Biology Concepts and Applications

Biology 1408

9th Edition

Starr | Evers | Starr

Australia • Brazil • Japan • Korea • Mexico • Singapore • Spain • United Kingdom • United States

Biology Concepts and Applications: Biology 1408, 9th edition

Senior Manager, Student Engagement:

Linda deStefano

Janey Moeller

Manager, Student Engagement:

Julie Dierig

Marketing Manager:

Rachael Kloos

Manager, Production Editorial:

Kim Fry

Manager, Intellectual Property Project Manager:

Brian Methe

Senior Manager, Production and Manufacturing:

Donna M. Brown

Manager, Production:

Terri Daley

Printed in the United States of America

Biology: Concepts and Applications, 9th Edition
Cecie Starr | Christine Evers | Lisa Starr

ISBN-13: 978-1-305-01734-4
ISBN-10: 1-305-01734-X

WCN: 01-100-101

Cengage Learning
5191 Natorp Boulevard
Mason, Ohio 45040
USA

Cengage Learning is a leading provider of customized learning solutions with office locations around the globe, including Singapore, the United Kingdom, Australia, Mexico, Brazil, and Japan. Locate your local office at: **international.cengage.com/region.**

Cengage Learning products are represented in Canada by Nelson Education, Ltd.
For your lifelong learning solutions, visit **www.cengage.com/custom.**
Visit our corporate website at **www.cengage.com.**

Brief Contents

INTRODUCTION

PRINCIPLES OF CELLULAR LIFE

GENETICS

PRINCIPLES OF EVOLUTION

APPENDICES

Near a tent serving as a makeshift laboratory, herpetologist Paul Oliver records the call of a frog on an expedition to New Guinea's Foja Mountains cloud forest.

1 INVITATION TO BIOLOGY

Links to Earlier Concepts

Whether or not you have studied biology, you already have an intuitive understanding of life on Earth because you are part of it. Every one of your experiences with the natural world—from the warmth of the sun on your skin to the love of your pet—contributes to that understanding.

KEY CONCEPTS

THE SCIENCE OF NATURE

We can understand life by studying it at many levels, starting with atoms that are components of all matter, and extending to interactions of organisms with their environment.

LIFE'S UNITY

All living things require ongoing inputs of energy and raw materials; all sense and respond to change; and all have DNA that guides their functioning.

LIFE'S DIVERSITY

Observable characteristics vary tremendously among organisms. Various classification systems help us keep track of the differences.

THE NATURE OF SCIENCE

Carefully designing experiments helps researchers unravel cause-and-effect relationships in complex natural systems.

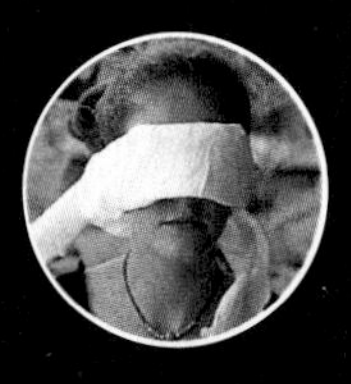

LIMITATIONS OF SCIENCE

Science addresses only testable ideas about observable events and processes. It does not address anything untestable, such as beliefs and opinions.

1.1 HOW DO LIVING THINGS DIFFER FROM NONLIVING THINGS?

LIFE IS MORE THAN THE SUM OF ITS PARTS

Biology is the study of life, past and present. What, exactly, is the property we call "life"? We may never actually come up with a good definition, because living things are too diverse, and they consist of the same basic components as nonliving things. When we try to define life, we end up only identifying properties that differentiate living from nonliving things.

Complex properties, including life, often emerge from the interactions of much simpler parts. To understand why, take a look at this drawing:

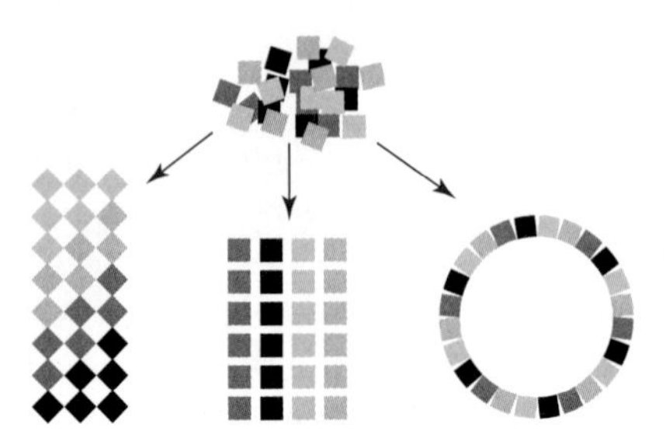

The property of "roundness" emerges when the parts are organized one way, but not other ways. Characteristics of a system that do not appear in any of the system's components are called **emergent properties**. The idea that structures with emergent properties can be assembled from the same basic building blocks is a recurring theme in our world, and also in biology.

LIFE'S ORGANIZATION

Through the work of biologists, we are beginning to understand an overall pattern in the way life is organized. We can look at life in successive levels of organization, with new emergent properties appearing at each level (**FIGURE 1.1**).

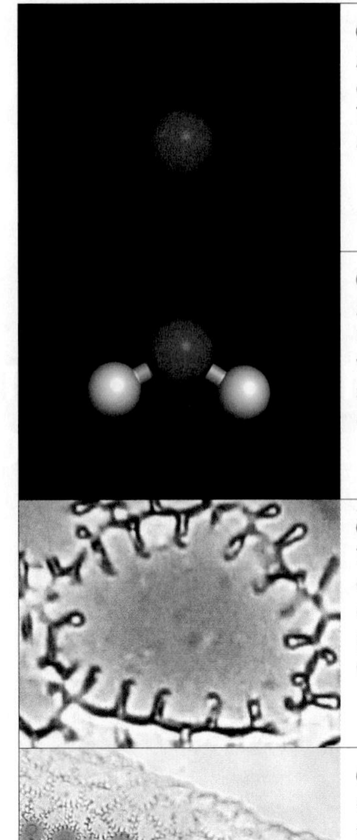

❶ atom
Atoms are fundamental units of all substances, living or not. This image shows a model of a single atom.

❷ molecule
Atoms join other atoms in molecules. This is a model of a water molecule. The molecules special to life are much larger and more complex than water.

❸ cell
The cell is the smallest unit of life. Some, like this plant cell, live and reproduce as part of a multicelled organism; others do so on their own.

❹ tissue
Organized array of cells that interact in a collective task. This is epidermal tissue on the outer surface of a flower petal.

❺ organ
Structural unit of interacting tissues. Flowers are the reproductive organs of many plants.

FIGURE 1.1 {Animated} An overall pattern in the way life is organized. New emergent properties appear at each successive level.

Life's organization starts with interactions between atoms. **Atoms** are fundamental building blocks of all substances ❶. Atoms join as **molecules** ❷. There are no atoms unique to living things, but there are unique molecules. In today's world, only living things make the "molecules of life," which are lipids, proteins, DNA, RNA, and complex carbohydrates. The emergent property of "life" appears at the next level, when many molecules of life become organized as a cell ❸. A **cell** is the smallest unit of life. Cells survive and reproduce themselves using energy, raw materials, and information in their DNA.

Some cells live and reproduce independently. Others do so as part of a multicelled organism. An **organism** is an individual that consists of one or more cells. A poppy plant is an example of a multicelled organism ❼.

In most multicelled organisms, cells are organized as tissues ❹. A **tissue** consists of specific types of cells organized in a particular pattern. The arrangement

6 organ system
A set of interacting organs. The shoot system of this poppy plant includes its aboveground parts: leaves, flowers, and stems.

7 multicelled organism
Individual that consists of more than one cell. Cells of this California poppy plant are part of its two organ systems: aboveground shoots and belowground roots.

8 population
Group of single-celled or multicelled individuals of a species in a given area. This population of California poppy plants is in California's Antelope Valley Poppy Reserve.

9 community
All populations of all species in a specified area. These plants are part of a community called the Antelope Valley Poppy Reserve.

10 ecosystem
A community interacting with its physical environment through the transfer of energy and materials. Sunlight and water sustain the community in the Antelope Valley.

11 biosphere
The sum of all ecosystems: every region of Earth's waters, crust, and atmosphere in which organisms live. No ecosystem in the biosphere is truly isolated from any other.

allows the cells to collectively perform a special function such as protection from injury (dermal tissue), movement (muscle tissue), and so on.

An **organ** is an organized array of tissues that collectively carry out a particular task or set of tasks 5. For example, a flower is an organ of reproduction in plants; a heart, an organ that pumps blood in animals. An **organ system** is a set of organs and tissues that interact to keep the individual's body working properly 6. Examples of organ systems include the aboveground parts of a plant (the shoot system), and the heart and blood vessels of an animal (the circulatory system).

A **population** is a group of individuals of the same type, or species, living in a given area 8. An example would be all of the California poppies that are living in California's Antelope Valley Poppy Reserve. At the next level, a **community** consists of all populations of all species in a given area. The Antelope Valley Reserve community includes California poppies and all other organisms—plants, animals, microorganisms, and so on—living in the reserve 9. Communities may be large or small, depending on the area defined.

The next level of organization is the **ecosystem**, which is a community interacting with its environment 10. The most inclusive level, the **biosphere**, encompasses all regions of Earth's crust, waters, and atmosphere in which organisms live 11.

atom Fundamental building block of all matter.
biology The scientific study of life.
biosphere All regions of Earth where organisms live.
cell Smallest unit of life.
community All populations of all species in a given area.
ecosystem A community interacting with its environment.
emergent property A characteristic of a system that does not appear in any of the system's component parts.
molecule An association of two or more atoms.
organ In multicelled organisms, a grouping of tissues engaged in a collective task.
organism Individual that consists of one or more cells.
organ system In multicelled organisms, set of organs engaged in a collective task that keeps the body functioning properly.
population Group of interbreeding individuals of the same species that live in a given area.
tissue In multicelled organisms, specialized cells organized in a pattern that allows them to perform a collective function.

TAKE-HOME MESSAGE 1.1

Biologists study life by thinking about it at different levels of organization, with new emergent properties appearing at each successive level.

All things, living or not, consist of the same building blocks: atoms. Atoms join as molecules.

The unique properties of life emerge as certain kinds of molecules become organized into cells.

Higher levels of life's organization include multicelled organisms, populations, communities, ecosystems, and the biosphere.

CREDITS: (l) 6: Lady Bird Johnson Wildflower Center; 7: Michael Szoenyi/Science Source; 8: Photographers Choice RF/SuperStock; 9: © Sergei Krupnov, www.flickr.com/photos/7969319@N03; 10: © Mark Koberg Photography; 11: NASA.

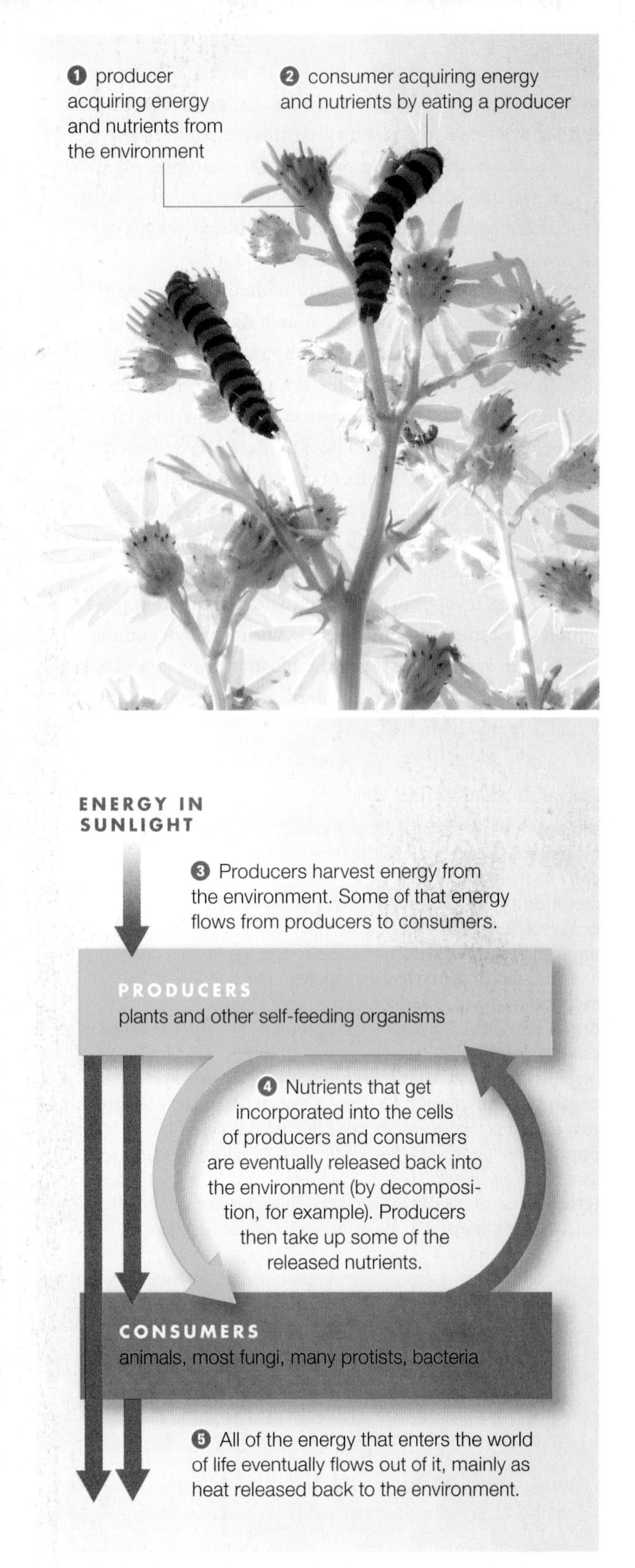

FIGURE 1.2 {Animated} The one-way flow of energy and cycling of materials through the world of life.

Even though we cannot precisely define "life," we can intuitively understand what it means because all living things share a set of key features. All require ongoing inputs of energy and raw materials; all sense and respond to change; and all pass DNA to offspring (**TABLE 1.1**).

TABLE 1.1

Three Key Features of Living Things

Requirement for energy and nutrients	Ongoing inputs of energy and nutrients sustain life.
Homeostasis	Each living thing has the capacity to sense and respond to change.
Use of DNA as hereditary material	DNA is passed to offspring during reproduction.

ORGANISMS REQUIRE ENERGY AND NUTRIENTS

Not all living things eat, but all require energy and nutrients on an ongoing basis. Both are essential to maintain the functioning of individual organisms and the organization of life. A **nutrient** is a substance that an organism needs for growth and survival but cannot make for itself.

Organisms spend a lot of time acquiring energy and nutrients (**FIGURE 1.2**). However, the source of energy and the type of nutrients required differ among organisms. These differences allow us to classify all living things into two categories: producers and consumers. **Producers** make their own food using energy and simple raw materials they get from nonbiological sources 1. Plants are producers that use the energy of sunlight to make sugars from water and carbon dioxide (a gas in air), a process called **photosynthesis**. By contrast, **consumers** cannot make their own food. They get energy and nutrients by feeding on other organisms 2. Animals are consumers. So are decomposers, which feed on the wastes or remains of other organisms. The leftovers from consumers' meals end up in the environment, where they serve as nutrients for producers. Said another way, nutrients cycle between producers and consumers.

Unlike nutrients, energy is not cycled. It flows through the world of life in one direction: from the environment 3, through organisms 4, and back to the environment 5. This flow maintains the organization of every living cell and body, and it also influences how individuals interact with one another and their environment. The energy flow is one-way, because with each transfer, some energy escapes as heat, and cells cannot use heat as an energy source. Thus, energy that enters the world of life eventually leaves it (we return to this topic in Chapter 5).

ORGANISMS SENSE AND RESPOND TO CHANGE

An organism cannot survive for very long in a changing environment unless it adapts to the changes. Thus, every living thing has the ability to sense and respond to change both inside and outside of itself (**FIGURE 1.3**). For example, after you eat, the sugars from your meal enter your bloodstream. The added sugars set in motion a series of events that causes cells throughout the body to take up sugar faster, so the sugar level in your blood quickly falls. This response keeps your blood sugar level within a certain range, which in turn helps keep your cells alive and your body functioning.

The fluid portion of your blood is a component of your internal environment, which is all of the body fluids outside of cells. Unless that internal environment is kept within certain ranges of temperature and other conditions, your body cells will die. By sensing and adjusting to change, you and all other organisms keep conditions in the internal environment within a range that favors survival. **Homeostasis** is the name for this process, and it is one of the defining features of life.

ORGANISMS USE DNA

With little variation, the same types of molecules perform the same basic functions in every organism. For example, information in an organism's **DNA** (deoxyribonucleic acid) guides ongoing functions that sustain the individual through its lifetime. Such functions include **development**: the process by which the first cell of a new individual gives rise to a multicelled adult; **growth**: increases in cell number, size, and volume; and **reproduction**: processes by which individuals produce offspring.

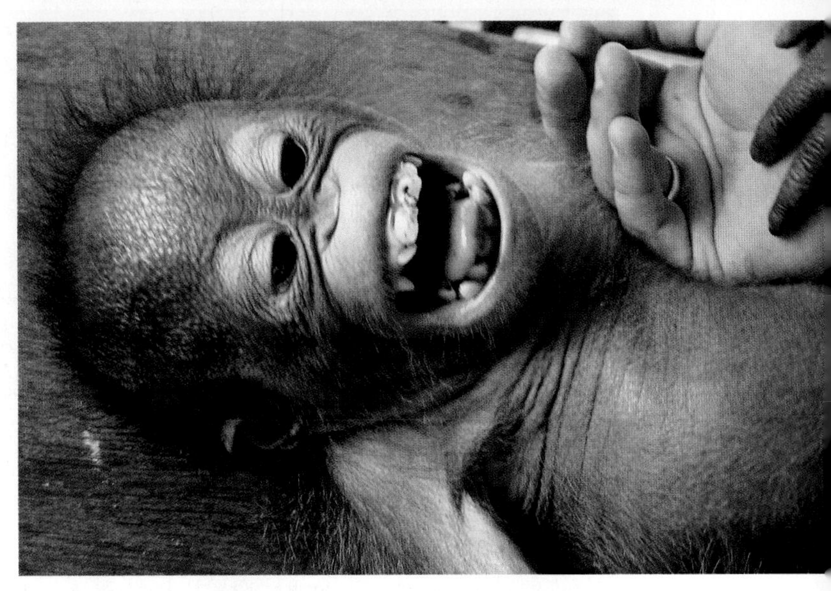

FIGURE 1.3 Living things sense and respond to their environment. This baby orangutan is laughing in response to being tickled. Apes and humans make different sounds when being tickled, but the airflow patterns are so similar that we can say apes really do laugh.

Individuals of every natural population are alike in certain aspects of their body form and behavior because their DNA is very similar: Orangutans look like orangutans and not like caterpillars because they inherited orangutan DNA, which differs from caterpillar DNA in the information it carries. **Inheritance** refers to the transmission of DNA to offspring. All organisms inherit their DNA from one or two parents.

DNA is the basis of similarities in form and function among organisms. However, the details of DNA molecules differ, and herein lies the source of life's diversity. Small variations in the details of DNA's structure give rise to differences among individuals, and also among types of organisms. As you will see in later chapters, these differences are the raw material of evolutionary processes.

TAKE-HOME MESSAGE 1.2

Continual inputs of energy and the cycling of materials maintain life's complex organization.

Organisms sense and respond to change inside and outside themselves. They make adjustments that keep conditions in their internal environment within a range that favors cell survival, a process called homeostasis.

All organisms use information in the DNA they inherited from their parent or parents to develop, grow, and reproduce. DNA is the basis of similarities and differences in form and function among organisms.

consumer Organism that gets energy and nutrients by feeding on tissues, wastes, or remains of other organisms.
development Multistep process by which the first cell of a new multicelled organism gives rise to an adult.
DNA Deoxyribonucleic acid; carries hereditary information that guides development and other activities.
growth In multicelled species, an increase in the number, size, and volume of cells.
homeostasis Process in which an organism keeps its internal conditions within tolerable ranges by sensing and responding to change.
inheritance Transmission of DNA to offspring.
nutrient Substance that an organism needs for growth and survival but cannot make for itself.
photosynthesis Process by which producers use light energy to make sugars from carbon dioxide and water.
producer Organism that makes its own food using energy and nonbiological raw materials from the environment.
reproduction Processes by which parents produce offspring.

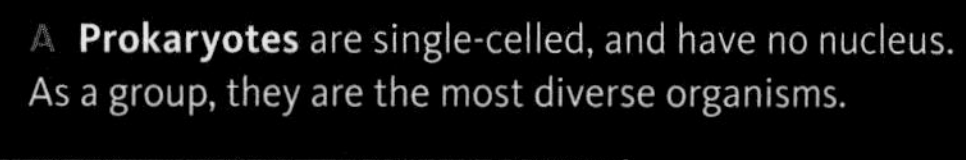

A **Prokaryotes** are single-celled, and have no nucleus. As a group, they are the most diverse organisms.

B **Eukaryotes** consist of cells that have a nucleus. Eukaryotic cells are typically larger and more complex than prokaryotes.

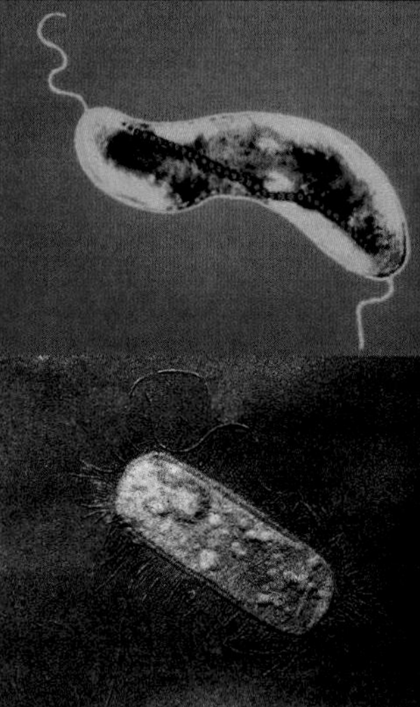

bacteria are the most numerous organisms on Earth. Top, this bacterium has a row of iron crystals that functions like a tiny compass; bottom, a resident of human intestines.

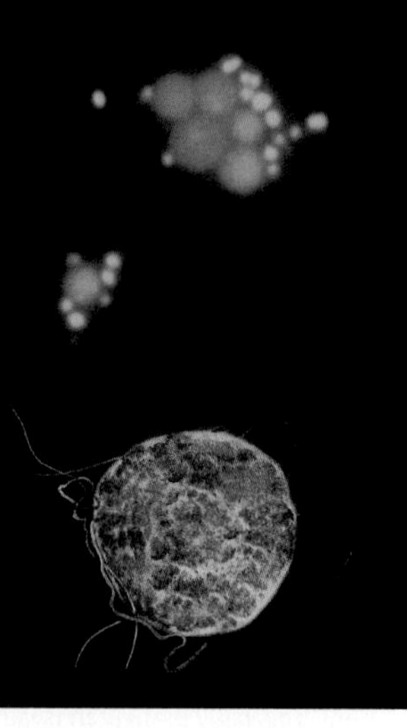

archaea resemble bacteria, but they are more closely related to eukaryotes. Top: two types from a hydrothermal vent on the seafloor. Bottom, a type that grows in sulfur hot springs.

protists are a group of extremely diverse eukaryotes that range from microscopic single cells (top) to giant multicelled seaweeds (bottom).

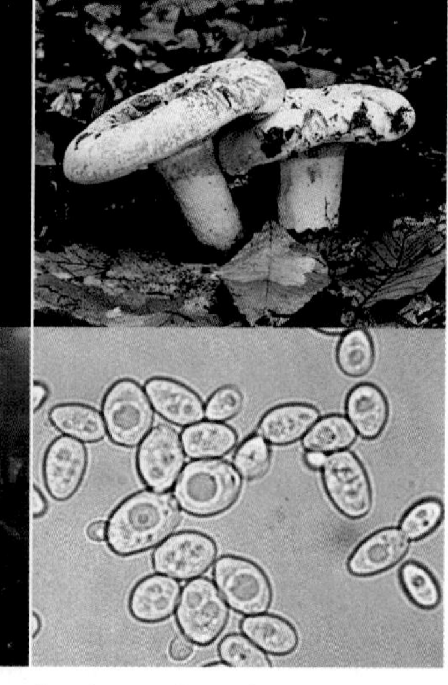

fungi are eukaryotic consumers that secrete substances to break down food outside their body. Most are multicelled (top), but some are single-celled (bottom).

FIGURE 1.4 A few representatives of life's diversity: **A** some prokaryotes; **B** some eukaryotes.

Living things differ tremendously in their observable characteristics. Various classification schemes help us organize what we understand about the scope of this variation, which we call Earth's **biodiversity**.

For example, organisms can be grouped on the basis of whether they have a **nucleus**, which is a sac with two membranes that encloses and protects a cell's DNA. **Bacteria** (singular, bacterium) and **archaea** (singular, archaeon) are organisms whose DNA is *not* contained within a nucleus. All bacteria and archaea are single-celled, which means each organism consists of one cell (**FIGURE 1.4A**). Collectively, these organisms are the most diverse representatives of life. Different kinds are producers or consumers in nearly all regions of Earth. Some inhabit such extreme environments as frozen desert rocks, boiling sulfurous lakes, and nuclear reactor waste. The first cells on Earth may have faced similarly hostile environments.

Traditionally, organisms without a nucleus have been called **prokaryotes**, but this designation is now used only informally. This is because, despite the similar appearance of bacteria and archaea, the two types of cells are less related to one another than we once thought. Archaea turned out to be more closely related to **eukaryotes**, which are organisms whose DNA is contained within a nucleus. Some eukaryotes live as individual cells; others are multicelled (**FIGURE 1.4B**). Eukaryotic cells are typically larger and more complex than bacteria or archaea.

Structurally, **protists** are the simplest eukaryotes, but as a group they vary dramatically, from single-celled consumers to giant, multicelled producers.

animal Multicelled consumer that develops through a series of stages and moves about during part or all of its life.
archaea Group of single-celled organisms that lack a nucleus but are more closely related to eukaryotes than to bacteria.
bacteria The most diverse and well-known group of single-celled organisms that lack a nucleus.
biodiversity Scope of variation among living organisms.
eukaryote Organism whose cells characteristically have a nucleus.
fungus Single-celled or multicelled eukaryotic consumer that breaks down material outside itself, then absorbs nutrients released from the breakdown.
nucleus Sac that encloses a cell's DNA; has two membranes.
plant A multicelled, typically photosynthetic producer.
prokaryote Single-celled organism without a nucleus.
protist Member of a diverse group of simple eukaryotes.

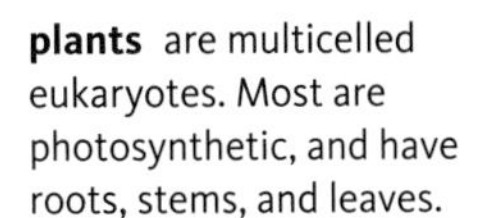

plants are multicelled eukaryotes. Most are photosynthetic, and have roots, stems, and leaves.

animals are multicelled eukaryotes that ingest tissues or juices of other organisms. All actively move about during at least part of their life.

Fungi (singular, fungus) are eukaryotic consumers that secrete substances to break down food externally, then absorb nutrients released by this process. Many fungi are decomposers. Most fungi, including those that form mushrooms, are multicellular. Fungi that live as single cells are called yeasts.

Plants are multicelled eukaryotes; the majority are photosynthetic producers that live on land. Besides feeding themselves, plants also serve as food for most other land-based organisms.

Animals are multicelled consumers that consume tissues or juices of other organisms. Unlike fungi, animals break down food inside their body. They also develop through a series of stages that lead to the adult form. All kinds actively move about during at least part of their lives.

TAKE-HOME MESSAGE 1.3

Organisms differ in their details; they show tremendous variation in observable characteristics, or traits.

We can divide Earth's biodiversity into broad groups based on traits such as having a nucleus or being multicellular.

PEOPLE MATTER

National Geographic Explorer
KRISTOFER HELGEN

Kristofer Helgen discovers new animals. Deep in a New Guinea rain forest. High on an Andean mountainside. Resting in a museum's specimen drawer. "Conventional wisdom would have it that we know all the mammals of the world," he notes. "In fact, we know so little. Unique species, profoundly different from anything ever discovered, are out there waiting to be found." His own efforts prove this. Helgen himself has discovered approximately 100 new species of mammals previously unknown to science. "Since I was three years old, I've been transfixed by animals," he recalls. "Even then, my excitement revolved around figuring out how many different kinds there were."

Helgen's search plunges him into the wild on almost every continent. Yet about three times as many new finds are made within the walls of museums. "An expert can go into any large natural history museum and identify kinds of animals no one knew existed," he explains. When only a few specimens of a species exist, and reside in museums scattered across the globe, sheer logistics often prevent researchers from connecting the dots and pinpointing a new find. "Collections build up over centuries," he says, "It's virtually impossible to fully interpret that wealth of material. Every day brings surprises." As Curator of Mammals for the Smithsonian Institution's National Museum of Natural History, he oversees not only the collection's use as an invaluable research resource, but also its continued expansion through exploration.

1.4 WHAT IS A SPECIES?

Each time we discover a new **species**, or unique kind of organism, we name it. **Taxonomy**, a system of naming and classifying species, began thousands of years ago, but naming species in a consistent way did not become a priority until the eighteenth century. At the time, European explorers who were just discovering the scope of life's diversity started having more and more trouble communicating with one another because species often had multiple names. For example, the dog rose (a plant native to Europe, Africa, and Asia) was alternately known as briar rose, witch's briar, herb patience, sweet briar, wild briar, dog briar, dog berry, briar hip, eglantine gall, hep tree, hip fruit, hip rose, hip tree, hop fruit, and hogseed—and those are only the English names! Species often had multiple scientific names too, in Latin that was descriptive but often cumbersome. The scientific name of the dog rose was *Rosa sylvestris inodora seu canina* (odorless woodland dog rose), and also *Rosa sylvestris alba cum rubore, folio glabro* (pinkish white woodland rose with smooth leaves).

An eighteenth-century naturalist, Carolus Linnaeus, standardized a naming system that we still use. By the Linnaean system, every species is given a unique two-part scientific name. The first part is the name of the **genus** (plural, genera), a group of species that share a unique set of features. The second part is the **specific epithet**. Together, the genus name and the specific epithet designate one species. Thus, the dog rose now has one official name, *Rosa canina*, that is recognized worldwide.

Genus and species names are always italicized. For example, *Panthera* is a genus of big cats. Lions belong to the species *Panthera leo*. Tigers belong to a different species in the same genus (*Panthera tigris*), and so do leopards (*P. pardus*). Note how the genus name may be abbreviated after it has been spelled out once.

A ROSE BY ANY OTHER NAME . . .

The individuals of a species share a unique set of inherited characteristics, or **traits**. For example, giraffes normally have very long necks, brown spots on white coats, and so on. These are morphological traits (*morpho–* means form). Individuals of a species also share biochemical traits (they make and use the same molecules) and behavioral traits (they respond the same way to certain stimuli, as when hungry giraffes feed on tree leaves).

We can rank species into ever more inclusive categories based on shared sets of traits. Each rank, or **taxon** (plural, taxa), is a group of organisms that share a unique set of traits. Each category above species—genus, family, order, class, phylum (plural, phyla), kingdom, and domain—consists of a group of the next lower taxon (**FIGURE 1.5**). Using this system, we can sort all life into a few categories (**FIGURE 1.6** and **TABLE 1.2**).

domain	Eukarya	Eukarya	Eukarya	Eukarya	Eukarya
kingdom	Plantae	Plantae	Plantae	Plantae	Plantae
phylum	Magnoliophyta	Magnoliophyta	Magnoliophyta	Magnoliophyta	Magnoliophyta
class	Magnoliopsida	Magnoliopsida	Magnoliopsida	Magnoliopsida	Magnoliopsida
order	Apiales	Rosales	Rosales	Rosales	Rosales
family	Apiaceae	Cannabaceae	Rosaceae	Rosaceae	Rosaceae
genus	*Daucus*	*Cannabis*	*Malus*	*Rosa*	*Rosa*
species	*carota*	*sativa*	*domestica*	*acicularis*	*canina*
common name	wild carrot	marijuana	apple	prickly rose	dog rose

FIGURE 1.5 Linnaean classification of five species that are related at different levels. Each species has been assigned to ever more inclusive groups, or taxa: in this case, from genus to domain.

FIGURE IT OUT: Which of the plants shown here are in the same order?

Answer: Marijuana, apple, prickly rose, and dog rose

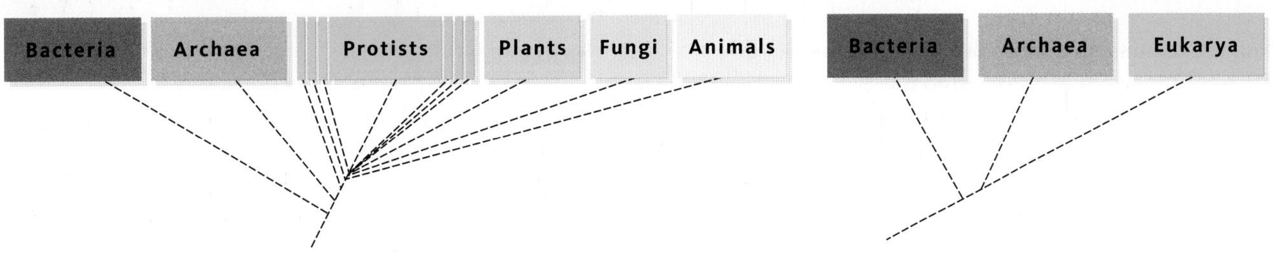

A **Six-kingdom classification system.** The protist kingdom includes the most ancient multicelled and all single-celled eukaryotes.

B **Three-domain classification system.** The Eukarya domain includes protists, plants, fungi, and animals.

FIGURE 1.6 {Animated} Two ways to see the big picture of life. The lines in such diagrams indicate evolutionary connections.

It is easy to tell that orangutans and caterpillars are different species because they appear very different. Distinguishing species that are more closely related may be much more challenging (**FIGURE 1.7**). In addition, traits shared by members of a species often vary a bit among individuals, such as eye color does among people. How do we decide if similar-looking organisms belong to different species or not? The short answer to that question is that we rely on whatever information we have. Early naturalists studied anatomy and distribution—essentially the only methods available at the time—so species were named and classified according to what they looked like and where they lived. Today's biologists are able to compare traits that the early naturalists did not even know about, including biochemical ones.

FIGURE 1.7 Four butterflies, two species: Which are which?

The top row shows two forms of the species *Heliconius melpomene*; the bottom row, two forms of *H. erato*.

H. melpomene and *H. erato* never cross-breed. Their alternate but similar patterns of coloration evolved as a shared warning signal to predatory birds that these butterflies taste terrible.

genus A group of species that share a unique set of traits.
species Unique type of organism.
specific epithet Second part of a species name.
taxon Group of organisms that share a unique set of traits.
taxonomy The science of naming and classifying species.
trait An observable characteristic of an organism or species.

The discovery of new information sometimes changes the way we distinguish a particular species or how we group it with others. For example, Linnaeus grouped plants by the number and arrangement of reproductive parts, a scheme that resulted in odd pairings such as castor-oil plants with pine trees. Having more information today, we place these plants in separate phyla.

Evolutionary biologist Ernst Mayr defined a species as one or more groups of individuals that potentially can interbreed, produce fertile offspring, and do not interbreed with other groups. This "biological species concept" is useful in many cases, but it is not universally applicable. For example, we may never know whether separate populations could interbreed even if they did get together. As another example, populations often continue to interbreed even as they diverge, so the exact moment at which two populations become two species is often impossible to pinpoint. We return to speciation and how it occurs in Chapter 17, but for now it is important to remember that a "species" is a convenient but artificial construct of the human mind.

TABLE 1.2

All of Life in Three Domains

Bacteria	Single cells, no nucleus. Most ancient lineage.
Archaea	Single cells, no nucleus. Evolutionarily closer to eukaryotes than bacteria.
Eukarya	Eukaryotic cells (with a nucleus). Single-celled and multicelled species of protists, plants, fungi, and animals.

TAKE-HOME MESSAGE 1.4

Each type of organism, or species, is given a unique, two-part scientific name.

Classification systems group species on the basis of shared, inherited traits.

Most of us assume that we do our own thinking, but do we, really? You might be surprised to find out how often we let others think for us. Consider how a school's job (which is to impart as much information to students as quickly as possible) meshes perfectly with a student's job (which is to acquire as much knowledge as quickly as possible). In this rapid-fire exchange of information, it is sometimes easy to forget about the quality of what is being exchanged. Anytime you accept information without questioning it, you let someone else think for you.

THINKING ABOUT THINKING

Critical thinking is the deliberate process of judging the quality of information before accepting it. "Critical" comes from the Greek *kriticos* (discerning judgment). When you use critical thinking, you move beyond the content of new information to consider supporting evidence, bias, and alternative interpretations. How does the busy student manage this? Critical thinking does not necessarily require extra time, just a bit of extra awareness. There are many ways to do it. For example, you might ask yourself some of the following questions while you are learning something new:

What message am I being asked to accept?
Is the message based on facts or opinion?
Is there a different way to interpret the facts?
What biases might the presenter have?
How do my own biases affect what I'm learning?

Such questions are a way of being conscious about learning. They can help you decide whether to allow new information to guide your beliefs and actions.

THE SCIENTIFIC METHOD

Critical thinking is a big part of **science**, the systematic study of the observable world and how it works (**FIGURE 1.8**). A scientific line of inquiry usually begins with curiosity about something observable, such as, say, a decrease in the number of birds in a particular area. Typically, a scientist will read about what others have discovered before making a **hypothesis**, a testable explanation for a natural phenomenon. An example of a hypothesis would be, "The number of birds is decreasing because the number of cats is increasing." Making a hypothesis this way is an example of **inductive reasoning**, which means arriving at a conclusion based on one's observations. Inductive reasoning is the way we come up with new ideas about groups of objects or events.

A **prediction**, or statement of some condition that should exist if the hypothesis is correct, comes next. Making predictions is called the if–then process, in which the "if" part is the hypothesis, and the "then" part is the prediction. Using a hypothesis to make a prediction is a form of **deductive reasoning**, the logical process of using a general premise to draw a conclusion about a specific case.

Next, a scientist will devise ways to test a prediction. Tests may be performed on a **model**, or analogous system, if working with an object or event directly is not possible. For example, animal diseases are often used as models of similar human diseases. Careful observations are one way to test predictions that flow from a hypothesis. So are **experiments**: tests designed to support or falsify a prediction. A typical experiment explores a cause-and-effect relationship.

Researchers often investigate causal relationships by changing and observing **variables**, characteristics or events that can differ among individuals or over time.

TABLE 1.3

The Scientific Method

1. **Observe** some aspect of nature.
2. **Think of an explanation** for your observation (in other words, form a hypothesis).
3. **Test the hypothesis.**
   - **a.** Make a prediction based on the hypothesis.
   - **b.** Test the prediction using experiments or surveys.
   - **c.** Analyze the results of the tests (data).
4. **Decide** whether the results of the tests support your hypothesis or not (form a conclusion).
5. **Report** your experiment, data, and conclusion to the scientific community.

control group Group of individuals identical to an experimental group except for the independent variable under investigation.
critical thinking Judging information before accepting it.
data Experimental results.
deductive reasoning Using a general idea to make a conclusion about a specific case.
dependent variable In an experiment, a variable that is presumably affected by an independent variable being tested.
experiment A test designed to support or falsify a prediction.
experimental group In an experiment, a group of individuals who have a certain characteristic or receive a certain treatment.
hypothesis Testable explanation of a natural phenomenon.
independent variable Variable that is controlled by an experimenter in order to explore its relationship to a dependent variable.
inductive reasoning Drawing a conclusion based on observation.
model Analogous system used for testing hypotheses.
prediction Statement, based on a hypothesis, about a condition that should exist if the hypothesis is correct.
science Systematic study of the observable world.
scientific method Making, testing, and evaluating hypotheses.
variable In an experiment, a characteristic or event that differs among individuals or over time.

FIGURE 1.8 Tierney Thys travels the world's oceans to study the giant sunfish (mola). This mola is carrying a satellite tracking device.

An **independent variable** is defined or controlled by the person doing the experiment. A **dependent variable** is an observed result that is supposed to be influenced by the independent variable. For example, an independent variable in an investigation of our observed decrease in the number of birds may be the removal of cats in the area. The dependent variable in this experiment would be the number of birds.

Biological systems are complex, with many interacting variables. It can be difficult to study one variable separately from the rest. Thus, biology researchers often test two groups of individuals simultaneously. An **experimental group** is a set of individuals that have a certain characteristic or receive a certain treatment. This group is tested side by side with a **control group**, which is identical to the experimental group except for one independent variable: the characteristic or the treatment being tested. Any differences in experimental results between the two groups is likely to be an effect of changing the variable.

Test results—**data**—that are consistent with the prediction are evidence in support of the hypothesis. Data inconsistent with the prediction are evidence that the hypothesis is flawed and should be revised.

A necessary part of science is reporting one's results and conclusions in a standard way, such as in a peer-reviewed journal article. The communication gives other scientists an opportunity to evaluate the information for themselves, both by checking the conclusions drawn and by repeating the experiments.

Forming a hypothesis based on observation, and then systematically testing and evaluating the hypothesis, are collectively called the **scientific method** (**TABLE 1.3**).

TAKE-HOME MESSAGE 1.5

Judging the quality of information before accepting it is called critical thinking.

The scientific method consists of making, testing, and evaluating hypotheses. It is a way of critical thinking.

Experiments measure how changing an independent variable affects a dependent variable.

1.6 WHY DO BIOLOGISTS PERFORM EXPERIMENTS?

There are many different ways to do research, particularly in biology. Some biologists make surveys; they observe without making hypotheses. Some make hypotheses and leave experimentation to others. Despite a broad range of approaches, however, researchers typically try to design experiments in a consistent way. They change one independent variable at a time, and carefully measure the effects of the change on a dependent variable.

To give you a sense of how biology experiments work, we summarize two published studies here.

POTATO CHIPS AND STOMACHACHES

In 1996 the U.S. Food and Drug Administration (the FDA) approved Olestra® (a fat replacement manufactured from sugar and vegetable oil) for use as a food additive. Potato chips were the first Olestra-containing food product on the market in the United States. Controversy soon raged. Many people complained of intestinal problems after eating the chips and thought that the Olestra was at fault.

Two years later, researchers at Johns Hopkins University School of Medicine designed an experiment to test the hypothesis that this food additive causes cramps. The researchers predicted *if* Olestra causes cramps, *then* people who eat Olestra will be more likely to get cramps than people who do not. To test their prediction, they used a Chicago theater as a "laboratory," and asked 1,100 people between the ages of thirteen and thirty-eight to eat potato chips while watching a movie. Each person got an unmarked bag that contained 13 ounces of chips. In this experiment, individuals who ate Olestra-containing potato chips constituted the experimental group, and individuals who ate regular chips were the control group. The independent variable was the presence or absence of Olestra in the chips.

A few days after the experiment was finished, the researchers contacted everyone and collected reports of any post-movie cramps (the dependent variable). Of the 563 people in the experimental (Olestra-eating) group, 89 (15.8 percent) complained about cramps. However, so did 93 of the 529 people (17.6 percent) making up the control group—who had eaten the regular chips. In this experiment, people were about as likely to get cramps whether or not they ate chips made with Olestra. These results did not support the prediction, so the researchers concluded that eating Olestra does not cause cramps (**FIGURE 1.9**).

A Hypothesis
Olestra® causes intestinal cramps.

B Prediction
People who eat potato chips made with Olestra will be more likely to get intestinal cramps than those who eat potato chips made without Olestra.

	Control Group	Experimental Group
C Experiment	Eats regular potato chips	Eats Olestra potato chips
D Results	93 of 529 people get cramps later (17.6%)	89 of 563 people get cramps later (15.8%)

E Conclusion
Percentages are about equal. People who eat potato chips made with Olestra are just as likely to get intestinal cramps as those who eat potato chips made without Olestra. These results do not support the hypothesis.

FIGURE 1.9 The steps in a scientific experiment to determine if Olestra causes cramps. A report of this study was published in the *Journal of the American Medical Association* in January 1998.

FIGURE IT OUT: What was the dependent variable in this experiment?

Answer: Whether or not a person got cramps

BUTTERFLIES AND BIRDS

A 2005 experiment investigated whether certain peacock butterfly behaviors defend these insects from predatory birds. The researchers performing this experiment began with two observations. First, when a peacock butterfly rests, it folds its wings, so only the dark underside shows (**FIGURE 1.10A**). Second, when a butterfly sees a predator approaching, it repeatedly flicks its wings open, while also moving them in a way that produces a hissing sound and a series of clicks (**FIGURE 1.10B**).

The researchers were curious about why the peacock butterfly flicks its wings. After they reviewed earlier studies, they came up with two hypotheses that might explain the wing-flicking behavior:

1. Although wing-flicking probably attracts predatory birds, it also exposes brilliant spots that resemble owl eyes. Anything that looks like owl eyes is known to startle small, butterfly-eating birds, so exposing the wing spots might scare off predators.
2. The hissing and clicking sounds produced when the peacock butterfly moves its wings may be an additional defense that deters predatory birds.

A With wings folded, a resting peacock butterfly resembles a dead leaf.

B When a bird approaches, a butterfly repeatedly flicks its wings open. This behavior exposes brilliant spots and also produces hissing and clicking sounds.

C Researchers tested whether peacock butterfly wing flicking and hissing reduce predation by blue tits.

FIGURE 1.10 Testing peacock butterfly defenses. Researchers painted out the spots of some butterflies, cut the sound-making part of the wings on others, and did both to a third group; then exposed each butterfly to a hungry blue tit. Results, listed below in Table 1.4, support the hypotheses that peacock butterfly spots and sounds can deter predatory birds.

FIGURE IT OUT: What was the dependent variable in this series of experiments?

Answer: Being eaten

TABLE 1.4

Results of Peacock Butterfly Experiment*

Wing Spots	Wing Sound	Total Number of Butterflies	Number Eaten	Number Survived
Spots	Sound	9	0	9 (100%)
No spots	Sound	10	5	5 (50%)
Spots	No sound	8	0	8 (100%)
No spots	No sound	10	8	2 (20%)

* *Proceedings of the Royal Society of London, Series B (2005) 272: 1203–1207.*

The researchers then used their hypotheses to make the following predictions:

1. If peacock butterflies startle predatory birds by exposing their brilliant wing spots, then individuals with wing spots will be less likely to get eaten by predatory birds than those without wing spots.
2. If peacock butterfly sounds deter predatory birds, then sound-producing individuals will be less likely to get eaten by predatory birds than silent individuals.

The next step was the experiment. The researchers used a marker to paint the wing spots of some butterflies black, and scissors to cut off the sound-making part of the wings of others. A third group had both treatments: Their wings were painted and cut. The researchers then put each butterfly into a large cage with a hungry blue tit (**FIGURE 1.10C**) and watched the pair for thirty minutes.

TABLE 1.4 lists the results of the experiment. All of the butterflies with unmodified wing spots survived, regardless of whether they made sounds. By contrast, only half of the butterflies that had spots painted out but could make sounds survived. Most of the silenced butterflies with painted-out spots were eaten quickly. The test results confirmed both predictions, so they support the hypotheses. Predatory birds are indeed deterred by peacock butterfly sounds, and even more so by wing spots.

TAKE-HOME MESSAGE 1.6

Natural processes are often influenced by many interacting variables.

Researchers unravel cause-and-effect relationships in complex natural processes by performing experiments in which they change one variable at a time.

SAMPLING ERROR

When researchers cannot directly observe all individuals of a population, all instances of an event, or some other aspect of nature, they may test or survey a subset. Results from the subset are then used to make generalizations about the whole. For example, a survey team may catalog the number of beetles in a given area of a very large forest. If that given area is one-thousandth of the forest, then an estimate of the number of beetles in the entire forest would be one thousand times their result. However, this type of generalization is risky because the subset may not be representative of the whole. In our beetle survey, for example, if the only nest of beetles in the entire forest happened to be located in the area that was surveyed, then the generalized result would be in error. **Sampling error** is a difference between results obtained from a subset, and results from the whole (**FIGURE 1.11A**).

A Natalie chooses a random jelly bean from a jar. She is blindfolded, so she does not know that the jar contains 120 green and 280 black jelly beans.

The jar is hidden from Natalie's view before she removes her blindfold. She sees one green jelly bean in her hand and assumes that the jar must hold only green jelly beans. This assumption is incorrect: 30 percent of the jelly beans in the jar are green, and 70 percent are black. The small sample size has resulted in sampling error.

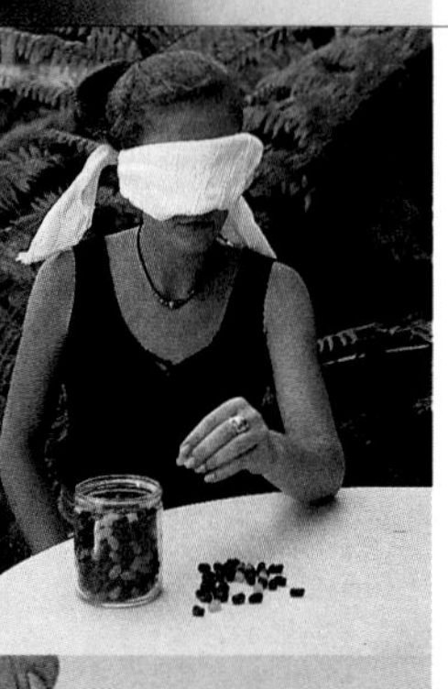

B Still blindfolded, Natalie randomly picks out 50 jelly beans from the jar. She ends up choosing 10 green and 40 black ones.

The larger sample leads Natalie to assume that one-fifth of the jar's jelly beans are green (20 percent) and four-fifths are black (80 percent). The larger sample more closely approximates the jar's actual green-to-black ratio of 30 percent to 70 percent.

The more times Natalie repeats the sampling, the greater the chance she has of guessing the actual ratio.

FIGURE 1.11 {Animated} Demonstration of sampling error, and the effect of sample size on it.

Sampling error may be unavoidable, but knowing how it can occur helps researchers design their experiments to minimize it. For example, sampling error can be a substantial problem with a small subset, so experimenters try to start with a relatively large sample, and they repeat their experiments (**FIGURE 1.11B**). To understand why these practices reduce the risk of sampling error, think about flipping a coin. There are two possible outcomes of each flip: The coin lands heads up, or it lands tails up. Thus, the chance that the coin will land heads up is one in two (1/2), which is a proportion of 50 percent. However, when you flip a coin repeatedly, it often lands heads up, or tails up, several times in a row. With just 3 flips, the proportion of times that heads actually land up may not even be close to 50 percent. With 1,000 flips, however, the overall proportion of times the coin lands heads up is much more likely to approach 50 percent.

In cases such as flipping a coin, it is possible to calculate **probability**, which is the measure, expressed as a percentage, of the chance that a particular outcome will occur. That chance depends on the total number of possible outcomes. For instance, if 10 million people enter a drawing, each has the same probability of winning: 1 in 10 million, or (an extremely improbable) 0.00001 percent.

Analysis of experimental data often includes probability calculations. If a result is very unlikely to have occurred by chance alone, it is said to be **statistically significant**. In this context, the word "significant" does not refer to the result's importance. It means that the result has been subjected to a rigorous statistical analysis that shows it has a very low probability (usually 5 percent or less) of being skewed by sampling error.

Variation in data is often shown as error bars on a graph (**FIGURE 1.12**). Depending on the graph, error bars may indicate variation around an average for one sample set, or the difference between two sample sets.

BIAS IN INTERPRETING RESULTS

Particularly when studying humans, changing a single variable apart from all others is not often possible. For example, remember that the people who participated in the Olestra experiment were chosen randomly. That means the study was not controlled for gender, age, weight, medications taken, and so on. Such variables may have influenced the results.

Human beings are by nature subjective, and scientists are no exception. Experimenters risk interpreting their results in terms of what they want to find out. That is why they often design experiments to yield quantitative results, which are counts or some other data that can be measured or gathered objectively. Such results minimize the potential for bias, and also give other scientists an opportunity to repeat the experiments and check the conclusions drawn from them.

This last point gets us back to the role of critical thinking in science. Scientists expect one another to recognize and put aside bias in order to test their hypotheses in ways that may prove them wrong. If a scientist does not, then others will, because exposing errors is just as useful as applauding insights. The scientific community consists of critically thinking people trying to poke holes in one another's ideas. Their collective efforts make science a self-correcting endeavor.

THE LIMITS OF SCIENCE

Science helps us be objective about our observations in part because of its limitations. For example, science does not address many questions, such as "Why do I exist?" Answers to such questions can only come from within as an integration of all the personal experiences and mental connections that shape our consciousness. This is not to say subjective answers have no value, because no human society can function for long unless its individuals share standards for making judgments, even if they are subjective. Moral, aesthetic, and philosophical standards vary from one society to the next, but all help people decide what is important and good. All give meaning to our lives.

Neither does science address the supernatural, or anything that is "beyond nature." Science neither assumes nor denies that supernatural phenomena occur, but scientists may cause controversy when they discover a natural explanation for something that was thought to have none. Such controversy often arises when a society's moral standards are interwoven with its understanding of nature. For example, Nicolaus Copernicus proposed in 1540 that Earth orbits the sun. Today that idea is generally accepted, but the prevailing belief system had Earth as the immovable center of the universe. In 1610, astronomer Galileo Galilei published evidence for the Copernican model of the solar system, an act that resulted in his imprisonment. He was publicly forced to recant his work, spent the rest of his life under house arrest, and was never allowed to publish again.

probability The chance that a particular outcome of an event will occur; depends on the total number of outcomes possible.
sampling error Difference between results derived from testing an entire group of events or individuals, and results derived from testing a subset of the group.
statistically significant Refers to a result that is statistically unlikely to have occurred by chance.

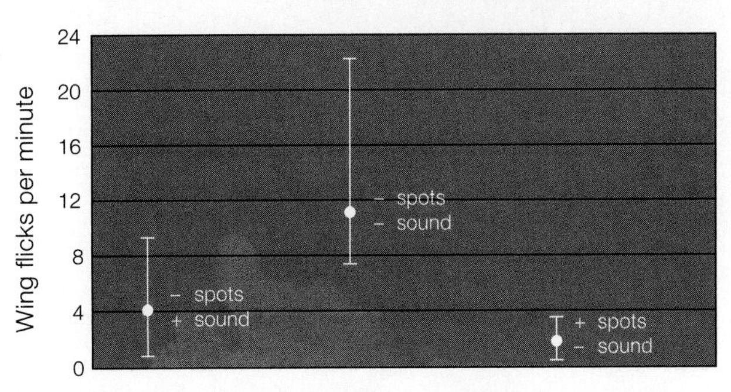

FIGURE 1.12 Example of error bars in a graph. This graph was adapted from the peacock butterfly research described in Section 1.6.

The researchers recorded the number of times each butterfly flicked its wings in response to an attack by a bird.

The dots represent average frequency of wing flicking for each sample set of butterflies. The error bars that extend above and below the dots indicate the range of values—the sampling error.

FIGURE IT OUT: What was the fastest rate at which a butterfly with no spots or sound flicked its wings?

Answer: 22 times per minute

As Galileo's story illustrates, exploring a traditional view of the natural world from a scientific perspective can be misinterpreted as a violation of morality. As a group, scientists are no less moral than anyone else, but they follow a particular set of rules that do not necessarily apply to others: Their work concerns only the natural world, and their ideas must be testable by other scientists.

Science helps us communicate our experiences without bias. As such, it may be as close as we can get to a universal language. We are fairly sure, for example, that the laws of gravity apply everywhere in the universe. Intelligent beings on a distant planet would likely understand the concept of gravity. We might well use gravity or another scientific concept to communicate with them, or anyone, anywhere. The point of science, however, is not to communicate with aliens. It is to find common ground here on Earth.

TAKE-HOME MESSAGE 1.7

- Checks and balances inherent in the scientific process help researchers to be objective about their observations.
- Researchers minimize sampling error by using large sample sizes and by repeating their experiments.
- Probability calculations can show whether a result is likely to have occurred by chance alone.
- Science is a self-correcting process because it is carried out by an aggregate community of people systematically checking one another's ideas.

1.8 WHAT IS A "THEORY"?

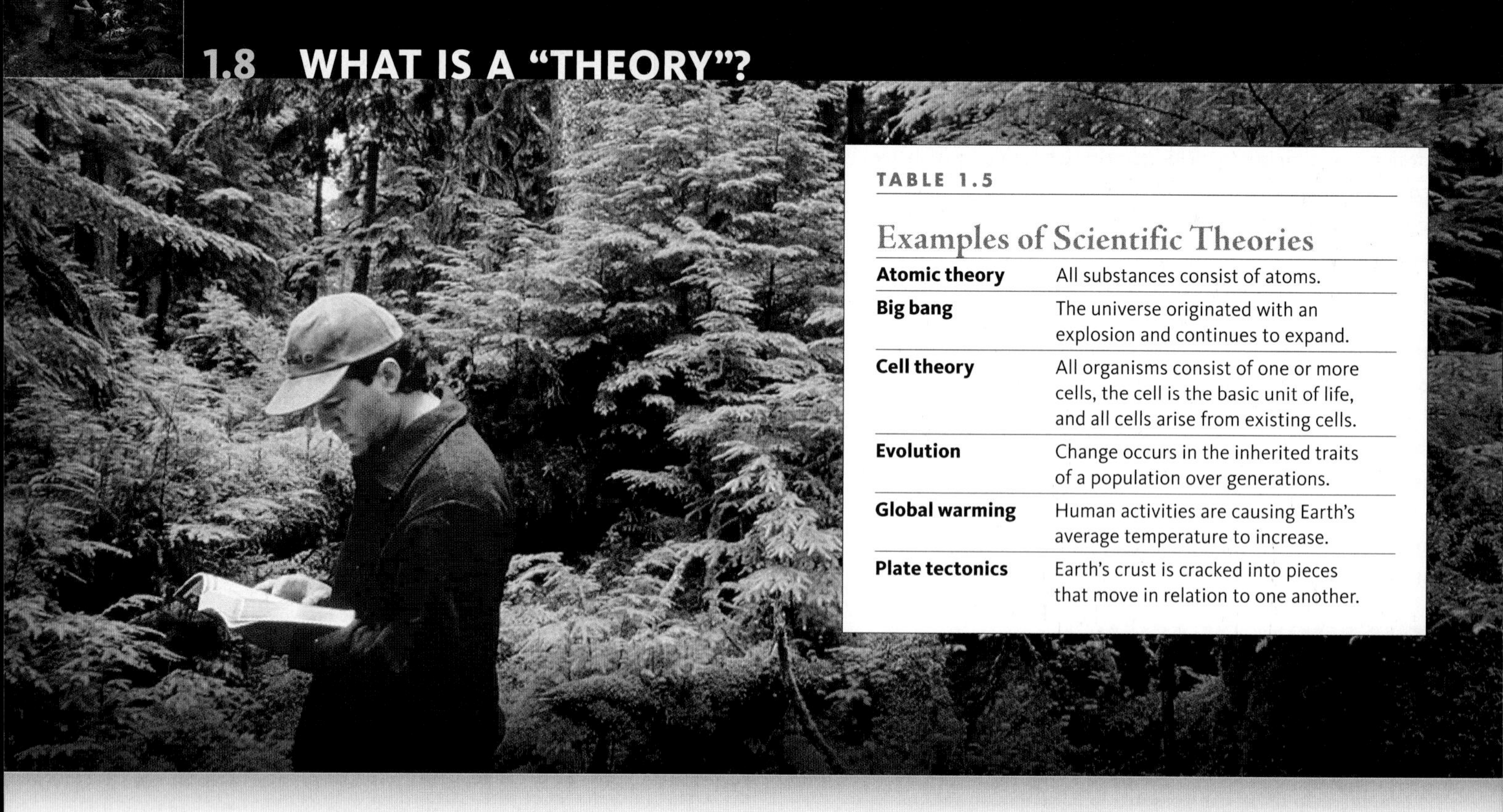

TABLE 1.5

Examples of Scientific Theories

Atomic theory	All substances consist of atoms.
Big bang	The universe originated with an explosion and continues to expand.
Cell theory	All organisms consist of one or more cells, the cell is the basic unit of life, and all cells arise from existing cells.
Evolution	Change occurs in the inherited traits of a population over generations.
Global warming	Human activities are causing Earth's average temperature to increase.
Plate tectonics	Earth's crust is cracked into pieces that move in relation to one another.

Suppose a hypothesis stands even after years of tests. It is consistent with all data ever gathered, and it has helped us make successful predictions about other phenomena. When a hypothesis meets these criteria, it is considered to be a **scientific theory** (TABLE 1.5). To give an example, all observations to date have been consistent with the hypothesis that matter consists of atoms. Scientists no longer spend time testing this hypothesis for the compelling reason that, since we started looking 200 years ago, no one has discovered matter that consists of anything else. Thus, scientists use the hypothesis, now called atomic theory, to make other hypotheses about matter.

Scientific theories are our best objective descriptions of the natural world, but they can never be proven absolutely because to do so would necessitate testing under every possible circumstance. For example, in order to prove atomic theory, the composition of all matter in the universe would have to be checked—an impossible task even if someone wanted to try.

Like all hypotheses, a scientific theory can be disproven by one observation or result that is inconsistent with it. For example, if someone discovers a form of matter that does not consist of atoms, atomic theory would be revised until no one could prove it to be incorrect. This potentially falsifiable nature of scientific theories is part of science's built-in system of checks and balances. The theory of evolution, which states that change occurs in a line of descent over time, still holds after a century of observations and testing. As with all other scientific theories, no one can be absolutely sure that it will hold under all possible conditions, but it has a very high probability of not being wrong. Few other theories have withstood as much scrutiny.

You may hear people apply the word "theory" to a speculative idea, as in the phrase "It's just a theory." This everyday usage of the word differs from the way it is used in science. Speculation is an opinion, belief, or personal conviction that is not necessarily supported by evidence. A scientific theory differs because it is supported by a large body of evidence, and it is consistent with all known facts.

A scientific theory also differs from a **law of nature**, which describes a phenomenon that has been observed to occur in every circumstance without fail, but for which we do not have a complete scientific explanation. The laws of thermodynamics, which describe energy, are examples. As you will see in Chapter 5, we understand *how* energy behaves, but not exactly *why* it behaves the way it does.

law of nature Generalization that describes a consistent natural phenomenon for which there is incomplete scientific explanation.
scientific theory Hypothesis that has not been disproven after many years of rigorous testing.

TAKE-HOME MESSAGE 1.8

A scientific theory is a time-tested hypothesis that is consistent with all known facts. It is our most objective way of describing the natural world.

1.9 Application: THE SECRET LIFE OF EARTH

Researcher Paul Oliver discovered this tiny tree frog perched on a sack of rice during a particularly rainy campsite lunch in New Guinea's Foja Mountains. The explorers dubbed the new species "Pinocchio frog" after the Disney character because the male frog's long nose inflates and points upward during times of excitement.

Exploration

IN THIS ERA OF DETAILED CELL PHONE GPS, could there possibly be any places left on Earth that humans have not yet explored? Actually, there are plenty. For example, a 2-million-acre cloud forest in New Guinea was only recently penetrated by explorers. How did the explorers know they had landed in uncharted territory? For one thing, the forest was filled with plants and animals unknown even to native peoples that have long inhabited other parts of the region. Team member Bruce Beehler remarked, "I was shouting. This trip was a once-in-a-lifetime series of shouting experiences." The team members discovered many new species, including a rhododendron plant with flowers the size of plates and a frog the size of a pea. They also came across hundreds of species that are on the brink of extinction in other parts of the world, and some that supposedly had been extinct for decades.

Each new species is a reminder that we do not yet know all of the organisms that share our planet. We don't even know how many to look for. Why does that matter? Understanding the scope of life on Earth gives us perspective on where we fit into it. For example, the current rate of extinctions is about 1,000 times faster than ever recorded, and we now know that human activities are responsible for the acceleration. At this rate, we will never know about most of the species that are alive today. Is that important? Biologists think so. Whether or not we are aware of it, humans are intimately connected with the world around us. Our activities are profoundly changing the entire fabric of life on Earth. The changes are, in turn, affecting us in ways we are only beginning to understand.

Ironically, the more we learn about the natural world, the more we realize we have yet to learn. But don't take our word for it. Find out what biologists know, and what they do not, and you will have a solid foundation upon which to base your own opinions about the human connection—your connection—with all life on Earth.

Summary

SECTION 1.1 **Biology** is the scientific study of life. Biologists think about life at different levels of organization, with **emergent properties** appearing at successive levels. All matter consists of **atoms**, which combine as **molecules. Organisms** are individuals that consist of one or more **cells**, the level at which life emerges. Cells of larger multicelled organisms are organized as **tissues**, **organs**, and **organ systems**. A **population** is a group of interbreeding individuals of a species in a given area; a **community** is all populations of all species in a given area. An **ecosystem** is a community interacting with its environment. The **biosphere** includes all regions of Earth that hold life.

SECTION 1.2 All organisms require energy and **nutrients** to sustain themselves. **Producers** harvest energy from the environment to make their own food by processes such as **photosynthesis**; **consumers** eat other organisms, their wastes, or remains. Organisms keep the conditions in their internal environment within ranges that their cells tolerate—a process called **homeostasis. DNA** contains information that guides an organism's **growth**, **development**, and **reproduction**. The passage of DNA from parents to offspring is called **inheritance**.

SECTION 1.3 The many types of organisms that currently exist on Earth differ greatly in details of body form and function. **Biodiversity** is the sum of differences among living things. **Bacteria** and **archaea** are both **prokaryotes**, single-celled organisms whose DNA is not contained within a **nucleus**. The DNA of single-celled or multicelled **eukaryotes** (**protists**, **plants**, **fungi**, and **animals**) is contained within a nucleus.

SECTION 1.4 Each **species** has a two-part name. The first part is the **genus** name. When combined with the **specific epithet**, it designates the particular species. With **taxonomy**, species are ranked into ever more inclusive **taxa** on the basis of shared **traits**.

SECTION 1.5 **Critical thinking**, the self-directed act of judging the quality of information as one learns, is an important part of **science**. Generally, a researcher observes something in nature, uses **inductive reasoning** to form a **hypothesis** (testable explanation) for it, then uses **deductive reasoning** to make a testable **prediction** about what might occur if the hypothesis is correct. **Experiments** with **variables** may be performed on an **experimental group** as compared with a **control group**, and sometimes on **models**. A researcher changes an **independent variable**, then observes the effects of the change on a **dependent variable**. Conclusions are drawn from the resulting **data**. The **scientific method** consists of making, testing, and evaluating hypotheses, and sharing results.

SECTION 1.6 Biological systems are usually influenced by many interacting variables. Research approaches differ, but experiments are typically designed in a consistent way, in order to study a single cause-and-effect relationship in a complex natural system.

SECTION 1.7 Small sample size increases the potential for **sampling error** in experimental results. In such cases, a subset may be tested that is not representative of the whole. Researchers design experiments carefully to minimize sampling error and bias, and they use **probability** rules to check the **statistical significance** of their results. Science is ideally a self-correcting process because scientists check and test one another's ideas. Science helps us be objective about our observations because it is only concerned with testable ideas about observable aspects of nature. Opinion and belief have value in human culture, but they are not addressed by science.

SECTION 1.8 A **scientific theory** is a long-standing hypothesis that is useful for making predictions about other phenomena. It is our best way of describing reality. A **law of nature** describes something that occurs without fail, but has an incomplete scientific explanation.

SECTION 1.9 We know about only a fraction of the organisms that live on Earth, in part because we have explored only a fraction of its inhabited regions.

Self-Quiz

Answers in Appendix VII

1. _______ are fundamental building blocks of all matter.
   a. Atoms
   b. Molecules
   c. Cells
   d. Organisms

2. The smallest unit of life is the _______ .
   a. atom
   b. molecule
   c. cell
   d. organism

3. Organisms require _______ and _______ to maintain themselves, grow, and reproduce.

4. By sensing and responding to change, organisms keep conditions in the internal environment within ranges that cells can tolerate. This process is called _______ .

5. DNA _______ .
   a. guides form and function
   b. is the basis of traits
   c. is transmitted from parents to offspring
   d. all of the above

6. A process by which an organism produces offspring is called _______ .

7. _______ is the transmission of DNA to offspring.
   a. Reproduction
   b. Development
   c. Homeostasis
   d. Inheritance

8. A butterfly is a(n) _______ (choose all that apply).
   a. organism
   b. domain
   c. species
   d. eukaryote
   e. consumer
   f. producer
   g. prokaryote
   h. trait

Data Analysis Activities

Peacock Butterfly Predator Defenses The photographs below represent experimental and control groups used in the peacock butterfly experiment discussed in Section 1.6. See if you can identify the experimental groups, and match them up with the relevant control group(s). ***Hint:*** Identify which variable is being tested in each group (each variable has a control).

A Wing spots painted out

B Wing spots visible; wings silenced

C Wing spots painted out; wings silenced

D Wings painted but spots visible

E Wings cut but not silenced

F Wings painted, spots visible; wings cut, not silenced

9. _______ move around for at least part of their life.

10. A bacterium is _______ (choose all that apply).
    a. an organism　　c. an animal
    b. single-celled　　d. a eukaryote

11. Bacteria, Archaea, and Eukarya are three _______ .

12. A control group is _______ .
    a. a set of individuals that have a certain characteristic or receive a certain treatment
    b. the standard against which an experimental group is compared
    c. the experiment that gives conclusive results

13. Fifteen randomly selected students are found to be taller than 6 feet. The researchers concluded that the average height of a student is greater than 6 feet. This is an example of _______ .
    a. experimental error　　c. a subjective opinion
    b. sampling error　　d. experimental bias

14. Science only addresses that which is _______ .
    a. alive　　c. variable
    b. observable　　d. indisputable

15. Match the terms with the most suitable description.

___ life	a. if–then statement
___ probability	b. unique type of organism
___ species	c. emerges with cells
___ hypothesis	d. testable explanation
___ prediction	e. measure of chance
___ producer	f. makes its own food

Critical Thinking

1. A person is declared dead upon the irreversible ceasing of spontaneous body functions: brain activity, blood circulation, and respiration. Only about 1% of a body's cells have to die in order for all of these things to happen. How can a person be dead when 99% of his or her cells are alive?

2. Explain the difference between a one-celled organism and a single cell of a multicelled organism.

3. Why would you think twice about ordering from a restaurant menu that lists the specific epithet but not the genus name of its offerings? *Hint:* Look up *Homarus americanus*, *Ursus americanus*, *Ceanothus americanus*, *Bufo americanus*, *Lepus americanus*, and *Nicrophorus americanus*.

4. Once there was a highly intelligent turkey that had nothing to do but reflect on the world's regularities. Morning always started out with the sky turning light, followed by the master's footsteps, which were always followed by the appearance of food. Other things varied, but food always followed footsteps. The sequence of events was so predictable that it eventually became the basis of the turkey's theory about the goodness of the world. One morning, after more than 100 confirmations of this theory, the turkey listened for the master's footsteps, heard them, and had its head chopped off.

Any scientific theory is modified or discarded upon discovery of contradictory evidence. The absence of absolute certainty has led some people to conclude that "theories are irrelevant because they can change." If that is so, should we stop doing scientific research? Why or why not?

5. In 2005, researcher Woo-suk Hwang reported that he had made immortal stem cells from human patients. His research was hailed as a breakthrough for people affected by degenerative diseases, because stem cells may be used to repair a person's own damaged tissues. Hwang published his results in a peer-reviewed journal. In 2006, the journal retracted his paper after other scientists discovered that Hwang's group had faked their data.

Does the incident show that results of scientific studies cannot be trusted? Or does it confirm the usefulness of a scientific approach, because other scientists discovered and exposed the fraud?

A unique set of properties makes water essential to life. These properties arise from interactions among individual water molecules.

2 LIFE'S CHEMICAL BASIS

Links to Earlier Concepts

In this chapter, you will explore the first level of life's organization--atoms—as you encounter the first example of how the same building blocks, arranged different ways, form different products (Section 1.1). You will also see one aspect homeostasis, the process by which organisms keep themselve in a state that favors cell survival (1.2).

KEY CONCEPTS

ATOMS AND ELEMENTS

Atoms, the building blocks of all matter, diffe in their numbers of protons, neutrons, and electrons. Atoms of an element have the sam number of protons.

WHY ELECTRONS MATTER

Whether an atom interacts with other atoms depends on the number of electrons it has. An atom with an unequal number of electrons and protons is an ion.

ATOMS BOND

Atoms of many elements interact by acquiring, sharing, and giving up electrons. Interacting atoms may form ionic, covalent, or hydrogen bonds.

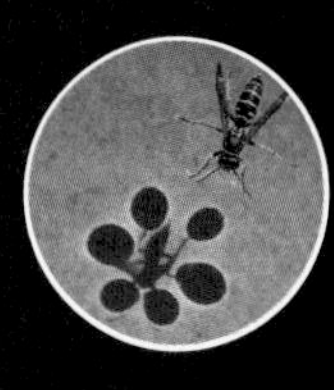

WATER

Hydrogen bonding among individual molecule gives water properties that make life possible temperature stabilization, cohesion, and the ability to dissolve many other substances.

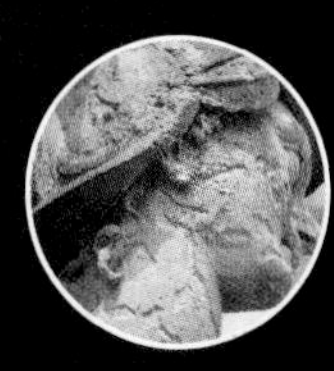

HYDROGEN POWER

Most of the chemistry of life occurs in a narr range of pH, so most fluids inside organisms buffered to stay within that range.

Photograph by Paul Nicklen, National Geographic Creative.

2.1 WHAT ARE THE BASIC BUILDING BLOCKS OF ALL MATTER?

Even though atoms are about 20 million times smaller than a grain of sand, they consist of even smaller subatomic particles. Positively charged **protons** (p^+) and uncharged **neutrons** occur in an atom's core, or **nucleus**. Negatively charged **electrons** (e^-) move around the nucleus (**FIGURE 2.1A**). **Charge** is an electrical property: Opposite charges attract, and like charges repel.

A typical atom has about the same number of electrons and protons. The negative charge of an electron is the same magnitude as the positive charge of a proton, so the two charges cancel one another. Thus, an atom with the same number of electrons and protons carries no charge.

All atoms have protons. The number of protons in the nucleus is called the **atomic number**, and it determines the type of atom, or element. **Elements** are pure substances, each consisting only of atoms with the same number of protons in their nucleus (**FIGURE 2.1B**). For example, the atomic number of carbon is 6, so all atoms with six protons in their nucleus are carbon atoms, no matter how many electrons or neutrons they have. Carbon, the substance, consists only of carbon atoms, and all of those atoms have six protons.

Knowing the numbers of electrons, protons, and neutrons in atoms helps us predict how elements will behave. In 1869, chemist Dmitry Mendeleyev arranged the elements known at the time by their chemical properties. The arrangement, which he called the **periodic table** (**FIGURE 2.1C**), turned out to be by atomic number, even though subatomic particles would not be discovered until the early 1900s.

In the periodic table, each element is represented by a symbol that is typically an abbreviation of the element's Latin or Greek name. For instance, Pb (lead) is short for *plumbum*; the word "plumbing" is related—ancient Romans made their water pipes with lead. Carbon's symbol, C, is from *carbo*, the Latin word for coal (which is mostly carbon).

FIGURE 2.1 {Animated} Atoms and elements.

A Atoms consist of electrons moving around a nucleus of protons and neutrons. Models such as this one do not show what atoms look like. Electrons move in defined, three-dimensional spaces about 10,000 times bigger than the nucleus.

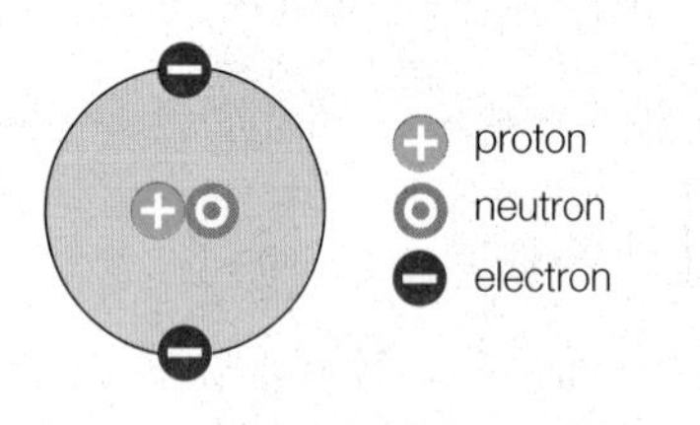

B Example of an element.

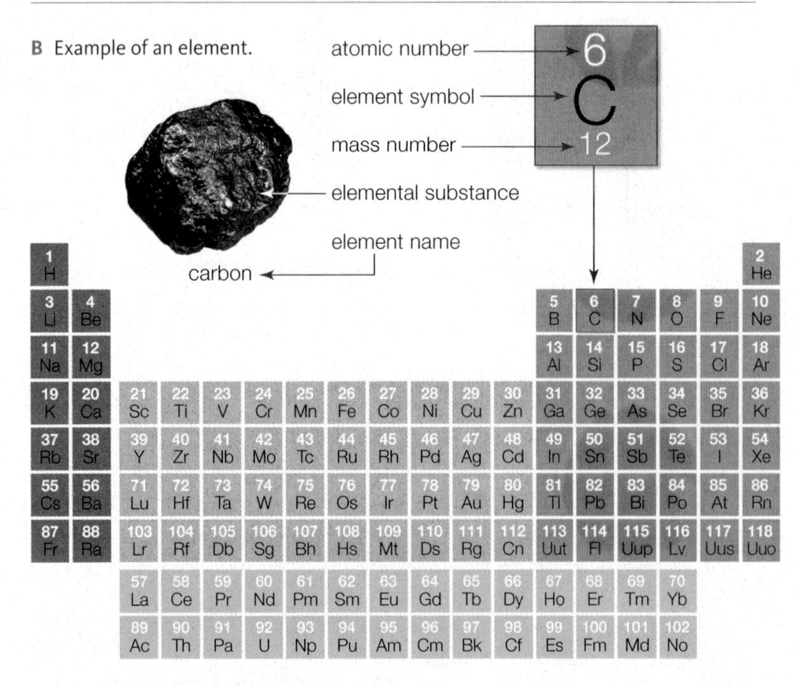

C The periodic table of the elements.

ISOTOPES AND RADIOISOTOPES

All atoms of an element have the same number of protons, but they can differ in the number of other subatomic particles. Those that differ in the number of neutrons are called **isotopes**. We define isotopes by their **mass number**, which is the total number of protons and neutrons in their nucleus. Mass number is written as a superscript to the left of an element's symbol. For example, hydrogen, the simplest atom, has one proton and no neutrons, so it is designated ^{1}H. The most common isotope of carbon has six protons and six neutrons, so it is ^{12}C, or carbon 12. The other naturally occurring isotopes of carbon are ^{13}C (six protons, seven neutrons), and ^{14}C (six protons, eight neutrons).

Carbon 14 is a **radioisotope**, or radioactive isotope. Atoms of a radioisotope have an unstable nucleus that breaks up

atomic number Number of protons in the atomic nucleus; determines the element.
charge Electrical property. Opposite charges attract, and like charges repel.
electron Negatively charged subatomic particle.
element A pure substance that consists only of atoms with the same number of protons.
isotopes Forms of an element that differ in the number of neutrons their atoms carry.
mass number Of an isotope, the total number of protons and neutrons in the atomic nucleus.
neutron Uncharged subatomic particle in the atomic nucleus.
nucleus Core of an atom; occupied by protons and neutrons.
periodic table Tabular arrangement of all known elements by their atomic number.
proton Positively charged subatomic particle that occurs in the nucleus of all atoms.
radioactive decay Process by which atoms of a radioisotope emit energy and/or subatomic particles when their nucleus spontaneously breaks up.
radioisotope Isotope with an unstable nucleus.
tracer A molecule with a detectable component.

spontaneously. As a nucleus breaks up, it emits radiation—subatomic particles, energy, or both—a process called **radioactive decay**. The atomic nucleus cannot be altered by ordinary means, so radioactive decay is unaffected by external factors such as temperature, pressure, or whether the atoms are part of molecules.

Each radioisotope decays at a predictable rate into predictable products. For example, when carbon 14 decays, one of its neutrons splits into a proton and an electron. The nucleus emits the electron as radiation. Thus, a carbon atom with eight neutrons and six protons (^{14}C) becomes a nitrogen atom, with seven neutrons and seven protons (^{14}N):

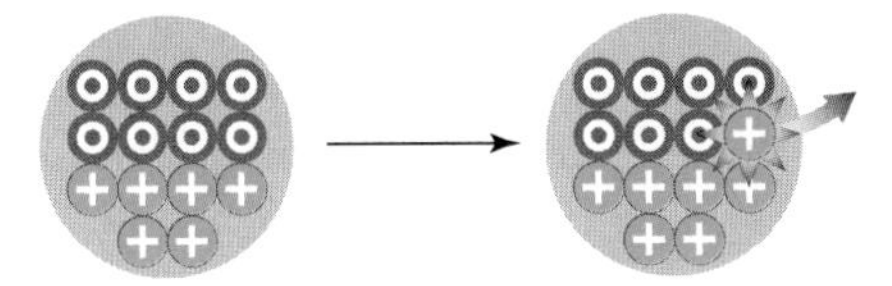

nucleus of ^{14}C, with 6 protons, 8 neutrons

nucleus of ^{14}N, with 7 protons, 7 neutrons

This process is so predictable that we can say with certainty that about half of the atoms in any sample of ^{14}C will be ^{14}N atoms after 5,730 years. The predictable rate of radioactive decay makes it possible for scientists to estimate the age of a rock or fossil by measuring its isotope content (we return to this topic in Section 16.4).

TRACERS

All isotopes of an element generally have the same chemical properties regardless of the number of neutrons in their atoms. This consistent chemical behavior means that organisms use atoms of one isotope the same way that they use atoms of another. Thus, radioisotopes can be used as tracers to study biological processes. A **tracer** is any substance with a detectable component. For example, a molecule in which an atom (such as ^{12}C) has been replaced with a radioisotope (such as ^{14}C) can be used as a radioactive tracer. When delivered into a biological system such as a cell, body, or ecosystem, this tracer may be followed as it moves through the system with instruments that detect radiation.

TAKE-HOME MESSAGE 2.1

All matter consists of atoms, tiny particles that in turn consist of electrons moving around a nucleus of protons and neutrons.

An element is a pure substance that consists only of atoms with the same number of protons. Isotopes are forms of an element that have different numbers of neutrons.

Unstable nuclei of radioisotopes break down spontaneously (decay) at a predictable rate to form predictable products.

PEOPLE MATTER

National Geographic Explorer
KENNETH SIMS

A volcano is a force of nature most of us prefer to observe from a very long and safe distance. Not volcanologist Ken Sims. The National Geographic explorer could never be satisfied with anything less than standing on the edge of an erupting volcano. In fact, even standing on the edge of an erupting volcano wasn't enough for Sims. As part of his research, he rappelled down into the mouth of Nyiragongo, a volcano in the Democratic Republic of the Congo, to gather fresh lava from a molten lake boiling at 1800°F.

Sims says, "While many think of me as a volcanologist, I am actually an isotope geochemist and, as such, I have pondered and written professional papers on a wide range of problems, including chemical oceanography; the Earth's paleo-climate; oceanic and continental crustal growth; continental crustal weathering; ground water transport; and, of course, the genesis and evolution of volcanic systems. The measurement of radioactive isotopes in natural systems allows me to quantify fundamental processes that would otherwise be limited to qualitative observation. This may make me sound like a nerd but to be able to quantify the Earth's processes is truly inspiring."

2.2 WHY DO ATOMS INTERACT?

ELECTRONS MATTER

Electrons are really, really small. How small are they? If they were as big as apples, you would be about 3.5 times taller than our solar system is wide. Simple physics explains the motion of, say, an apple falling from a tree, but electrons are so tiny that such everyday physics cannot explain their behavior. For example, electrons carry energy, but only in incremental amounts. An electron gains energy only by absorbing the exact amount needed to boost it to the next energy level. Likewise, it loses energy only by emitting the exact difference between two energy levels. This concept will be important to remember when you learn how cells harvest and release energy.

A lot of electrons may be zipping around in the same atom. Despite moving really fast (around 3 million meters per second), they never collide. Why not? For one reason, electrons in an atom occupy different orbitals, which are defined volumes of space around the atomic nucleus.

To understand how orbitals work, imagine that an atom is a multilevel apartment building, with the nucleus in the basement. Each "floor" of the building corresponds to a certain energy level, and each has a certain number of "rooms" (orbitals) available for rent. Two electrons can occupy each room. Pairs of electrons populate rooms from the ground floor up; in other words, they fill orbitals from lower to higher energy levels. The farther an electron is from the nucleus in the basement, the greater its energy. An electron can move to a room on a higher floor if an energy input gives it a boost, but it immediately emits the extra energy and moves back down.

A **shell model** helps us visualize how electrons populate atoms (**FIGURE 2.2**). In this model, nested "shells" correspond to successively higher energy levels. Thus, each shell includes all of the rooms (orbitals) on one floor (energy level) of our atomic apartment building.

We draw a shell model of an atom by filling it with electrons (represented as balls or dots), from the innermost shell out, until there are as many electrons as the atom has protons. There is only one room on the first floor, one orbital at the lowest energy level. It fills up first. In hydrogen, the simplest atom, a single electron occupies that room (**FIGURE 2.2A**). Helium, with two protons, has two electrons that fill the room—and the first shell. In larger atoms, more electrons rent the second-floor rooms (**FIGURE 2.2B**). When the second floor fills, more electrons rent third-floor rooms (**FIGURE 2.2C**), and so on.

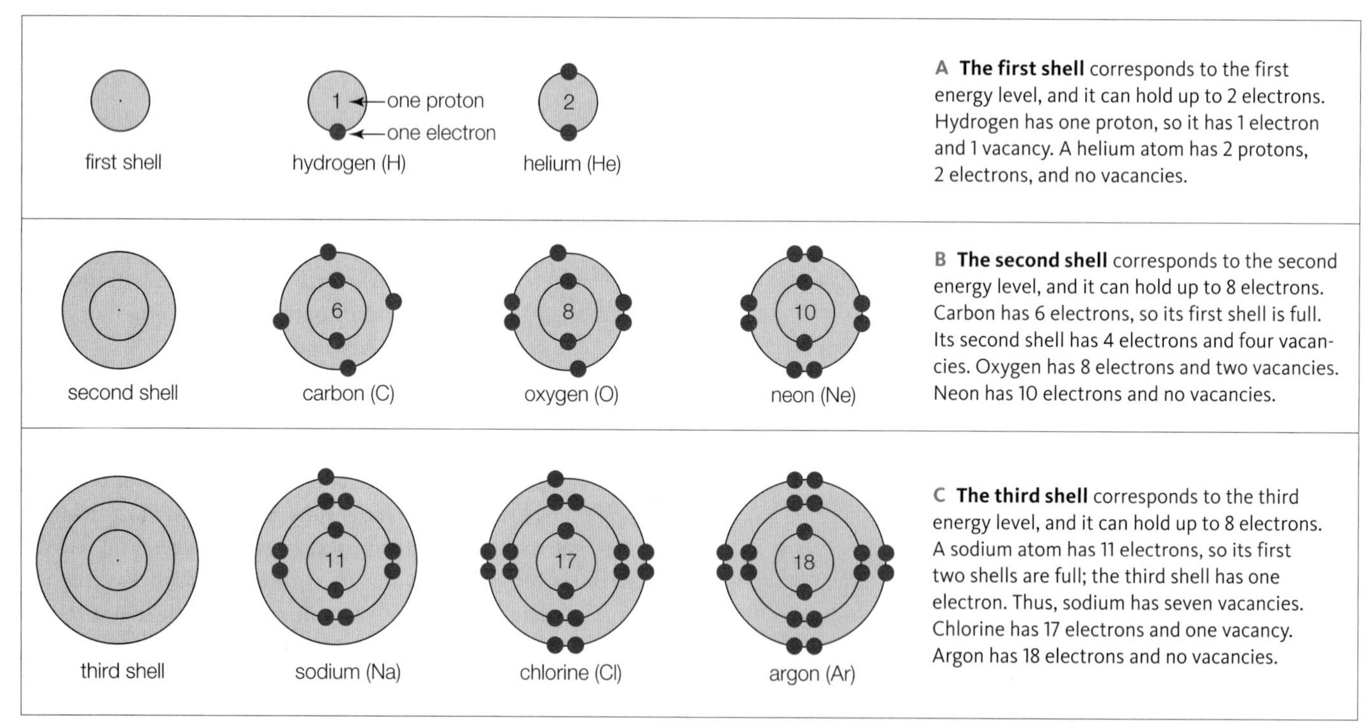

FIGURE 2.2 {Animated} Shell models. Each circle (shell) represents one energy level. To make these models, we fill the shells with electrons from the innermost shell out, until there are as many electrons as the atom has protons. The number of protons in each model is indicated.

FIGURE IT OUT: Which of these models have unpaired electrons in their outer shell? Answer: Hydrogen, carbon, oxygen, sodium, and chlorine

ABOUT VACANCIES

When an atom's outermost shell is filled with electrons, we say that it has no vacancies. Any atom is in its most stable state when it has no vacancies. Helium, neon, and argon are examples of elements with no vacancies. Atoms of these elements are chemically stable, which means they have no tendency to interact with other atoms. Thus, these elements occur most frequently in nature as solitary atoms.

By contrast, when an atom's outermost shell has room for another electron, it has a vacancy. Atoms with vacancies tend to get rid of them by interacting with other atoms; in other words, they are chemically active. For example, the sodium atom (Na) depicted in **FIGURE 2.2C** has one electron in its outer (third) shell, which can hold eight. With seven vacancies, we can predict that this atom is chemically active.

In fact, this particular sodium atom is not just active, it is extremely so. Why? The shell model shows that a sodium atom has an unpaired electron, but in the real world, electrons really like to be in pairs when they occupy orbitals. Solitary atoms that have unpaired electrons are called **free radicals**. With some exceptions, free radicals are very unstable, easily forcing electrons upon other atoms or ripping electrons away from them. This property makes free radicals dangerous to life (we return to this topic in Section 5.5).

A free radical sodium atom can easily evict its one unpaired electron, so that its second shell—which is full of electrons—becomes its outermost, and no vacancies remain. This is the atom's most stable state. The vast majority of sodium atoms on Earth are like this one, with 11 protons and 10 electrons.

Atoms with an unequal number of protons and electrons are called **ions**. Ions carry a net (or overall) charge. Sodium ions (Na^+) offer an example of how atoms gain a positive charge by losing an electron (**FIGURE 2.3A**). Other atoms gain a negative charge by accepting an electron. For example, an uncharged chlorine atom has 17 protons and 17 electrons. The outermost shell of this atom can hold eight electrons, but it has only seven. With one vacancy and one unpaired electron, we can predict—correctly—that this atom is chemically very active. An uncharged chlorine atom easily fills its third shell by accepting an electron. When that happens, the atom becomes a chloride ion (Cl^-) with 17 protons, 18 electrons, and a net negative charge (**FIGURE 2.3B**).

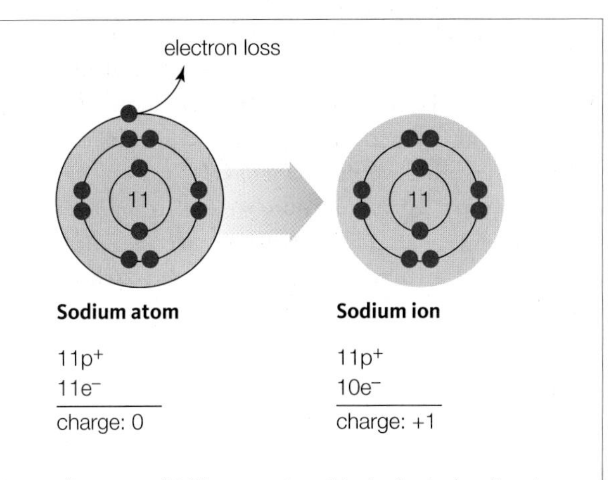

A A sodium atom (Na) becomes a positively charged sodium ion (Na^+) when it loses the single electron in its third shell. The atom's full second shell is now its outermost, so it has no vacancies.

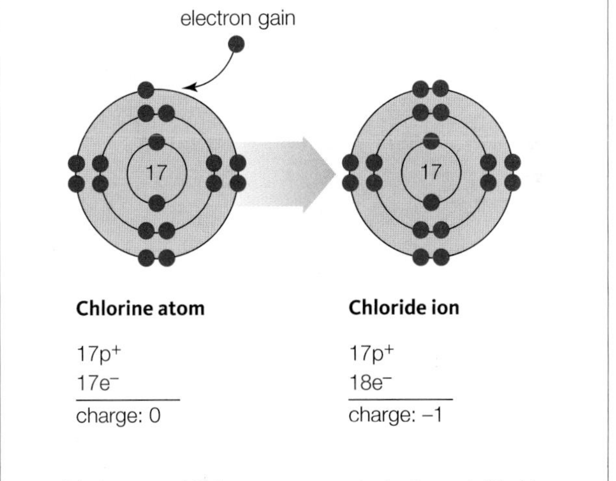

B A chlorine atom (Cl) becomes a negatively charged chloride ion (Cl^-) when it gains an electron and fills the vacancy in its third, outermost shell.

FIGURE 2.3 Ion formation.

TAKE-HOME MESSAGE 2.2

An atom's electrons are the basis of its chemical behavior.

Shells represent all electron orbitals at one energy level in an atom. When the outermost shell is not full of electrons, the atom has a vacancy.

Atoms with vacancies tend to interact with other atoms.

free radical Atom with an unpaired electron.
ion Charged atom.
shell model Model of electron distribution in an atom.

2.3 HOW DO ATOMS INTERACT IN CHEMICAL BONDS?

An atom can get rid of vacancies by participating in a chemical bond with another atom. A **chemical bond** is an attractive force that arises between two atoms when their electrons interact. Chemical bonds link atoms into molecules. In other words, each molecule consists of atoms held together in a particular number and arrangement by chemical bonds. For example, a water molecule consists of three atoms: two hydrogen atoms bonded to the same oxygen atom (**FIGURE 2.4**). A water molecule is also a **compound**, which means it has atoms of two or more elements. Other molecules, including molecular oxygen (a gas in air), have atoms of one element only.

The term "bond" applies to a continuous range of atomic interactions. However, we can categorize most bonds into distinct types based on their properties. Which type forms depends on the atoms taking part in the molecule.

FIGURE 2.4 The water molecule. Each water molecule has two hydrogen atoms bonded to the same oxygen atom.

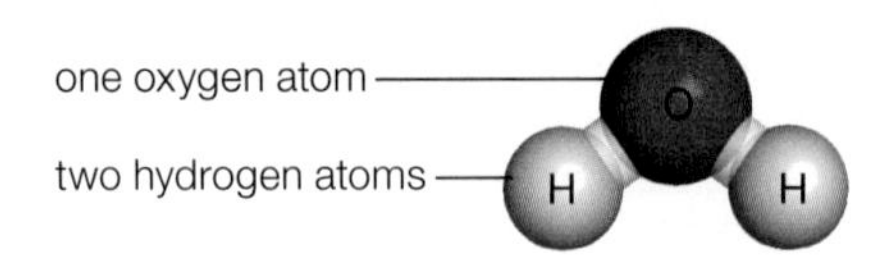

A Each crystal of table salt consists of many sodium and chloride ions locked together in a cubic lattice by ionic bonds.

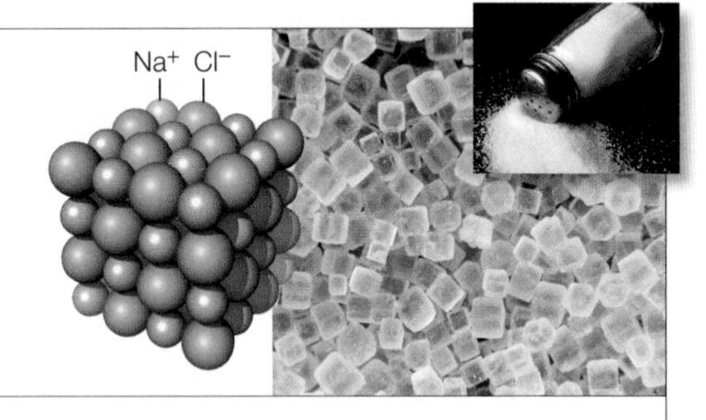

B The strong mutual attraction of opposite charges holds a sodium ion and a chloride ion together in an ionic bond.

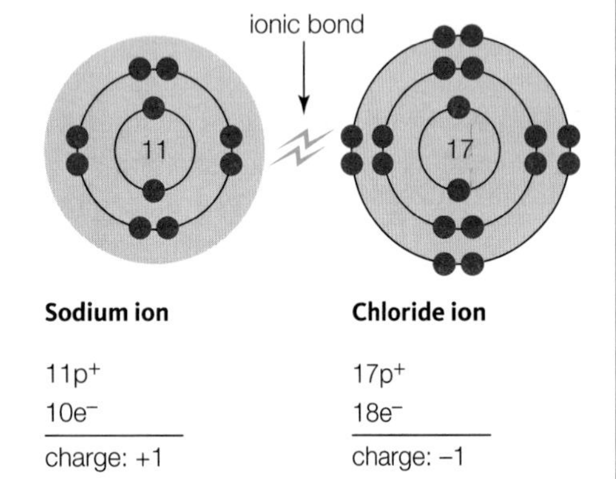

C Ions taking part in an ionic bond retain their charge, so the molecule itself is polar. One side is positively charged (represented by a blue overlay); the other side is negatively charged (red overlay).

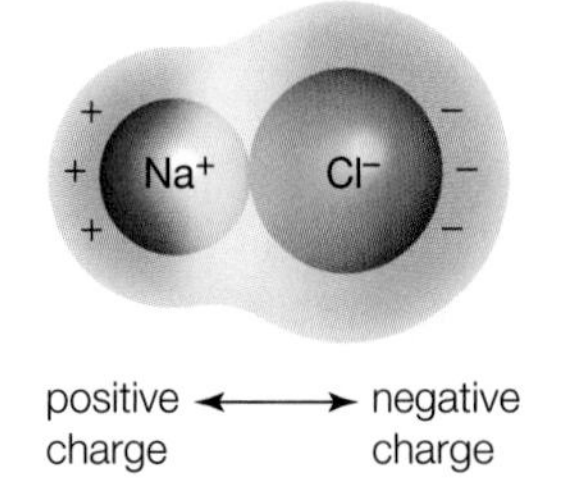

FIGURE 2.5 Ionic bonds in table salt, or NaCl.

IONIC BONDS

Two ions may stay together by the mutual attraction of their opposite charges, an association called an **ionic bond**. Ionic bonds can be quite strong. Ionically bonded sodium and chloride ions make up sodium chloride (NaCl), which we know as common table salt. A crystal of this substance consists of a cubic lattice of sodium and chloride ions interacting in ionic bonds (**FIGURE 2.5A**).

Ions retain their respective charges when participating in an ionic bond (**FIGURE 2.5B**). Thus, one "end" of an ionic bond has a positive charge, and the other "end" has a negative charge. Any such separation of charge into distinct positive and negative regions is called **polarity** (**FIGURE 2.5C**). A sodium chloride molecule is polar because the chloride ion keeps a very strong hold on its extra electron. In other words, it is strongly electronegative. **Electronegativity** is a measure of an atom's ability to pull electrons away from another atom. Electronegativity is not the same thing as charge. Rather, an atom's electronegativity depends on its size, how many vacancies it has, and what other atoms it is interacting with.

An ionic bond is very polar because the atoms that are participating in it have a very large difference in electronegativity. When atoms with a lower difference in electronegativity interact, they tend to form chemical bonds that are less polar than ionic bonds.

COVALENT BONDS

Covalent bonds form between atoms with a small difference in electronegativity or none at all. In a **covalent bond**, two atoms share a pair of electrons, so each atom's vacancy is partially filled (**FIGURE 2.6**). Sharing electrons links the two atoms, just as sharing a pair of earphones links two friends (*above*). Covalent bonds can be stronger than ionic bonds, but they are not always so.

TABLE 2.1 shows different ways of representing covalent bonds. In structural formulas, a line between two

TABLE 2.1

Ways of Representing Molecules

Common name:	Water	Familiar term.
Chemical name:	Dihydrogen monoxide	Describes elemental composition.
Chemical formula:	H_2O	Indicates unvarying proportions of elements. Subscripts show number of atoms of an element per molecule. The absence of a subscript means one atom.
Structural formula:	H—O—H	Represents each covalent bond as a single line between atoms.
Structural model:		Shows relative sizes and positions of atoms in three dimensions.
Shell model:		Shows how pairs of electrons are shared in covalent bonds.

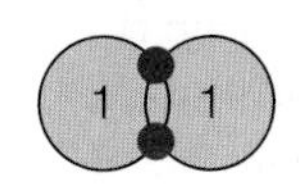

MOLECULAR HYDROGEN (H—H)

Two hydrogen atoms, each with one proton, share two electrons in a nonpolar covalent bond.

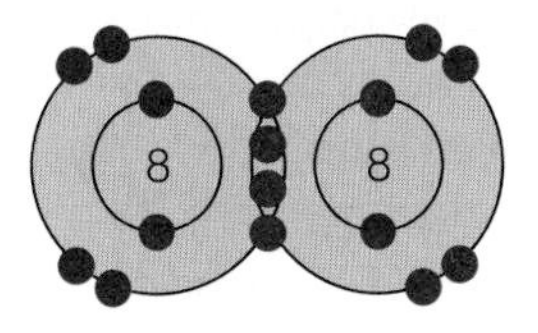

MOLECULAR OXYGEN (O=O)

Two oxygen atoms, each with eight protons, share four electrons in a double covalent bond.

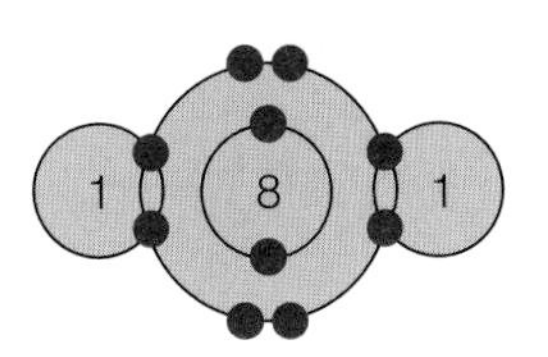

WATER (H—O—H)

Two hydrogen atoms share electrons with an oxygen atom in two covalent bonds. The bonds are polar because the oxygen exerts a greater pull on the shared electrons than the hydrogens do.

FIGURE 2.6 {Animated} Covalent bonds, in which atoms fill vacancies by sharing electrons. Two electrons are shared in each covalent bond. When sharing is equal, the bond is nonpolar. When one atom exerts a greater pull on the electrons, the bond is polar.

atoms represents a single covalent bond, in which two atoms share one pair of electrons. For example, molecular hydrogen (H_2) has one covalent bond between hydrogen atoms (H—H). Two, three, or even four covalent bonds may form between atoms when they share multiple pairs of electrons. For example, two atoms sharing two pairs of electrons are connected by two covalent bonds, which are represented by a double line between the atoms. A double bond links the two oxygen atoms in molecular oxygen (O=O). Three lines indicate a triple bond, in which two atoms share three pairs of electrons. A triple covalent bond links the two nitrogen atoms in molecular nitrogen (N≡N). Comparing bonds between the same two atoms: A triple bond is stronger than a double bond, which is stronger than a single bond.

Double and triple bonds are not distinguished from single bonds in structural models, which show positions and relative sizes of the atoms in three dimensions. The bonds are shown as one stick connecting two balls, which represent atoms. Elements are usually coded by color:

carbon

hydrogen

oxygen

nitrogen

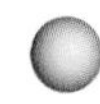
phosphorus

Atoms share electrons unequally in a polar covalent bond. A bond between an oxygen atom and a hydrogen atom in a water molecule is an example. One atom (the oxygen, in this case) is a bit more electronegative. It pulls the electrons a little more toward its side of the bond, so that atom bears a slight negative charge. The atom at the other end of the bond (the hydrogen) bears a slight positive charge. Covalent bonds in compounds are usually polar. By contrast, atoms participating in a nonpolar covalent bond share electrons equally. There is no difference in charge between the two ends of such bonds. The bonds in molecular hydrogen (H_2), oxygen (O_2), and nitrogen (N_2) are nonpolar.

chemical bond An attractive force that arises between two atoms when their electrons interact.
compound Molecule that has atoms of more than one element.
covalent bond Chemical bond in which two atoms share a pair of electrons.
electronegativity Measure of the ability of an atom to pull electrons away from other atoms.
ionic bond Type of chemical bond in which a strong mutual attraction links ions of opposite charge.
polarity Separation of charge into positive and negative regions.

TAKE-HOME MESSAGE 2.3

A chemical bond forms between atoms when their electrons interact. A chemical bond may be ionic or covalent depending on the atoms taking part in it.

An ionic bond is a strong mutual attraction between two ions of opposite charge. Ionic bonds are very polar.

Atoms share a pair of electrons in a covalent bond. When the atoms share electrons unequally, the bond is polar.

2.4 WHAT ARE LIFE-SUSTAINING PROPERTIES OF WATER?

HYDROGEN BONDING IN WATER

Water has unique properties that arise from the two polar covalent bonds in each water molecule. Overall, the molecule has no charge, but the oxygen atom carries a slight negative charge; the hydrogen atoms, a slight positive charge. Thus, the molecule itself is polar (**FIGURE 2.7A**).

The polarity of individual water molecules attracts them to one another. The slight positive charge of a hydrogen atom in one water molecule is drawn to the slight negative charge of an oxygen atom in another. This type of interaction is called a hydrogen bond. A **hydrogen bond** is an attraction between a covalently bonded hydrogen atom and another atom taking part in a separate polar covalent bond (**FIGURE 2.7B**). Like ionic bonds, hydrogen bonds form by the mutual attraction of opposite charges. However, unlike ionic bonds, hydrogen bonds do not make molecules out of atoms, so they are not chemical bonds.

Hydrogen bonds are on the weaker end of the spectrum of atomic interactions, and they form and break much more easily than covalent or ionic bonds. Even so, many of them form, and collectively they are quite strong. As you will see, hydrogen bonds stabilize the characteristic structures of biological molecules such as DNA and proteins. They also form in tremendous numbers among water molecules (**FIGURE 2.7C**). Extensive hydrogen bonding among water molecules gives liquid water several special properties that make life possible.

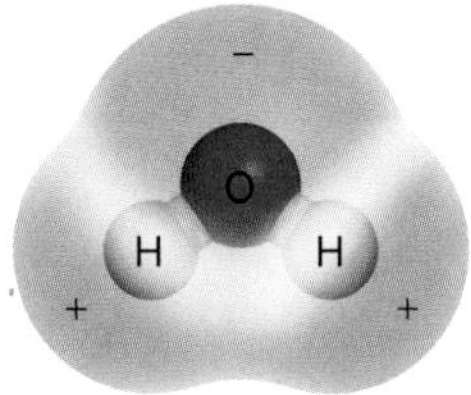

A Polarity of the water molecule. Each of the hydrogen atoms in a water molecule bears a slight positive charge (represented by a blue overlay). The oxygen atom carries a slight negative charge (red overlay).

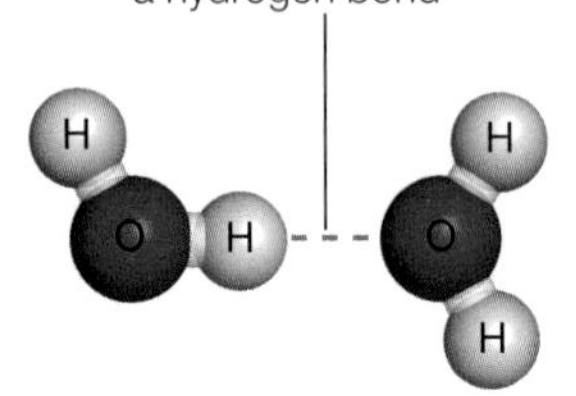

B A hydrogen bond is an attraction between a hydrogen atom and another atom taking part in a separate polar covalent bond.

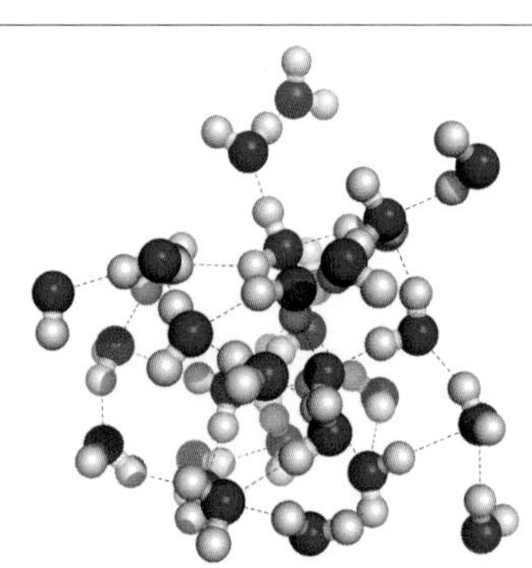

C The many hydrogen bonds that form among water molecules impart special properties to liquid water.

FIGURE 2.7 {Animated} Hydrogen bonds in water.

WATER'S SPECIAL PROPERTIES

Water Is an Excellent Solvent The polarity of the water molecule and its ability to form hydrogen bonds make water an excellent **solvent**, which means that many other substances easily dissolve in it. Substances that dissolve easily in water are **hydrophilic** (water-loving). Ionic solids such as sodium chloride (NaCl) dissolve in water because the slight positive charge on each hydrogen atom in a water molecule attracts negatively charged ions (Cl^-), and the slight negative charge on the oxygen atom attracts positively charged ions (Na^+). Hydrogen bonds among many water molecules are collectively stronger than an ionic bond between two ions, so the solid dissolves as water molecules tug the ions apart and surround each one (*right*).

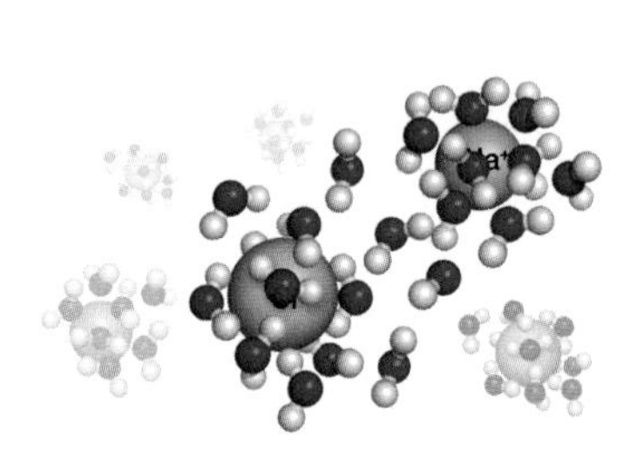

Sodium chloride is called a **salt** because it releases ions other than H^+ and OH^- when it dissolves in water (more about these ions in the next section). When a substance such as NaCl dissolves, its component ions disperse uniformly among the molecules of liquid, and it becomes a **solute**. A uniform mixture such as salt dissolved in water is called a **solution**. Chemical bonds do not form between molecules of solute and solvent, so the proportions of the two substances in a solution can vary.

Nonionic solids such as sugars dissolve easily in water because their molecules can form hydrogen bonds with water molecules. Hydrogen bonding with water does not break the covalent bonds of such molecules; rather, it dissolves the substance by pulling individual molecules away from one another and keeping them apart.

Water does not interact with **hydrophobic** (water-dreading) substances such as oils. Oils consist of nonpolar molecules, and hydrogen bonds do not form between nonpolar molecules and water. When you mix oil and water, the water breaks into small droplets, but quickly begins to cluster into larger drops as new hydrogen bonds form among its molecules. The bonding excludes molecules of oil and pushes them together into drops that rise to the surface of the water. The same interactions occur at the thin, oily membrane that separates the watery fluid inside cells from the watery fluid outside of them. As you will see in Chapter 3, such interactions give rise to the structure of cell membranes.

Water Has Cohesion Molecules of some substances resist separating from one another, and the resistance gives rise to a property called **cohesion**. Water has cohesion because hydrogen bonds collectively exert a continuous pull on its individual molecules. You can see cohesion in water as surface tension, which means that the surface of liquid water behaves a bit like a sheet of elastic (*left*).

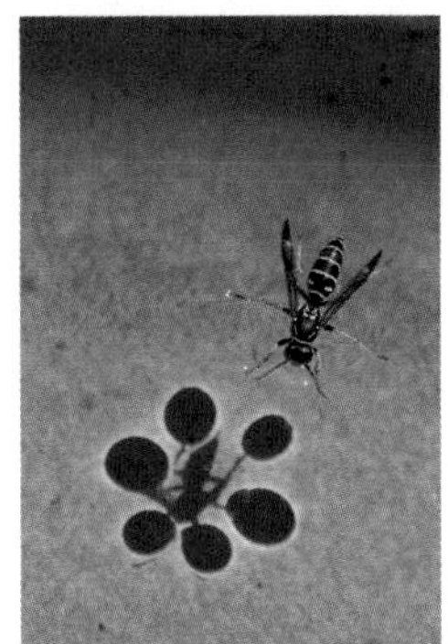

Cohesion plays a role in many processes that sustain multicelled bodies. As one example, water molecules constantly escape from the surface of liquid water as vapor, a process called **evaporation**. Evaporation is resisted by hydrogen bonding among water molecules. In other words, overcoming water's cohesion takes energy. Thus, evaporation sucks energy (in the form of heat) from liquid water, and this lowers the water's surface temperature. Evaporative water loss helps you and some other mammals cool off when you sweat in hot, dry weather. Sweat, which is about 99 percent water, cools the skin as it evaporates.

Cohesion works inside organisms, too. Consider how plants absorb water from soil as they grow. Water molecules evaporate from leaves, and replacements are pulled upward from roots. Cohesion makes it possible for columns of liquid water to rise from roots to leaves inside narrow pipelines of vascular tissue. In some trees, these pipelines extend hundreds of feet above the soil (Section 26.4 returns to this topic).

FIGURE 2.8 {Animated} Hydrogen bonds lock water molecules in a rigid lattice in ice. The molecules in this lattice pack less densely than in liquid water, which is why ice floats on water. A covering of ice can insulate water underneath it, thus keeping aquatic organisms from freezing during harsh winters.

Water Stabilizes Temperature All atoms jiggle nonstop, so the molecules they make up jiggle too. We measure the energy of this motion as degrees of **temperature**. Adding energy (in the form of heat, for example) makes the jiggling faster, so the temperature rises.

Hydrogen bonding keeps water molecules from jiggling as much as they would otherwise, so it takes more heat to raise the temperature of water compared with other liquids. Temperature stability is an important part of homeostasis, because most of the molecules of life function properly only within a certain range of temperature.

Below 0°C (32°F), water molecules do not jiggle enough to break hydrogen bonds, and they become locked in the rigid, lattice-like bonding pattern of ice (**FIGURE 2.8**). Individual water molecules pack less densely in ice than they do in water, which is why ice floats on water. Sheets of ice that form on the surface of ponds, lakes, and streams can insulate the water under them from subfreezing air temperatures. Such "ice blankets" protect aquatic organisms during cold winters.

TAKE-HOME MESSAGE 2.4

Extensive hydrogen bonding among water molecules, which arises from the polarity of the individual molecules, gives water special properties.

Liquid water is an excellent solvent. Hydrophilic substances such as salts and sugars dissolve easily in water to form solutions. Hydrophobic substances do not dissolve in water.

Water also has cohesion, and it stabilizes temperature.

cohesion Property of a substance that arises from the tendency of its molecules to resist separating from one another.
evaporation Transition of a liquid to a vapor.
hydrogen bond Attraction between a covalently bonded hydrogen atom and another atom taking part in a separate covalent bond.
hydrophilic Describes a substance that dissolves easily in water.
hydrophobic Describes a substance that resists dissolving in water.
salt Compound that releases ions other than H^+ and OH^- when it dissolves in water.
solute A dissolved substance.
solution Uniform mixure of solute completely dissolved in solvent.
solvent Liquid that can dissolve other substances.
temperature Measure of molecular motion.

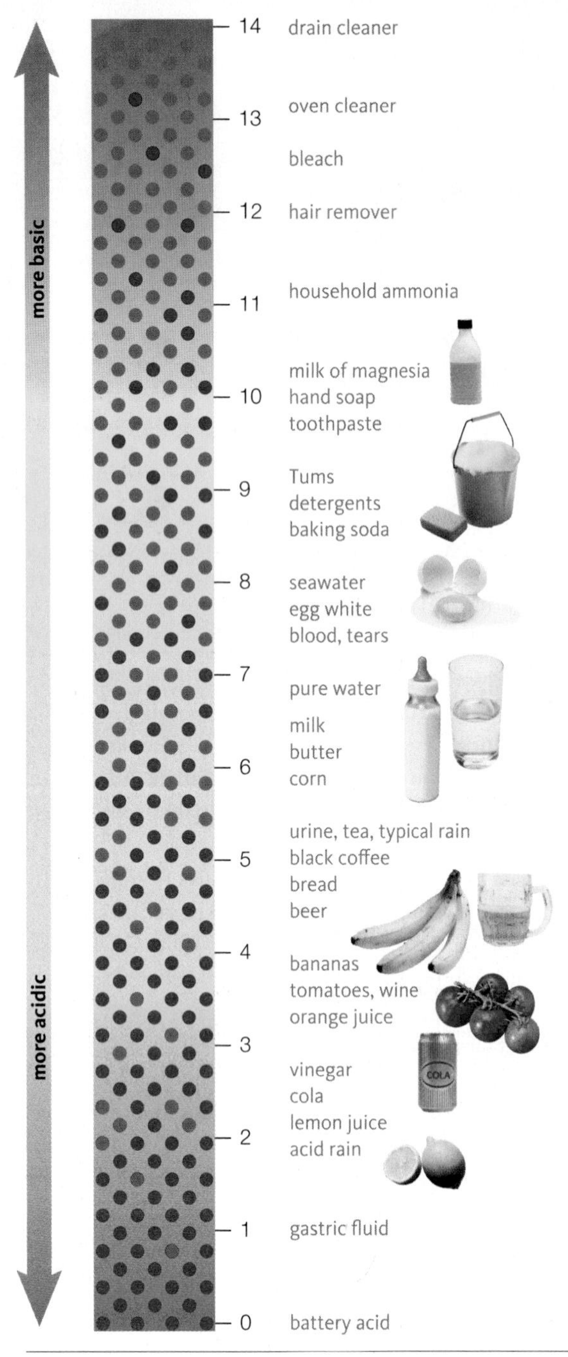

FIGURE 2.9 A pH scale. Here, red dots signify hydrogen ions (H^+) and blue dots signify hydroxyl ions (OH^-). Also shown are the approximate pH values for some common solutions.

This pH scale ranges from 0 (most acidic) to 14 (most basic). A change of one unit on the scale corresponds to a tenfold change in the amount of H^+ ions.

FIGURE IT OUT: What is the approximate pH of cola?

Answer: 2.5

When water is liquid, some of its molecules spontaneously separate into hydrogen ions (H^+) and hydroxide ions (OH^-). These ions can combine again to form water:

$$\underset{\text{water}}{H_2O} \longrightarrow \underset{\text{hydrogen ions}}{H^+} + \underset{\text{hydroxide ions}}{OH^-} \longrightarrow \underset{\text{water}}{H_2O}$$

Concentration refers to the amount of a particular solute dissolved in a given volume of fluid. Hydrogen ion (H^+) concentration is a special case. We measure the amount of hydrogen ions in a solution using a value called **pH**. When the number of H^+ ions equals the number of OH^- ions in the liquid, the pH is 7, or neutral. The higher the number of hydrogen ions, the lower the pH. A one-unit decrease in pH corresponds to a tenfold increase in the number of H^+ ions, and a one-unit increase corresponds to a tenfold decrease in the number of H^+ ions (**FIGURE 2.9**).

One way to get a sense of the pH scale is to taste dissolved baking soda (pH 9), pure water (pH 7), and lemon juice (pH 2). Nearly all of life's chemistry occurs near pH 7. Most of your body's internal environment (tissue fluids and blood) stays between pH 7.3 and 7.5.

Substances called **bases** accept hydrogen ions, so they can raise the pH of fluids and make them basic, or alkaline (above pH 7). **Acids** give up hydrogen ions when they dissolve in water, so they lower the pH of fluids and make them acidic (below pH 7).

Strong acids ionize completely in water to give up all of their H^+ ions; weak acids give up only some of them. Hydrochloric acid (HCl) is an example of a strong acid: its H^+ and Cl^- ions stay separated in water. Inside your stomach, the H^+ from HCl makes gastric fluid acidic (pH 1–2). Carbonic acid is an example of a weak acid. It forms when carbon dioxide gas (CO_2) dissolves in the watery, fluid portion of human blood:

$$\underset{\text{carbon dioxide}}{CO_2} + H_2O \longrightarrow \underset{\text{carbonic acid}}{H_2CO_3}$$

A carbonic acid molecule can break apart into a hydrogen ion and a bicarbonate ion, which in turn can recombine to form carbonic acid again:

$$\underset{\text{carbonic acid}}{H_2CO_3} \longrightarrow H^+ + \underset{\text{bicarbonate}}{HCO_3^-} \longrightarrow \underset{\text{carbonic acid}}{H_2CO_3}$$

Together, carbonic acid and bicarbonate constitute a **buffer**, a set of chemicals that can keep the pH of a solution stable by alternately donating and accepting ions that contribute to pH. For example, when a base is added to an unbuffered fluid, the number of OH^- ions increases, so the pH rises. However, if the fluid is buffered, the addition of base causes the buffer to release H^+ ions. These combine with OH^- ions to form water, which has no effect on pH. Excess hydrogen ions combine with the buffer, so they do not contribute to pH. Thus, the pH of a buffered fluid stays the same when base or acid is added.

Under normal circumstances, the fluids inside cells (as well as those inside bodies) stay within a consistent range of pH because they are buffered. For example, excess OH^- in blood combines with the H^+ from carbonic acid to form water, which does not contribute to pH. Excess H^+ in blood combines with bicarbonate, so it does not affect pH. This exchange of ions keeps the blood pH stable, but only up to a certain point. A buffer can neutralize only so many ions; even slightly more than that limit and the pH of the fluid will change dramatically.

Most biological molecules can function properly only within a narrow range of pH. Even a slight deviation from that range can halt cellular processes, so buffer failure can be catastrophic in a biological system. For instance, when breathing is impaired suddenly, carbon dioxide gas accumulates in tissues, so too much carbonic acid forms in blood. The resulting decline in blood pH may cause the person to enter a coma (a dangerous level of unconsciousness). By contrast, hyperventilation (sustained rapid breathing) causes the body to lose too much CO_2. The loss results in a rise in blood pH. If blood pH rises too much, prolonged muscle spasm (tetany) or coma may occur.

Burning fossil fuels such as coal releases sulfur and nitrogen compounds that affect the pH of rain and other forms of precipitation. Rainwater is not buffered, so the addition of acids or bases has a dramatic effect. In places with a lot of fossil fuel emissions, the rain and fog can be more acidic than vinegar. The corrosive effect of this acid rain is visible in urban areas (*left*). Acid rain also drastically changes the pH of water in soil, lakes, and streams. Such changes can overwhelm the buffering capacity of fluids inside organisms, with lethal effects. We return to acid rain in Section 44.4.

acid Substance that releases hydrogen ions in water.
base Substance that accepts hydrogen ions in water.
buffer Set of chemicals that can keep the pH of a solution stable by alternately donating and accepting ions that contribute to pH.
concentration Amount of solute per unit volume of solution.
pH Measure of the number of hydrogen ions in a fluid.

TAKE-HOME MESSAGE 2.5

The number of hydrogen ions in a fluid determines its pH. Most biological systems function properly only within a narrow range of pH.

Acids release hydrogen ions in water; bases accept them. Salts release ions other than H^+ and OH^-.

Buffers help keep pH stable. Inside organisms, they are part of homeostasis.

2.6 MERCURY RISING

FIGURE 2.10 Mercury that falls on Earth's oceans accumulates in the bodies of tuna and other large predatory fish.

MERCURY IS A TOXIC ELEMENT. Most of it is safely locked away in rocks, but volcanic activity and other geologic processes release it into the atmosphere. So do human activities, especially burning coal. Airborne mercury can drift long distances before settling to Earth's surface, where microbes combine it with carbon to form a substance called methylmercury.

Unlike mercury alone, methylmercury easily crosses skin and mucous membranes. In water, it ends up in the tissues of aquatic organisms. All fish and shellfish contain it. Humans contain it too, mainly as a result of eating seafood. When mercury enters the body, it damages the nervous system, brain, kidneys, and other organs. An average-sized adult who ingests as little as 200 micrograms of methylmercury may experience blurred vision, tremors, itching or burning sensations, and loss of coordination. Exposure to larger amounts can result in thought and memory impairment, coma, and death. Methylmercury in a pregnant woman's blood passes to her unborn child, along with a legacy of permanent developmental problems.

It takes months or even years for mercury to be cleared from the body, so the toxin can build up to high levels if even small amounts are ingested on a regular basis. That is why large predatory fish have a lot of mercury in their tissues (FIGURE 2.10). It is also why the U.S. Environmental Protection Agency recommends that adult humans ingest less than 0.1 microgram of mercury per kilogram of body weight per day. For an average-sized person, that limit works out to be about 7 micrograms per day, which is not a big amount if you eat seafood. A typical 6-ounce can of albacore tuna contains about 60 micrograms of mercury, and the occasional can has many times that amount. It does not matter if the fish is canned, grilled, or raw, because methylmercury is unaffected by cooking. Eat a medium-sized tuna steak, and you could be getting more than 700 micrograms of mercury along with it.

CREDITS: (in text) W. K. Fletcher/Science Source; (10) Brian J. Skerry/National Geographic Creative.

Summary

SECTION 2.1 Atoms consist of **electrons**, which carry a negative **charge**, moving about a **nucleus** of positively charged **protons** and uncharged **neutrons** (TABLE 2.2). The **periodic table** lists **elements** in order of **atomic number**. **Isotopes** of an element differ in the number of neutrons. The total number of protons and neutrons is the **mass number**. **Tracers** can be made with **radioisotopes**, which, by a process called **radioactive decay**, emit particles and energy when their nucleus spontaneously breaks up.

SECTION 2.2 Which atomic orbital an electron occupies depends on its energy. A **shell model** represents successive energy levels as concentric circles. Atoms tend to get rid of vacancies. Many do so by gaining or losing electrons, thereby becoming **ions**. Unpaired electrons make **free radicals** chemically active.

SECTION 2.3 A **chemical bond** is an attractive force that unites two atoms as a molecule. A **compound** consists of two or more elements. Atoms form different types of bonds depending on their **electronegativity**. The mutual attraction of opposite charges can hold atoms together in an **ionic bond**, which is completely polar (**polarity** is separation of charge). Atoms share a pair of electrons in a **covalent bond**, which is nonpolar if the sharing is equal, and polar if it is not.

SECTION 2.4 Two polar covalent bonds give each water molecule an overall polarity. **Hydrogen bonds** that form among water molecules in tremendous numbers are the basis of water's unique properties. Water has **cohesion** and a capacity to act as a **solvent** for **salts** and other polar **solutes**; and it resists **temperature** changes. **Hydrophilic** substances dissolve easily in water to form **solutions**; **hydrophobic** substances do not. **Evaporation** is the transition of liquid to vapor.

SECTION 2.5 A solute's **concentration** refers to the amount of solute in a given volume of fluid; **pH** reflects the number of hydrogen ions (H^+). **Acids** release hydrogen ions in water; **bases** accept them. A **buffer** can keep a solution within a consistent range of pH. Most cell and body fluids are buffered because most molecules of life work only within a narrow range of pH.

SECTION 2.6 Interactions between atoms make the molecules that sustain life, and also some that destroy it. Mercury in air pollution ends up in the bodies of fish, and in turn, in the bodies of humans.

TABLE 2.2

Players in the Chemisty of Life

Atoms	Particles that are basic building blocks of all matter.
Proton (p^+)	Positively charged particle of an atom's nucleus.
Electron (e^-)	Negatively charged particle that can occupy a defined volume of space (orbital) around an atom's nucleus.
Neutron	Uncharged particle of an atom's nucleus.
Element	Pure substance that consists entirely of atoms with the same, characteristic number of protons.
Isotopes	Atoms of an element that differ in the number of neutrons.
Radioisotope	Unstable isotope that emits particles and energy when its nucleus breaks up.
Tracer	Molecule that has a detectable component such as a radioisotope. Used to track the movement or destination of the molecule in a biological system.
Ion	Atom that carries a charge after it has gained or lost one or more electrons. A single proton without an electron is a hydrogen ion (H^+).
Molecule	Two or more atoms joined in a chemical bond.
Compound	Molecule of two or more different elements in unvarying proportions (for example, water: H_2O).
Solute	Substance dissolved in a solvent.
Hydrophilic	Refers to a substance that dissolves easily in water. Such substances consist of polar molecules.
Hydrophobic	Refers to a substance that resists dissolving in water. Such substances consist of nonpolar molecules.
Acid	Compound that releases H^+ when dissolved in water.
Base	Compound that accepts H^+ when dissolved in water.
Salt	Ionic compound that releases ions other than H^+ or OH^- when dissolved in water.
Solvent	Substance that can dissolve other substances.

Self-Quiz

Answers in Appendix VII

1. What atom has only one proton?
   a. hydrogen
   b. an isotope
   c. a free radical
   d. a radioisotope

2. A molecule into which a radioisotope has been incorporated can be used as a(n) ________ .
   a. compound
   b. tracer
   c. salt
   d. acid

3. Which of the following statements is incorrect?
   a. Isotopes have the same atomic number and different mass numbers.
   b. Atoms have about the same number of electrons as protons.
   c. All ions are atoms.
   d. Free radicals are dangerous because they emit energy.

4. In the periodic table, symbols for the elements are arranged according to ________ .
   a. size
   b. charge
   c. mass number
   d. atomic number

5. An ion is an atom that has ________ .
   a. the same number of electrons and protons
   b. a different number of electrons and protons
   c. electrons, protons, and neutrons

Data Analysis Activities

Radioisotopes in PET Scans Positron-emission tomography (PET) helps us "see" a functional process inside the body. By this procedure, a radioactive sugar or other tracer is injected into a patient, who is then moved into a scanner. Inside the patient's body, cells with differing rates of activity take up the tracer at different rates. The scanner detects radioactive decay wherever the tracer is, then translates that data into an image.

1. What conclusion does **FIGURE 2.11** and its caption suggest about the behavior of a smoker compared with that of a nonsmoker?
2. What is an alternate interpretation of the differences between the results of these two PET scans?
3. This experiment compares two human individuals, who undoubtedly differ in factors other than smoking. What would be an appropriate control for this study?

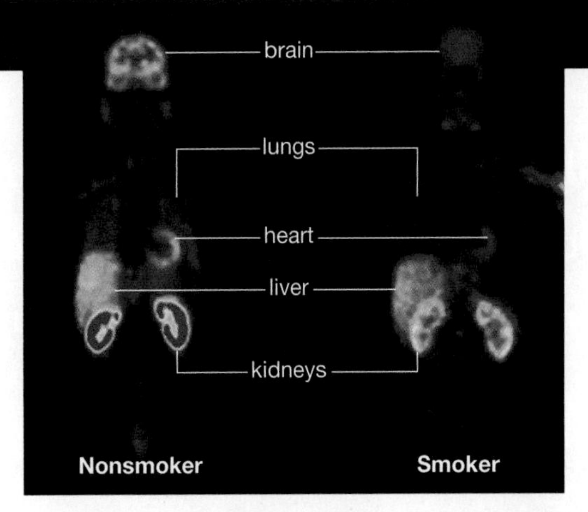

FIGURE 2.11 Two PET scans showing the activity of a molecule called MAO-B in the body of a nonsmoker (*left*) and a smoker (*right*). The activity is color-coded from red (highest activity) to purple (lowest). Low MAO-B activity is associated with violence, impulsiveness, and other behavioral problems.

6. The measure of an atom's ability to pull electrons away from another atom is called _______ .
   a. electronegativity b. charge c. polarity

7. The mutual attraction of opposite charges holds atoms together as molecules in a(n) _______ bond.
   a. ionic c. polar covalent
   b. hydrogen d. nonpolar covalent

8. Atoms share electrons unequally in a(n) _______ bond.
   a. ionic c. polar covalent
   b. hydrogen d. nonpolar covalent

9. A(n) _______ substance repels water.
   a. acidic c. hydrophobic
   b. basic d. polar

10. A salt does not release _______ in water.
    a. ions b. energy c. H^+

11. Hydrogen ions (H^+) are _______ .
    a. indicated by a pH scale c. in blood
    b. protons d. all of the above

12. When dissolved in water, a(n) _______ donates H^+; a(n) _______ accepts H^+.
    a. acid; base c. buffer; solute
    b. base; acid d. base; buffer

13. A(n) _______ can help keep the pH of a solution stable.
    a. covalent bond c. buffer
    b. hydrogen bond d. pH

14. A _______ is dissolved in a solvent.
    a. molecule b. solute c. salt

15. Match the terms with their most suitable description.

___ hydrophilic	a. protons > electrons
___ atomic number	b. number of protons in nucleus
___ hydrogen bonds	c. polar; dissolves easily in water
___ positive charge	d. collectively strong
___ temperature	e. protons < electrons
___ negative charge	f. measure of molecular motion

Critical Thinking

1. Alchemists were medieval scholars and philosophers who were the forerunners of modern-day chemists. Many spent their lives trying to transform lead (atomic number 82) into gold (atomic number 79). Explain why they never did succeed in that endeavor.

2. Draw a shell model of a lithium atom (Li), which has 3 protons, then predict whether the majority of lithium atoms on Earth are uncharged, positively charged, or negatively charged.

3. Polonium is a rare element with 33 radioisotopes. The most common one, ^{210}Po, has 82 protons and 128 neutrons. When ^{210}Po decays, it emits an alpha particle, which is a helium nucleus (2 protons and 2 neutrons). ^{210}Po decay is tricky to detect because alpha particles do not carry very much energy compared to other forms of radiation. They can be stopped by, for example, a sheet of paper or a few inches of air. This property is one reason why authorities failed to discover toxic amounts of ^{210}Po in the body of former KGB agent Alexander Litvinenko until after he died suddenly and mysteriously in 2006. What element does an atom of ^{210}Po change into after it emits an alpha particle?

4. Some undiluted acids are not as corrosive as when they are diluted with water. That is why lab workers are told to wipe off splashes with a towel before washing. Explain.

Rice has been cultivated for thousands of years. Carbohydrate-packed seeds make this grain the most important food source for humans worldwide.

Links to Earlier Concepts

Having learned about atomic interactions (Section 2.3), you are now in a position to understand the structure of the molecules of life. Keep the big picture in mind by reviewing Section 1.1. You will be building on your knowledge of covalent bonding (2.3), acids and bases (2.5), and the effects of hydrogen bonds (2.4).

MOLECULES OF LIFE

KEY CONCEPTS

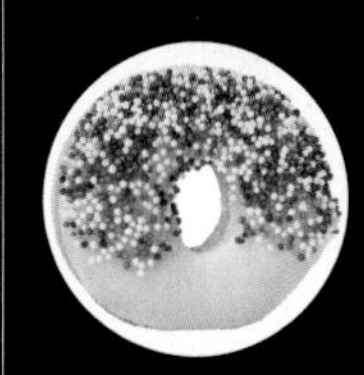

STRUCTURE DICTATES FUNCTION

Complex carbohydrates and lipids, proteins, and nucleic acids are assembled from simpler molecules. Functional groups add chemical character to a backbone of carbon atoms.

CARBOHYDRATES

Cells use carbohydrates as structural materials, for fuel, and to store and transport energy. They can build different complex carbohydrates from the same simple sugars.

LIPIDS

Lipids are the main structural component of all cell membranes. Cells use them to make other compounds, to store energy, and as waterproofing or lubricating substances.

PROTEINS

Proteins are the most diverse molecules of life. They include enzymes and structural materials. A protein's function arises from and depends on its structure.

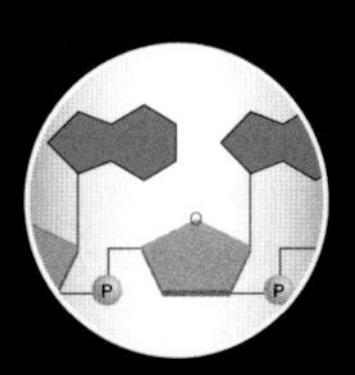

NUCLEIC ACIDS

Nucleotides are building blocks of nucleic acids; some have additional roles in metabolism. DNA stores a cell's heritable information. RNA helps put that information to use.

Photograph by Alex Treadway, National Geographic Creative.

THE STUFF OF LIFE: CARBON

The same elements that make up a living body also occur in nonliving things, but their proportions differ. For example, compared to sand or seawater, a human body contains a much larger proportion of carbon atoms. Why? Unlike sand or seawater, a body consists of a very high proportion of the molecules of life—complex carbohydrates and lipids, proteins, and nucleic acids—which in turn consist of a high proportion of carbon atoms. Molecules that have primarily hydrogen and carbon atoms are said to be **organic**. The term is a holdover from a time when these molecules were thought to be made only by living things, as opposed to the "inorganic" molecules that formed by nonliving processes.

Carbon's importance to life arises from its versatile bonding behavior. Carbon has four vacancies (Section 2.2), so it can form four covalent bonds with other atoms, including other carbon atoms. Many organic molecules have a backbone—a chain of carbon atoms—to which other atoms attach. The ends of a backbone may join to form a carbon ring structure (**FIGURE 3.1**). Carbon's ability to form chains and rings, and also to bond with many other elements, means that atoms of this element can be assembled into a wide variety of organic compounds.

We represent organic molecules in several ways. The structure of many organic molecules is quite complex (**FIGURE 3.2A**). For clarity, we may omit some of the bonds in a structural formula. Hydrogen atoms bonded to a carbon backbone may also be omitted. Carbon rings are often represented as polygons (**FIGURE 3.2B**). If no atom is shown at a corner or at the end of a bond, a carbon is implied there. Ball-and-stick models are useful for representing smaller organic compounds (**FIGURE 3.2C**). Space-filling models show a molecule's overall shape (**FIGURE 3.2D**). Proteins and nucleic acids are often represented as ribbon structures, which, as you will see in Section 3.4, show how the backbone folds and twists.

FIGURE 3.1 Carbon rings.

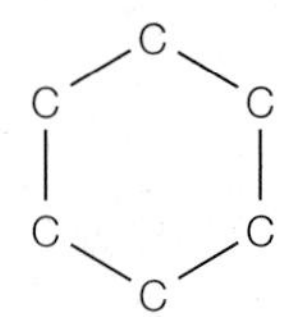

A Carbon's versatile bonding behavior allows it to form a variety of structures, including rings.

B Carbon rings form the framework of many sugars, starches, and fats (such as those found in doughnuts).

A A structural formula for an organic molecule—even a simple one—can be very complicated. The overall structure is obscured by detail.

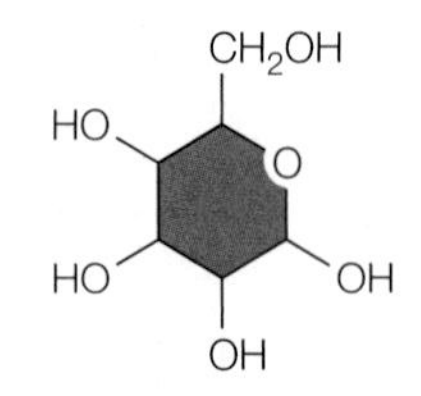

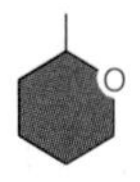

B Structural formulas of organic molecules are typically simplified by using polygons as symbols for rings, omitting some bonds and element labels.

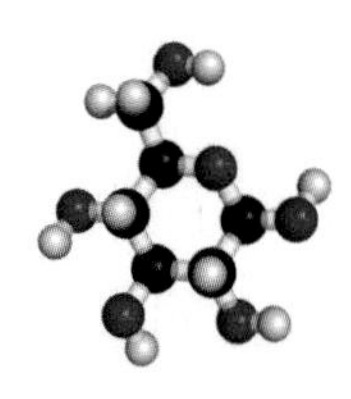

C A ball-and-stick model is often used to show the arrangement of atoms and bonds in three dimensions.

D A space-filling model can be used to show a molecule's overall shape. Individual atoms are visible in this model. Space-filling models of larger molecules often show only the surface contours.

FIGURE 3.2 Modeling an organic molecule. All of these models represent the same molecule: glucose.

FROM STRUCTURE TO FUNCTION

An organic molecule that consists only of hydrogen and carbon atoms is called a **hydrocarbon**. Hydrocarbons are generally nonpolar. Methane, the simplest kind, is one carbon atom bonded to four hydrogen atoms. Other organic

condensation Chemical reaction in which an enzyme builds a large molecule from smaller subunits; water also forms.
enzyme Organic molecule that speeds up a reaction without being changed by it.
functional group An atom (other than hydrogen) or a small molecular group bonded to a carbon of an organic compound; imparts a specific chemical property.
hydrocarbon Compound or region of one that consists only of carbon and hydrogen atoms.
hydrolysis Water-requiring chemical reaction in which an enzyme breaks a molecule into smaller subunits.
metabolism All of the enzyme-mediated chemical reactions by which cells acquire and use energy as they build and break down organic molecules.
monomers Molecules that are subunits of polymers.
organic Describes a molecule that consists mainly of carbon and hydrogen atoms.
polymer Molecule that consists of multiple monomers.
reaction Process of molecular change.

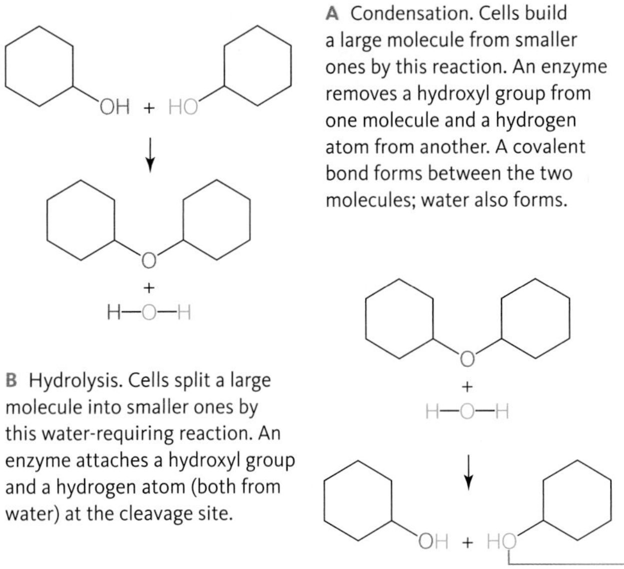

A Condensation. Cells build a large molecule from smaller ones by this reaction. An enzyme removes a hydroxyl group from one molecule and a hydrogen atom from another. A covalent bond forms between the two molecules; water also forms.

B Hydrolysis. Cells split a large molecule into smaller ones by this water-requiring reaction. An enzyme attaches a hydroxyl group and a hydrogen atom (both from water) at the cleavage site.

FIGURE 3.3 {Animated} Two common metabolic processes by which cells build and break down organic molecules.

molecules, including the molecules of life, have at least one functional group. A **functional group** is an atom (other than hydrogen) or small molecular group covalently bonded to a carbon atom of an organic compound. These groups impart chemical properties such as acidity or polarity (**TABLE 3.1**). The chemical behavior of the molecules of life arises mainly from the number, kind, and arrangement of their functional groups.

All biological systems are based on the same organic molecules, a similarity that is one of many legacies of life's common origin. However, the details of those molecules differ among organisms. Just as atoms bonded in different numbers and arrangements form different molecules, simple organic building blocks bonded in different numbers and arrangements form different versions of the molecules of life. These small organic molecules—simple sugars, fatty acids, amino acids, and nucleotides—are called **monomers** when they are used as subunits of larger molecules. Molecules that consist of multiple monomers are called **polymers**.

Cells build polymers from monomers, and break down polymers to release monomers. These and any other processes of molecular change are called chemical **reactions**. Cells constantly run reactions as they acquire and use energy to stay alive, grow, and reproduce—activities that are collectively called **metabolism**. Metabolism requires **enzymes**, which are organic molecules (usually proteins) that speed up reactions without being changed by them.

In many metabolic reactions, large organic molecules are assembled from smaller ones. With **condensation**, an enzyme covalently bonds two molecules together. Water (H—O—H) usually forms as a product of condensation

TABLE 3.1

Some Functional Groups in Biological Molecules

Group	Structure	Character	Formula	Found in:
acetyl	O=C(—)—CH_3	polar, acidic	—$COCH_3$	some proteins, coenzymes
aldehyde	—C—C(=O)—H	polar, reactive	—CHO	simple sugars
amide	—C(=O)—N<	weakly basic, stable, rigid	—C(O)N—	proteins nucleotide bases
amine	—N(H)H	very basic	—NH_2	nucleotide bases amino acids
carboxyl	—C—C(=O)—OH	very acidic	—COOH	fatty acids amino acids
hydroxyl	—O—H	polar	—OH	alcohols sugars
ketone	—C—C(=O)—C—	polar, acidic	—CO—	simple sugars nucleotide bases
methyl	—CH_3	nonpolar	—CH_3	fatty acids some amino acids
sulfhydryl	—S—H	forms rigid disulfide bonds	—SH	cysteine many cofactors
phosphate	—O—P(=O)(OH)—OH	polar, reactive	—PO_4	nucleotides DNA, RNA phospholipids proteins

when a hydroxyl group (—OH) from one of the molecules combines with a hydrogen atom (—H) from the other molecule (**FIGURE 3.3A**). With **hydrolysis**, the reverse of condensation, an enzyme breaks apart a large organic molecule into smaller ones. During hydrolysis, a bond between two atoms breaks when a hydroxyl group gets attached to one of the atoms, and a hydrogen atom gets attached to the other (**FIGURE 3.3B**). The hydroxyl group and hydrogen atom come from a water molecule, so this reaction requires water.

We will revisit enzymes and metabolic reactions in Chapter 5. The remainder of this chapter introduces the different types of biological molecules and the monomers from which they are built.

TAKE-HOME MESSAGE 3.1

The molecules of life are organic, which means they consist mainly of carbon and hydrogen atoms. Functional groups bonded to their carbon backbone impart chemical characteristics to these molecules.

Cells assemble large polymers from smaller monomer molecules. They also break apart polymers into monomers.

3.2 WHAT IS A CARBOHYDRATE?

Molecules of glycolaldehyde, a simple sugar, were recently discovered floating in gas surrounding a young, sunlike star. The finding is important because glycolaldehyde can react with other molecules found in space gas to form ribose, the five-carbon monosaccharide component of RNA. "What is really exciting about our findings is that the sugar molecules are falling in towards one of the stars of the system," says team member Cécile Favre. "The sugar molecules are not only in the right place to find their way onto a planet, but they are also going in the right direction." The discovery does not prove that life has developed elsewhere in the universe—but it implies that there is no reason it could not. It shows that the carbon-rich molecules that are the building blocks of life can be present even before planets have begun forming.

FIGURE 3.4 Astronomers made a sweet discovery in 2012.

Carbohydrates are organic compounds that consist of carbon, hydrogen, and oxygen in a 1:2:1 ratio. Cells use different kinds as structural materials, for fuel, and for storing and transporting energy. The three main types of carbohydrates in living systems are monosaccharides, oligosaccharides, and polysaccharides.

SIMPLE SUGARS

"Saccharide" is from *sacchar*, a Greek word that means sugar. Monosaccharides (one sugar) are the simplest type of carbohydrate. These molecules have extremely important biological roles. Common monosaccharides have a backbone of five or six carbon atoms (carbon atoms of sugars are numbered in a standard way: 1′, 2′, 3′, and so on, as illustrated in the model of glucose on the *right*).

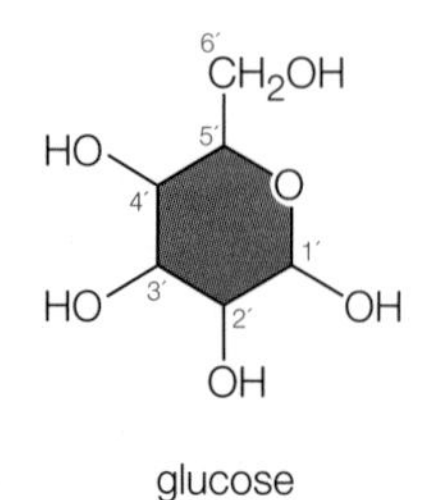

Glucose has six carbon atoms. Five-carbon monosaccharides are components of the nucleotide monomers of DNA and RNA (**FIGURE 3.4**). Two or more hydroxyl (—OH) groups impart solubility to a sugar molecule, which means that monosaccharides move easily through the water-based internal environments of all organisms.

Cells use monosaccharides for cellular fuel, because breaking the bonds of sugars releases energy that can be harnessed to power other cellular processes (we return to this important metabolic process in Chapter 7). Monosaccharides are also used as precursors, or parent molecules, that are remodeled into other molecules; and as structural materials to build larger molecules.

POLYMERS OF SIMPLE SUGARS

Oligosaccharides are short chains of covalently bonded monosaccharides (*oligo*– means a few). Disaccharides consist of two monosaccharide monomers. The lactose in milk, with one glucose and one galactose, is a disaccharide. Sucrose, the most plentiful sugar in nature, has a glucose and a fructose unit (sucrose extracted from sugarcane or sugar beets is our table sugar). Oligosaccharides attached to lipids or proteins have important functions in immunity.

Foods that we call "complex" carbohydrates consist mainly of polysaccharides: chains of hundreds or thousands of monosaccharide monomers. The chains may be straight or branched, and can consist of one or many types of monosaccharides. The most common polysaccharides are cellulose, starch, and glycogen. All consist only of glucose monomers, but as substances their properties are very different. Why? The answer begins with differences in patterns of covalent bonding that link their monomers.

Cellulose, the major structural material of plants, is the most abundant biological molecule on Earth. Its long, straight chains are locked into tight, sturdy bundles by hydrogen bonds (**FIGURE 3.5A**). The bundles form tough fibers that act like reinforcing rods inside stems and other plant parts, helping these structures resist wind and other forms of mechanical stress. Cellulose does not dissolve in water, and it is not easily broken down. Some bacteria and fungi make enzymes that can break it apart into its component sugars, but humans and other mammals do not. Dietary fiber, or "roughage," usually refers to the cellulose in our vegetable foods. Bacteria that live in the guts of termites and grazers such as cattle and sheep help these animals digest the cellulose in plants.

In starch, a different covalent bonding pattern between glucose monomers makes a chain that coils up into a spiral (**FIGURE 3.5B**). Like

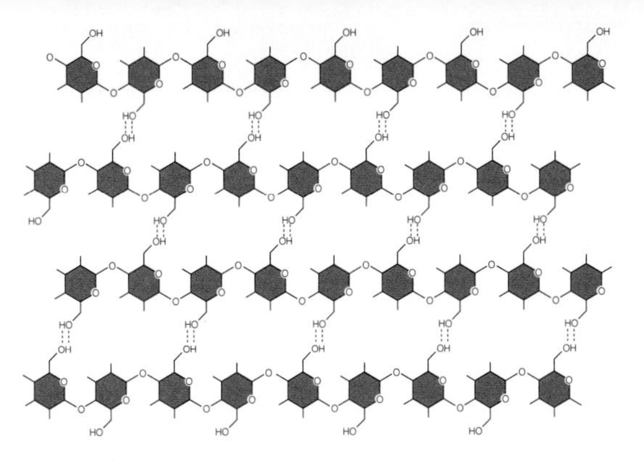

A Cellulose
Cellulose is the main structural component of plants. Above, in cellulose, chains of glucose monomers stretch side by side and hydrogen-bond at many —OH groups. The hydrogen bonds stabilize the chains in tight bundles that form long fibers. Very few types of organisms can digest this tough, insoluble material.

B Starch
Starch is the main energy reserve in plants, which store it in their roots, stems, leaves, seeds, and fruits. Below, in starch, a series of glucose monomers form a chain that coils up.

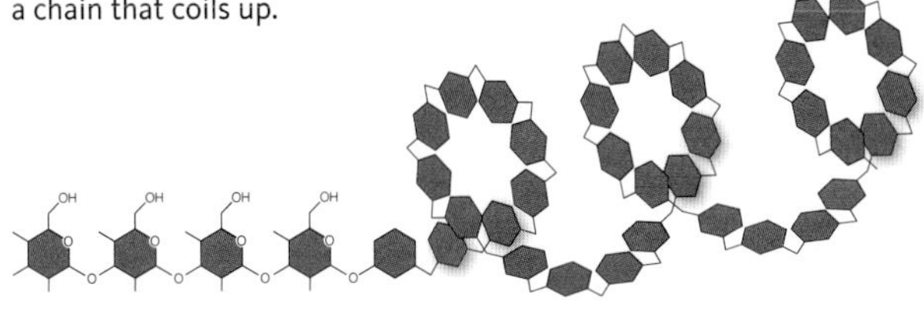

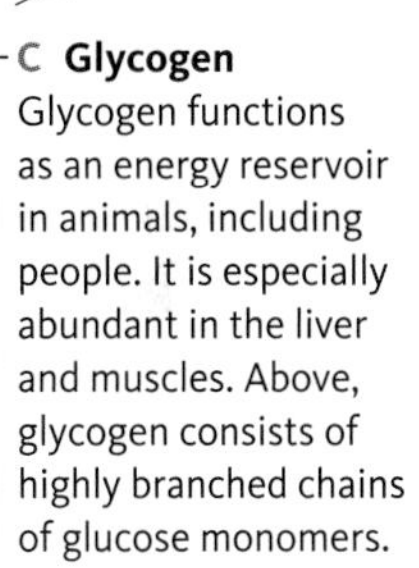

C Glycogen
Glycogen functions as an energy reservoir in animals, including people. It is especially abundant in the liver and muscles. Above, glycogen consists of highly branched chains of glucose monomers.

FIGURE 3.5 {Animated} Three of the most common complex carbohydrates and their locations in a few organisms. Each polysaccharide consists only of glucose subunits, but different bonding patterns result in substances with very different properties.

cellulose, starch does not dissolve easily in water, but it is more easily broken down than cellulose. These properties make starch ideal for storing sugars in the watery, enzyme-filled interior of plant cells. Most plant leaves make glucose during the day, and their cells store it by building starch. At night, hydrolysis enzymes break the bonds between starch's glucose monomers. The released glucose can be broken down immediately for energy, or converted to sucrose that is transported to other parts of the plant. Humans also have hydrolysis enzymes that break down starch, so this carbohydrate is an important component of our food.

Animals store their sugars in the form of glycogen. The covalent bonding pattern between glucose monomers in glycogen forms highly branched chains (**FIGURE 3.5C**). Muscle and liver cells contain most of the body's stored glycogen. When the sugar level in blood falls, liver cells break down stored glycogen, and the released glucose subunits enter the blood.

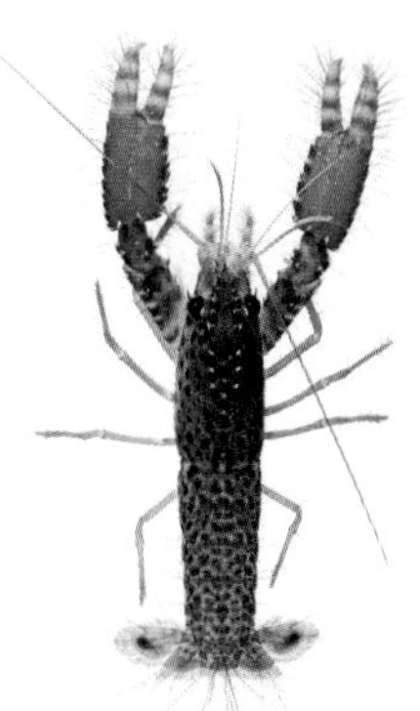

In chitin, a polysaccharide similar to cellulose, long, unbranching chains of nitrogen-containing monomers are linked by hydrogen bonds. As a structural material, chitin is durable, translucent, and flexible. It strengthens hard parts of many animals, including the outer cuticle of lobsters (*left*), and it reinforces the cell wall of many fungi.

TAKE-HOME MESSAGE 3.2

Cells use simple carbohydrates (sugars) for energy and to build other molecules.

Glucose monomers, bonded in different ways, form complex carbohydrates, including cellulose, starch, and glycogen.

carbohydrate Molecule that consists primarily of carbon, hydrogen, and oxygen atoms in a 1:2:1 ratio.
cellulose Tough, insoluble carbohydrate that is the major structural material in plants.

3.3 WHAT ARE LIPIDS?

Lipids are fatty, oily, or waxy organic compounds. Many lipids incorporate **fatty acids**, which are small organic molecules that consist of a long hydrocarbon "tail" with a carboxyl group "head" (**FIGURE 3.6**). The tail is hydrophobic; the carboxyl group makes the head hydrophilic (and acidic). You are already familiar with the properties of fatty acids because these molecules are the main component of soap. The hydrophobic tails of fatty acids in soap attract oily dirt, and the hydrophilic heads dissolve the dirt in water.

Saturated fatty acids have only single bonds linking the carbons in their tails. In other words, their carbon chains are fully saturated with hydrogen atoms (**FIGURE 3.6A**). Saturated fatty acid tails are flexible and they wiggle freely. Double bonds between carbons of unsaturated fatty acid tails limit their flexibility (**FIGURE 3.6B,C**).

FATS

The carboxyl group head of a fatty acid can easily form a covalent bond with another molecule. When it bonds to a glycerol, a type of alcohol, it loses its hydrophilic character and becomes part of a fat. **Fats** are lipids with one, two, or three fatty acids bonded to the same glycerol. A fat with three fatty acid tails is called a **triglyceride** (**FIGURE 3.7A**). Triglycerides are entirely hydrophobic, so they do not dissolve in water. Most "neutral" fats, such as butter and vegetable oils, are examples. Triglycerides are the most abundant and richest energy source in vertebrate bodies. Gram for gram, fats store more energy than carbohydrates.

Butter, cream, and other high-fat animal products have a high proportion of **saturated fats**, which means they consist mainly of triglycerides with three saturated fatty acid tails. Saturated fats tend to be solid at room temperature because their floppy saturated tails can pack tightly. Most vegetable oils are **unsaturated fats**, which means they consist mainly of triglycerides with one or more unsaturated fatty acid tails. Each double bond in a fatty acid tail makes a rigid kink. Kinky tails do not pack tightly, so unsaturated fats are typically liquid at room temperature.

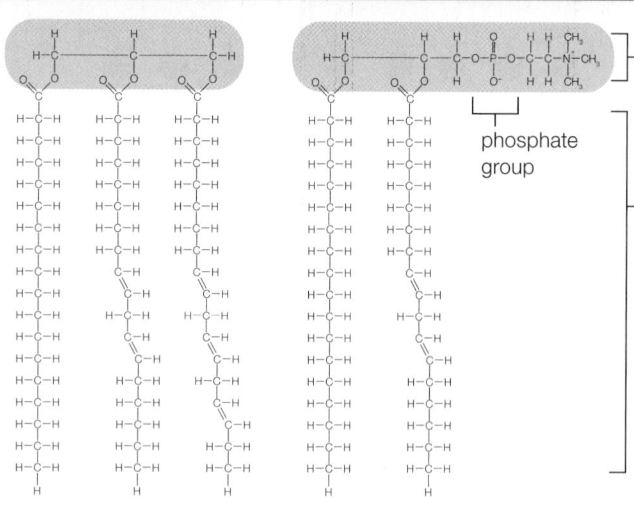

A The three fatty acid tails of a triglyceride are attached to a glycerol head.

B The two fatty acid tails of this phospholipid are attached to a phosphate-containing head.

FIGURE 3.7 {Animated} Lipids with fatty acid tails.

FIGURE 3.6 {Animated} Fatty acids. **A** The tail of stearic acid is fully saturated with hydrogen atoms. **B** Linoleic acid, with two double bonds, is unsaturated. The first double bond occurs at the sixth carbon from the end, so linoleic acid is called an omega-6 fatty acid. Omega-6 and **C** omega-3 fatty acids are "essential fatty acids." Your body does not make them, so they must come from food.

hydrophilic "head" (acidic carboxyl group)

hydrophobic "tail"

A stearic acid (saturated)

B linoleic acid (omega-6)

C linolenic acid (omega-3)

fat Lipid that consists of a glycerol molecule with one, two, or three fatty acid tails.
fatty acid Organic compound that consists of a chain of carbon atoms with an acidic carboxyl group at one end.
lipid Fatty, oily, or waxy organic compound.
lipid bilayer Double layer of lipids arranged tail-to-tail; structural foundation of cell membranes.
phospholipid A lipid with a phosphate group in its hydrophilic head, and two nonpolar tails typically derived from fatty acids.
saturated fat Triglyceride that has three saturated fatty acid tails.
steroid Type of lipid with four carbon rings and no tails.
triglyceride A fat with three fatty acid tails.
unsaturated fat Triglyceride that has one or more unsaturated fatty acid tails.
wax Water-repellent mixture of lipids with long fatty acid tails bonded to long-chain alcohols or carbon rings.

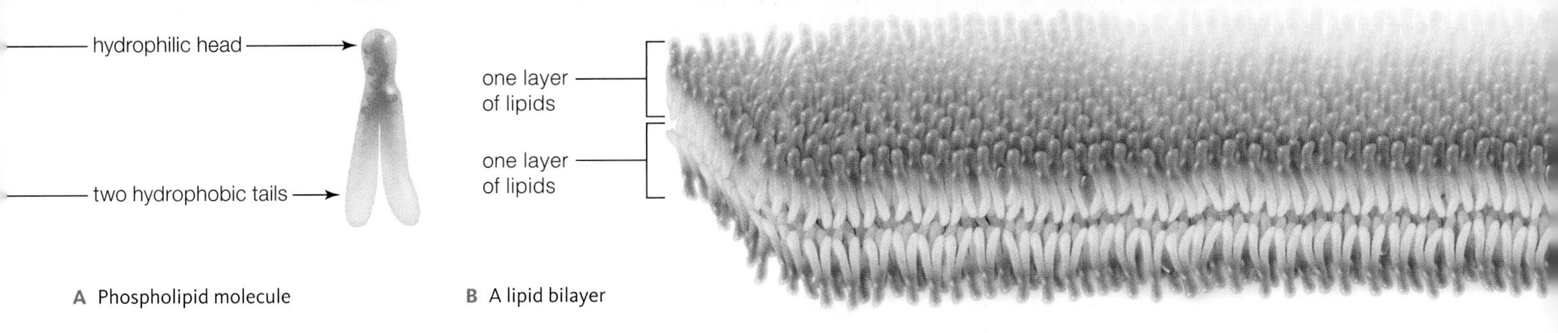

FIGURE 3.8 {Animated} Phospholipids as components of cell membranes. A double layer of phospholipids—the lipid bilayer—is the structural foundation of all cell membranes. You will read more about the structure of cell membranes in Chapter 4.

FIGURE 3.9 Steroids. Estrogen and testosterone are steroid hormones that govern reproduction and secondary sexual traits. The two hormones are the source of gender-specific traits in many species, including wood ducks.

PHOSPHOLIPIDS

A **phospholipid** consists of a phosphate-containing head with two long hydrocarbon tails that are typically derived from fatty acids (**FIGURE 3.7B**). The tails are hydrophobic, but the highly polar phosphate group makes the head hydrophilic. These opposing properties give rise to the basic structure of cell membranes, which consist mainly of phospholipids. In a cell membrane, phospholipids are arranged in two layers—a **lipid bilayer** (**FIGURE 3.8**). The heads of one layer are dissolved in the cell's watery interior, and the heads of the other layer are dissolved in the cell's fluid surroundings. All of the hydrophobic tails are sandwiched between the hydrophilic heads.

WAXES

A **wax** is a complex, varying mixture of lipids with long fatty acid tails bonded to alcohols or carbon rings. The molecules pack tightly, so waxes are firm and water-repellent. Plants secrete waxes onto their exposed surfaces to restrict water loss and keep out parasites and other pests. Other types of waxes protect, lubricate, and soften skin and hair. Waxes, together with fats and fatty acids, make feathers waterproof. Bees store honey and raise new generations of bees inside a honeycomb of secreted beeswax.

STEROIDS

Steroids are lipids with no fatty acid tails; they have a rigid backbone that consists of twenty carbon atoms arranged in a characteristic pattern of four rings (**FIGURE 3.9**). Functional groups attached to the rings define the type of steroid. These molecules serve varied and important physiological functions in plants, fungi, and animals. Cells remodel cholesterol, the most common steroid in animal tissue, to produce many other molecules, including bile salts (which help digest fats), vitamin D (required to keep teeth and bones strong), and steroid hormones.

TAKE-HOME MESSAGE 3.3

- Lipids are fatty, waxy, or oily organic compounds.
- Fats have one, two, or three fatty acid tails; triglyceride fats are an important energy reservoir in vertebrate animals.
- Phospholipids arranged in a lipid bilayer are the main component of cell membranes.
- Waxes have complex, varying structures. They are components of water-repelling and lubricating secretions.
- Steroids serve varied and important physiological roles in plants, fungi, and animals.

3.4 WHAT ARE PROTEINS?

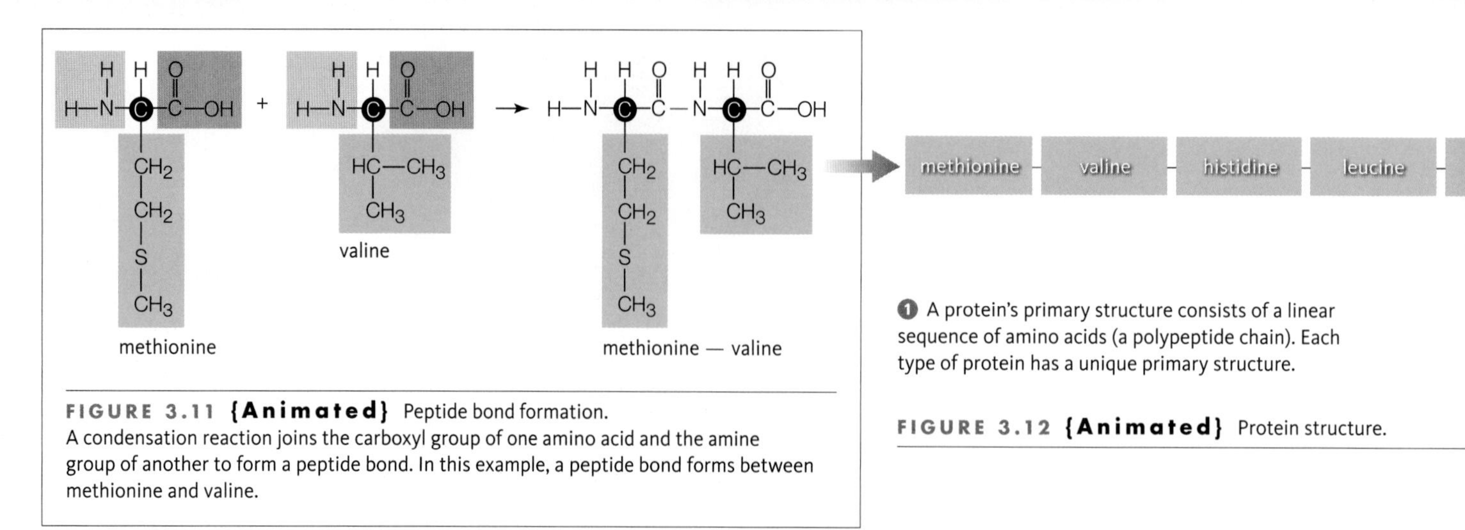

FIGURE 3.11 {Animated} Peptide bond formation.
A condensation reaction joins the carboxyl group of one amino acid and the amine group of another to form a peptide bond. In this example, a peptide bond forms between methionine and valine.

❶ A protein's primary structure consists of a linear sequence of amino acids (a polypeptide chain). Each type of protein has a unique primary structure.

FIGURE 3.12 {Animated} Protein structure.

AMINO ACID SUBUNITS

With a few exceptions, cells can make all of the thousands of different proteins they need from only twenty kinds of monomers called amino acids. An **amino acid** is a small organic compound with an amine group (—NH_2), a carboxyl group (—COOH, the acid), and a side chain called an "R group" that defines the kind of amino acid. In most amino acids, all three groups are attached to the same carbon atom (**FIGURE 3.10**).

FIGURE 3.10 Generalized structure of an amino acid.
See Appendix V for the complete structures of the twenty most common amino acids found in eukaryotic proteins.

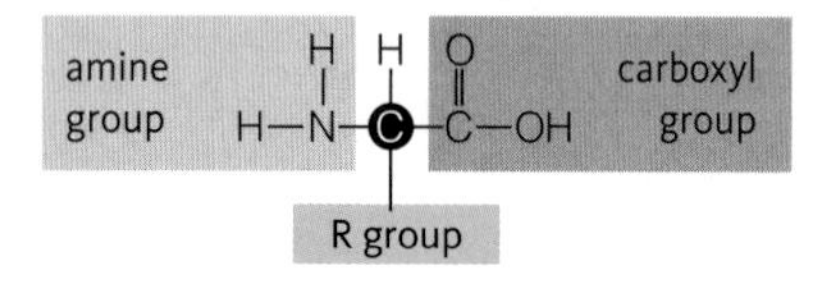

The covalent bond that links amino acids in a protein is called a **peptide bond**. During protein synthesis, a peptide bond forms between the carboxyl group of the first amino acid and the amine group of the second (**FIGURE 3.11**). Another peptide bond links a third amino acid to the second, and so on (you will learn more about the details of protein synthesis in Chapter 9). A short chain of amino acids is called a **peptide**; as the chain lengthens, it becomes a **polypeptide**. **Proteins** consist of polypeptides that are hundreds or even thousands of amino acids long.

STRUCTURE DICTATES FUNCTION

Of all biological molecules, proteins are the most diverse. Structural proteins support cell parts and, as part of tissues, multicelled bodies. Feathers, hooves, and hair, as well as tendons and other body parts, consist mainly of structural proteins. A tremendous number of different proteins, including some structural types, participate in all processes that sustain life. Most enzymes that help cells carry out metabolic reactions are proteins. Proteins also function in movement, defense, and cellular communication.

One of the fundamental ideas in biology is that structure dictates function. This idea is particularly appropriate for proteins, because a protein's biological activity arises from and depends on its structure (**FIGURE 3.12**).

The linear series of amino acids in a polypeptide chain is called primary structure ❶, which defines the type of protein. The protein's three-dimensional shape begins to arise during synthesis, when hydrogen bonds that form among amino acids cause the lengthening polypeptide chain to twist and fold. Parts of the polypeptide form loops, helixes (coils), or flat sheets, and these patterns constitute secondary structure ❷. The primary structure of each type of protein is unique, but most proteins have similar patterns of secondary structure.

Much as an overly twisted rubber band coils back upon itself, hydrogen bonding between nonadjacent regions of a protein makes its loops, helices, and sheets fold up into even more compact domains (**FIGURE 3.13A**). These domains are called tertiary structure ❸. Tertiary structure makes a protein a working molecule. For example, the helices and loops in a globin chain fold up together to form a pocket that can hold a heme, which is a small compound essential to the finished protein's function. Sheets, loops, and helices of other proteins roll up into complex structures that resemble barrels, propellers, sandwiches, and so on. Large proteins typically have several domains, each contributing a particular structural or functional property to the molecule. For example, some barrel domains rotate like motors in

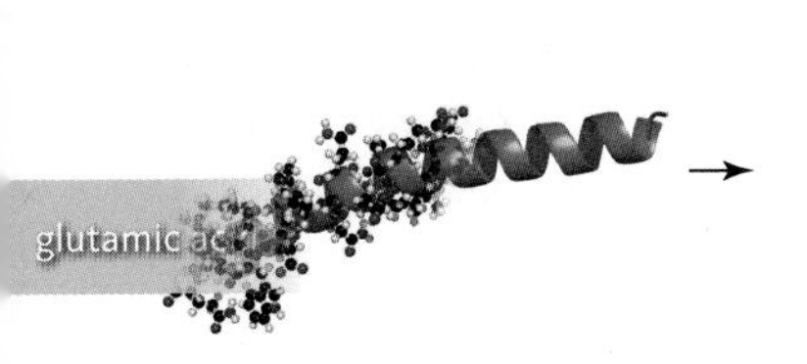

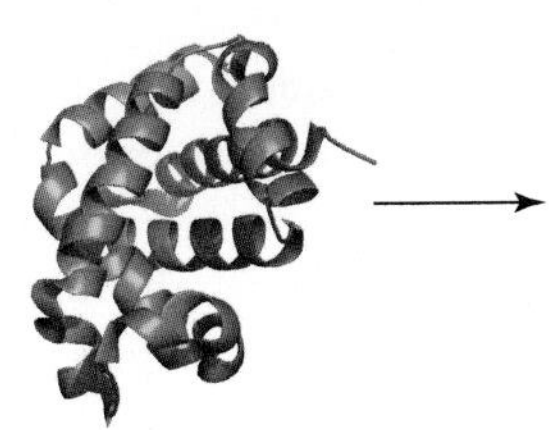

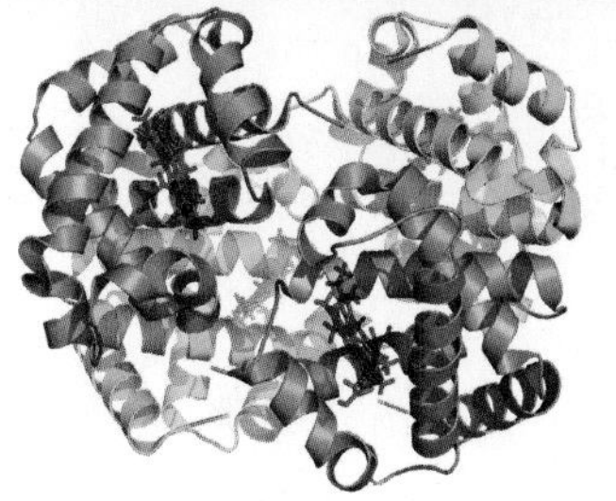

❷ Secondary structure arises as a polypeptide chain twists into a helix (coil), loop, or sheet held in place by hydrogen bonds.

❸ Tertiary structure occurs when loops, helices, and sheets fold up into a domain. In this example, the helices of a globin chain form a pocket.

❹ Many proteins have two or more polypeptide chains (quaternary structure). Hemoglobin, shown here, consists of four globin chains (green and blue). Each globin pocket now holds a heme group (red).

❺ Some types of proteins aggregate into much larger structures. As an example, organized arrays of keratin, a fibrous protein, compose filaments that make up your hair.

small molecular machines (**FIGURE 3.13B**). Other barrels function as tunnels for small molecules, allowing them to pass, for example, through a cell membrane.

Many proteins also have quaternary structure, which means they consist of two or more polypeptide chains that are closely associated or covalently bonded together ❹. Most enzymes are like this, with multiple polypeptide chains that collectively form a roughly spherical shape.

Fibrous proteins aggregate by many thousands into much larger structures, with their polypeptide chains organized into strands or sheets. The keratin in your hair is an example ❺. Some fibrous proteins contribute to the structure and organization of cells and tissues. Others help cells, cell parts, and multicelled bodies move.

Enzymes often attach sugars or lipids to proteins. A glycoprotein forms when oligosaccharides are attached to a polypeptide. The molecules that allow a tissue or a body to recognize its own cells are glycoproteins, as are other molecules that help cells interact in immunity.

Some lipoproteins form when enzymes covalently bond lipids to a protein. Other lipoproteins are aggregate structures that consist of variable amounts and types of proteins and lipids (**FIGURE 3.14**).

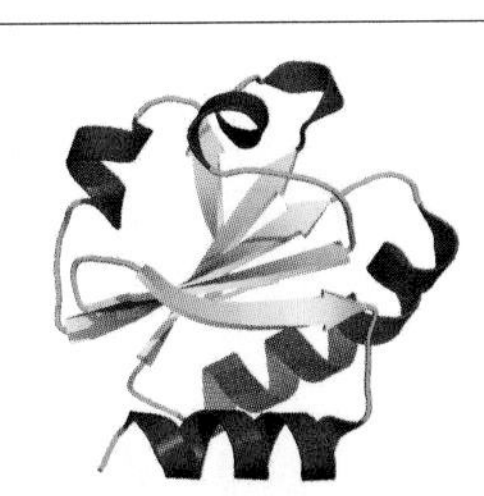

A In this protein, loops (green), coils (red), and a sheet (yellow) fold up together into a chemically active pocket. The pocket gives this protein the ability to transfer electrons from one molecule to another. Many other proteins have the same pocket structure.

B This barrel domain is part of a rotary mechanism in a larger protein. The protein functions as a molecular motor that pumps hydrogen ions through cell membranes.

FIGURE 3.13 Examples of protein domains.

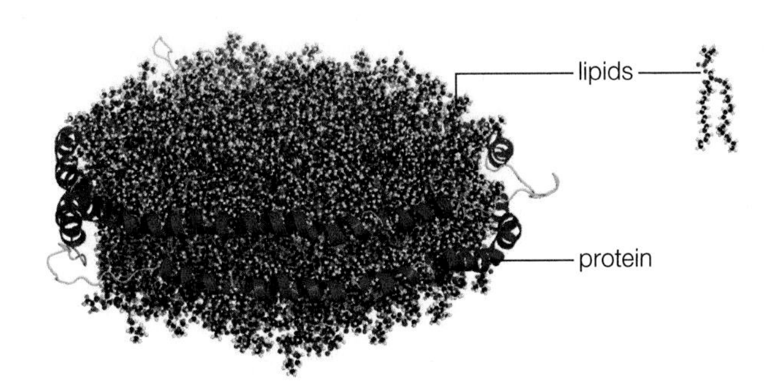

FIGURE 3.14 A lipoprotein particle. The one depicted here (HDL, often called "good" cholesterol) consists of thousands of lipids lassoed into a clump by two protein chains.

amino acid Small organic compound that is a subunit of proteins. Consists of a carboxyl group, an amine group, and a characteristic side group (R), all typically bonded to the same carbon atom.
peptide bond A bond between the amine group of one amino acid and the carboxyl group of another. Joins amino acids in proteins.
peptide Short chain of amino acids linked by peptide bonds.
polypeptide Long chain of amino acids linked by peptide bonds.
protein Organic molecule that consists of one or more polypeptides.

TAKE-HOME MESSAGE 3.4

Proteins are chains of amino acids. The order of amino acids in a polypeptide chain dictates the type of protein.

Polypeptide chains twist and fold into coils, sheets, and loops, which fold and pack further into functional domains.

A protein's function arises from its three-dimensional shape.

CREDITS: (13A, 14) Castrignanò T, De Meo PD, Cozzetto D, Talamo IG, Tramontano A. (2006). The PMDB Protein Model Database. Nucleic Acids Research, 34: D306-D309. (13B) pdb ID2W5J, Vollmar, M., Shlieper, D., Winn M., Buechner, C., Groth, G. "Structure of the C14 rotor ring of the proton translocating chloroplast ATP synthase." (2009) *J. Biol. Chem.* 284:18228.

3.5 WHY IS PROTEIN STRUCTURE IMPORTANT?

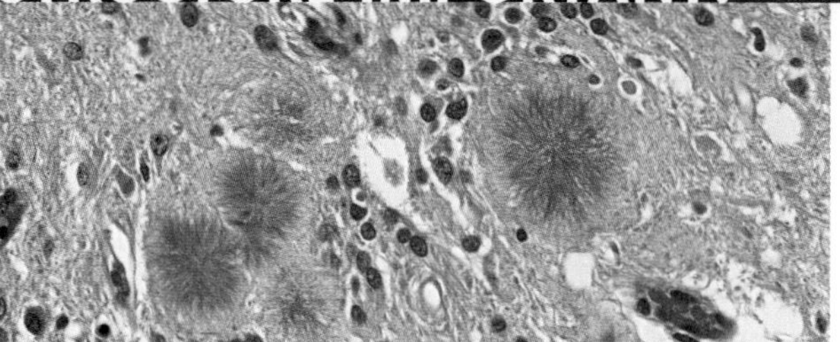

FIGURE 3.15 Variant Creutzfeldt–Jakob disease (vCJD). Characteristic holes and prion protein fibers radiating from several deposits are visible in this slice of brain tissue from a person with vCJD.

Protein shape depends on hydrogen bonding, which can be disrupted by heat, some salts, shifts in pH, or detergents. Such disruption causes proteins to **denature**, which means they lose their three-dimensional shape. Once a protein's shape unravels, so does its function.

Consider three fatal diseases: scrapie in sheep, mad cow disease (BSE, bovine spongiform encephalopathy), and variant Creutzfeldt–Jakob disease (vCJD) in humans. All begin with a glycoprotein called PrPC that occurs normally in cell membranes of the mammalian body. Sometimes, a PrPC protein spontaneously misfolds. A single misfolded protein molecule should not pose much of a threat, but when this particular protein misfolds it becomes a **prion**, or infectious protein. The altered shape of a misfolded PrPC protein causes normally folded PrPC proteins to misfold too. Because each protein that misfolds becomes infectious, the number of prions increases exponentially.

The shape of misfolded PrPC proteins allows them to align tightly into long fibers. In the brain, these fibers accumulate in water-repellent patches that disrupt brain cell function, resulting in relentlessly worsening symptoms of confusion, memory loss, and lack of coordination. Holes form in the brain as its cells die (**FIGURE 3.15**).

In the mid-1980s, an epidemic of mad cow disease in Britain was followed by an outbreak of vCJD in humans. The cattle became infected by the prion after eating feed prepared from the remains of scrapie-infected sheep, and people became infected by eating beef from infected cattle. The use of animal parts in livestock feed is now banned in many countries, and the number of cases of BSE and vCJD has since declined.

denature To unravel the shape of a protein or other large biological molecule.
prion Infectious protein.

TAKE-HOME MESSAGE 3.5

Protein shape can unravel if hydrogen bonds are disrupted.

A protein's function depends on its structure, so conditions that alter a protein's structure also alter its function.

3.6 WHAT ARE NUCLEIC ACIDS?

A **nucleotide** is a small organic molecule that consists of a sugar with a five-carbon ring bonded to a nitrogen-containing base and one, two, or three phosphate groups (**FIGURE 3.16A**). When the third phosphate group of a nucleotide is transferred to another molecule, energy is transferred along with it. The nucleotide **ATP** (adenosine triphosphate) serves an especially important role as an energy carrier in cells.

Nucleic acids are polymers, chains of nucleotides in which the sugar of one nucleotide is bonded to the phosphate group of the next (**FIGURE 3.16B**). An example is ribonucleic acid, or **RNA**, named after the ribose sugar of its component nucleotides. An RNA molecule is a chain of four kinds of nucleotide monomers, one of which is ATP. RNA molecules carry out protein synthesis. Deoxyribonucleic acid, or **DNA**, is a nucleic acid named after the deoxyribose sugar of its component nucleotides. A DNA molecule consists of two chains of nucleotides twisted into a double helix. Hydrogen bonds between the nucleotides hold the chains together. Each cell's DNA holds all information necessary to build a new cell and, in the case of multicelled organisms, a new individual.

FIGURE 3.16 Nucleic acids.

A ATP
(a nucleotide)

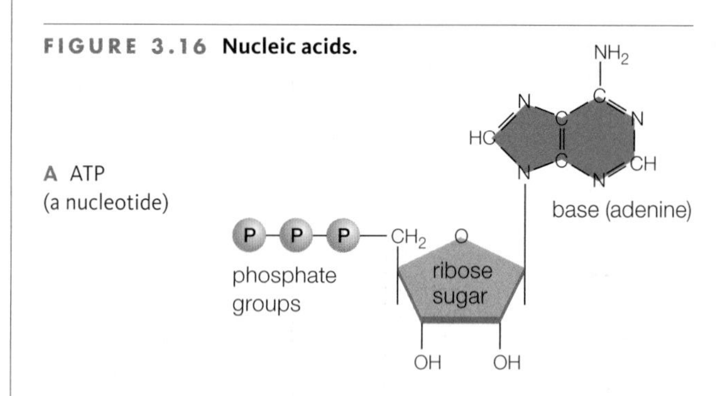

B RNA
(a nucleic acid)

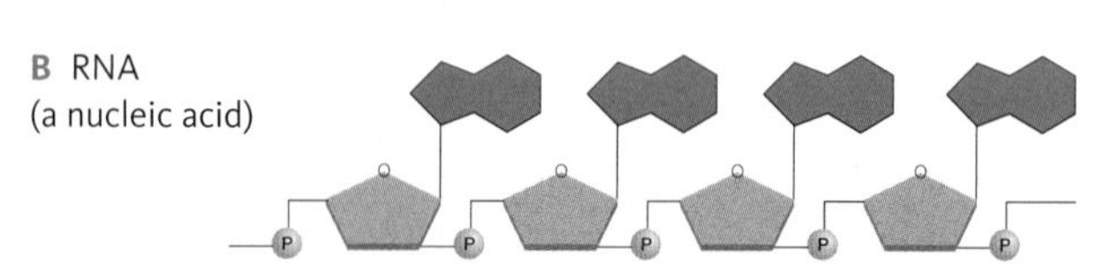

ATP Adenosine triphosphate. Nucleotide that serves an important role as an energy carrier in cells.
DNA Deoxyribonucleic acid. Consists of two chains of nucleotides twisted into a double helix.
nucleic acid Polymer of nucleotides; DNA or RNA.
nucleotide Monomer of nucleic acids; has a five-carbon sugar, a nitrogen-containing base, and one, two, or three phosphate groups.
RNA Ribonucleic acid. Single-stranded chain of nucleotides.

TAKE-HOME MESSAGE 3.6

Nucleotides are monomers of nucleic acids. ATP has an important metabolic role as an energy carrier.

RNA carries out protein synthesis. DNA holds information necessary to build cells and multicelled individuals.

FIGURE 3.17 *Trans* fats, an unhealthy food. Double bonds in the tail of most naturally occurring fatty acids are *cis*, which means that the two hydrogen atoms flanking the bond are on the same side of the carbon backbone. Hydrogenation creates abundant *trans* bonds, with hydrogen atoms on opposite sides of the tail.

FATS ARE NOT INERT MOLECULES THAT SIMPLY ACCUMULATE IN STRATEGIC AREAS OF OUR BODIES. They are major constituents of cell membranes, and as such they have powerful effects on cell function. As you learned in Section 3.3, the long carbon backbone of fatty acid tails can vary a bit in structure. *Trans* fats have unsaturated fatty acid tails with a particular arrangement of hydrogen atoms around the double bonds (FIGURE 3.17). Small amounts of *trans* fats occur naturally, but the main source of these fats in the American diet is an artificial food product called partially hydrogenated vegetable oil. Hydrogenation is a manufacturing process that adds hydrogen atoms to oils in order to change them into solid fats. In 1908, Procter & Gamble Co. developed partially hydrogenated soybean oil as a substitute for the more expensive solid animal fats they had been using to make candles. However, the demand for candles began to wane as more households in the United States became wired for electricity, and P&G looked for another way to sell its proprietary fat. Partially hydrogenated vegetable oil looks like lard, so the company began aggressively marketing it as a revolutionary new food: a solid cooking fat with a long shelf life, mild flavor, and lower cost than lard or butter. By the mid-1950s, hydrogenated vegetable oil had become a major part of the American diet, and it is still found in many manufactured and fast foods. For decades, it was considered healthier than animal fats, but we now know otherwise. *Trans* fats raise the level of cholesterol in our blood more than any other fat, and they directly alter the function of our arteries and veins. The effects of such changes are quite serious. Eating as little as 2 grams a day (about 0.4 teaspoon) of hydrogenated vegetable oil measurably increases a person's risk of atherosclerosis (hardening of the arteries), heart attack, and diabetes. A small serving of french fries made with hydrogenated vegetable oil contains about 5 grams of *trans* fat.

Summary

SECTION 3.1 Complex carbohydrates and lipids, proteins, and nucleic acids are **organic**, which means they consist mainly of carbon and hydrogen atoms. **Hydrocarbons** have only carbon and hydrogen atoms.

Carbon chains or rings form the backbone of the molecules of life. **Functional groups** attached to the backbone influence the chemical character of these compounds, and thus their function.

Metabolism includes chemical **reactions** and all other processes by which cells acquire and use energy as they make and break the bonds of organic compounds. In reactions such as **condensation**, **enzymes** build **polymers** from **monomers** of simple sugars, fatty acids, amino acids, and nucleotides. Reactions such as **hydrolysis** release the monomers by breaking apart the polymers.

SECTION 3.2 Enzymes build complex **carbohydrates** such as **cellulose**, glycogen, and starch from simple carbohydrate (sugar) subunits. Cells use carbohydrates for energy, and as structural materials.

SECTION 3.3 **Lipids** are fatty, oily, or waxy compounds. All are nonpolar. **Fats** have **fatty acid** tails; **triglycerides** have three. **Saturated fats** are mainly triglycerides with three saturated fatty acid tails (only single bonds link their carbons). **Unsaturated fats** are mainly triglycerides with one or more unsaturated fatty acids.

A **lipid bilayer** (that consists primarily of **phospholipids**) is the basic structure of all cell membranes. **Waxes** are part of water-repellent and lubricating secretions. **Steroids** occur in cell membranes, and some are remodeled into other molecules such as hormones.

SECTION 3.4 Structurally and functionally, **proteins** are the most diverse molecules of life. The shape of a protein is the source of its function. Protein structure begins as a series of **amino acids** (primary structure) linked by **peptide bonds** into a **peptide**, then a **polypeptide**. Polypeptides twist into helices, sheets, and coils (secondary structure) that can pack further into functional domains (tertiary structure). Many proteins, including most enzymes, consist of two or more polypeptides (quaternary structure). Fibrous proteins aggregate into much larger structures.

SECTION 3.5 A protein's structure dictates its function, so changes in a protein's structure may also alter its function. A protein's shape may be disrupted by shifts in pH or temperature, or exposure to detergent or some salts. If that happens, the protein unravels, or **denatures**, and so loses its function. **Prion** diseases are a fatal consequence of misfolded proteins.

SECTION 3.6 **Nucleotides** are small organic molecules that consist of a five-carbon sugar, a nitrogen-containing base, and one, two, or three phosphate groups. Nucleotides are monomers of **DNA** and **RNA**, which are **nucleic acids**. Some, especially **ATP**, have additional functions such as carrying energy. DNA encodes information necessary to build cells and multicelled individuals. RNA molecules carry out protein synthesis.

SECTION 3.7 All organisms consist of the same kinds of molecules. Seemingly small differences in the way those molecules are put together can have big effects inside a living organism.

Self-Quiz

Answers in Appendix VII

1. Organic molecules consist mainly of _______ atoms.
   a. carbon
   b. carbon and oxygen
   c. carbon and hydrogen
   d. carbon and nitrogen

2. Each carbon atom can bond with as many as _______ other atom(s).

3. _______ groups are the "acid" part of amino acids and fatty acids.
   a. Hydroxyl (—OH)
   b. Carboxyl (—COOH)
   c. Methyl (—CH_3)
   d. Phosphate (—PO_4)

4. _______ is a simple sugar (a monosaccharide).
   a. Glucose
   b. Sucrose
   c. Ribose
   d. Starch
   e. both a and c
   f. a, b, and c

5. Unlike saturated fats, the fatty acid tails of unsaturated fats incorporate one or more _______ .
   a. phosphate groups
   b. glycerols
   c. double bonds
   d. single bonds

6. Is this statement true or false? Unlike saturated fats, all unsaturated fats are beneficial to health because their fatty acid tails kink and do not pack together.

7. Steroids are among the lipids with no _______ .
   a. double bonds
   b. fatty acid tails
   c. hydrogens
   d. carbons

8. Name three kinds of carbohydrates that can be built using only glucose monomers.

9. Which of the following is a class of molecules that encompasses all of the other molecules listed?
   a. triglycerides
   b. fatty acids
   c. waxes
   d. steroids
   e. lipids
   f. phospholipids

10. ______ are to proteins as ______ are to nucleic acids.
    a. Sugars; lipids
    b. Sugars; proteins
    c. Amino acids; hydrogen bonds
    d. Amino acids; nucleotides

11. A denatured protein has lost its _______ .
    a. hydrogen bonds
    b. shape
    c. function
    d. all of the above

12. _______ consist(s) of nucleotides.
    a. Sugars b. DNA c. RNA d. b and c

Data Analysis Activities

Effects of Dietary Fats on Lipoprotein Levels Cholesterol that is made by the liver or that enters the body from food cannot dissolve in blood, so it is carried through the bloodstream by lipoproteins. Low-density lipoprotein (LDL) carries cholesterol to body tissues such as artery walls, where it can form deposits associated with cardiovascular disease. Thus, LDL is often called "bad" cholesterol. High-density lipoprotein (HDL) carries cholesterol away from tissues to the liver for disposal, so HDL is often called "good" cholesterol.

In 1990, Ronald Mensink and Martijn Katan published a study that tested the effects of different dietary fats on blood lipoprotein levels. Their results are shown in **FIGURE 3.18**.

1. In which group was the level of LDL ("bad" cholesterol) highest?
2. In which group was the level of HDL ("good" cholesterol) lowest?
3. An elevated risk of heart disease has been correlated with increasing LDL-to-HDL ratios. Which group had the highest LDL-to-HDL ratio?
4. Rank the three diets from best to worst according to their potential effect on heart disease.

	Main Dietary Fats			
	cis fatty acids	*trans* fatty acids	saturated fats	optimal level
LDL	103	117	121	<100
HDL	55	48	55	>40
ratio	1.87	2.44	2.2	<2

FIGURE 3.18 Effect of diet on lipoprotein levels. Researchers placed 59 men and women on a diet in which 10 percent of their daily energy intake consisted of *cis* fatty acids, *trans* fatty acids, or saturated fats. Blood LDL and HDL levels were measured after three weeks on the diet; averaged results are shown in mg/dL (milligrams per deciliter of blood). All subjects were tested on each of the diets. The ratio of LDL to HDL is also shown.

13. In the following list, identify the carbohydrate, the fatty acid, the amino acid, and the polypeptide:
    a. $NH_2—CH_2—COOH$
    b. $C_6H_{12}O_6$
    c. $(\text{methionine})_{20}$
    d. $CH_3(CH_2)_{16}COOH$

14. Match the molecules with the best description.
    ___ wax — a. sugar storage in plants
    ___ starch — b. richest energy source
    ___ triglyceride — c. water-repellent secretions

15. Match each polymer with the appropriate monomer(s).
    ___ protein — a. phosphate, fatty acids
    ___ phospholipid — b. amino acids, sugars
    ___ glycoprotein — c. glycerol, fatty acids
    ___ fat — d. nucleotides
    ___ nucleic acid — e. glucose only
    ___ wax — f. sugar, phosphate, base
    ___ nucleotide — g. amino acids
    ___ lipoprotein — h. glucose, fructose
    ___ sucrose — i. lipids, amino acids
    ___ glycogen — j. fatty acids, carbon rings

Critical Thinking

1. Lipoproteins are relatively large, spherical clumps of protein and lipid molecules (see **FIGURE 3.14**) that circulate in the blood of mammals. They are like suitcases that move cholesterol, fatty acid remnants, triglycerides, and phospholipids from one place to another in the body. Given what you know about the insolubility of lipids in water, which of the four kinds of lipids would you predict to be on the outside of a lipoprotein clump, bathed in the water-based fluid portion of blood?

2. In 1976, a team of chemists in the United Kingdom was developing new insecticides by modifying sugars with chlorine (Cl_2), phosgene (Cl_2CO), and other toxic gases. One young member of the team misunderstood his verbal instructions to "test" a new molecule. He thought he had been told to "taste" it. Luckily for him, the molecule was not toxic, but it was very sweet. It became the food additive sucralose.

Sucralose has three chlorine atoms substituted for three hydroxyl groups of sucrose (table sugar). It binds so strongly to the sweet-taste receptors on the tongue that the human brain perceives it as 600 times sweeter than sucrose. Sucralose was originally marketed as an artificial sweetener called Splenda®, but it is now available under several other brand names.

sucrose

Researchers investigated whether the body recognizes sucralose as a carbohydrate by feeding sucralose labeled with ^{14}C to volunteers. Analysis of the radioactive molecules in the volunteers' urine and feces showed that 92.8 percent of the sucralose passed through the body without being altered. Many people are worried that the chlorine atoms impart toxicity to sucralose. How would you respond to that concern?

sucralose

CENGAGE brain.com To access course materials, please visit www.cengagebrain.com.

CREDITS: (18) Source, Mensink RP, Katan MB., "Effect of dietary trans fatty acids on high-density and low-density lipoprotein cholesterol levels in healthy subjects." NEJM 323(7):439-45, 1990; From Starr/Taggart/Evers/Starr, Biology, 13E. © Cengage Learning; (insets) © Cengage Learning.

Each cell making up this seedling contains a nucleus (orange spots), which is the defining characteristic of eukaryotes. Rigid walls surround but do not isolate plant cells from one another.

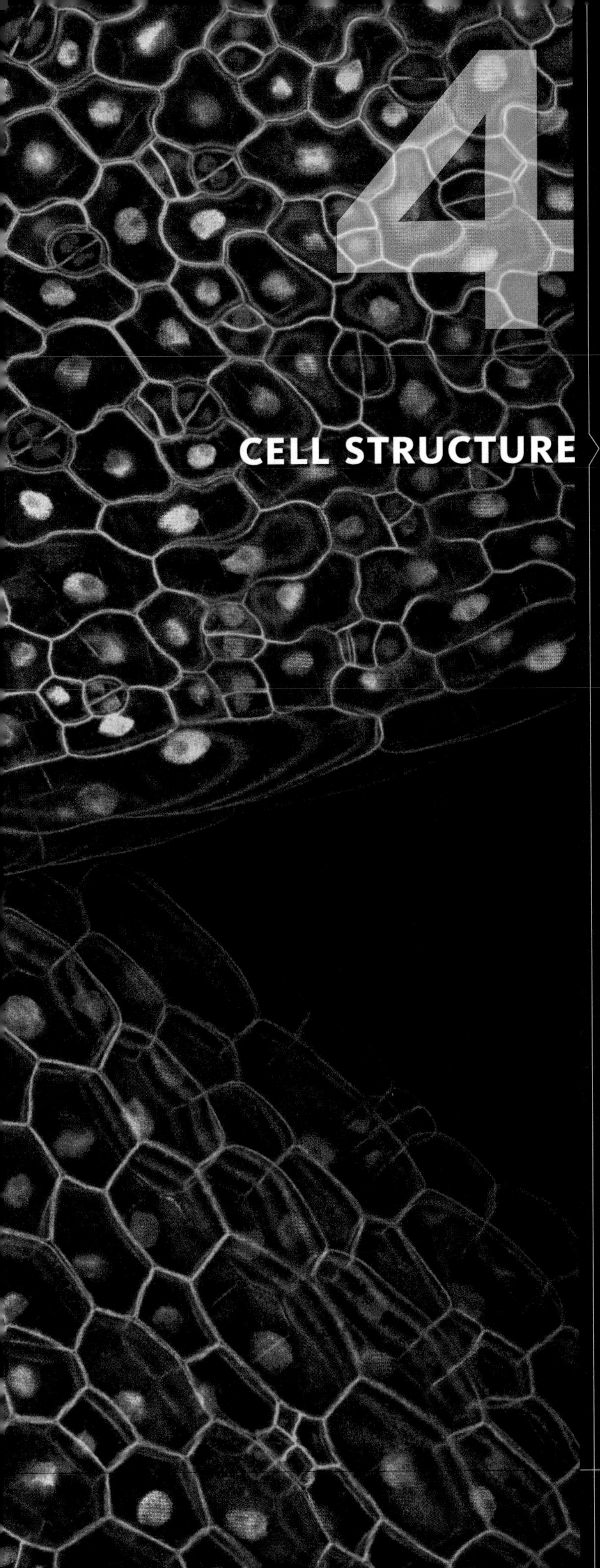

4 CELL STRUCTURE

Links to Earlier Concepts

Reflect on the overview of life's levels of organization in Section 1.1. In this chapter, you will see how the properties of lipids (3.3) give rise to cell membranes; consider the location of DNA (3.6) and the sites where carbohydrates are built and broken apart (3.1, 3.2); and expand your understanding of the vital roles of proteins in cell function (3.4, 3.5). You will also revisit the philosophy of science (1.5, 1.8) and tracers (2.1).

KEY CONCEPTS

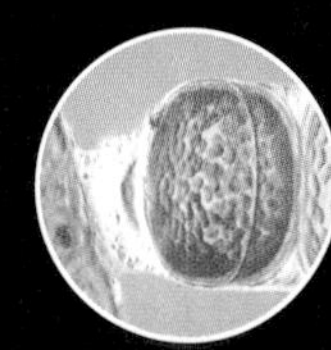

COMPONENTS OF ALL CELLS

Every cell has a plasma membrane separating its interior from the exterior environment. Its interior contains cytoplasm, DNA, and other structures.

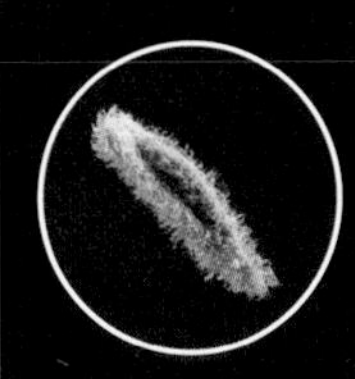

THE MICROSCOPIC WORLD

Most cells are too small to see with the naked eye. We use different types of microscopes to reveal different details of their structure.

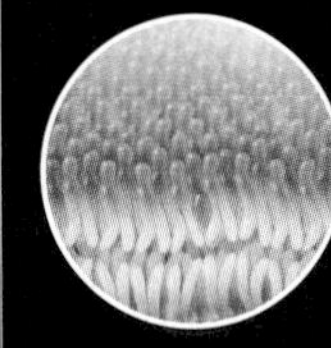

CELL MEMBRANES

All cell membranes consist of a lipid bilayer with various proteins embedded in it and attached to its surfaces. A membrane controls the kinds and amounts of substances that cross it.

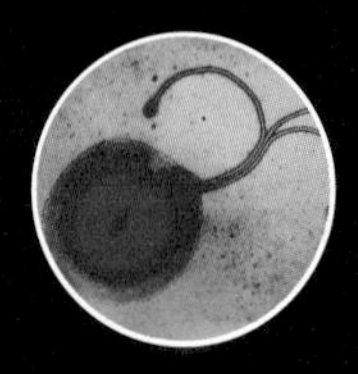

PROKARYOTIC CELLS

Archaea and bacteria have no nucleus. In general, they are smaller and structurally more simple than eukaryotic cells, but they are by far the most numerous and diverse organisms.

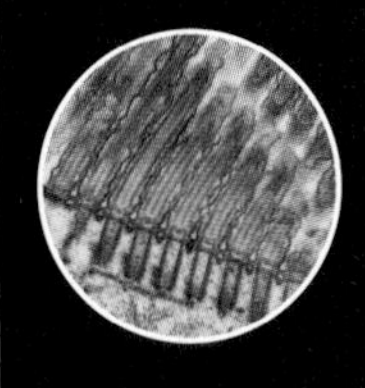

EUKARYOTIC CELLS

Protists, plants, fungi, and animals are eukaryotes. Cells of these organisms differ in internal parts and surface specializations, but all start out life with a nucleus.

CELL THEORY

No one knew cells existed until after the first microscopes were invented. By the mid-1600s, Antoni van Leeuwenhoek had constructed a crude instrument, and was writing about the tiny moving organisms he spied in rainwater, insects, fabric, sperm, feces, and other samples. In scrapings of tartar from his teeth, Leeuwenhoek saw "many very small animalcules, the motions of which were very pleasing to behold." He (incorrectly) assumed that movement defined life, and (correctly) concluded that the moving "beasties" he saw were alive. Leeuwenhoek might have been less pleased to behold his animalcules if he had grasped the implications of what he saw: Our world, and our bodies, teem with microbial life.

Today we know that a cell carries out metabolism and homeostasis, and reproduces either on its own or as part of a larger organism. By this definition, each cell is alive even if it is part of a multicelled body, and all living organisms consist of one or more cells. We also know that cells reproduce by dividing, so it follows that all existing cells must have arisen by division of other cells (later chapters discuss the processes by which cells divide). As a cell divides, it passes its hereditary material—its DNA—to offspring. Taken together, these generalizations constitute the **cell theory**, which is one of the foundations of modern biology (TABLE 4.1).

TABLE 4.1

Cell Theory

Cell Theory
1. Every living organism consists of one or more cells.
2. The cell is the structural and functional unit of all organisms. A cell is the smallest unit of life, individually alive even as part of a multicelled organism.
3. All living cells arise by division of preexisting cells.
4. Cells contain hereditary material, which they pass to their offspring when they divide.

COMPONENTS OF ALL CELLS

Cells vary in shape and function, but all have at least three components in common: a plasma membrane, cytoplasm, and DNA (FIGURE 4.1). A cell's **plasma membrane** is its outermost, separating the cell's contents from the external environment. Like all other cell membranes, a plasma membrane is selectively permeable, which means that only certain materials can cross it. Thus, a plasma membrane controls exchanges between the cell and its environment.

The plasma membrane encloses a jellylike mixture of water, sugars, ions, and proteins called **cytoplasm.** A major part of a cell's metabolism occurs in the cytoplasm, and the cell's internal components, including organelles, are suspended in it. **Organelles** are structures that carry out special functions inside a cell. Membrane-enclosed organelles allow a cell to compartmentalize activities.

All cells start out life with DNA, though a few types lose it as they mature. In nearly all bacteria and archaea, the DNA is suspended directly in cytoplasm. By contrast, all eukaryotic cells start out life with a **nucleus** (plural, nuclei), an organelle with a double membrane that contains the cell's DNA. All protists, fungi, plants, and animals are eukaryotes. Some of these organisms are independent, free-living cells; others consist of many cells working together as a body.

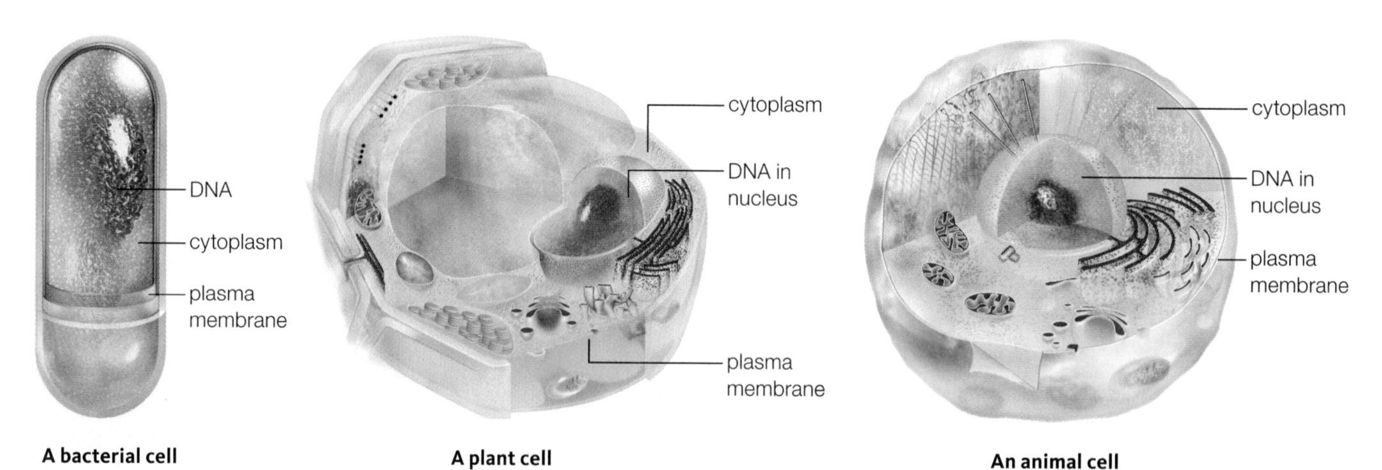

FIGURE 4.1 {Animated} All cells start out life with a plasma membrane, cytoplasm, and DNA. Archaea are similar to bacteria in overall structure; both are typically much smaller than eukaryotic cells. If the cells depicted here had been drawn to the same scale, the bacterium would be about this big:

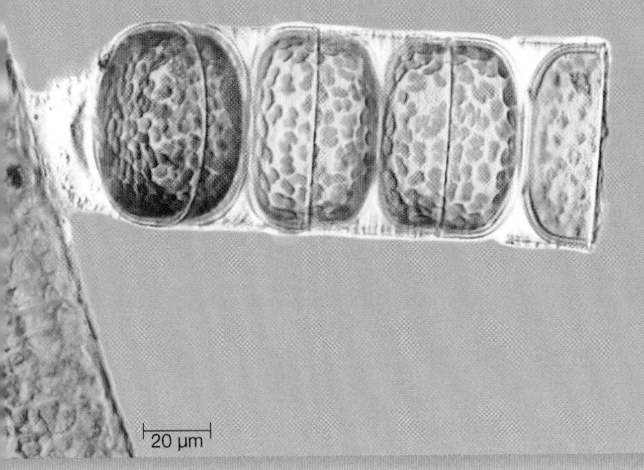

Sticky secretions hold these pill-shaped cells together end to end, forming a long strand of algae. The arrangement allows each algal cell to exchange substances directly with the surrounding water. Secretions also anchor the strand to a solid surface (such as the plant on the left).

FIGURE 4.3 An example of a colonial algae.

Diameter (cm)	2	3	6
Surface area (cm^2)	12.6	28.2	113
Volume (cm^3)	4.2	14.1	113
Surface-to-volume ratio	3:1	2:1	1:1

FIGURE 4.2 Examples of surface-to-volume ratio. This physical relationship between increases in volume and surface area limits the size and influences the shape of cells.

CONSTRAINTS ON CELL SIZE

Almost all cells are too small to see with the naked eye. Why? The answer begins with the processes that keep a cell alive. A living cell must exchange substances with its environment at a rate that keeps pace with its metabolism. These exchanges occur across the plasma membrane, which can handle only so many exchanges at a time. The rate of exchange across a plasma membrane depends on its surface area: the bigger it is, the more substances can cross it during a given interval. Thus, cell size is limited by a physical relationship called the **surface-to-volume ratio**. By this ratio, an object's volume increases with the cube of its diameter, but its surface area increases only with the square.

Apply the surface-to-volume ratio to a round cell. As **FIGURE 4.2** shows, when a cell expands in diameter, its volume increases faster than its surface area does. Imagine that a round cell expands until it is four times its original diameter. The volume of the cell has increased 64 times (4^3), but its surface area has increased only 16 times (4^2). Each unit of plasma membrane must now handle exchanges with four times as much cytoplasm ($64 \div 16 = 4$). If the cell gets too big, the inward flow of nutrients and the outward flow of wastes across that membrane will not be fast enough to keep the cell alive.

Surface-to-volume limits also affect the form of colonial types and multicelled ones too. For example, small cells attach end to end to form strandlike algae, so each can interact directly with the environment (**FIGURE 4.3**). Muscle cells in your thighs are as long as the muscle in which they occur, but each is thin, so it exchanges substances efficiently with fluids in the surrounding tissue.

cell theory Theory that all organisms consist of one or more cells, which are the basic unit of life; all cells come from division of preexisting cells; and all cells pass hereditary material to offspring.
cytoplasm Semifluid substance enclosed by a cell's plasma membrane.
nucleus Of a eukaryotic cell, organelle with a double membrane that holds the cell's DNA.
organelle Structure that carries out a specialized metabolic function inside a cell.
plasma membrane A cell's outermost membrane.
surface-to-volume ratio A relationship in which the volume of an object increases with the cube of the diameter, and the surface area increases with the square.

TAKE-HOME MESSAGE 4.1

Observations of cells led to the cell theory: All organisms consist of one or more cells; the cell is the smallest unit of life; each new cell arises from another cell; and a cell passes hereditary material to its offspring.

All cells start life with a plasma membrane, cytoplasm, and a region of DNA, which, in eukaryotic cells only, is enclosed by a nucleus.

The surface-to-volume ratio limits cell size and influences cell shape.

4.2 HOW DO WE SEE CELLS?

Most cells are 10–20 micrometers in diameter, about fifty times smaller than the unaided human eye can perceive (**FIGURE 4.4**). One micrometer (μm) is one-thousandth of a millimeter, which is one-thousandth of a meter (**TABLE 4.2**). We use microscopes to observe cells and other objects in the micrometer range of size.

Light microscopes use visible light to illuminate samples. As you will learn in Chapter 6, all light travels in waves, a property that makes it bend when it passes through a curved glass lens. Curved lenses inside a light microscope focus light that passes through a specimen, or bounces off of one, into a magnified image (**FIGURE 4.5A**). Photographs of images enlarged with a microscope are called micrographs. Microscopes that use polarized light can yield images in which the edges of some structures appear in three-dimensional relief (**FIGURE 4.5B**).

Most cells are nearly transparent, so their internal details may not be visible unless they are first stained (exposed to dyes that only some cell parts soak up). Parts that absorb the most dye appear darkest. Staining results in an increase in contrast (the difference between light and dark parts) that allows us to see a greater range of detail.

Researchers often use light-emitting tracers (Section 2.1) to pinpoint the location of a molecule of interest within a cell. When illuminated with laser light, the tracer fluoresces (emits light), and an image of the emitted light can be captured with a fluorescence microscope (**FIGURE 4.5C**).

Other microscopes can reveal even finer details. For example, electron microscopes use magnetic fields to focus a beam of electrons onto a sample; these instruments resolve details thousands of times smaller than light microscopes do. Transmission electron microscopes direct electrons through a thin specimen, and the specimen's internal details appear as shadows in the resulting image—a transmission electron micrograph, or TEM (**FIGURE 4.5D**). Scanning electron microscopes direct a beam of electrons across the surface of a specimen that has been coated with a thin layer of metal. The irradiated metal emits electrons and x-rays, which can be converted into an image (a scanning electron micrograph, or SEM) of the surface (**FIGURE 4.5E**). SEMs and TEMs are always black and white; colored versions have been digitally altered to highlight specific details.

TABLE 4.2

Equivalent Units of Length

	Equivalent	
Unit	**Meter**	**Inch**
centimeter (cm)	1/100	0.4
millimeter (mm)	1/1000	0.04
micrometer (μm)	1/1,000,000	0.00004
nanometer (nm)	1/1,000,000,000	0.00000004
meter (m)	100 cm 1,000 mm 1,000,000 μm 1,000,000,000 nm	

TAKE-HOME MESSAGE 4.2

Most cells are visible only with the help of microscopes.

We use different microscopes and techniques to reveal different aspects of cell structure.

FIGURE 4.4 Relative sizes. Most cells are between 1 and 100 micrometers in diameter. See also Units of Measure, Table 4.2 and Appendix VI.

FIGURE IT OUT: Which one is smallest: a protein, a virus, or a bacterium?

Answer: A protein

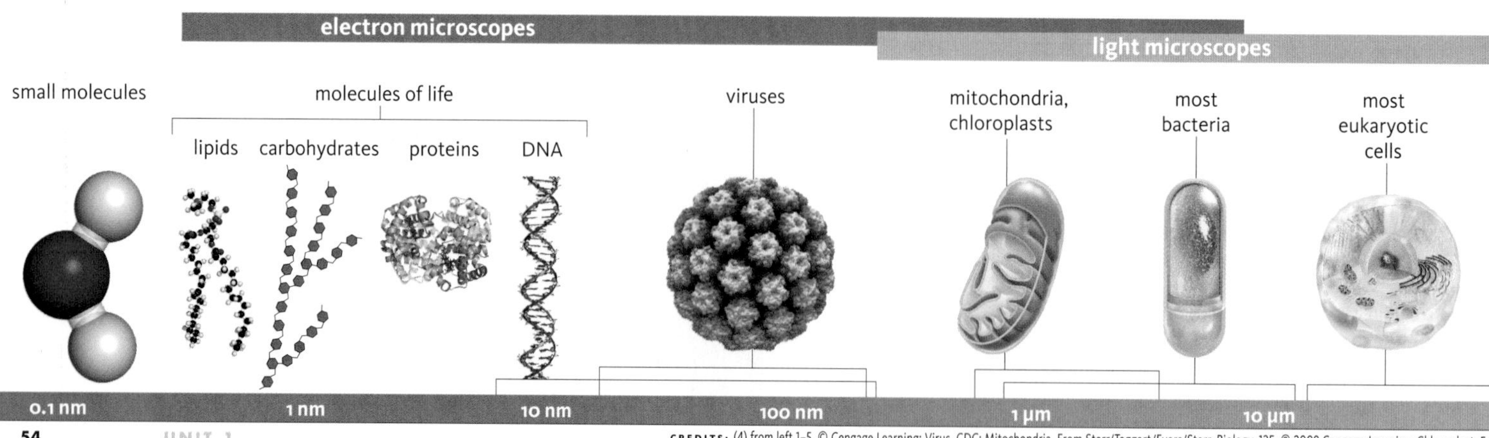

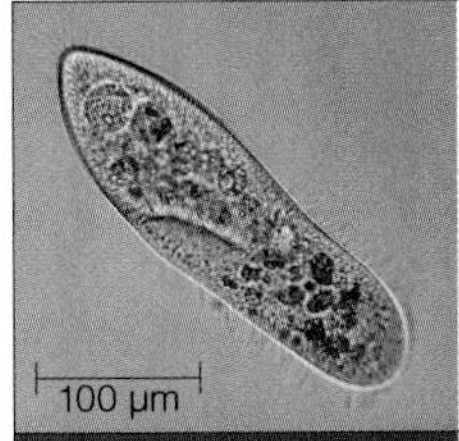

A The green blobs visible in this light micrograph of a living cell are ingested algal cells (also visible in B and D). Fine, hairlike structures on the cell's surface (also visible in E) are waving cilia that propel this motile organism through its fluid surroundings.

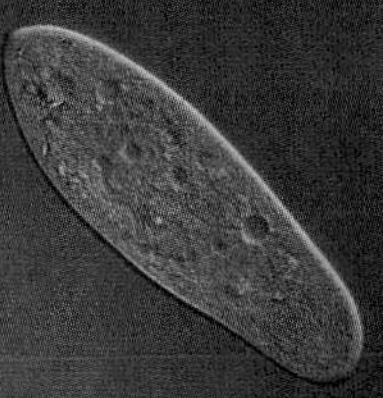

B A light micrograph taken with polarized light shows edges in relief. This technique reveals some internal structures not visible in A.

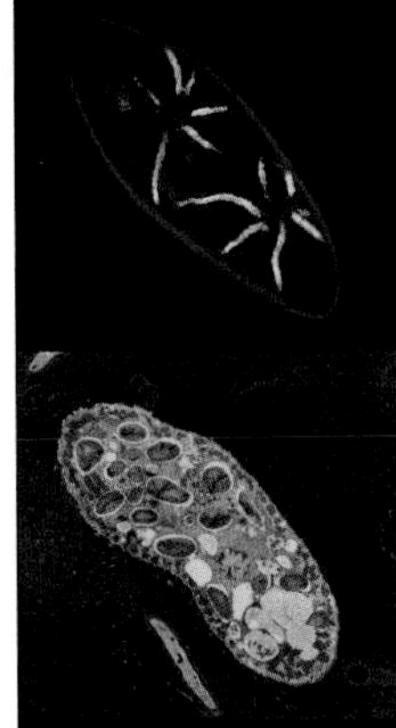

C In this fluorescence micrograph, yellow pinpoints the location of a particular type of protein in the membrane of organelles called contractile vacuoles. These organelles are also visible in B.

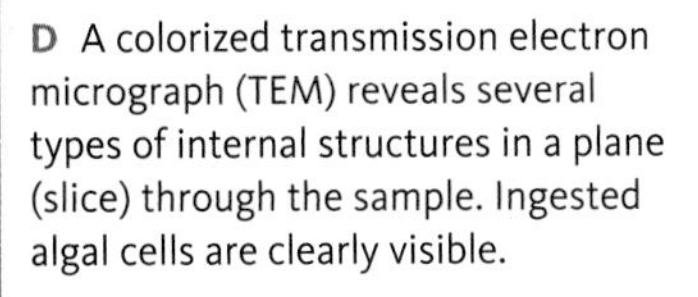

D A colorized transmission electron micrograph (TEM) reveals several types of internal structures in a plane (slice) through the sample. Ingested algal cells are clearly visible.

E A colorized scanning electron micrograph (SEM) shows details of the cell's surface. The cell ingests its food via the indentation (also visible in A).

FIGURE 4.5 Different microscopy techniques reveal different characteristics of the same organism, a protist (*Paramecium*).

FIGURE IT OUT: About how big are these cells?

Answer: About 250 µm long

PEOPLE MATTER

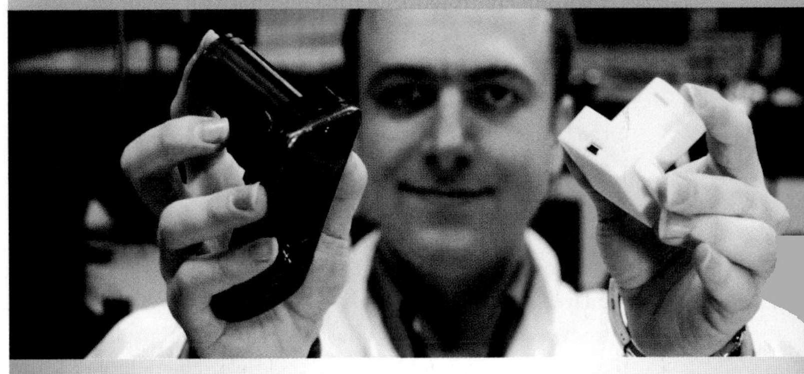

National Geographic Explorer
DR. AYDOGAN OZCAN

Aydogan Ozcan is developing a revolutionary global health solution using one of the most common forms of technology available—the smart phone. Using readily available parts that cost less than $50, Ozcan builds adapters that transform a smart phone into a mobile medical lab with the capability to test and diagnose diseases like HIV, malaria, and tuberculosis in remote communities.

Conventional microscopes, the mainstay of diagnosis for centuries, are impractical on a global level. "They are too heavy and powerful to be cost-effectively miniaturized. They also can't quickly capture and screen the large number of cells needed for statistically viable diagnoses," he says. What's more, because technicians in remote areas may be poorly trained, they often interpret images inaccurately. "In some parts of Africa, 70 percent of malaria diagnoses are incorrect false-positives." Ozcan's invention solves these problems by arming cell phones with sophisticated algorithms that do the interpreting. To tackle the most expensive part of microscopes—lenses—his team simply eliminated them. Ozcan's modified phone uses a special light source and the phone's camera to capture an image of a blood sample, essentially turning the phone into a lens-free microscope able to resolve structures smaller than one micron.

4.3 WHAT IS A MEMBRANE?

A plasma membrane physically separates a cell's external environment from its internal one, but that is not its only function. For example, you learned in Section 4.1 that a cell's plasma membrane allows some substances, but not others, to cross it. Membranes around organelles do this too. We return to membrane functions in Chapter 5; here, we explore the structure that gives rise to these functions.

THE FLUID MOSAIC MODEL

The foundation of almost all cell membranes is a lipid bilayer that consists mainly of phospholipids. Remember from Section 3.3 that a phospholipid has a phosphate-containing head and two fatty acid tails. The polar head is hydrophilic, which means that it interacts with water molecules. The nonpolar tails are hydrophobic, so they do not interact with water molecules. As a result of these opposing properties, phospholipids swirled into water will spontaneously organize themselves into lipid bilayer sheets or bubbles (*left*), with hydrophobic tails together, hydrophilic heads facing the watery surroundings (**FIGURE 4.6A**).

Other molecules, including cholesterol, proteins, glycoproteins, and glycolipids, are embedded in or attached to the lipid bilayer of a cell membrane. Many of these molecules move around the membrane more or less freely. We describe a eukaryotic or bacterial cell membrane as a **fluid mosaic** because it behaves like a two-dimensional liquid of mixed composition. The "mosaic" part of the name comes from the many different types of molecules in the membrane. A cell membrane is fluid because its phospholipids are not chemically bonded to one another; they stay organized in a bilayer as a result of collective hydrophobic and hydrophilic attractions. These interactions are, on an individual basis, relatively weak. Thus, individual phospholipids in the bilayer drift sideways and spin around their long axis, and their tails wiggle.

A cell membrane's properties vary depending on the types and proportions of molecules composing it. For example, membrane fluidity decreases with increasing cholesterol content. A membrane's fluidity also depends on the length and saturation of its phospholipids' fatty acid tails (Section 3.3). Archaea do not even use fatty acids to build their phospholipids. Instead, they use molecules with reactive side chains, so the tails of archaeal phospholipids form covalent bonds with one another. As a result of this rigid crosslinking, archaeal phospholipids do not drift, spin, or wiggle in a bilayer. Thus, membranes of archaea are stiffer than those of bacteria or eukaryotes, a characteristic that may help these cells survive in extreme habitats.

A In a watery fluid, phospholipids spontaneously line up into two layers: the hydrophobic tails cluster together, and the hydrophilic heads face outward, toward the fluid. This lipid bilayer forms the framework of all cell membranes. Many types of proteins intermingle among the lipids; a few that are typical of plasma membranes are shown *opposite*.

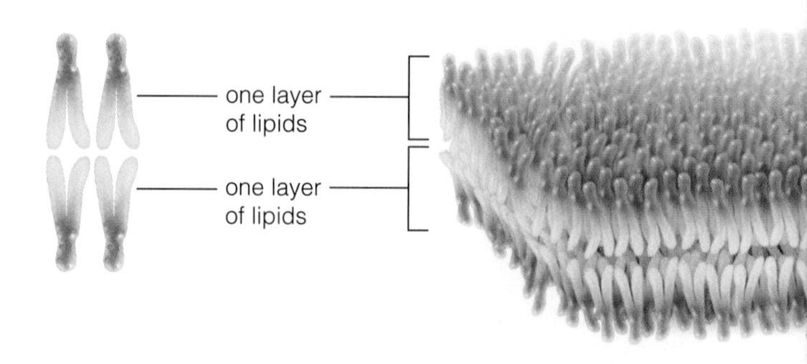

FIGURE 4.6 {Animated} Cell membrane structure.
A Organization of phospholipids in cell membranes.
B–E Examples of common membrane proteins.

TABLE 4.3

Common Membrane Proteins

Category	Function	Examples
Passive transport protein	Allows ions or small molecules to cross a membrane to the side where they are less concentrated.	Porin; glucose transporter
Active transport protein	Pumps ions or molecules through membranes to the side where they are more concentrated. Requires energy input, as from ATP.	Calcium pump; serotonin transporter
Receptor	Initiates change in a cell activity by responding to an outside signal (e.g., by binding a signaling molecule or absorbing light energy).	Insulin receptor; B cell receptor
Adhesion protein	Helps cells stick to one another, to cell junctions, and to extracellular matrix.	Integrins; cadherins
Recognition protein	Identifies a cell as self (belonging to one's own body or tissue) or nonself (foreign to the body).	MHC molecule
Enzyme	Speeds a specific reaction. Membranes provide a relatively stable reaction site for enzymes that work in series with other molecules.	Cytochrome *c* oxidase

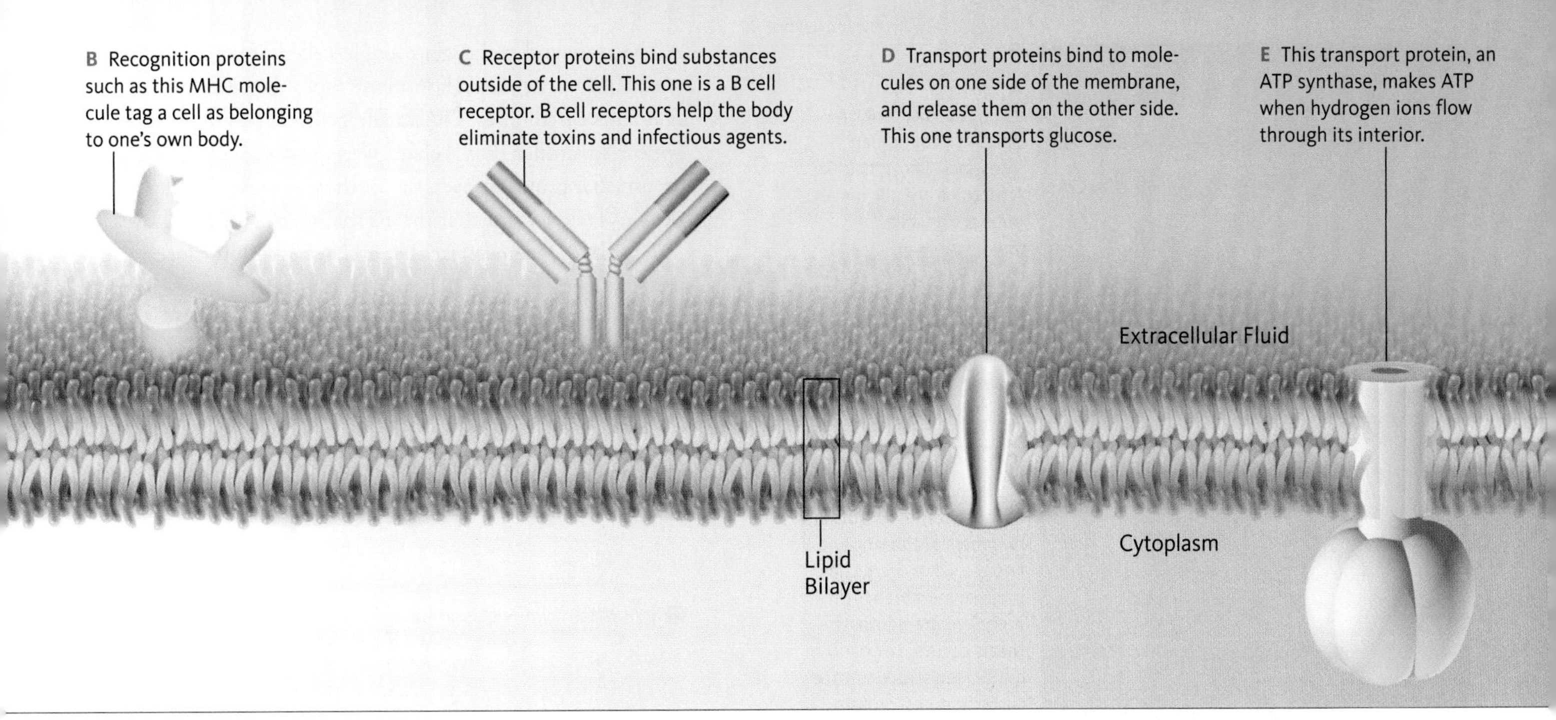

PROTEINS ADD FUNCTION

Many types of proteins are associated with a cell membrane (**TABLE 4.3**). Some are temporarily or permanently attached to one of the lipid bilayer's surfaces. Others have a hydrophobic domain that anchors the protein in the bilayer. Filaments inside the cell fasten some membrane proteins in place, including those that cluster as rigid pores.

Each type of protein in a membrane imparts a specific function to it. Thus, different cell membranes can have different functions depending on which proteins are associated with them. A plasma membrane has certain proteins that no internal cell membrane has. For example, cells in some animal tissues are fastened together by **adhesion proteins** in their plasma membranes, an arrangement that strengthens these tissues. **Recognition proteins** in the plasma membrane function as unique identity tags for an individual or a species (**FIGURE 4.6B**). As you will see in Chapter 34, being able to recognize "self" imparts the potential ability to distinguish nonself (foreign) cells or particles.

Plasma membranes and some internal membranes incorporate **receptor proteins**, which trigger a change in the cell's activities upon binding a particular substance (**FIGURE 4.6C**). Different receptors bind to hormones or other signaling molecules, toxins, or molecules on another cell. The response triggered may involve metabolism, movement, division, or even cell death.

All cell membranes have some types of proteins, including enzymes. **Transport proteins** move specific substances across a membrane, typically by forming a channel through it (**FIGURE 4.6D,E**). These proteins are important because lipid bilayers are impermeable to most substances, including ions and polar molecules. Some transport proteins are open channels through which a substance moves on its own across a membrane. Others use energy to actively pump a substance across.

adhesion protein Protein that helps cells stick together in animal tissues.
fluid mosaic Model of a cell membrane as a two-dimensional fluid of mixed composition.
receptor protein Membrane protein that triggers a change in cell activity after binding to a particular substance.
recognition protein Plasma membrane protein that identifies a cell as belonging to self (one's own body or species).
transport protein Protein that passively or actively assists specific ions or molecules across a membrane.

TAKE-HOME MESSAGE 4.3

A cell membrane selectively controls exchanges between the cell and its surroundings.

The foundation of almost all cell membranes is the lipid bilayer—two layers of lipids (mainly phospholipids), with tails sandwiched between heads.

Proteins embedded in or attached to a lipid bilayer add specific functions to each type of cell membrane.

4.4 HOW ARE BACTERIA AND ARCHAEA ALIKE?

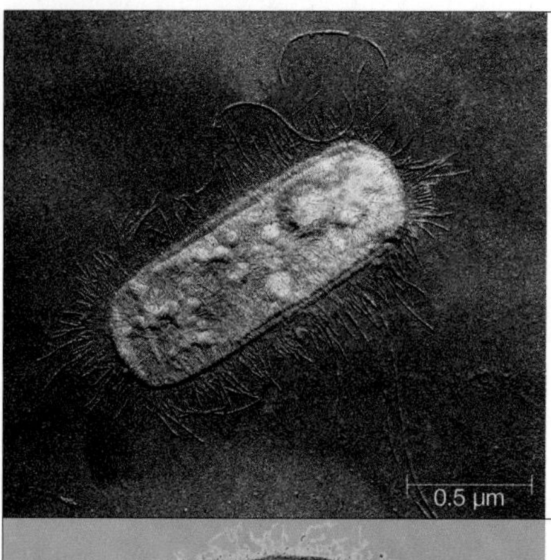

A *Escherichia coli*, a common bacterial inhabitant of human intestines. Short, hairlike structures are pili; longer ones are flagella.

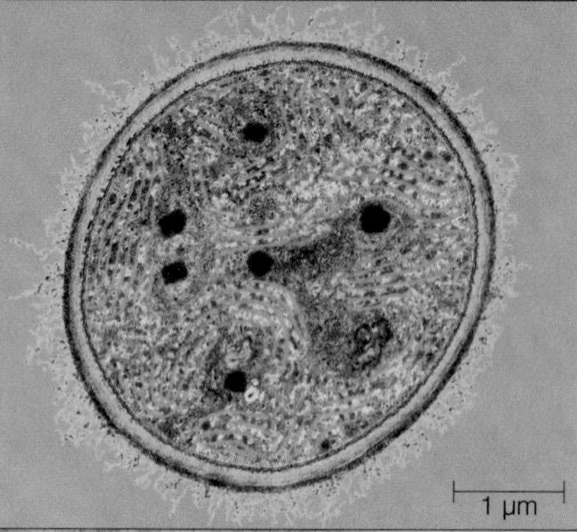

B *Dermocarpa*, a type of cyanobacteria. Like other members of this ancient lineage, *Dermocarpa* has internal membranes (in green) where photosynthesis occurs. The dark, multisided structures are carboxysomes, protein-enclosed organelles that assist photosynthesis.

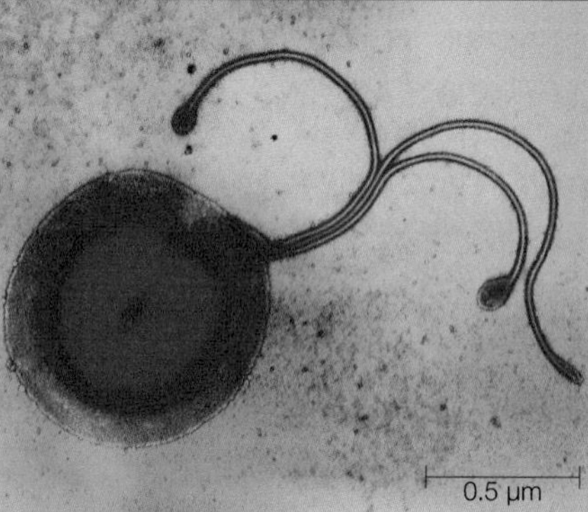

C *Helicobacter pylori*, a bacterium that can cause stomach ulcers when it infects the lining of the stomach. In unfavorable conditions, this species takes on a ball-shaped form (shown) that may offer the cells protection from environmental challenges such as antibiotic treatment.

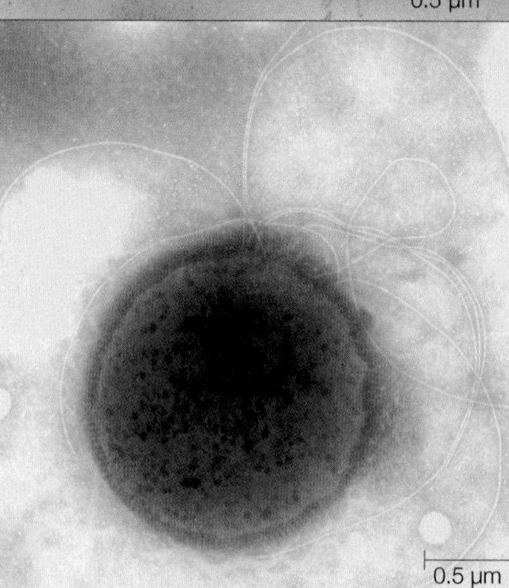

D *Thermococcus gammatolerans*, an archaeon discovered at a deep-sea hydrothermal vent, where it lives under extreme conditions of salt, temperature, and pressure. It is by far the most radiation-resistant organism ever discovered, capable of withstanding thousands of times more radiation than humans can.

FIGURE 4.7 Some representatives of bacteria (A–C) and an archaeon (D).

All bacteria and archaea are single-celled organisms (**FIGURE 4.7**), though in many types the cells form filaments or colonies. Outwardly, cells of the two groups appear so similar that archaea were once thought to be an unusual group of bacteria. Both were classified as prokaryotes, a word that means "before the nucleus." By 1977, it had become clear that archaea are more closely related to eukaryotes than to bacteria, so they were given their own separate domain. The term "prokaryote" is now an informal designation.

FIGURE 4.8 {Animated} Generalized body plan of a prokaryote (a bacterium or archaeon).

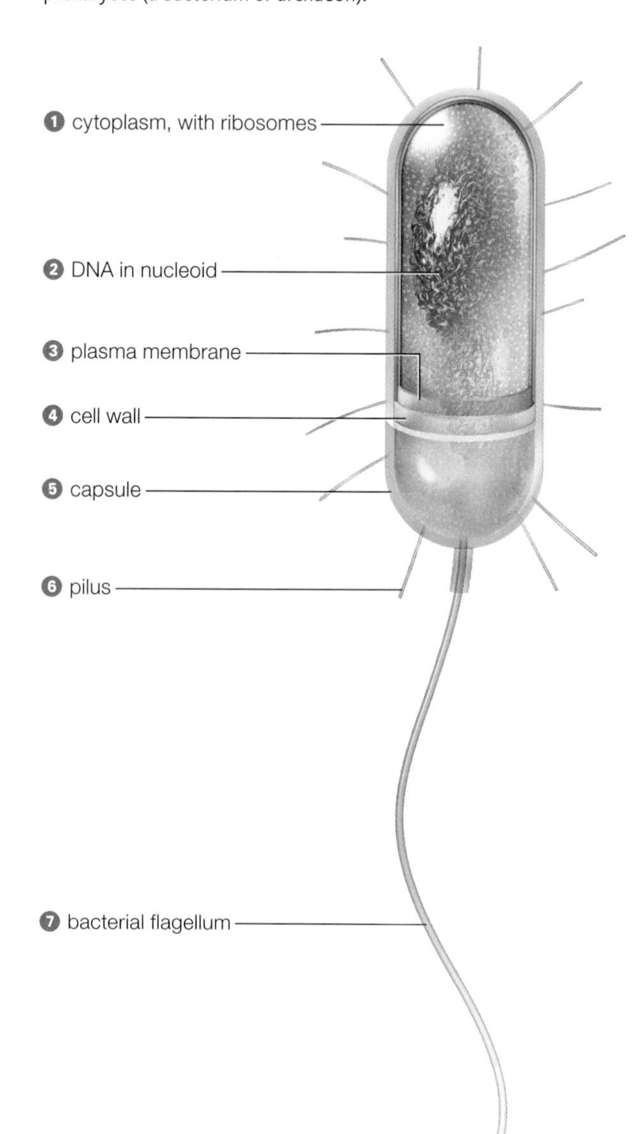

Bacteria and archaea are the smallest and most metabolically diverse forms of life that we know about. They inhabit nearly all of Earth's environments, including some extremely hostile places. The two kinds of cells differ in structure and metabolism. Chapter 19 revisits them in more detail; here we present an overview of structures shared by both groups (**FIGURE 4.8**).

Compared with eukaryotic cells, prokaryotes have little in the way of internal framework, but they do have protein filaments under the plasma membrane that reinforce the cell's shape and act as scaffolding for internal structures. The cytoplasm of these cells ❶ contains many **ribosomes** (organelles upon which polypeptides are assembled), and in some species, additional organelles. Cytoplasm also contains **plasmids**, small circles of DNA that carry a few genes (units of inheritance) that can provide advantages, such as resistance to antibiotics. The cell's remaining genes typically occur on one large circular molecule of DNA located in an irregularly shaped region of cytoplasm called the **nucleoid** ❷. In a few species, the nucleoid is enclosed by a membrane. Other internal membranes carry out special metabolic processes such as photosynthesis in some prokaryotes (**FIGURE 4.7B**).

Like all cells, bacteria and archaea have a plasma membrane ❸. In nearly all prokaryotes, a rigid **cell wall** ❹ surrounding the plasma membrane protects the cell and supports its shape. Most archaeal cell walls consist of proteins; most bacterial cell walls consist of a polymer of peptides and polysaccharides. Both types are permeable to water, so dissolved substances easily cross.

Polysaccharides form a slime layer or capsule ❺ around the wall of many types of bacteria. These sticky structures help the cells adhere to many types of surfaces, and they also offer protection against some predators and toxins.

Protein filaments called **pili** (singular, pilus) ❻ project from the surface of some prokaryotes. Pili help these cells move across or cling to surfaces. Many prokaryotes also have one or more **flagella** (singular, flagellum) ❼, which are long, slender cellular structures used for motion. A bacterial flagellum rotates like a propeller that drives the cell through fluid habitats.

BIOFILMS

Bacterial cells often live so close together that an entire community shares a layer of secreted polysaccharides and proteins. A communal living arrangement in which single-celled organisms live in a shared mass of slime is called a **biofilm**. A biofilm is often attached to a solid surface, and may include bacteria, algae, fungi, protists, and/or archaea. Participating in a biofilm allows the cells to linger in a favorable spot rather than be swept away by fluid currents, and to reap the benefits of living communally. For example, rigid or netlike secretions of some species serve as permanent scaffolding for others; species that break down toxic chemicals allow more sensitive ones to thrive in habitats that they could not withstand on their own; and waste products of some serve as raw materials for others. Later chapters discuss medical implications of biofilms, including the dental plaque that forms on teeth (**FIGURE 4.9**).

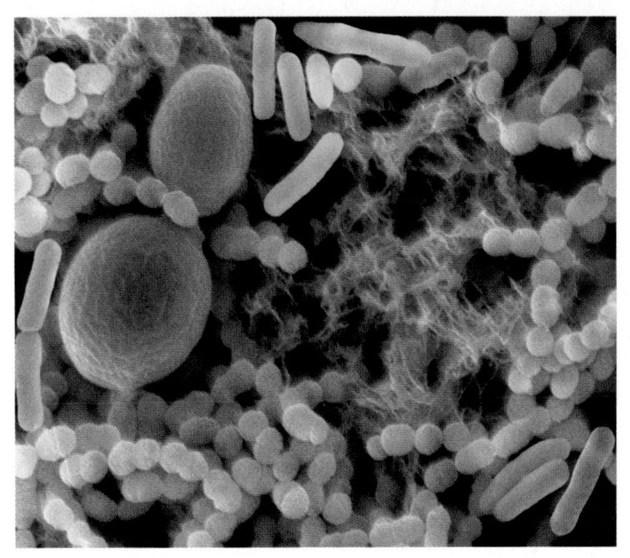

FIGURE 4.9 Oral bacteria in dental plaque, a biofilm. This micrograph shows two species of bacteria (tan, green) and a yeast (red) sticking to one another and to teeth via a gluelike mass of shared, secreted polysaccharides (pink). Other secretions of these organisms cause cavities and periodontal disease.

biofilm Community of microorganisms living within a shared mass of secreted slime.
cell wall Rigid but permeable structure that surrounds the plasma membrane of some cells.
flagellum Long, slender cellular structure used for motility.
nucleoid Of a bacterium or archaeon, region of cytoplasm where the DNA is concentrated.
pilus A protein filament that projects from the surface of some prokaryotic cells.
plasmid Small circle of DNA in some bacteria and archaea.
ribosome Organelle of protein synthesis.

TAKE-HOME MESSAGE 4.4

Bacteria and archaea do not have a nucleus. Most kinds have a cell wall around their plasma membrane. The permeable wall reinforces and imparts shape to the cell body.

The structure of bacteria and archaea is relatively simple, but as a group these organisms are the most diverse forms of life.

4.5 WHAT DO EUKARYOTIC CELLS HAVE IN COMMON?

In addition to the nucleus, a typical eukaryotic cell has many other organelles, including endoplasmic reticulum, Golgi bodies, ribosomes, and at least one mitochondrion (**TABLE 4.4** and **FIGURE 4.10**). Organelles with membranes can regulate the types and amounts of substances that enter and exit. Such control maintains a special internal environment that allows the organelle to carry out its particular function—for example, isolating toxic or sensitive substances from the rest of the cell, moving substances through cytoplasm, maintaining fluid balance, or providing a favorable environment for a special process.

In this section, we detail the nucleus, which is the defining characteristic of eukaryotes. The remaining sections of the chapter introduce the functions of other organelles typical of eukaryotic cells.

TABLE 4.4

Some Organelles in Eukaryotic Cells

Organelles with membranes	
Nucleus	Protecting and controlling access to DNA
Endoplasmic reticulum (ER)	Making, modifying new polypeptides and lipids; other tasks
Golgi body	Modifying and sorting new polypeptides and lipids
Vesicle	Transporting, storing, or breaking down substances
Mitochondrion	Making ATP by glucose breakdown
Chloroplast	Making sugars in plants, some protists
Lysosome	Intracellular digestion
Peroxisome	Breaking down fatty acids, amino acids, toxins
Vacuole	Storage, breaking down food or waste
Organelles without membranes	
Ribosome	Assembling polypeptides
Centriole	Anchor for cytoskeleton
Other components	
Cytoskeleton	Contributes to cell shape, internal organization, movement

THE NUCLEUS

A cell nucleus (**FIGURE 4.11**) serves multiple functions. First, it keeps the cell's genetic material—its one and only copy of DNA—safe from metabolic processes that might damage it. Isolated in its own compartment, the DNA stays separated from the bustling activity of the cytoplasm. The nucleus also allows some molecules, but not others, to access the DNA. The nuclear membrane, which is called

FIGURE 4.10 Some components of eukaryotic cells.

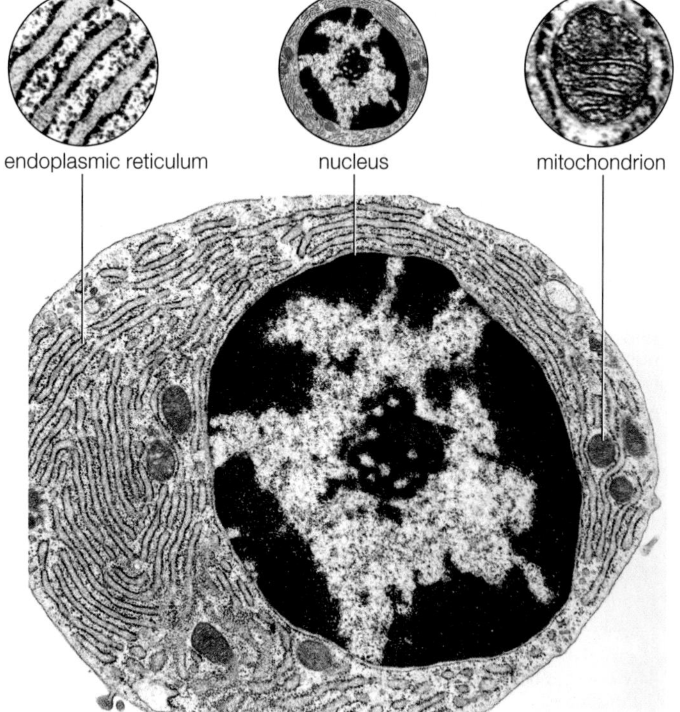

An animal cell (a white blood cell of a guinea pig)

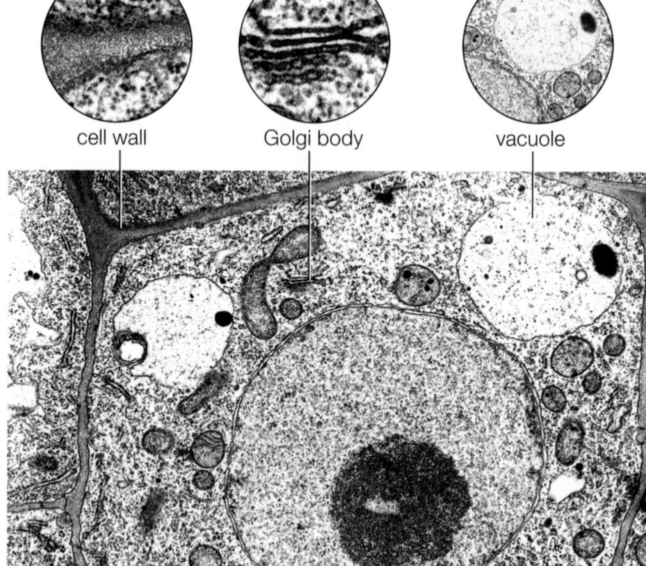

A plant cell (from a root of thale cress)

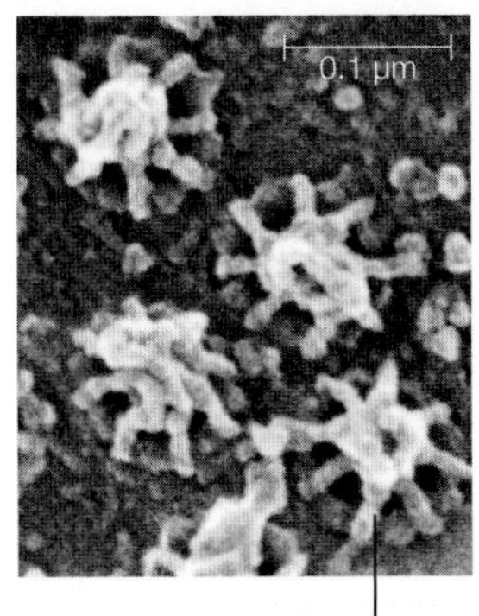

the **nuclear envelope**, carries out this function. A nuclear envelope consists of two lipid bilayers folded together as a single membrane. Membrane proteins aggregate into thousands of tiny pores (*left*) that span the nuclear envelope. The pores are anchored by the nuclear lamina, a dense mesh of fibrous proteins that supports the inner surface of the membrane. Some bacteria have membranes around their DNA, but we do not consider the bacteria to have nuclei because there are no pores in these membranes.

As you will see in Chapter 5, large molecules, including RNA and proteins, cannot cross a lipid bilayer on their own. Nuclear pores function as gateways for these molecules to enter and exit a nucleus. Protein synthesis offers an example of why this movement is important. Protein synthesis occurs in cytoplasm, and it requires the participation of many molecules of RNA. RNA is produced in the nucleus. Thus, RNA molecules must move from nucleus to cytoplasm, and they do so through nuclear pores. Proteins that carry out RNA synthesis must move in the opposite direction, because this process occurs in the nucleus. A cell can regulate the amounts and types of proteins it makes at a given time by selectively restricting the passage of certain molecules through nuclear pores. (Later chapters return to details of protein synthesis and controls over it.)

The nuclear envelope encloses **nucleoplasm**, a viscous fluid similar to cytoplasm, in which the cell's DNA is suspended. The nucleus contains at least one **nucleolus** (plural, nucleoli), a dense, irregularly shaped region of proteins and nucleic acid where subunits of ribosomes are produced.

---

nuclear envelope A double membrane that constitutes the outer boundary of the nucleus. Pores in the membrane control which substances can cross.
nucleolus In a cell nucleus, a dense, irregularly shaped region where ribosomal subunits are assembled.
nucleoplasm Viscous fluid enclosed by the nuclear envelope.

TAKE-HOME MESSAGE 4.5

All eukaryotic cells start life with a nucleus and other membrane-enclosed organelles.

A nucleus protects and controls access to a eukaryotic cell's DNA.

The nuclear envelope is a double lipid bilayer. Proteins embedded in the bilayer form pores that control the passage of molecules between the nucleus and cytoplasm.

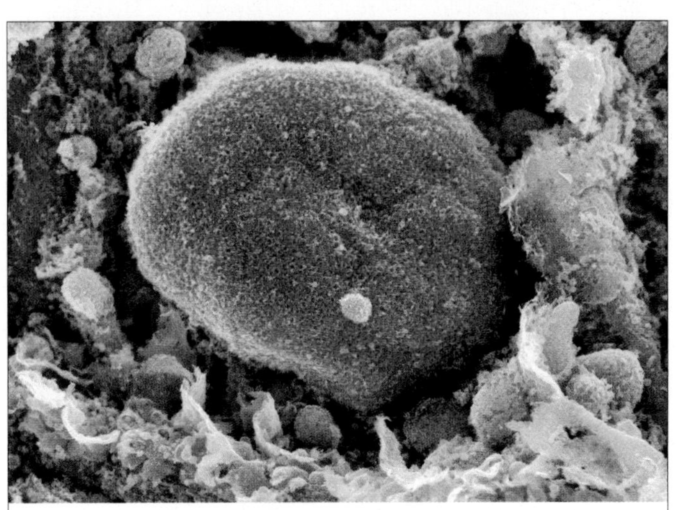

A Nucleus of a liver cell (in pink).

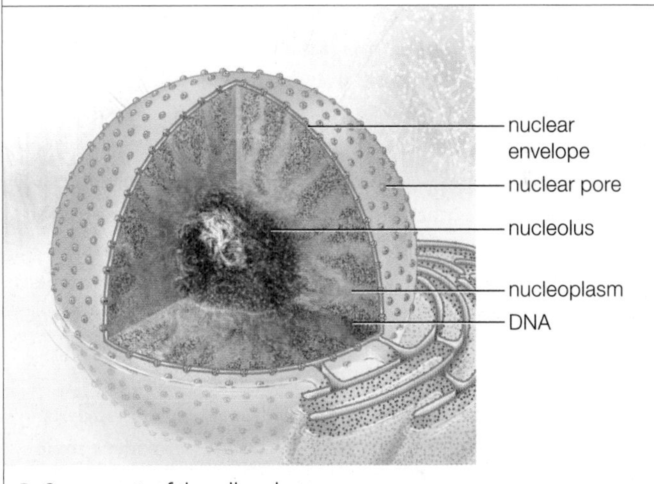

B Components of the cell nucleus.

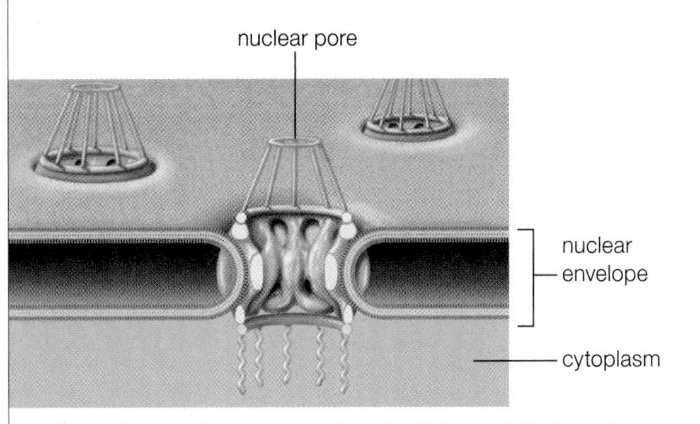

C The nuclear envelope consists of two lipid bilayers folded together as a single membrane and studded with thousands of nuclear pores. Each nuclear pore is an organized cluster of membrane proteins that selectively allows certain substances to cross it on their way into and out of the nucleus.

FIGURE 4.11 {Animated} The cell nucleus.

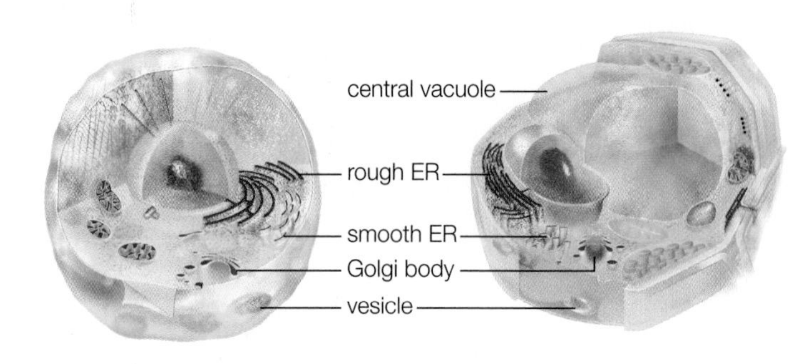

The **endomembrane system** is a series of interacting organelles between the nucleus and the plasma membrane (*above*). Its main function is to make lipids, enzymes, and proteins for insertion into the cell's membranes or secretion to the external environment. The endomembrane system also destroys toxins, recycles wastes, and has other special functions. Components of the system vary among different types of cells, but here we present an overview of the most common ones (**FIGURE 4.12**).

A VARIETY OF VESICLES

Small, membrane-enclosed sacs called **vesicles** form by budding from other organelles or when a patch of plasma membrane sinks into the cytoplasm ❶. Vesicles have a variety of functions. Many transport substances from one organelle to another, or to and from the plasma membrane. Some are a bit like trash cans that collect and dispose of waste, debris, or toxins. Enzymes in **peroxisomes** break down fatty acids, amino acids, and poisons such as alcohol. They also break down hydrogen peroxide, a toxic by-product of fatty acid metabolism. **Lysosomes** take part in intracellular digestion. They contain powerful enzymes that can break down cellular debris and wastes (carbohydrates, proteins, nucleic acids, and lipids). Vesicles in cells such as amoebas or white blood cells deliver ingested bacteria, cell parts, and other debris to lysosomes for breakdown.

Vacuoles form by the fusion of multiple vesicles. They have different functions in different kinds of cells. Many isolate or break down waste, debris, toxins, or food (**FIGURE 4.13**). Amino acids, sugars, ions, wastes, and toxins accumulate in the water-filled interior of a plant cell's large **central vacuole**. Fluid pressure in a central vacuole keeps plant cells plump, so stems, leaves, and other plant parts stay firm.

---

central vacuole Fluid-filled vesicle in many plant cells.
endomembrane system Series of interacting organelles (endoplasmic reticulum, Golgi bodies, vesicles) between nucleus and plasma membrane; produces lipids, proteins.
endoplasmic reticulum (ER) Organelle that is a continuous system of sacs and tubes extending from the nuclear envelope. Smooth ER makes lipids and breaks down carbohydrates and fatty acids; ribosomes on the surface of rough ER synthesize proteins.
Golgi body Organelle that modifies proteins and lipids, then packages the finished products into vesicles.
lysosome Enzyme-filled vesicle that breaks down cellular wastes and debris.
peroxisome Enzyme-filled vesicle that breaks down amino acids, fatty acids, and toxic substances.
vacuole A fluid-filled organelle that isolates or disposes of waste, debris, or toxic materials.
vesicle Small, membrane-enclosed organelle; different kinds store, transport, or break down their contents.

FIGURE 4.12 Some interactions among components of the endomembrane system.

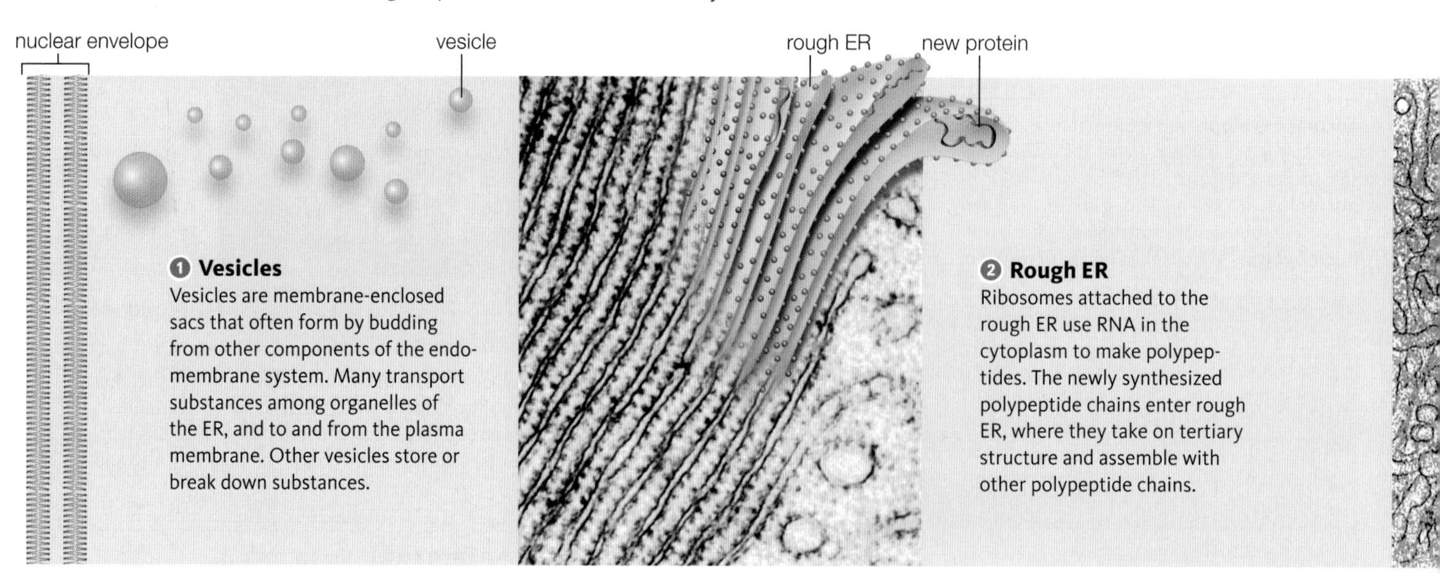

ENDOPLASMIC RETICULUM

The membrane of the **endoplasmic reticulum (ER)** is an extension of the nuclear envelope. Its interconnected tubes and flattened sacs form a single compartment that houses many enzymes. Two kinds of ER, rough and smooth, are named for their appearance. Thousands of ribosomes that attach to the outer surface of rough ER give this organelle its "rough" appearance. These ribosomes make polypeptides that thread into the interior of the ER as they are assembled ❷. Inside the ER, the polypeptide chains fold and take on their tertiary structure, and many assemble with other polypeptide chains (Section 3.4). Cells that make, store, and secrete proteins have a lot of rough ER. For example, ER-rich cells in the pancreas make digestive enzymes that they secrete into the small intestine.

Some proteins made in rough ER become part of its membrane. Others migrate through the ER compartment to smooth ER. Smooth ER has no ribosomes, so it does not make its own proteins ❸. Some proteins that arrive in smooth ER are immediately packaged into vesicles for delivery elsewhere. Others are enzymes that stay and become part of the smooth ER. Some of these enzymes break down carbohydrates, fatty acids, and some drugs and poisons. Others make lipids for the cell's membranes.

GOLGI BODIES

A **Golgi body** has a folded membrane that often looks like a stack of pancakes ❹. Enzymes inside of it put finishing touches on proteins and lipids that have been delivered from ER. These enzymes attach phosphate groups or carbohydrates, and cleave certain proteins. The finished products (such as membrane proteins, proteins for secretion, and enzymes) are sorted and packaged in new vesicles. Some of the vesicles deliver their cargo to the plasma membrane; others become lysosomes.

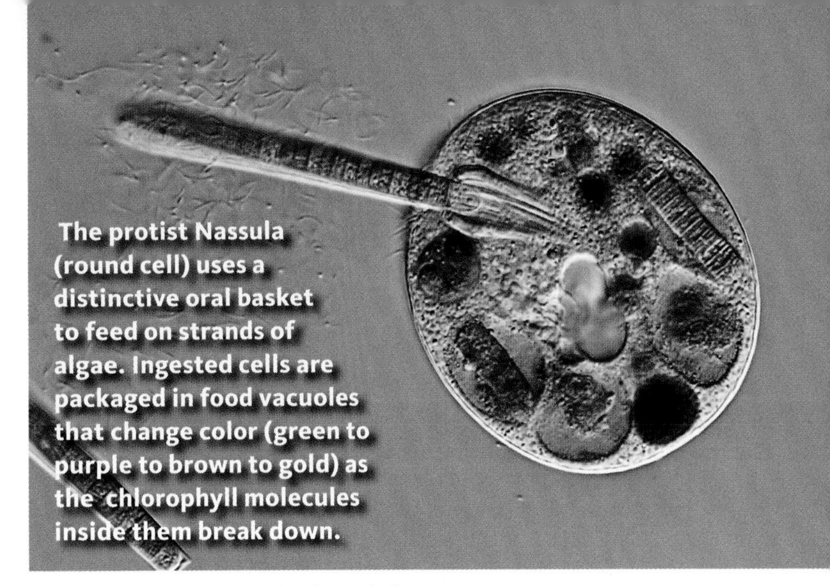

FIGURE 4.13 An example of vacuole function.

> **TAKE-HOME MESSAGE 4.6**
>
> Rough ER produces enzymes, membrane proteins, and secreted proteins. Smooth ER produces lipids and breaks down carbohydrates, fatty acids, and toxins.
>
> Golgi bodies modify proteins and lipids.

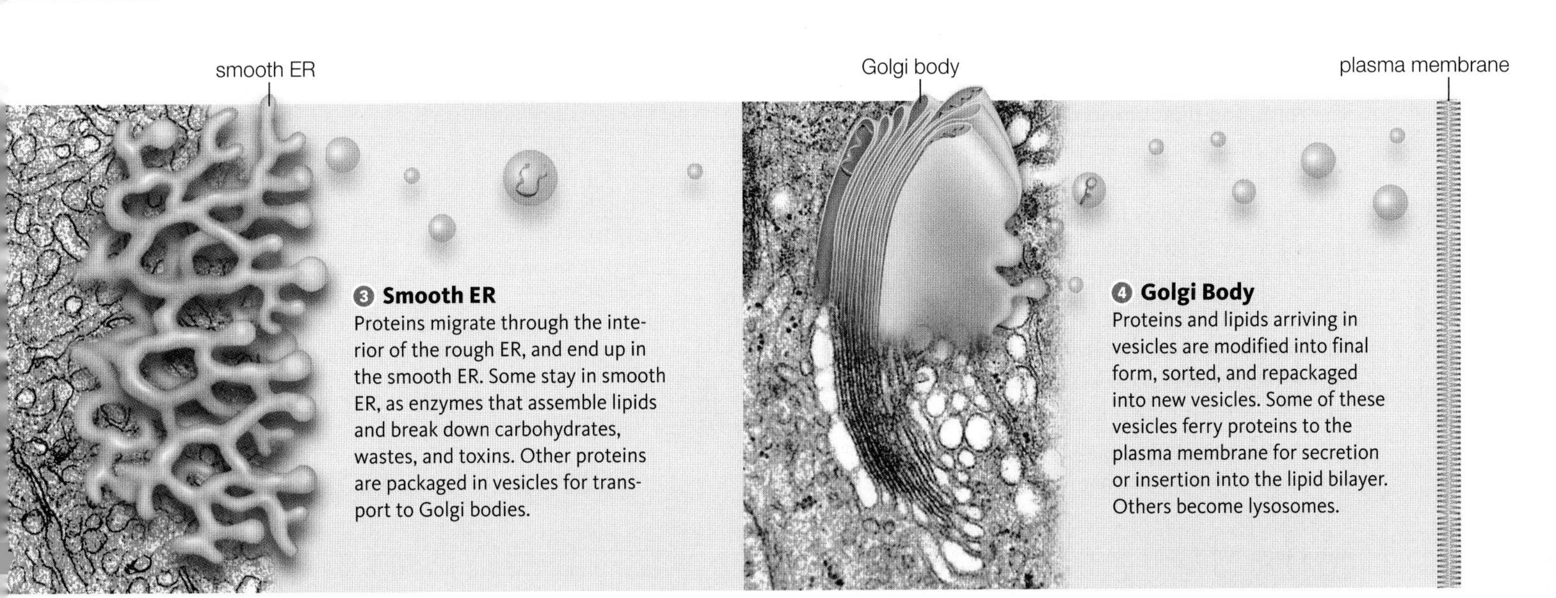

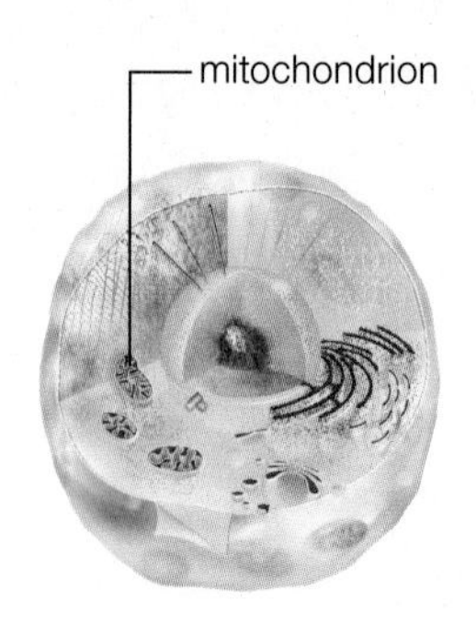

As you will see in Chapter 5, biologists think of the nucleotide ATP as a type of cellular currency because it carries energy between reactions. Cells require a lot of ATP. The most efficient way they can produce it is by aerobic respiration, a series of oxygen-requiring reactions that harvests the energy in sugars by breaking their bonds. In eukaryotes, aerobic respiration occurs inside organelles called **mitochondria** (singular, mitochondrion). With each breath, you are taking in oxygen mainly for the mitochondria in your trillions of aerobically respiring cells.

The structure of a mitochondrion is specialized for carrying out reactions of aerobic respiration. Each mitochondrion has two membranes, one highly folded inside the other (**FIGURE 4.14**). This arrangement creates two compartments: an outer one (between the two membranes), and an inner one (inside the inner membrane). Hydrogen ions accumulate in the outer compartment. The buildup pushes the ions across the inner membrane, into the inner compartment, and this flow drives ATP formation. Chapter 7 returns to the details of aerobic respiration.

Nearly all eukaryotic cells (including plant cells) have mitochondria, but the number varies by the type of cell and by the organism. For example, single-celled organisms such as yeast often have only one mitochondrion, but human skeletal muscle cells have a thousand or more. In general, cells that have the highest demand for energy tend to have the most mitochondria.

Typical mitochondria are between 1 and 4 micrometers in length. These organelles can change shape, split in two, branch, or fuse together. They resemble bacteria in size, form, and biochemistry. They have their own DNA, which is circular and otherwise similar to bacterial DNA. They divide independently of the cell, and have their own ribosomes. Such clues led to a theory that mitochondria evolved from aerobic bacteria that took up permanent residence inside a host cell (we return to this topic in Section 18.5).

Some eukaryotes that live in oxygen-free environments have modified mitochondria that produce hydrogen in addition to ATP. Like mitochondria, these organelles have two membranes, but they have lost the ability to divide independently because they lack their own DNA.

mitochondrion Double-membraned organelle that produces ATP by aerobic respiration in eukaryotes.

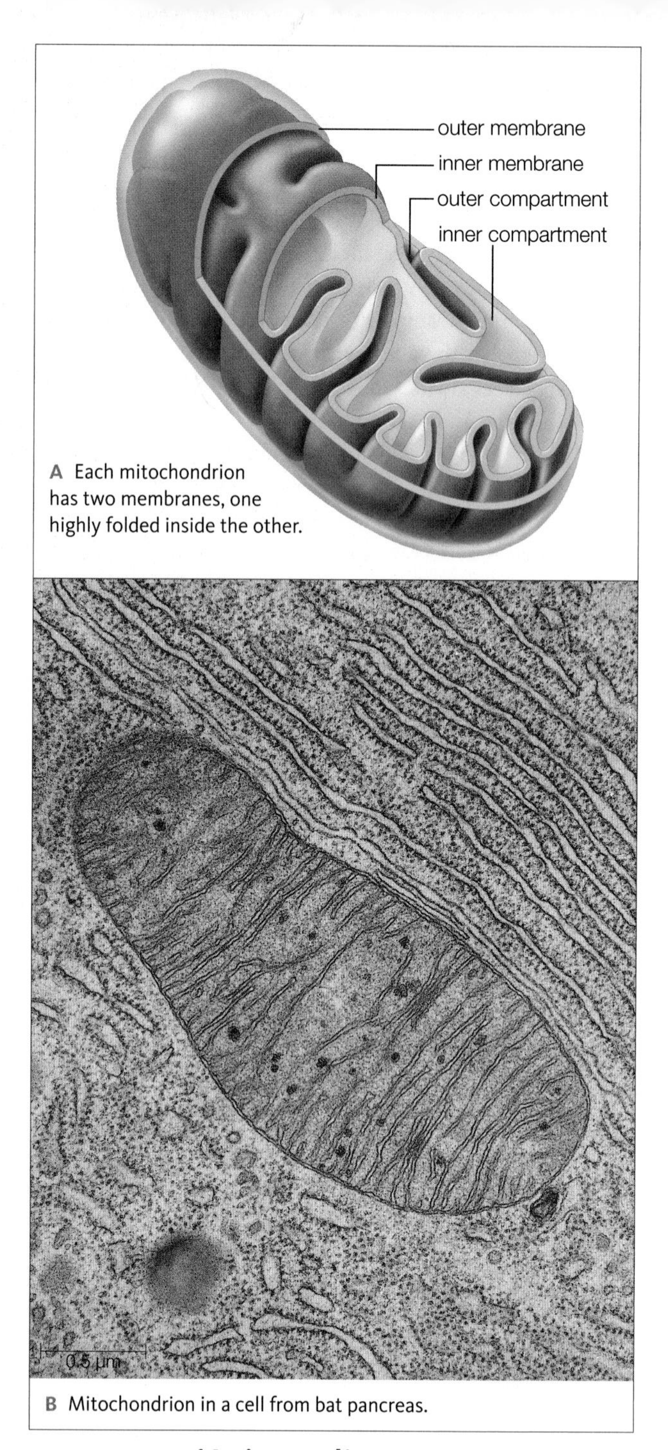

A Each mitochondrion has two membranes, one highly folded inside the other.

B Mitochondrion in a cell from bat pancreas.

FIGURE 4.14 {Animated} The mitochondrion, a eukaryotic organelle that specializes in producing ATP.

FIGURE IT OUT: What organelle is visible in the upper right-hand corner of the TEM?

Answer: Rough ER

TAKE-HOME MESSAGE 4.7

Mitochondria are eukaryotic organelles specialized to produce ATP by aerobic respiration.

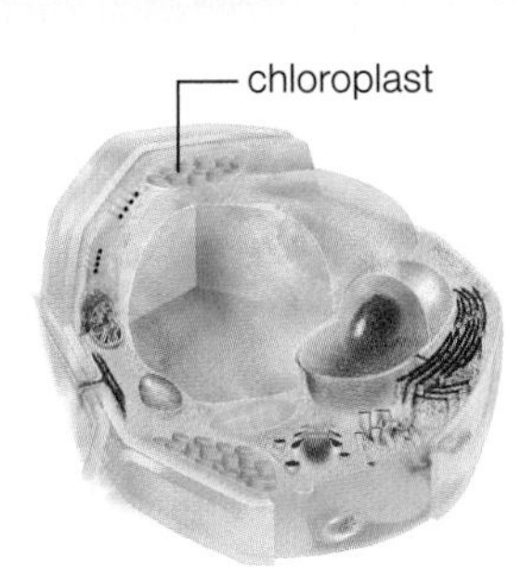

Plastids are double-membraned organelles that function in photosynthesis, storage, or pigmentation in plant and algal cells. Photosynthetic cells of plants and many protists contain **chloroplasts**, which are plastids specialized for photosynthesis (**FIGURE 4.15**). Most chloroplasts are oval or disk-shaped. Each has two outer membranes enclosing a semifluid interior, the stroma, that contains enzymes and the chloroplast's own DNA. In the stroma, a third, highly folded membrane forms a single, continuous compartment. Photosynthesis occurs at this inner membrane.

The innermost membrane of a chloroplast incorporates many pigments, including a green one called chlorophyll (the abundance of chlorophyll in plant cell chloroplasts is the reason most plants are green). During photosynthesis, these pigments capture energy from sunlight, and pass it to other molecules that require energy to make ATP. The resulting ATP is used inside the stroma to build sugars from carbon dioxide and water. (Chapter 6 returns to details of these processes.) In many ways, chloroplasts resemble the photosynthetic bacteria that they evolved from.

Chromoplasts are plastids that make and store pigments other than chlorophylls. They often contain red or orange carotenoids that color flowers, leaves, roots, and fruits (**FIGURE 4.16**). Chromoplasts are related to chloroplasts, and the two types of plastids are interconvertible. For example, as fruits such as tomatoes ripen, green chloroplasts in their cells are converted to red chromoplasts, so the color of the fruit changes.

Amyloplasts are unpigmented plastids that make and store starch grains. They are notably abundant in cells of stems, tubers (underground stems), fruits, and seeds. Like chromoplasts, amyloplasts are related to chloroplasts, and one type can change into the other. Starch-packed amyloplasts are dense and heavy compared to cytoplasm; in some plant cells, they function as gravity-sensing organelles (we return to this topic in Chapter 27).

chloroplast Organelle of photosynthesis in the cells of plants and photosynthetic protists.
plastid One of several types of double-membraned organelles in plants and algal cells; for example, a chloroplast or amyloplast.

TAKE-HOME MESSAGE 4.8

Plastids occur in plants and some protists; they function in photosynthesis, storage, and pigmentation.

Chloroplasts are plastids that carry out photosynthesis.

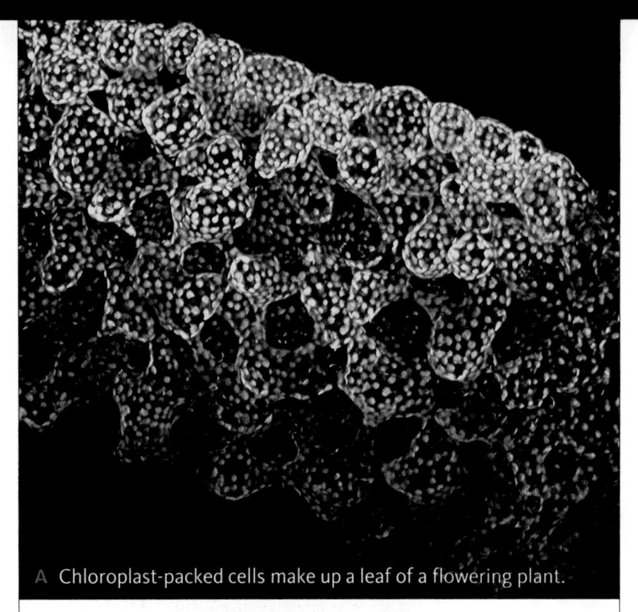

A Chloroplast-packed cells make up a leaf of a flowering plant.

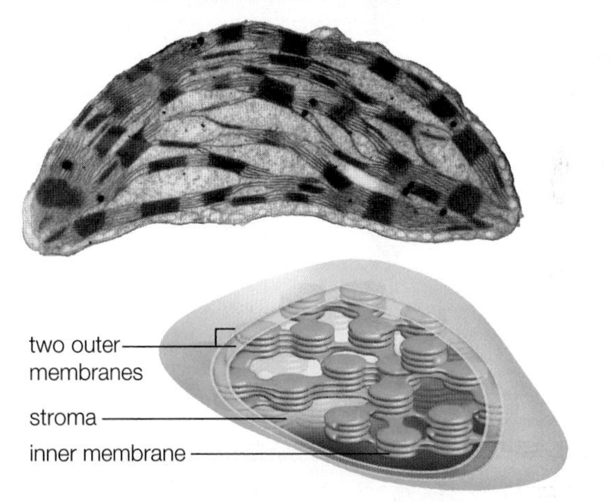

B Each chloroplast has two outer membranes. Photosynthesis occurs at a much-folded inner membrane. The electron micrograph shows a chloroplast from a leaf of corn.

FIGURE 4.15 {Animated} The chloroplast.

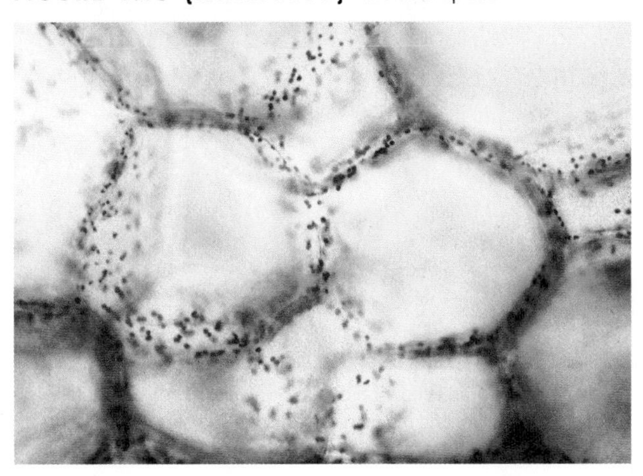

FIGURE 4.16 Chromoplasts. The color of a red bell pepper arises from chromoplasts in its cells.

CREDITS: (in text) © Cengage Learning; (15A) Heiti Paves/Science Photo Library; (15B) top, Dr. George Chapman/Visuals Unlimited, Inc.; bottom, © Cengage Learning 2015; (16) © David T. Webb.

Between the nucleus and plasma membrane of all eukaryotic cells is a system of interconnected protein filaments collectively called the **cytoskeleton.** Elements of the cytoskeleton reinforce, organize, and move cell structures, and often the whole cell. Some are permanent; others form only at certain times.

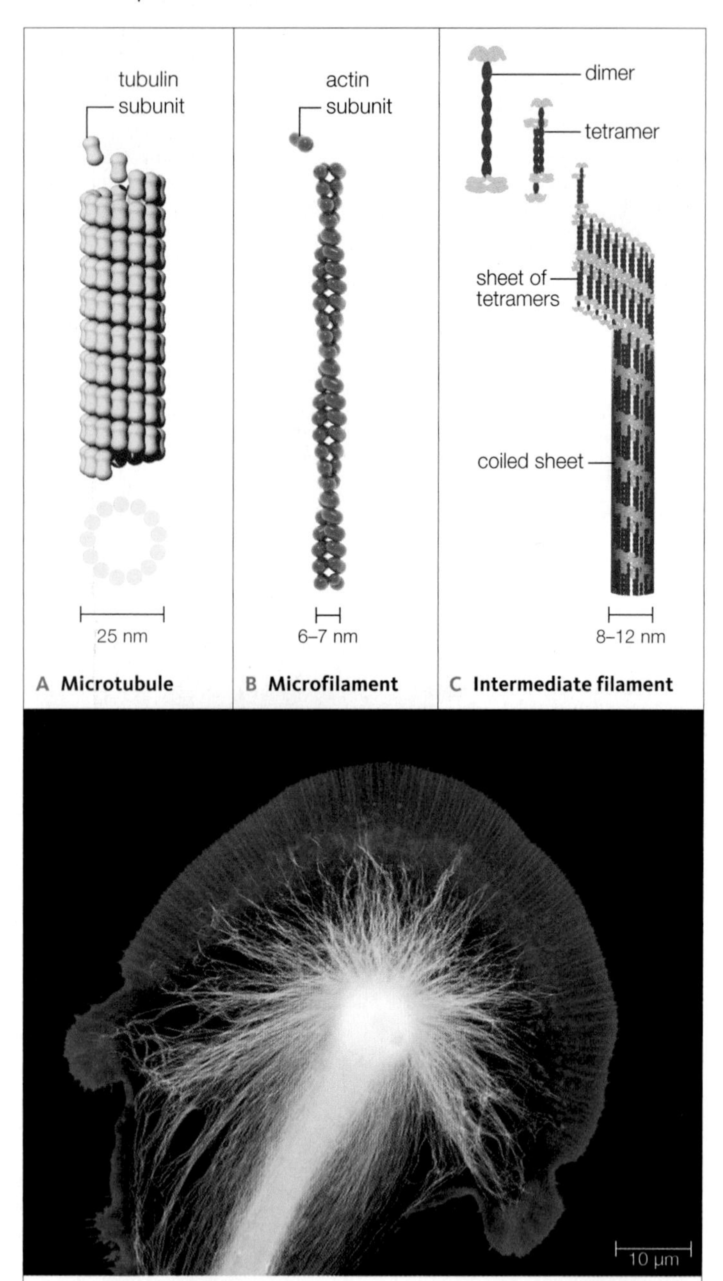

D A fluorescence micrograph shows microtubules (yellow) and microfilaments (blue) in the growing end of a nerve cell. These cytoskeletal elements support and guide the cell's lengthening in a particular direction.

FIGURE 4.17 **{Animated}** Cytoskeletal elements.

Microtubules are long, hollow cylinders that consist of subunits of the protein tubulin (FIGURE 4.17A). They form a dynamic scaffolding for many cellular processes, rapidly assembling when they are needed, disassembling when they are not. For example, before a eukaryotic cell divides, microtubules assemble, separate the cell's duplicated DNA molecules, then disassemble. As another example, microtubules that form in the growing end of a young nerve cell support its lengthening in a particular direction (FIGURE 4.17D).

Microfilaments are fibers that consist primarily of subunits of the globular protein actin (FIGURE 4.17B). These fine fibers strengthen or change the shape of eukaryotic cells, and have a critical function in cell migration, movement, and contraction. Crosslinked, bundled, or gel-like arrays of them make up the **cell cortex**, a reinforcing mesh under the plasma membrane. Microfilaments also connect plasma membrane proteins to other proteins inside the cell.

Intermediate filaments are the most stable elements of the cytoskeleton, forming a framework that lends structure and resilience to cells and tissues in multicelled organisms. Several types of intermediate filaments are assembled from different proteins (FIGURE 4.17C). For example, intermediate filaments that make up your hair consist of keratin, a fibrous protein (Section 3.4). Intermediate filaments that form the nuclear lamina consist of lamins, another type of fibrous protein.

Motor proteins that associate with cytoskeletal elements move cell parts when energized by a phosphate-group transfer from ATP (Section 3.6). A cell is like a bustling train station, with molecules and structures being moved continuously throughout its interior. Motor proteins are like freight trains, dragging cellular cargo along tracks of microtubules and microfilaments (FIGURE 4.18). The motor protein myosin interacts with microfilaments to bring about muscle cell contraction. Another motor protein, dynein, interacts with microtubules to bring about movement of flagella and cilia in eukaryotes. Eukaryotic flagella whip back and forth to propel cells such as sperm (*right*) through fluid. **Cilia** (singular, cilium) are short, hairlike structures that project from the surface of some cells. The coordinated

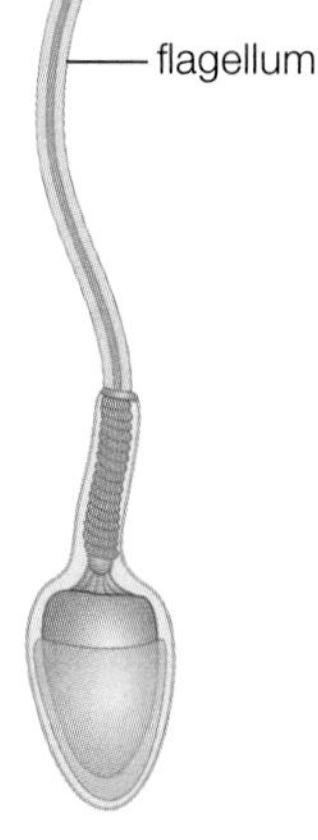

FIGURE 4.18 {Animated} Motor proteins. Here, kinesin (tan) drags a pink vesicle as it inches along a microtubule.

waving of many cilia propels some cells through fluid, and stirs fluid around other cells that are stationary. The waving movement of eukaryotic flagella and cilia, which differs from the propeller-like rotation of prokaryotic flagella, arises from their internal architecture. Microtubules extend lengthwise through them, in what is called a 9+2 array (**FIGURE 4.19**). The array consists of nine pairs of microtubules ringing another pair in the center. The microtubules grow from a barrel-shaped organelle called the **centriole**, which remains below the finished array as a **basal body**.

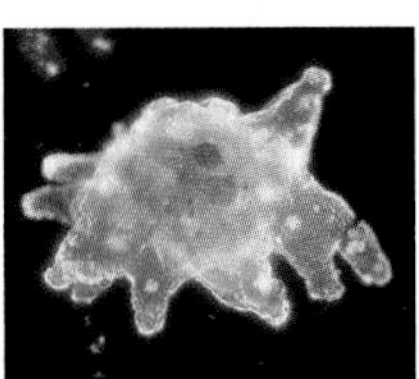

Amoebas (*left*) and other types of eukaryotic cells form **pseudopods**, or "false feet." As these temporary, irregular lobes bulge outward, they move the cell and engulf a target such as prey. Elongating microfilaments force the lobe to advance in a steady direction. Motor proteins that are attached to the microfilaments drag the plasma membrane along with them.

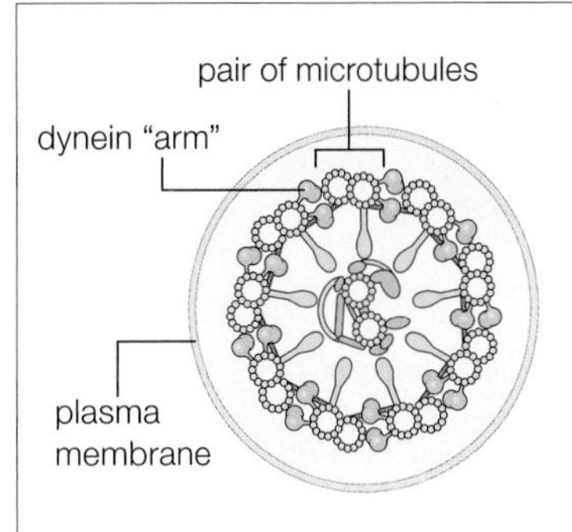

A A 9+2 array, which consists of a ring of nine pairs of microtubules plus one pair at their core, runs lengthwise through a eukaryotic flagellum or cilium. Stabilizing spokes and linking elements connect the microtubules and keep them aligned in this pattern. Projecting from each pair of microtubules in the outer ring are "arms" of the motor protein dynein.

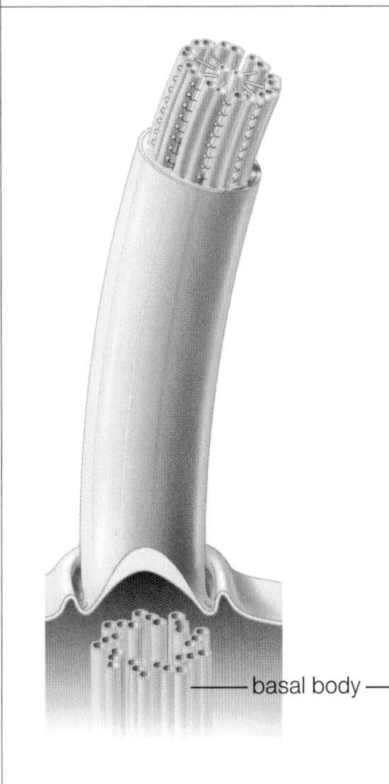

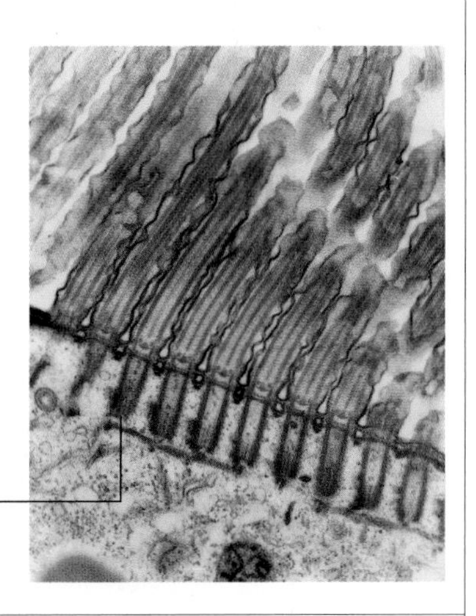

B Microtubules of a developing 9+2 array grow from a centriole, which remains below the finished array as a basal body. The micrograph *below* shows basal bodies underlying cilia of the protist pictured in **FIGURE 4.5**.

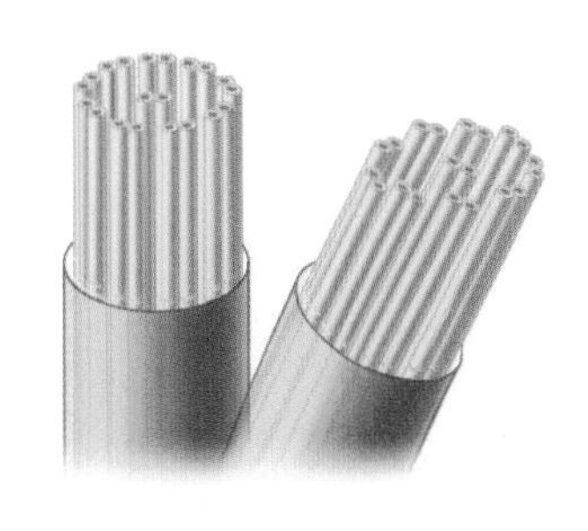

C Phosphate-group transfers from ATP cause the dynein arms in a 9+2 array to repeatedly bind the adjacent pair of microtubules, bend, and then disengage. The dynein arms "walk" along the microtubules, so adjacent microtubule pairs slide past one another. The short, sliding strokes of the dynein arms occur in a coordinated sequence around the ring, down the length of the microtubules. The movement causes the entire structure to bend.

FIGURE 4.19 {Animated} How eukaryotic flagella and cilia move.

basal body Organelle that develops from a centriole.
cell cortex Mesh of cytoskeletal elements under a plasma membrane.
centriole Barrel-shaped organelle from which microtubules grow.
cilium Short, movable structure that projects from the plasma membrane of some eukaryotic cells.
cytoskeleton Network of interconnected protein filaments that support, organize, and move eukaryotic cells and their parts.
intermediate filament Stable cytoskeletal element that structurally supports cell membranes and tissues.
microfilament Cytoskeletal element that is a fiber of actin subunits. Reinforces cell membranes; functions in muscle contractions.
microtubule Cytoskeletal element involved in movement; hollow filament of tubulin subunits.
motor protein Type of energy-using protein that interacts with cytoskeletal elements to move the cell's parts or the whole cell.
pseudopod A temporary protrusion that helps some eukaryotic cells move and engulf prey.

TAKE-HOME MESSAGE 4.9

A cytoskeleton of protein filaments is the basis of eukaryotic cell shape, internal structure, and movement.

Microtubules organize eukaryotic cells and help move their parts. Networks of microfilaments reinforce cell shape and function in movement. Intermediate filaments strengthen and maintain the shape of cell membranes and tissues, and form external structures such as hair.

When energized by ATP, motor proteins move along tracks of microtubules and microfilaments. As part of cilia, flagella, and pseudopods, they can move the whole cell.

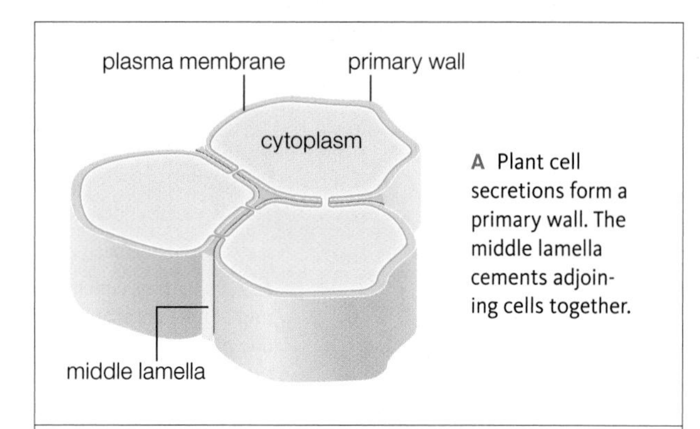

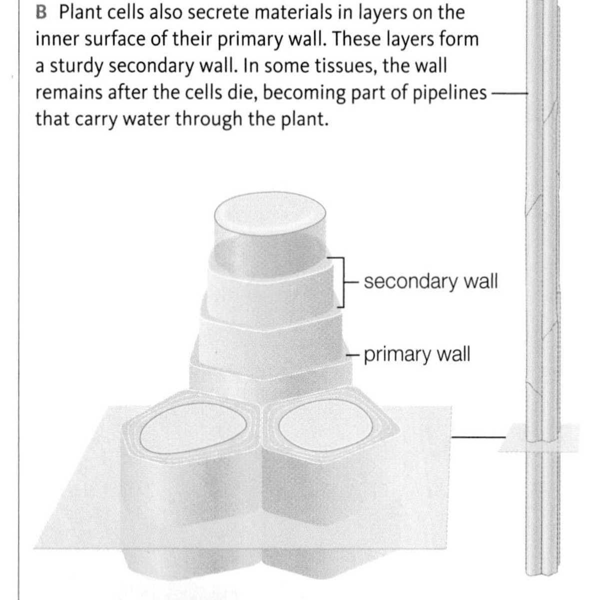

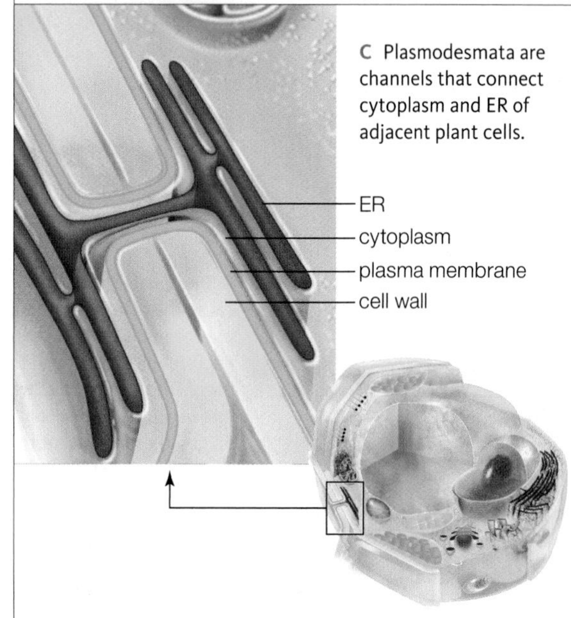

FIGURE 4.20 {Animated} Plant cell walls.

CELL MATRIXES

Many cells secrete an **extracellular matrix (ECM)**, a complex mixture of molecules that often includes polysaccharides and fibrous proteins. The composition and function of ECM vary by the type of cell that secretes it.

A cell wall is an example of ECM. Among eukaryotes, fungi and some protists have walls, as do plant cells (as shown in this chapter's opening photo). The composition of the wall differs among these groups. Like a prokaryotic cell wall, a eukaryotic cell wall is porous: Water and solutes easily cross it on the way to and from the plasma membrane.

In plants, the cell wall forms as a young cell secretes pectin and other polysaccharides onto the outer surface of its plasma membrane. The sticky coating is shared between adjacent cells, and it cements them together. Each cell then forms a **primary wall** by secreting strands of cellulose into the coating. Some of the pectin coating remains as the middle lamella, a sticky layer in between the primary walls of abutting plant cells (**FIGURE 4.20A**).

Being thin and pliable, a primary wall allows a growing plant cell to enlarge and change shape. In some plants, mature cells secrete material onto the primary wall's inner surface. These deposits form a firm **secondary wall** (**FIGURE 4.20B**). One of the materials deposited is **lignin**, an organic compound that makes up as much as 25 percent of the secondary wall of cells in older stems and roots. Lignified plant parts are stronger, more waterproof, and less susceptible to plant-attacking organisms than younger tissues.

Animal cells have no walls, but some types secrete an extracellular matrix called basement membrane. Despite the name, basement membrane is not a cell membrane because it does not consist of a lipid bilayer. Rather, it is a sheet of fibrous material that structurally supports and organizes

FIGURE 4.21 A plant ECM. Section through a plant leaf showing cuticle, a protective covering secreted by living cells.

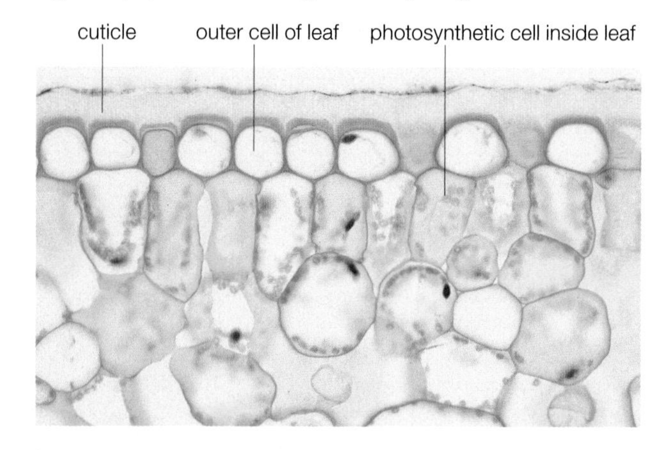

tissues, and it has roles in cell signaling. Bone is an ECM composed mostly of the fibrous protein collagen, and hardened by deposits of calcium and phosphorus.

A **cuticle** is a type of ECM secreted by cells at a body surface. In plants, a cuticle of waxes and proteins helps stems and leaves fend off insects and retain water (FIGURE 4.21). Crabs, spiders, and other arthropods have a cuticle that consists mainly of chitin (Section 3.2).

CELL JUNCTIONS

In multicelled species, cells can interact with one another and their surroundings by way of cell junctions. **Cell junctions** are structures that connect a cell directly to other cells and to its environment. Cells send and receive substances and signals through some junctions. Other junctions help cells recognize and stick to each other and to ECM.

Three types of cell junctions are common in animal tissues (FIGURE 4.22A). In tissues that line body surfaces and internal cavities, rows of adhesion proteins form **tight junctions** between the plasma membranes of adjacent cells. These junctions prevent body fluids from seeping between the cells (FIGURE 4.22B). For example, the lining of the stomach is leak-proof because tight junctions seal its cells together. These junctions keep gastric fluid, which contains acid and destructive enzymes, safely inside the stomach. If a bacterial infection damages the stomach lining, gastric fluid leaks into and damages the underlying layers. A painful peptic ulcer is the result.

Adhering junctions, which fasten cells to one another and to basement membrane, also consist of adhesion proteins. These junctions make a tissue quite strong because they connect to cytoskeletal elements inside the cells. Contractile tissues (such as heart muscle) have a lot of adhering junctions, as do tissues subject to abrasion or stretching (such as skin).

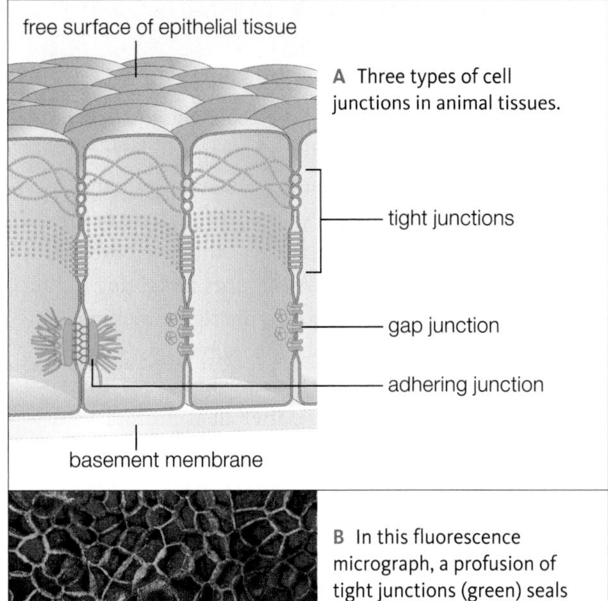

A Three types of cell junctions in animal tissues.

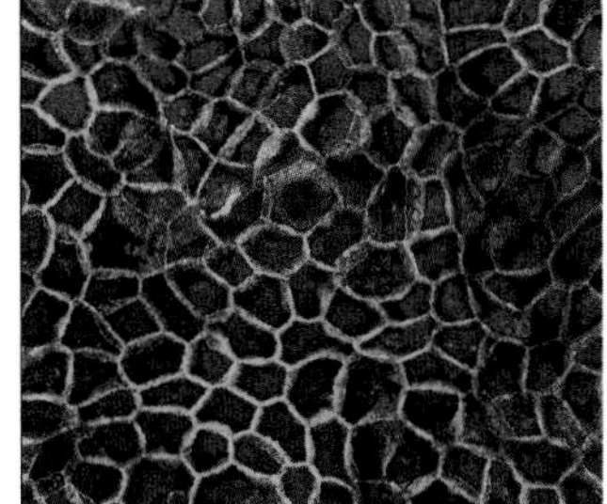

B In this fluorescence micrograph, a profusion of tight junctions (green) seals abutting surfaces of kidney cell membranes and forms a leak-proof tissue. The DNA in each cell nucleus appears red.

FIGURE 4.22 **{Animated}** Cell junctions.

Gap junctions are closable channels that connect the cytoplasm of adjoining animal cells. When open, they permit water, ions, and small molecules to pass directly from the cytoplasm of one cell to another. These channels allow entire regions of cells to respond to a single stimulus. Heart muscle and other tissues in which the cells perform a coordinated action have many gap junctions.

In plants, open channels called **plasmodesmata** (singular, plasmodesma) extend across plant cell walls to connect the cytoplasm of adjacent cells. Like gap junctions, plasmodesmata also allow substances to flow quickly from cell to cell.

TAKE-HOME MESSAGE 4.10

Many cells secrete an extracellular matrix (ECM). ECM varies in composition and function depending on the cell type.

Plant cells, fungi, and some protists have a porous wall around their plasma membrane. Animal cells do not have walls.

Cell junctions structurally and functionally connect cells in tissues. In animal tissues, cell junctions also connect cells with basement membrane.

adhering junction Cell junction composed of adhesion proteins that connect to cytoskeletal elements. Fastens cells to each other and basement membrane.
cell junction Structure that connects a cell to another cell or to extracellular matrix.
cuticle Secreted covering at a body surface.
extracellular matrix (ECM) Complex mixture of cell secretions; its composition and function vary by cell type.
gap junction Cell junction that forms a closable channel across the plasma membranes of adjoining animal cells.
lignin Material that strengthens cell walls of vascular plants.
plasmodesmata Cell junctions that form an open channel between the cytoplasm of adjacent plant cells.
primary wall The first cell wall of young plant cells.
secondary wall Lignin-reinforced wall that forms inside the primary wall of a plant cell.
tight junctions Arrays of adhesion proteins that join epithelial cells and collectively prevent fluids from leaking between them.

You learned in Section 1.1 that the cell is the smallest unit with the properties of life. In this chapter, you learned that a living cell has at minimum a plasma membrane, cytoplasm, and a region of DNA; most cells have many other components in addition to these things. So what is it, exactly, that makes it alive? A cell does not spring to life from cellular components mixed in the right amounts and proportions. According to evolutionary biologist Gerald Joyce, the simplest definition of life might well be "that which is squishy." He says, "Life, after all, is protoplasmic and cellular. It is made up of cells and organic stuff and is undeniably squishy."

However, defining life more unambiguously than "squishy" is challenging, if not impossible. We can more easily describe what sets the living apart from the nonliving, but even that can be tricky. For example, living things have a high proportion of the organic molecules of life, but so do the remains of dead organisms in seams of coal. Living things use energy to reproduce themselves, but computer viruses, which are arguably not alive, can do that too.

So how do biologists, who study life as a profession, define it? The short answer is that their best definition is a long list of properties that collectively describe living things. You already know about two of these properties:

1. They make and use the organic molecules of life.
2. They consist of one or more cells.

The remainder of this book details the others:

3. They engage in self-sustaining biological processes such as metabolism and homeostasis.
4. They change over their lifetime, for example by growing, maturing, and aging.
5. They use DNA as their hereditary material when they reproduce.
6. They have the collective capacity to change over successive generations, for example by adapting to environmental pressures.

Collectively, these properties characterize living things as different from nonliving things.

TAKE-HOME MESSAGE 4.11

We describe the characteristic of "life" in terms of a set of properties. The set is unique to living things.

In living things, the molecules of life are organized as one or more cells that engage in self-sustaining biological processes.

Organisms make and use the organic molecules of life.

Living things change over lifetimes, and over generations.

National Geographic Explorer
DR. KEVIN PETER HAND

Today's weather forecast for Europa, Jupiter's fourth-largest moon, is –280°F. A layer of ice several miles thick coats its fractured surface, with 1,000-foot ice cliffs piercing a pitch-black sky. It is devoid of atmosphere, bombarded by fierce radiation—and National Geographic Explorer Kevin Hand can hardly wait to get there.

Hand works at the Jet Propulsion Laboratory (JPL), where he is helping NASA plan a mission to Jupiter's moons—an orbiting probe that will give Earthlings a closer look at Europa. Beneath Europa's icy shell lies a vast global liquid water ocean that Hand thinks could be a great place for life. "I want to know if DNA is the only game in town. Are there different biochemical pathways that could lead to other kinds of life? That's at the heart of why I want to go to Europa—to find something living in that ocean we can poke at and use to understand and define life in a much more comprehensive way."

For the first time, we have the technological capability of taking our search for life to distant worlds. "Nevertheless, our understanding of life as a phenomenon remains largely qualitative and poorly constrained," Hand says. In other words, without an exact definition of "life," how do we determine whether it exists on Europa? "Biology preferentially uses specific organic subunits to build larger compounds while abiotic organic chemistry proceeds randomly," says Hand. "The structures of life arise from a relatively small set of universal building blocks; thus, when we search for life we look for patterns indicative of life's structural biases."

CREDITS: (in text) left, © R. Llewellyn/Superstock, Inc.; right, Photo courtesy of Kevin Hand.

4.12 Application: FOOD FOR THOUGHT

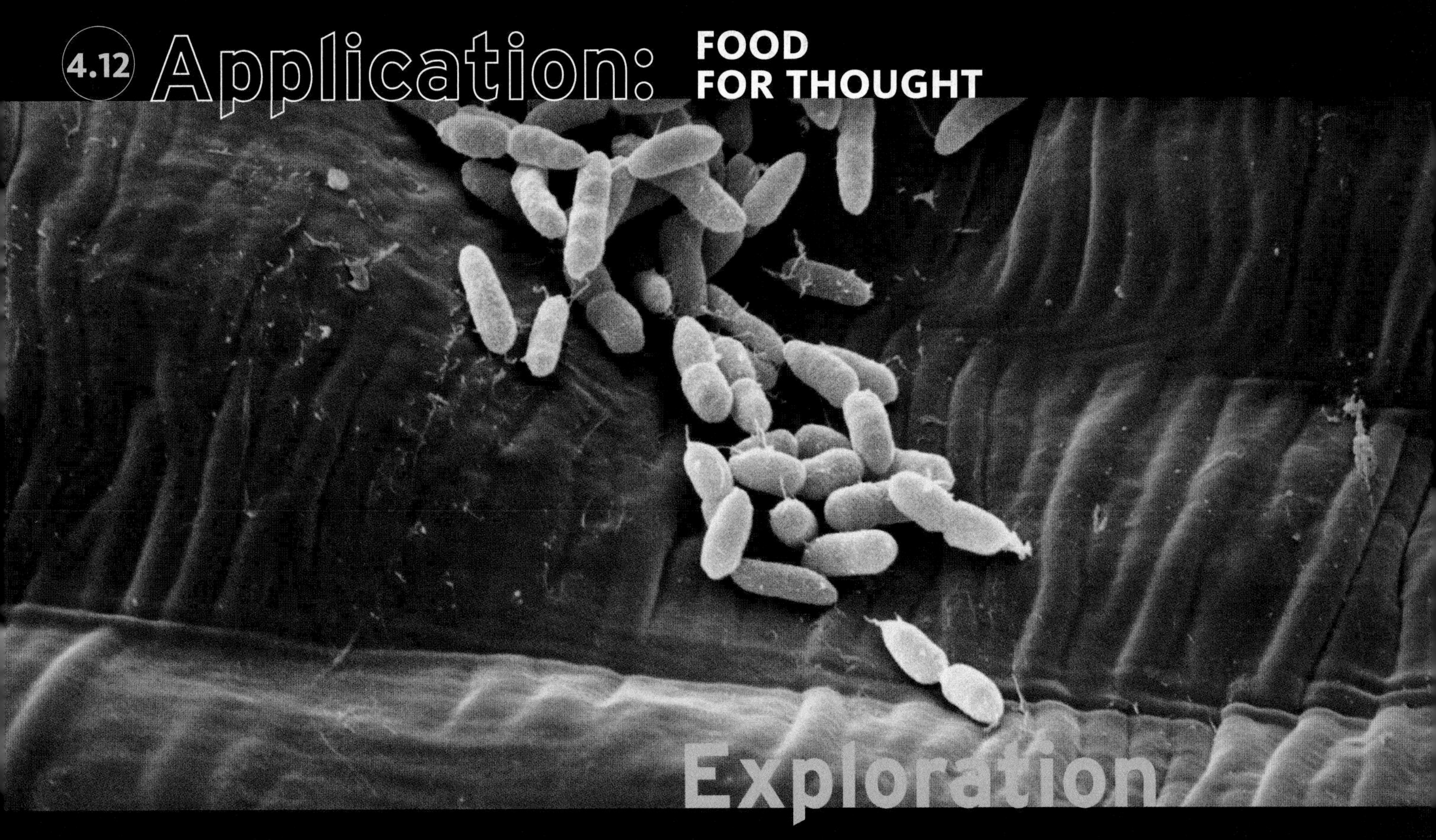

FIGURE 4.23 *Escherichia coli* cells sticking to the surface of a lettuce leaf. Some strains of this bacteria can cause a serious intestinal illness when they contaminate human food.

CELL FOR CELL, BACTERIA THAT LIVE IN AND ON A HUMAN BODY OUTNUMBER THE PERSON'S OWN CELLS BY ABOUT TEN TO ONE. One of the most common intestinal bacteria of warm-blooded animals (including humans) is *Escherichia coli*. Most of the hundreds of types, or strains, of *E. coli* are harmless, but a few strains make a toxic protein that can severely damage the lining of the intestine. After ingesting as few as ten cells of a toxic strain, a person may become ill with severe cramps and bloody diarrhea that lasts up to ten days. In some people, complications of infection result in kidney failure, blindness, paralysis, and death. Each year, about 265,000 people in the United States become infected with toxin-producing *E. coli*.

Strains of *E. coli* that are toxic to people live in the intestines of other animals—mainly cattle, deer, goats, and sheep—apparently without sickening them. Humans are exposed to the bacteria when they come into contact with feces of animals that harbor it, for example, by eating contaminated ground beef. During slaughter, meat can come into contact with feces. Bacteria in the feces stick to the meat, then get thoroughly mixed into it during the grinding process. Unless contaminated meat is cooked to at least 71°C (160°F), live bacteria will enter the digestive tract of whoever eats it.

People also become infected with toxic *E. coli* by eating fresh fruits and vegetables that have come into contact with animal feces. Washing produce with water does not remove all of the bacteria because they are sticky (FIGURE 4.23). In June 2011, more than 4,000 people in Germany and France were sickened after eating sprouts, and 49 of them died. The outbreak was traced to a single shipment of contaminated sprout seeds from Egypt.

The impact of such outbreaks, which occur with unfortunate regularity, extends beyond casualties. The contaminated sprouts cost growers in the European Union at least $600 million in lost sales. In 2011 alone, the United States Department of Agriculture (USDA) recalled 36.7 million pounds of ground meat products contaminated with toxic bacteria, at a cost in the billions of dollars. Such costs are eventually passed to taxpayers and consumers.

Food growers and processors are implementing new procedures intended to reduce the number and scope of these outbreaks. Meat and produce are being tested for some bacteria before sale, and improved documentation should allow a source of contamination to be pinpointed more quickly.

Summary

SECTION 4.1 **Cell theory** is the foundation of modern biology. By this theory, all organisms consist of one or more cells; the cell is the smallest unit of life; each new cell arises from another, preexisting cell; and a cell passes hereditary material to its offspring.

All cells start out life with **cytoplasm**, DNA, and a **plasma membrane** that controls the types and kinds of substances that cross it. Most cells have many additional components (TABLE 4.5 and FIGURE 4.24). In eukaryotes, a cell's DNA is contained within a **nucleus**, which is a membrane-enclosed **organelle**.

A cell's surface area increases with the square of its diameter, while its volume increases with the cube. This **surface-to-volume ratio** limits cell size and influences cell (and body) shape.

SECTION 4.2 Most cells are far too small to see with the naked eye, so we use microscopes to observe them. Different types of microscopes and techniques reveal different internal and external details of cells.

SECTION 4.3 A cell membrane is a mosaic of proteins and lipids (mainly phospholipids) organized as a lipid bilayer. The membranes of bacteria and eukaryotic cells can be described as a **fluid mosaic**; those of archaea are not fluid. Proteins contribute to membrane function. All cell membranes have enzymes, and all have **transport proteins** that help substances move across the membrane. Plasma membranes also incorporate **receptor proteins** that bind specific substances, **adhesion proteins** that lock cells together in tissues, and **recognition proteins** that identify a cell as belonging to a tissue or body.

SECTION 4.4 Bacteria and archaea, informally grouped as prokaryotes, are the most diverse forms of life that we know about. These single-celled organisms have no nucleus, but they do have **nucleoids** and **ribosomes**. Many also have a protective, rigid **cell wall** and a sticky capsule, and some have motile structures (**flagella**) and other projections (**pili**). There are often **plasmids** in addition to the single circular molecule of DNA. Bacteria and other microbial organisms may live together in a shared mass of slime as **biofilms**.

SECTION 4.5 All eukaryotic cells start out life with a nucleus and other membrane-enclosed organelles. Membranes allow organelles to compartmentalize tasks and substances that may be sensitive or dangerous to the rest of the cell. A nucleus protects and controls access to a eukaryotic cell's DNA. A double membrane studded with pores constitutes the **nuclear envelope**. The pores serve as gateways for molecules passing into and out of the nucleus.

TABLE 4.5

Summary of Typical Components of Cells

Cell Component	Main Function(s)	Bacteria, Archaea	Eukaryotes			
			Protists	Fungi	Plants	Animals
Cell wall	Protection, structural support	✔	✔	✔	✔	None
Plasma membrane	Control of substances moving into and out of cell	✔	✔	✔	✔	✔
Nucleus	Protecting and controlling access to DNA	None	✔	✔	✔	✔
DNA	Encoding of hereditary information	✔	✔	✔	✔	✔
RNA	Protein synthesis	✔	✔	✔	✔	✔
Ribosome	Protein synthesis	✔	✔	✔	✔	✔
Endoplasmic reticulum (ER)	Protein, lipid synthesis; carbohydrate and fatty acid breakdown	None	✔	✔	✔	✔
Golgi body	Final modification of proteins; lipid assembly	None	✔	✔	✔	✔
Lysosome	Intracellular digestion	None	✔	✔	✔	✔
Peroxisome	Breakdown of fatty acids, amino acids, and toxins	None	✔	✔	✔	✔
Mitochondrion	Production of ATP by aerobic respiration	None	✔	✔	✔	✔
Photosynthetic pigments	Capturing light for photosynthesis	✔	✔	None	✔	None
Chloroplast	Photosynthesis; starch storage	None	✔	None	✔	None
Vacuole	Isolation and breakdown of food, wastes, toxins	None	✔	✔	✔	None
Vesicle	Storage, transport, or breakdown of contents	✔	✔	✔	✔	✔
Flagellum	Locomotion through fluid surroundings	✔	✔	✔	✔	✔
Cilium	Movement through (and of) fluid	✔	✔	None	✔	✔
Cytoskeleton	Physical reinforcement; internal organization; movement of the cell and its parts	✔	✔	✔	✔	✔

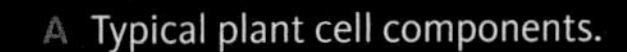

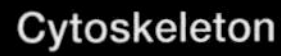

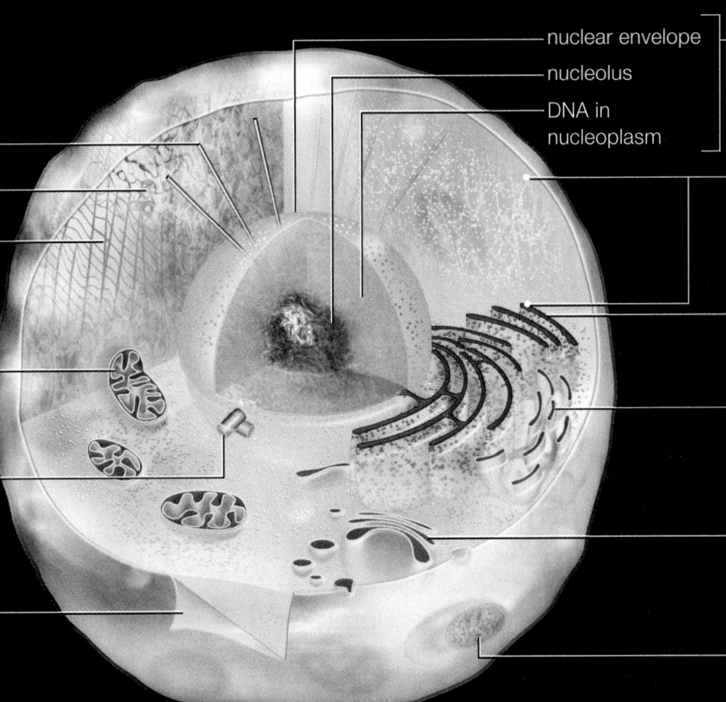

FIGURE 4.24 **{Animated}** Organelles and structures typical of A plant cells and B animal cells.

Summary continued

Inside the nuclear envelope, the cell's DNA is suspended in viscous **nucleoplasm**. Also inside the nucleus, ribosome subunits are assembled in dense, irregularly shaped areas called **nucleoli**.

SECTION 4.6 The **endomembrane system** is a series of organelles (endoplasmic reticulum, Golgi bodies, vesicles) that interact mainly to make lipids, enzymes, and proteins for insertion into membranes or secretion. **Endoplasmic reticulum (ER)** is a continuous system of sacs and tubes extending from the nuclear envelope. Ribosome-studded rough ER makes proteins; smooth ER makes lipids and breaks down carbohydrates and fatty acids. **Golgi bodies** modify proteins and lipids before sorting them into vesicles. Different types of **vesicles** store, break down, or transport substances through the cell. Enzymes in **peroxisomes** break down substances such as amino acids, fatty acids, and toxins. **Lysosomes** contain enzymes that break down cellular wastes and debris. Fluid-filled **vacuoles** store or break down waste, food, and toxins. Fluid pressure inside a **central vacuole** keeps plant cells plump, thus keeping plant parts firm.

SECTION 4.7 Double-membraned **mitochondria** specialize in making ATP by breaking down organic compounds in the oxygen-requiring metabolic pathway of aerobic respiration.

SECTION 4.8 Different types of **plastids** are specialized for photosynthesis or storage in plants and algal cells. In eukaryotes, photosynthesis takes place inside **chloroplasts**. Pigment-filled chromoplasts and starch-filled amyloplasts are used for storage; many of these plastids serve additional roles.

SECTION 4.9 Elements of a **cytoskeleton** reinforce, organize, and move cell structures, and often the whole cell. Cytoskeletal elements include **microtubules**, **microfilaments**, and **intermediate filaments**.

Interactions between ATP-driven **motor proteins** and hollow, dynamically assembled microtubules bring about the movement of cell parts. A microfilament mesh called the **cell cortex** reinforces plasma membranes. Elongating microfilaments bring about movement of **pseudopods**. Intermediate filaments lend structural support to cells and tissues, and they help support the nuclear membrane. **Centrioles** give rise to a special 9+2 array of microtubules inside **cilia** and eukaryotic flagella, then remain beneath these motile structures as **basal bodies**.

SECTION 4.10 Many cells secrete a complex mixture of fibrous proteins and polysaccharides onto their surfaces. The secretions form an **extracellular matrix (ECM)** that has different functions depending on the cell type. In animals, a secreted basement membrane supports and organizes cells in tissues. Among the eukaryotes, plant cells, fungi, and many protists secrete a cell wall around their plasma membrane. Older plant cells secrete a rigid, **lignin-containing secondary wall** inside their pliable **primary wall**. Many eukaryotic cell types also secrete a protective **cuticle**.

Plasmodesmata are open **cell junctions** that connect the cytoplasm of adjacent plant cells. In animals, **gap junctions** are closable channels between adjacent cells. **Adhering junctions** that connect to cytoskeletal elements fasten cells to one another and to basement membrane. **Tight junctions** form a waterproof seal between cells.

SECTION 4.11 We describe the quality of "life" as a set of properties that are collectively unique to living things. Living things consist of cells that engage in self-sustaining biological processes, pass their hereditary material (DNA) to offspring by mechanisms of reproduction, and have the capacity to change over successive generations.

SECTION 4.12 Bacteria are found in all parts of the biosphere, including the human body. Huge numbers inhabit our intestines, but most of these are beneficial. A few can cause disease. Contamination of food with disease-causing bacteria can result in food poisoning that is sometimes fatal.

Self-Quiz Answers in Appendix VII

1. Despite the diversity of cell type and function, all cells have these three things in common:
   a. cytoplasm, DNA, and organelles with membranes.
   b. a plasma membrane, DNA, and a nuclear envelope.
   c. cytoplasm, DNA, and a plasma membrane.
   d. a cell wall, cytoplasm, and DNA.

2. Every cell is descended from another cell. This idea is part of _______ .
   a. evolution
   b. the theory of heredity
   c. the cell theory
   d. cell biology

3. Unlike eukaryotic cells, prokaryotic cells _______ .
   a. have no plasma membrane
   b. have RNA but not DNA
   c. have no nucleus
   d. a and c

4. The surface-to-volume ratio _______ .
   a. does not apply to prokaryotic cells
   b. constrains cell size
   c. is part of the cell theory
   d. b and c

5. Cell membranes consist mainly of _______ and _______ .
   a. lipids; carbohydrates
   b. phospholipids; protein
   c. lipids; carbohydrates
   d. phospholipids; ECM

6. In a lipid bilayer, the _______ of all the lipid molecules are sandwiched between all of the _______ .
   a. hydrophilic tails; hydrophobic heads
   b. hydrophilic heads; hydrophilic tails
   c. hydrophobic tails; hydrophilic heads
   d. hydrophobic heads; hydrophilic tails

Data Analysis Activities

Abnormal Motor Proteins Cause Kartagener Syndrome An abnormal form of a motor protein called dynein causes Kartagener syndrome, a genetic disorder characterized by chronic sinus and lung infections. Biofilms form in the thick mucus that collects in the airways, and the resulting bacterial activities and inflammation damage tissues.

Affected men can produce sperm but are infertile (**FIGURE 4.25**). They can become fathers after a doctor injects their sperm cells directly into eggs. Review **FIGURE 4.20**, then explain how abnormal dynein could cause these observed effects.

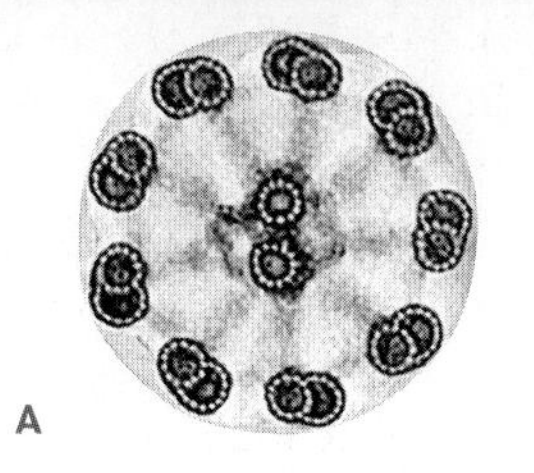

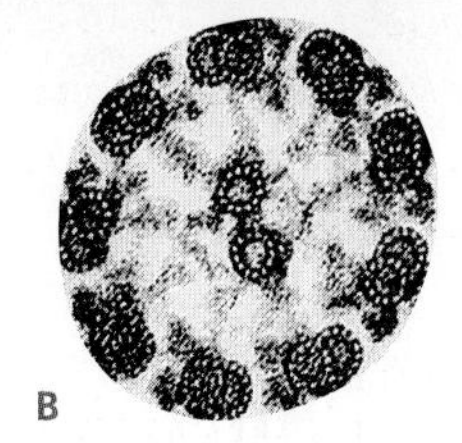

FIGURE 4.25 Cross-section of the flagellum of a sperm cell from **A** a man affected by Kartagener syndrome and **B** an unaffected man.

7. Most of a membrane's diverse functions are carried out by _______ .
   a. proteins
   b. phospholipids
   c. nucleic acids
   d. hormones

8. What controls the passage of molecules into and out of the nucleus?
   a. endoplasmic reticulum, an extension of the nucleus
   b. nuclear pores, which consist of membrane proteins
   c. nucleoli, in which ribosome subunits are made

9. The main function of the endomembrane system is _______ .
   a. building and modifying proteins and lipids
   b. isolating DNA from toxic substances
   c. secreting extracellular matrix onto the cell surface
   d. producing ATP by aerobic respiration

10. Which of the following statements is correct?
    a. Ribosomes are only found in bacteria and archaea.
    b. Some animal cells are prokaryotic.
    c. Only eukaryotic cells have mitochondria.
    d. The plasma membrane is the outermost boundary of all cells.

11. Enzymes contained in _______ break down worn-out organelles, bacteria, and other particles.
    a. lysosomes
    b. mitochondria
    c. endoplasmic reticulum
    d. peroxisomes

12. Put the following structures in order according to the pathway of a secreted protein:
    a. plasma membrane
    b. Golgi bodies
    c. endoplasmic reticulum
    d. post-Golgi vesicles

13. No animal cell has a _______ .
    a. plasma membrane
    b. flagellum
    c. lysosome
    d. cell wall

14. _______ connect the cytoplasm of plant cells.
    a. Plasmodesmata
    b. Adhering junctions
    c. Tight junctions
    d. Adhesion proteins

15. Match each cell component with its function.
    ___ mitochondrion
    ___ chloroplast
    ___ ribosome
    ___ nucleus
    ___ cell junction
    ___ flagellum
    ___ cell membrane

    a. connects cells
    b. movement
    c. ATP production
    d. protects DNA
    e. protein synthesis
    f. maintains internal environment
    g. photosynthesis

Critical Thinking

1. In a classic episode of *Star Trek*, a gigantic amoeba engulfs an entire starship. Spock blows the cell to bits before it can reproduce. Think of at least one inaccuracy that a biologist would identify in this scenario.

2. In plants, the cell wall forms as a young plant cell secretes polysaccharides onto the outer surface of its plasma membrane. Being thin and pliable, this primary wall allows the cell to enlarge and change shape. At maturity, cells in some plant tissues deposit material onto the primary wall's inner surface. Why doesn't this secondary wall form on the outer surface of the primary wall?

3. Which structures can you identify in the organism *below*? Is it prokaryotic or eukaryotic? How can you tell?

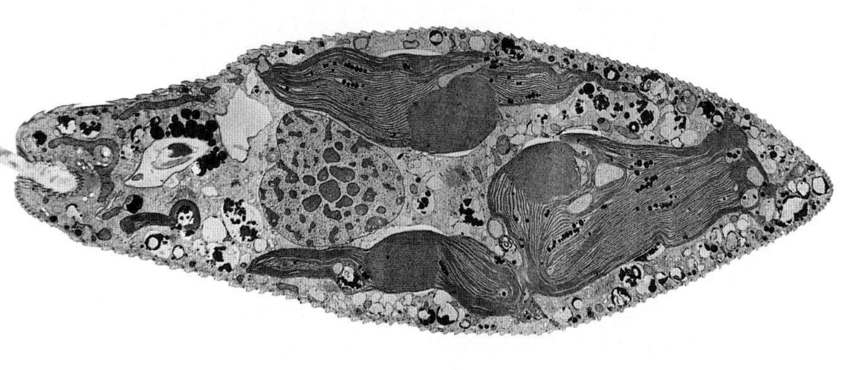

CREDITS: (25) From "Tissue & Cell", Vol. 27, pp.421–427, Courtesy of Bjorn Afzelius, Stockholm University; (in text) P.L. Walne and J. H. Arnott, *Planta*, 77:325–354, 1967.

A single-celled protist with the common name Sea Sparkle glows blue when agitated—for example, by breaking waves. The light is energy released from chemical reactions that run inside these cells. Light emitted by a living organism is called bioluminescence.

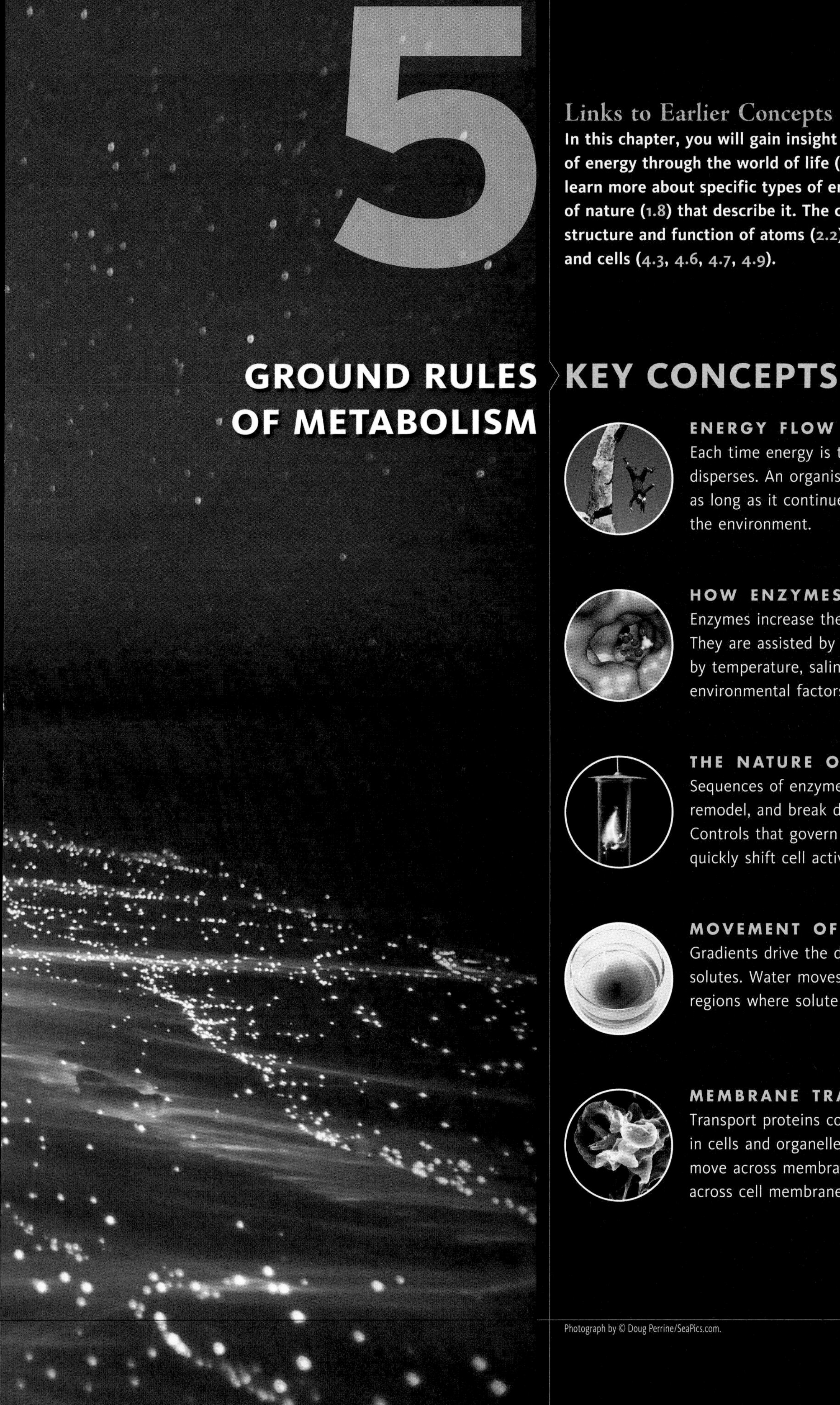

5

GROUND RULES OF METABOLISM

Links to Earlier Concepts

In this chapter, you will gain insight into the one-way flow of energy through the world of life (Sections 1.1, 1.2) as you learn more about specific types of energy (2.4) and the laws of nature (1.8) that describe it. The chapter also revisits the structure and function of atoms (2.2), molecules (2.3, 3.1–3.6), and cells (4.3, 4.6, 4.7, 4.9).

KEY CONCEPTS

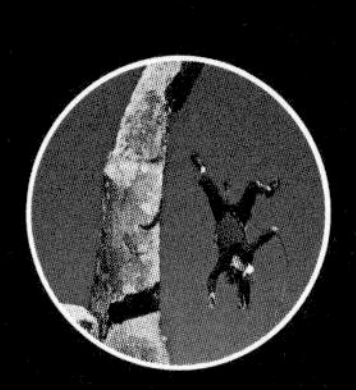

ENERGY FLOW

Each time energy is transferred, some of it disperses. An organism can sustain its life only as long as it continues to harvest energy from the environment.

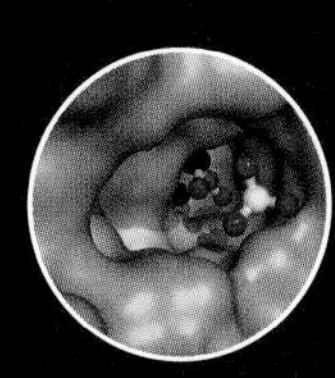

HOW ENZYMES WORK

Enzymes increase the rate of chemical reactions. They are assisted by cofactors, and affected by temperature, salinity, pH, and other environmental factors.

THE NATURE OF METABOLISM

Sequences of enzyme-mediated reactions build, remodel, and break down organic molecules. Controls that govern steps in these pathways quickly shift cell activities.

MOVEMENT OF FLUIDS

Gradients drive the directional movements of solutes. Water moves across cell membranes to regions where solute concentration is higher.

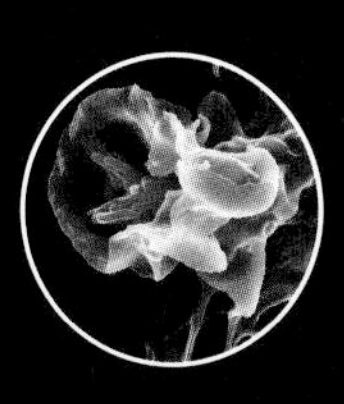

MEMBRANE TRANSPORT

Transport proteins control solute concentrations in cells and organelles by helping substances move across membranes. Substances also move across cell membranes inside vesicles.

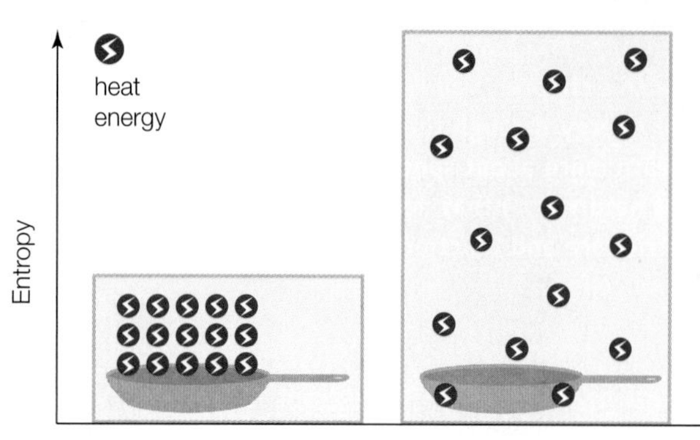

FIGURE 5.1 Entropy. Entropy tends to increase, which means that energy tends to spread out spontaneously.

FIGURE 5.2 It takes more than 10,000 pounds of soybeans and corn to raise a 1,000-pound steer. Where do the other 9,000 pounds go? About half of the steer's food is indigestible and passes right through it. The animal's body breaks down molecules in the remaining half to access energy stored in chemical bonds. Only about 15% of that energy goes toward building body mass. The rest is lost during energy conversions, as metabolic heat.

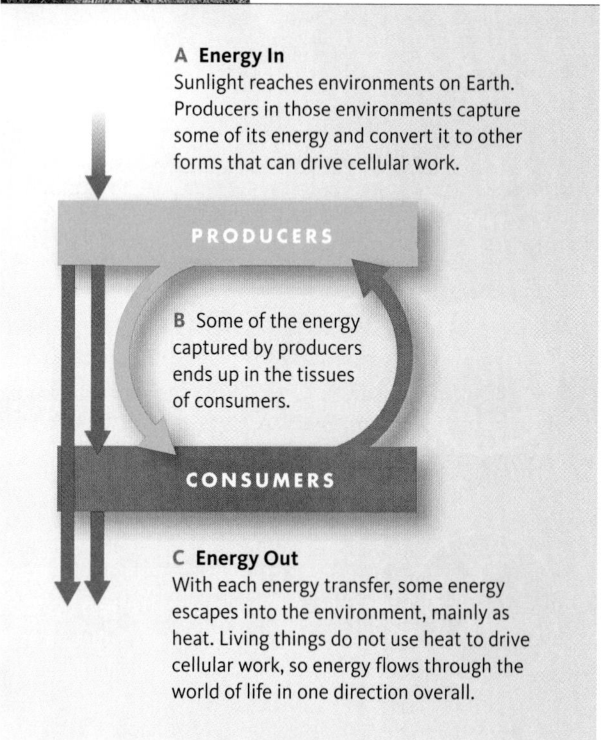

A Energy In
Sunlight reaches environments on Earth. Producers in those environments capture some of its energy and convert it to other forms that can drive cellular work.

PRODUCERS

B Some of the energy captured by producers ends up in the tissues of consumers.

CONSUMERS

C Energy Out
With each energy transfer, some energy escapes into the environment, mainly as heat. Living things do not use heat to drive cellular work, so energy flows through the world of life in one direction overall.

FIGURE 5.3 {Animated} Energy flows from the environment into living organisms, and then back to the environment. The flow drives a cycling of materials among producers and consumers.

ENERGY DISPERSES

Energy is formally defined as the capacity to do work, but this definition is not very satisfying. Even brilliant physicists who study energy cannot say exactly what it is. However, we do have an intuitive understanding of energy just by thinking about familiar forms of it, such as light, heat, electricity, and motion. We also understand intuitively that one form of energy can be converted to another. Think about how a lightbulb changes electricity into light, or how an automobile changes gasoline into the energy of motion, which is also called **kinetic energy**.

The formal study of heat and other forms of energy is thermodynamics (*therm* is a Greek word for heat; *dynam* means energy). By making careful measurements, thermodynamics researchers discovered that the total amount of energy before and after every conversion is always the same. In other words, energy cannot be created or destroyed—a phenomenon that is the **first law of thermodynamics**. Remember, a law of nature describes something that occurs without fail, but our explanation of why it occurs is incomplete (Section 1.8).

Energy also tends to spread out, or disperse, until no part of a system holds more than another part. In a kitchen, for example, heat always flows from a hot pan to cool air until the temperature of both is the same. We never see cool air raising the temperature of a hot pan. **Entropy** is a measure of how much the energy of a particular system has become dispersed. We can use the hot pan in a cool kitchen as an example of a system. As heat flows from the pan into the air, the entropy of the system increases (**FIGURE 5.1**). Entropy continues to increase until the heat is evenly distributed throughout the kitchen, and there is no longer a net (or overall) flow of heat from one area to another. Our system has now reached its maximum entropy with respect to heat. The tendency of entropy to increase is the **second law of thermodynamics**. This is the formal way of saying that energy tends to spread out spontaneously.

Biologists use the concept of entropy as it applies to chemical bonding, because energy flow in living things occurs mainly by the making and breaking of chemical bonds. How is entropy related to chemical bonding? Think about it just in terms of motion. Two unbound atoms can vibrate, spin, and rotate in every direction, so they are at high entropy with respect to motion. A covalent bond between the atoms restricts their movement, so they are able to move in fewer ways than they did before bonding. Thus, the entropy of two atoms decreases when a bond forms between them. Such entropy changes are part of the reason why some reactions occur spontaneously and others require an energy input, as you will see in the next section.

ENERGY'S ONE-WAY FLOW

Work occurs as a result of energy transfers. Consider how it takes work to push a box across a floor. In this case, a body (you) transfers energy to another body (the box) to make it move. Similarly, a plant cell works to make sugars. Inside the cell, one set of molecules harvests energy from light, then transfers it to another set of molecules. The second set of molecules uses the energy to build the sugars from carbon dioxide and water. This particular energy transfer involves the conversion of light energy to chemical energy. Most other types of cellular work occur by the transfer of chemical energy from one molecule to another.

As you learn about such processes, remember that every time energy is transferred, a bit of it disperses. Energy lost from a transfer is usually in the form of heat. As a simple example, a typical incandescent lightbulb converts only about 5 percent of the energy of electricity into light. The remaining 95 percent of the energy ends up as heat that disperses from the bulb.

Dispersed heat is not very useful for doing work, and it is not easily converted to a more useful form of energy (such as electricity). Because some of the energy in every transfer disperses as heat, and heat is not useful for doing work, we can say that the total amount of energy available for doing work in the universe is always decreasing.

Is life an exception to this inevitable flow? An organized body is hardly dispersed. Energy becomes concentrated in each new organism as the molecules of life organize into cells. Even so, living things constantly use energy—to grow, to move, to acquire nutrients, to reproduce, and so on—and some energy is lost in every one of these processes (**FIGURE 5.2**). Unless those losses are replenished with energy from another source, the complex organization of life will end.

The energy that fuels most life on Earth comes from the sun. That energy flows through producers such as plants, then consumers such as animals (**FIGURE 5.3**). During this journey, the energy is transferred many times. With each transfer, some energy escapes as heat until, eventually, all of it is permanently dispersed. However, the second law of thermodynamics does not say how quickly the dispersal has to happen. Energy's spontaneous dispersal is resisted by chemical bonds. The energy in chemical bonds is a type of **potential energy**, which is energy stored in the position or arrangement of objects in a system (**FIGURE 5.4**). Think of all the bonds in the countless molecules that make up your skin, heart, liver, fluids, and other body parts. Those bonds hold the molecules, and you, together—at least for the time being.

FIGURE 5.4 Illustration of potential energy. By opposing the downward pull of gravity, the rope attached to the rock prevents the man from falling. Similarly, a chemical bond keeps two atoms from moving apart.

TAKE-HOME MESSAGE 5.1

Energy, which is the capacity to do work, cannot be created or destroyed.

Energy disperses spontaneously.

Energy can be transferred between systems or converted from one form to another, but some is lost (as heat, typically) during every such exchange.

Sustaining life's organization requires ongoing energy inputs to counter energy loss. Organisms stay alive by replenishing themselves with energy they harvest from someplace else.

energy The capacity to do work.
entropy Measure of how much the energy of a system is dispersed.
first law of thermodynamics Energy cannot be created or destroyed.
kinetic energy The energy of motion.
potential energy Stored energy.
second law of thermodynamics Energy disperses spontaneously.

5.2 HOW DO CELLS USE ENERGY?

Remember from Section 3.1 that chemical reactions change molecules into other molecules. During a reaction, one or more **reactants** (molecules that enter a reaction and become changed by it) become one or more **products** (molecules that are produced by the reaction). Intermediate molecules may form between reactants and products.

We show a chemical reaction as an equation in which an arrow points from reactants to products:

$$\underset{\text{(hydrogen)}}{2H_2} + \underset{\text{(oxygen)}}{O_2} \longrightarrow \underset{\text{(water)}}{2H_2O}$$

A number before a chemical formula in such equations indicates the number of molecules; a subscript indicates the number of atoms of that element per molecule. Note that atoms shuffle around in a reaction, but they never disappear: The same number of atoms that enter a reaction remain at the reaction's end (**FIGURE 5.5**).

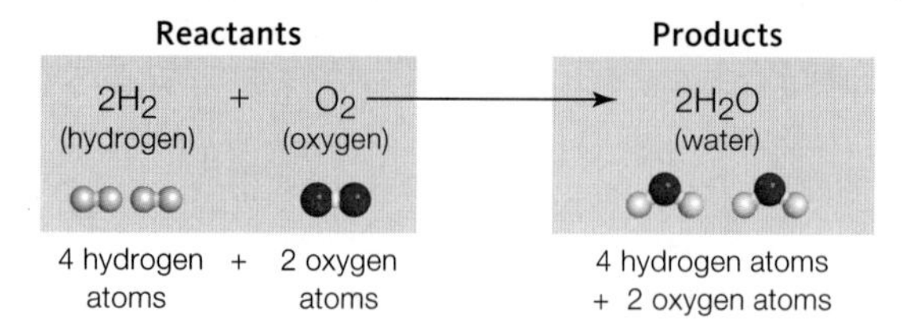

FIGURE 5.5 Chemical bookkeeping. In equations that represent chemical reactions, reactants are written to the left of an arrow that points to the products. A number before a formula indicates the number of molecules. Atoms may shuffle around in a reaction, but the same number of atoms that enter the reaction remain at the reaction's end.

CHEMICAL BOND ENERGY

Every chemical bond holds a certain amount of energy. That is the amount of energy required to break the bond, and it is also the amount of energy released when the bond forms. The particular amount of energy held by a bond depends on which elements are taking part in it. For example, two covalent bonds—one between an oxygen and a hydrogen atom in a water molecule, the other between two oxygen atoms in molecular oxygen (O_2)—both hold energy, but different amounts of it.

Bond energy and entropy both contribute to a molecule's free energy, which is the amount of energy that is available ("free") to do work. In most reactions, the free energy of reactants differs from the free energy of products. If the reactants have less free energy than the products, the reaction will not proceed without a net energy input. Such reactions are **endergonic**, which means "energy in" (**FIGURE 5.6A**). If the reactants have more free energy than the products, the reaction will end with a net release of energy. Such reactions are **exergonic**, which means "energy out" (**FIGURE 5.6B**).

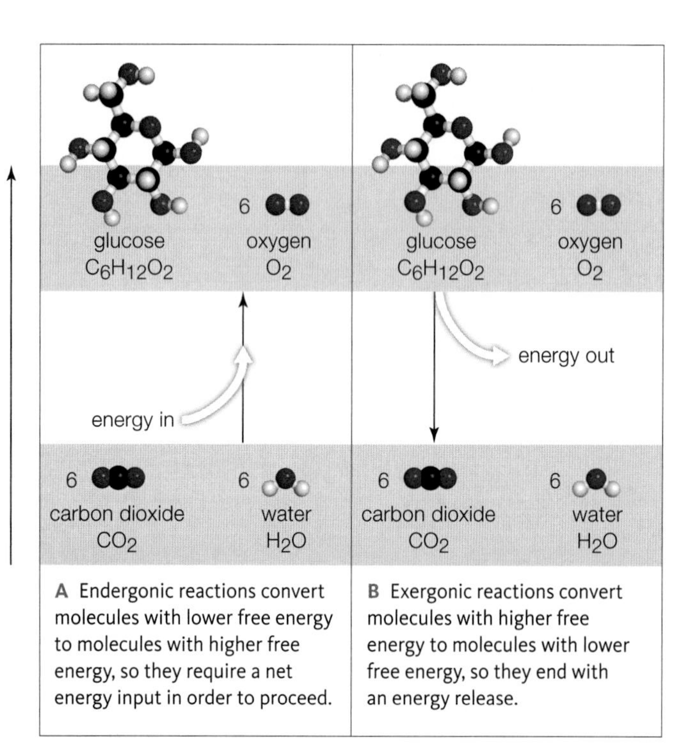

A Endergonic reactions convert molecules with lower free energy to molecules with higher free energy, so they require a net energy input in order to proceed.

B Exergonic reactions convert molecules with higher free energy to molecules with lower free energy, so they end with an energy release.

FIGURE 5.6 Energy inputs and outputs in chemical reactions.

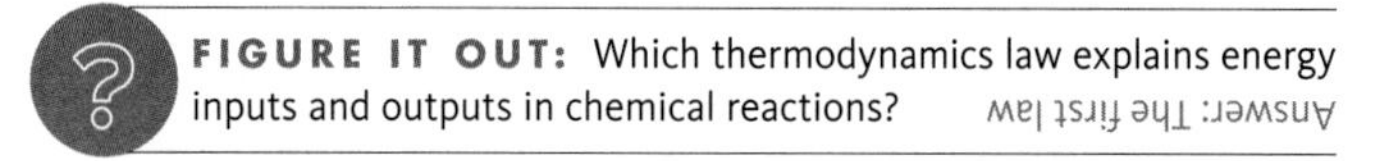

FIGURE IT OUT: Which thermodynamics law explains energy inputs and outputs in chemical reactions?

Answer: The first law

WHY EARTH DOES NOT GO UP IN FLAMES

The molecules of life release energy when they combine with oxygen. For example, think of how a spark ignites wood. Wood is mostly cellulose, which consists of long chains of repeating glucose monomers (Section 3.2). A spark starts a reaction that converts cellulose (in wood) and oxygen (in air) to water and carbon dioxide. The reaction is highly exergonic, which means it releases a lot of energy—enough to initiate the same reaction with other cellulose and oxygen molecules. That is why wood keeps burning after it has been lit.

Earth is rich in oxygen—and in potential exergonic reactions. Why doesn't it burst into flames? Luckily, chemical bonds do not break without at least a small input of energy, even in an energy-releasing reaction. We call this input activation energy. **Activation energy**, the minimum amount of energy required to get a chemical reaction started, is a bit like a hill that reactants must climb before they can coast down the other side to become products (**FIGURE 5.7**).

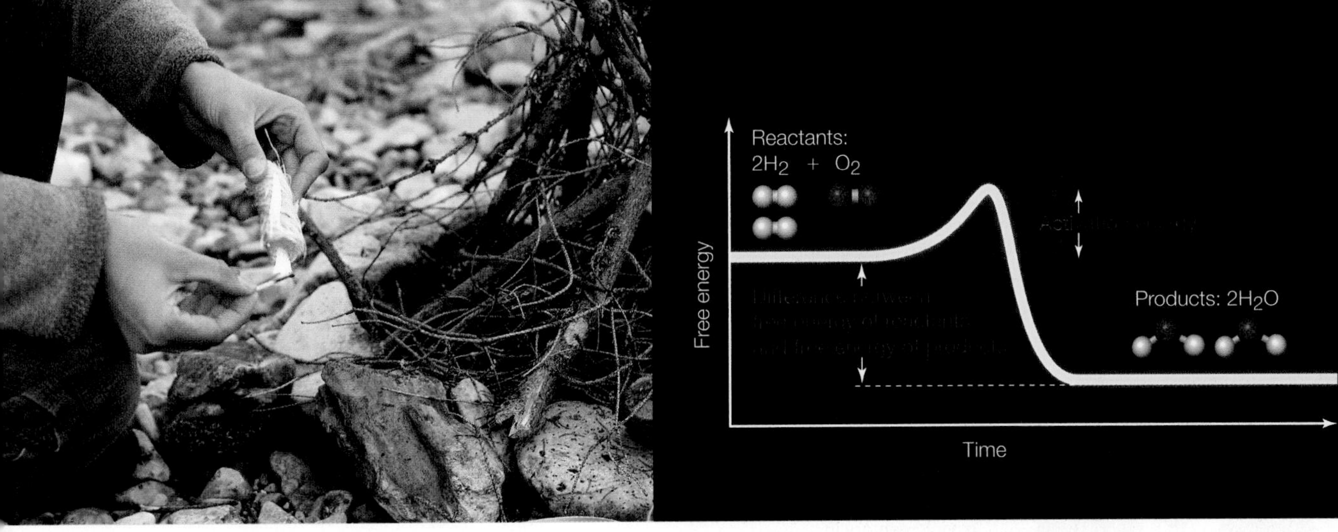

FIGURE 5.7 {Animated} Activation energy. Most reactions will not begin without an input of activation energy, which is shown in the graph as a bump in a free energy hill. Reactants in this example have more energy than the products. Activation energy keeps this and other reactions, including exergonic ones such as burning wood cellulose, from starting spontaneously.

Both endergonic and exergonic reactions have activation energy, but the amount varies with the reaction. Consider guncotton (nitrocellulose), a highly explosive derivative of cellulose. Christian Schönbein accidentally discovered a way to manufacture it when he used his wife's cotton apron to wipe up a nitric acid spill on his kitchen table, then hung it up to dry next to the oven. The apron exploded. Being a chemist in the 1800s, Schönbein immediately thought of marketing guncotton as a firearm explosive, but it proved to be too unstable to manufacture. So little activation energy is needed to make guncotton react with oxygen that it tends to explode unexpectedly. Several manufacturing plants burned to the ground before guncotton was abandoned for use as a firearm explosive. The substitute? Gunpowder, which has a higher activation energy for a reaction with oxygen.

ENERGY IN, ENERGY OUT

Cells store energy by running endergonic reactions that build organic compounds (**FIGURE 5.8A**). For example, light energy drives the overall reactions of photosynthesis, which produce sugars such as glucose from carbon dioxide and water. Unlike light, glucose can be stored in a cell. Cells harvest energy by running exergonic reactions that break the bonds of organic compounds (**FIGURE 5.8B**). Most cells do this when they carry out the overall reactions of aerobic respiration, which releases the energy of glucose by breaking the bonds between its carbon atoms. You will see in the next few sections how cells use energy released from some reactions to drive others.

FIGURE 5.8 {Animated} Cells store and retrieve energy in the chemical bonds of organic molecules.

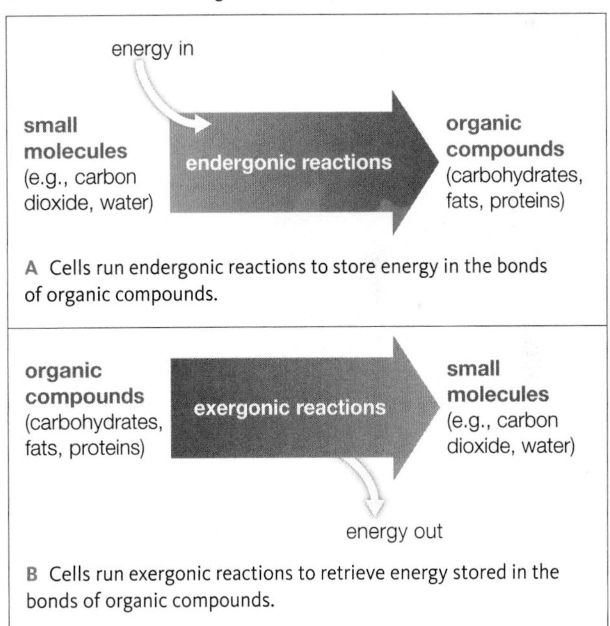

A Cells run endergonic reactions to store energy in the bonds of organic compounds.

B Cells run exergonic reactions to retrieve energy stored in the bonds of organic compounds.

activation energy Minimum amount of energy required to start a reaction.
endergonic Describes a reaction that requires a net input of free energy to proceed.
exergonic Describes a reaction that ends with a net release of free energy.
product A molecule that is produced by a reaction.
reactant A molecule that enters a reaction and is changed by participating in it.

TAKE-HOME MESSAGE 5.2

Endergonic reactions will not run without a net input of energy. Exergonic reactions end with a net release of energy.

Both endergonic and exergonic reactions require an input of activation energy to begin.

Cells store energy in chemical bonds by running endergonic reactions that build organic compounds. To release this stored energy, they run exergonic reactions that break the bonds.

THE NEED FOR SPEED

Metabolism requires enzymes. Why? Consider that sugar can break down to carbon dioxide and water on its own, but it might take decades. That same conversion takes just seconds inside your cells. Enzymes make the difference.

FIGURE 5.9 How an active site works.

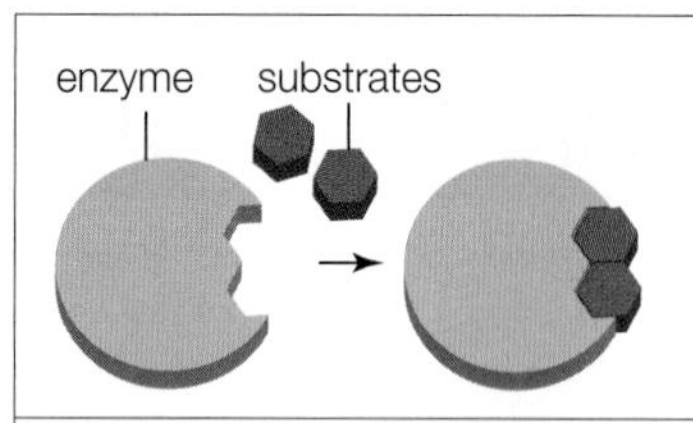

A An active site binds substrates that are complementary in shape, size, polarity, and charge.

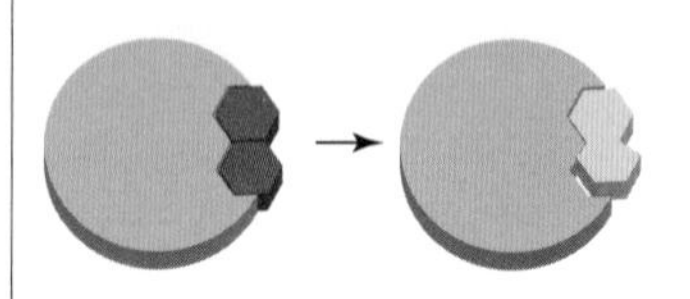

B The binding squeezes substrates together, influences their charge, or causes some change that lowers activation energy, so the reaction proceeds.

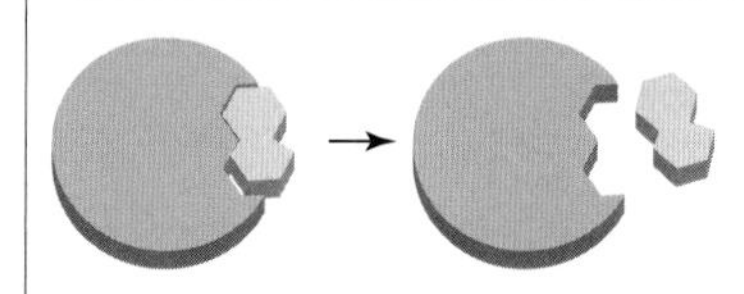

C The product leaves the active site after the reaction is finished. The enzyme is unchanged, so it can work again.

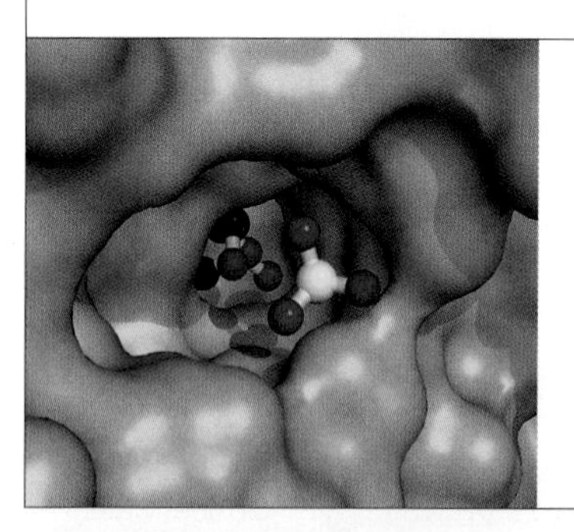

D For simplicity, enzymes are often depicted as blobs or geometric shapes. This model shows the actual contours of an active site in an enzyme (hexokinase) that adds a phosphate group to a six-carbon sugar. Both substrates are shown.

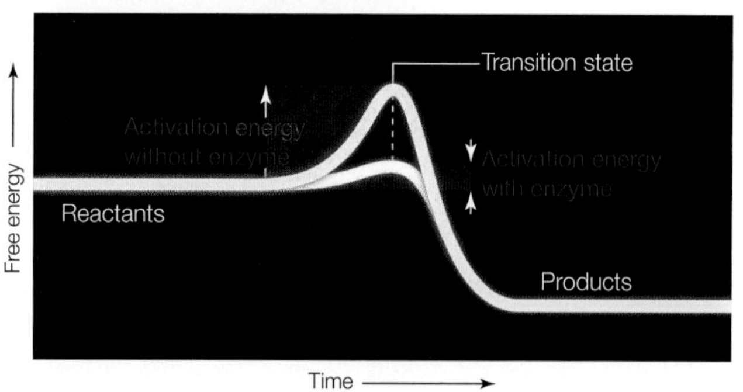

FIGURE 5.10 {Animated} The transition state. An enzyme enhances the rate of a reaction by lowering activation energy.

FIGURE IT OUT: Is the reaction shown in this graph endergonic or exergonic?

Answer: Exergonic

In a process called **catalysis**, an enzyme makes a reaction run much faster than it would on its own. The enzyme is unchanged by participating in the reaction, so it can work again and again.

Some enzymes are RNAs, but most are proteins. Each kind of enzyme recognizes specific reactants, or **substrates**, and alters them in a specific way. For instance, the enzyme hexokinase adds a phosphate group to the hydroxyl group on the sixth carbon of glucose. Such specificity occurs because an enzyme's polypeptide chains fold up into one or more **active sites**, which are pockets where substrates bind and where reactions proceed (**FIGURE 5.9**). An active site is complementary in shape, size, polarity, and charge to the enzyme's substrate. This fit is the reason why each enzyme acts in a specific way on a specific substrate.

When we talk about activation energy, we are really talking about the energy required to bring reactant bonds to their breaking point. At that point, which is called the transition state, the reaction can run without any additional energy input. Enzymes help bring on the transition state by lowering activation energy (**FIGURE 5.10**). They do so by the following four mechanisms.

Forcing Substrates Together Binding at an active site brings substrates together. The closer the substrates are to one another, the more likely they are to react.

Orienting Substrates Substrate molecules in a solution collide from random directions. By contrast, binding at an active site positions substrates optimally for reaction.

Inducing Fit By the **induced-fit model**, an enzyme's active site is not quite complementary to its substrate. Interacting with a substrate molecule causes the enzyme to change shape so that the fit between them improves. The improved fit may result in a stronger bond between enzyme and substrate, or it may better bring on the transition state.

Shutting Out Water Metabolism occurs in water-based fluids, but water molecules can interfere with certain reactions. The active sites of some enzymes repel water, and keep it away from the reactions.

active site Pocket in an enzyme where substrates bind and a reaction occurs.
catalysis The acceleration of a reaction rate by a molecule that is unchanged by participating in the reaction.
induced-fit model Substrate binding to an active site improves the fit between the two.
substrate Of an enzyme, a reactant that is specifically acted upon by the enzyme.

CREDITS: (9A–C, 10) © Cengage Learning; (9D) PDB ID: 1GZX; Paoli, M., Liddington, R., Tame, J., Wilinson, A., Dodson, G.; Crystal Structure of T state hemoglobin with oxygen bound at all four haems. *J. Mol.Bio.*, v256, pp. 775–792, 1996.

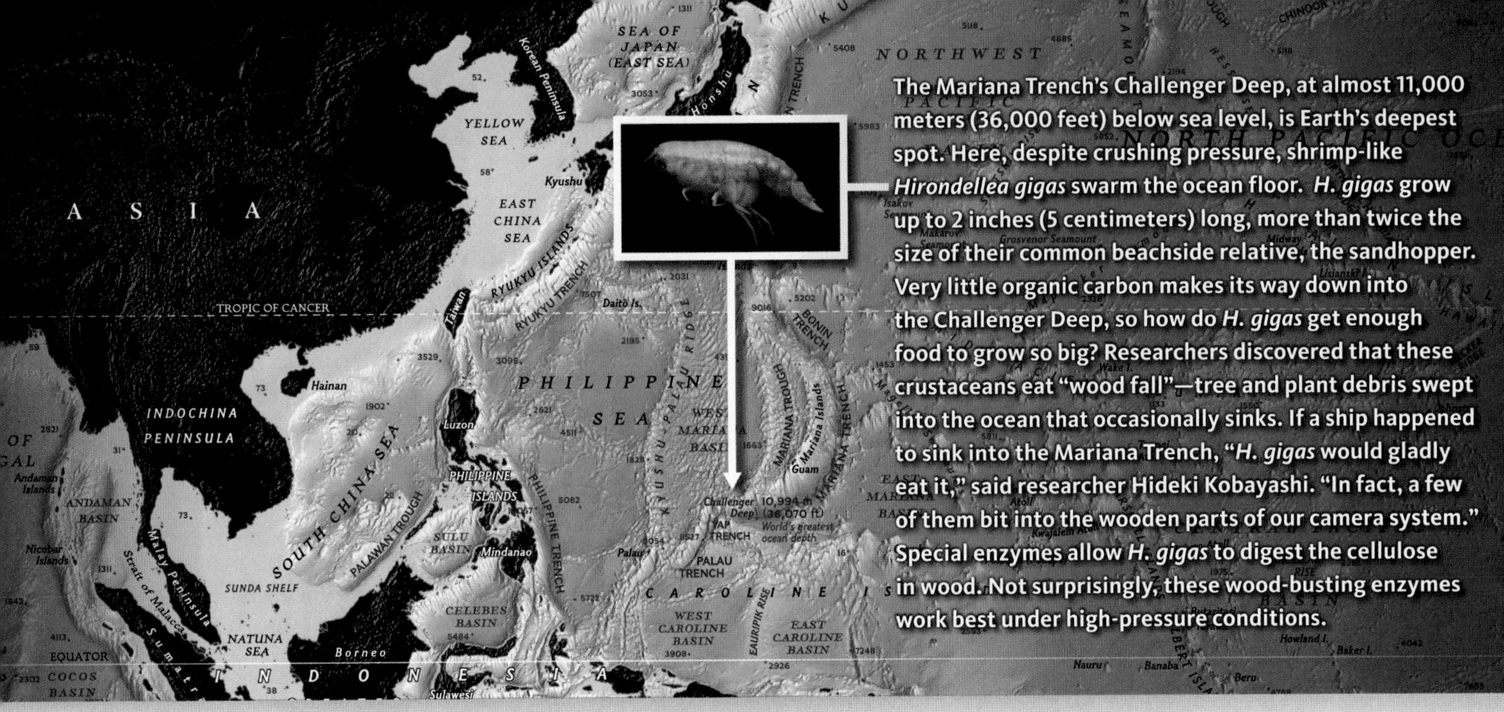

FIGURE 5.11 An organism (and its enzymes) adapted to life in a particular environment.

ENZYME ACTIVITY

Environmental factors such as pH, temperature, and salt influence an enzyme's shape, which in turn influences its function (Sections 3.4 and 3.5). Each enzyme functions best in a particular range of conditions that reflect the environment in which it evolved (**FIGURE 5.11**).

Consider pepsin, a digestive enzyme that works best at low pH (**FIGURE 5.12A**). Pepsin begins the process of protein digestion in the very acidic environment of the stomach (pH 2). During digestion, the stomach's contents pass into the small intestine, where the pH rises to about 9. Pepsin denatures (unfolds) above pH 5.5, so this enzyme becomes inactivated in the small intestine. Here, protein digestion continues with the assistance of trypsin, an enzyme that functions well at the higher pH.

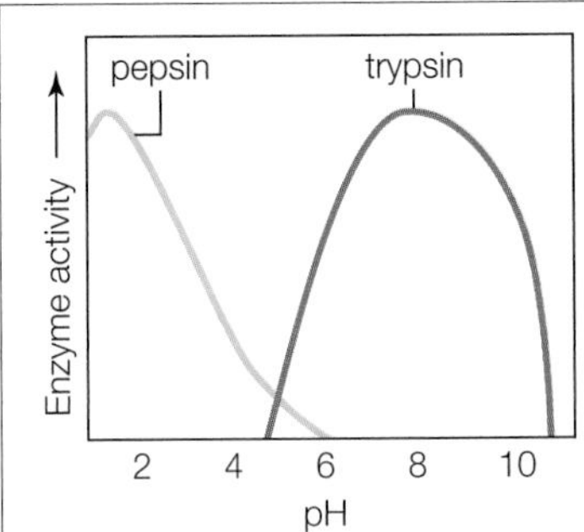

A The pH-dependent activity of two digestive enzymes, pepsin and trypsin. Pepsin acts in the stomach, where the normal pH is 2. Trypsin acts in the small intestine, where the normal pH is 9.

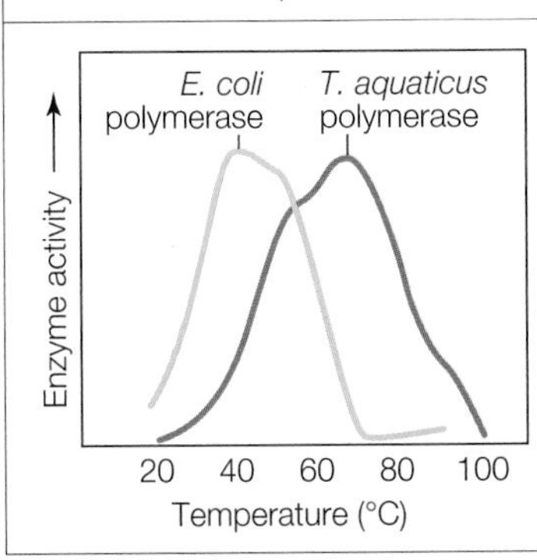

B Comparison of temperature-dependent activity of a DNA synthesis enzyme from two species of bacteria: *E. coli*, which inhabits the gut (normally 37°C); and *Thermus aquaticus*, which lives in hot springs around 70°C.

FIGURE 5.12 Each enzyme functions best within a characteristic range of conditions—generally, the same conditions that occur in the environment in which the enzyme evolved.

Adding heat boosts free energy, which is why molecular motion increases with temperature. The greater the free energy of reactants, the closer they are to activation energy. Thus, the rate of an enzymatic reaction typically increases with temperature—but only up to a point. An enzyme denatures above a characteristic temperature. Then, the reaction rate falls sharply as the shape of the enzyme changes and it stops working (**FIGURE 5.12B**). Body temperatures above 42°C (107.6°F) adversely affect the function of many of your enzymes, which is why severe fevers are dangerous.

The activity of many enzymes is also influenced by the amount of salt in the surrounding fluid. Too little salt, and polar parts of the enzyme attract one another so strongly that the enzyme's shape changes. Too much salt interferes with the hydrogen bonds that hold the enzyme in its characteristic shape, and the enzyme denatures.

TAKE-HOME MESSAGE 5.3

Enzymes greatly enhance the rate of specific reactions.

Binding at an enzyme's active site causes a substrate to reach its transition state. In this state, the substrate's bonds are at the breaking point, and the reaction can run spontaneously.

Each enzyme works best at certain environmental conditions that include temperature, pH, and salt concentration.

CREDITS: (11) inset, JAMSTEC; map, Courtesy National Geographic Maps; (12) © Cengage Learning.

Metabolism, remember, refers to the activities by which cells acquire and use energy as they build, break down, or remodel organic molecules. These activities often occur stepwise, in a series of enzymatic reactions called a **metabolic pathway**. Some metabolic pathways are linear, meaning that the reactions run straight from reactant to product (**FIGURE 5.13A**), and others are cyclic. In a cyclic pathway, the last step regenerates a reactant for the first step (**FIGURE 5.13B**). Both linear and cyclic pathways are common in cells; both can involve thousands of molecules and be quite complex. Later chapters detail the steps in some important pathways.

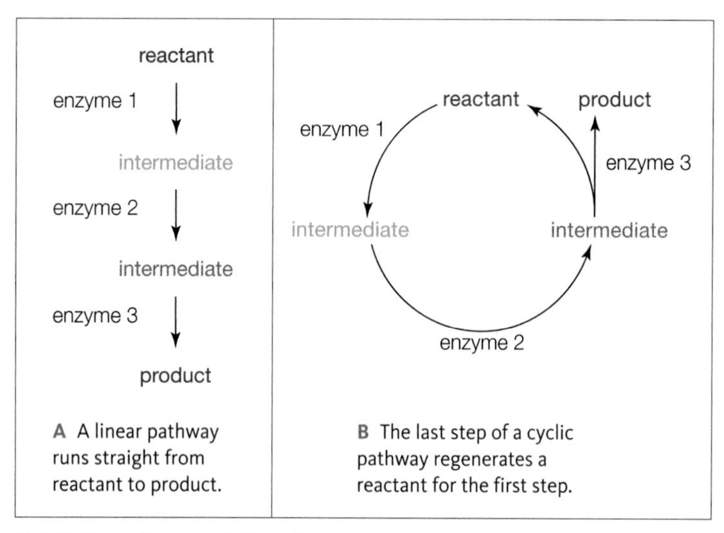

FIGURE 5.13 Metabolic pathways.

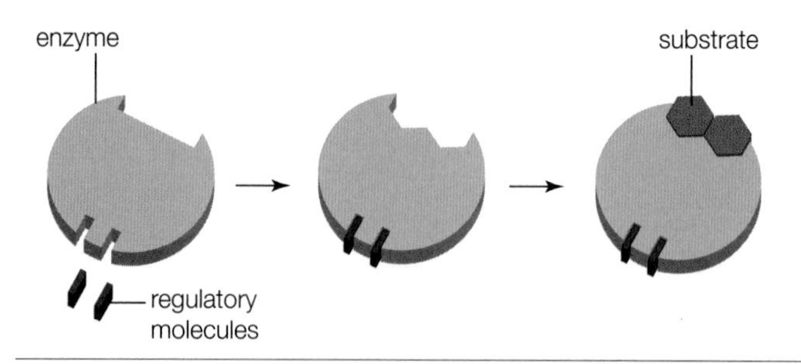

FIGURE 5.14 {Animated} Allosteric regulation, in which regulatory molecules bind to a region of an enzyme that is not the active site. The binding changes the shape of the enzyme, and thus alters its activity.

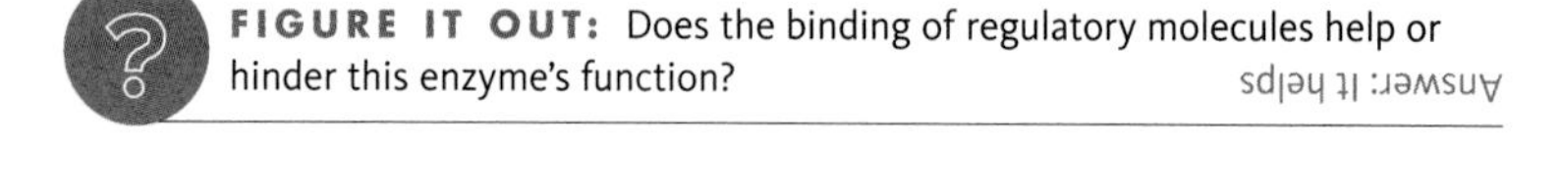

FIGURE IT OUT: Does the binding of regulatory molecules help or hinder this enzyme's function?

Answer: It helps

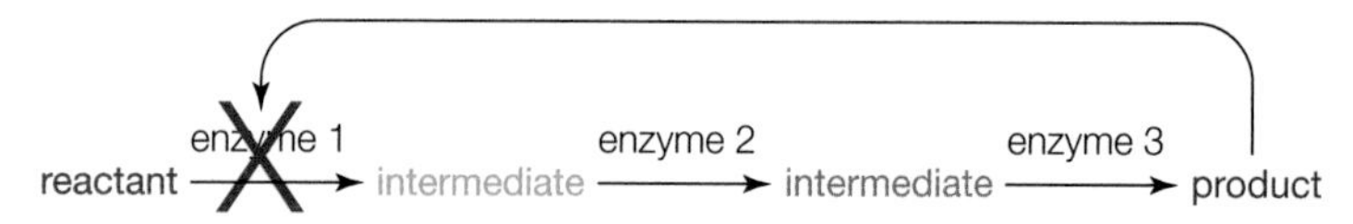

FIGURE 5.15 {Animated} Feedback inhibition. In this example, three different enzymes act in sequence to convert a substrate to a product. The product inhibits the activity of the first enzyme.

FIGURE IT OUT: Is this metabolic pathway cyclic or linear?

Answer: Linear

CONTROLS OVER METABOLISM

Cells conserve energy and resources by making only what they need at any given moment—no more, no less. Several mechanisms help a cell maintain, raise, or lower its production of thousands of different substances. Consider that reactions do not only run from reactants to products. Many also run in reverse at the same time, with some of the products being converted back to reactants. The rates of the forward and reverse reactions often depend on the concentrations of reactants and products: A high concentration of reactants pushes the reaction in the forward direction, and a high concentration of products pushes it in the reverse direction.

Other mechanisms more actively regulate enzymatic reactions. Certain substances—regulatory molecules or ions—can influence enzyme activity. In some cases, a regulatory substance activates or inhibits an enzyme by binding directly to the active site. In other cases, the regulatory substance binds outside of the active site, a mechanism called **allosteric regulation** (*allo*– means other; steric means structure). Binding of an allosteric regulator alters the shape of the enzyme in a way that enhances or inhibits its function (**FIGURE 5.14**).

Regulation of a single enzyme can affect an entire metabolic pathway. For example, the end product of a series of enzymatic reactions often inhibits the activity of one of the enzymes in the series (**FIGURE 5.15**). This regulatory mechanism is an example of **feedback inhibition**, in which a change that results from an activity decreases or stops the activity.

allosteric regulation Control of enzyme activity by a regulatory molecule or ion that binds to a region outside the enzyme's active site.
electron transfer chain Array of enzymes and other molecules that accept and give up electrons in sequence, thus releasing the energy of the electrons in steps.
feedback inhibition Regulatory mechanism in which a change that results from some activity decreases or stops the activity.
metabolic pathway Series of enzyme-mediated reactions by which cells build, remodel, or break down an organic molecule.
redox reaction Oxidation–reduction reaction, in which one molecule accepts electrons (it becomes reduced) from another molecule (which becomes oxidized). Also called electron transfer.

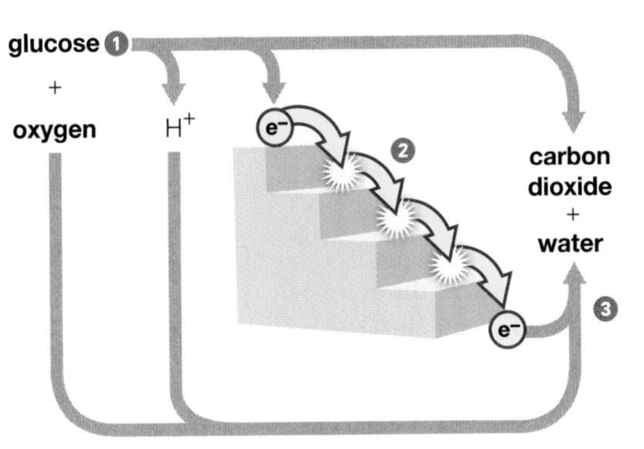

A Left, glucose in a metal spoon reacts (burns) with oxygen inside a glass jar. Energy in the form of light and heat is released all at once as CO_2 and water form.

B In cells, the same overall reaction occurs in a stepwise fashion that involves an electron transfer chain, represented here by a staircase. Energy is released in amounts that cells are able to use.

1 An input of activation energy splits glucose into carbon dioxide, electrons, and hydrogen ions (H^+).
2 Electrons lose energy as they move through an electron transfer chain. Energy released by electrons — is harnessed for cellular work.
3 Electrons, hydrogen ions, and oxygen combine to form water.

FIGURE 5.16 {Animated} Comparing uncontrolled and controlled energy release.

ELECTRON TRANSFERS

The bonds of organic molecules hold a lot of energy that can be released in a reaction with oxygen. Burning is one type of reaction with oxygen, and it releases the energy of organic molecules all at once—explosively (**FIGURE 5.16A**). Cells use oxygen to break the bonds of organic molecules, but they have no way to harvest the explosive burst of energy that occurs during burning. Instead, they break the molecules apart in pathways that release the energy in small, manageable steps. Most of these steps are oxidation–reduction reactions, or redox reactions for short. In a typical **redox reaction**, one molecule accepts electrons (it becomes reduced) from another molecule (which becomes oxidized). To remember what reduced means, think of how the negative charge of an electron "reduces" the charge of a recipient molecule.

An oxidation always occurs together with a reduction (**FIGURE 5.17**). In the next two chapters, you will learn about the importance of this concept in electron transfer chains. An **electron transfer chain** is a series of membrane-bound enzymes and other molecules that give up and accept electrons in turn. Electrons are at a higher energy level (Section 2.2) when they enter a chain than when they leave. Energy given off by an electron as it drops to a lower energy level is harvested by molecules of the electron transfer chain (**FIGURE 5.16B**).

Many electrons are delivered to electron transfer chains in photosynthesis and aerobic respiration. Energy released at certain steps in those chains helps drive the synthesis of ATP. These pathways will occupy our attention in chapters to come.

FIGURE 5.17 Visible evidence of oxidation–reduction: a glowing protist, *Noctiluca scintillans* (*below*). The pathway that produces the glow involves an enzyme, luciferase, and its substrate, luciferin. It occurs when the cells are mechanically stimulated, as by waves (shown in the chapter opening photo) or an attack by a protist-eating predator.

$$\text{luciferin—H}_2 + O_2 \xrightarrow{\text{luciferase}} \text{luciferin=O} + H_2O + \textbf{light}$$

Luciferin is oxidized during the reactions, which are summarized above. At the same time, an oxygen atom is reduced when it combines it with electrons and hydrogen, so water forms.

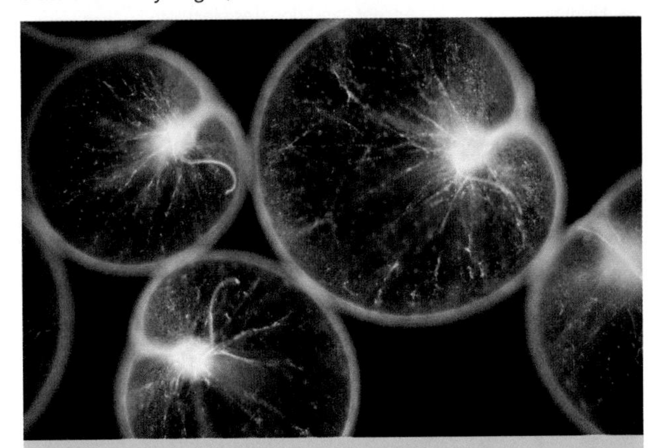

TAKE-HOME MESSAGE 5.4

A metabolic pathway is a series of enzyme-mediated reactions that builds, breaks down, or remodels an organic molecule.

Cells conserve energy and resources by producing only what they require at a given time. This metabolic control arises from regulatory molecules and other mechanisms that influence metabolic pathways and individual reactions.

Many metabolic pathways involve electron transfers. Electron transfer chains are important sites of energy exchange.

5.5 HOW DO COFACTORS WORK?

Most enzymes cannot function properly without assistance from metal ions or small organic molecules. Such enzyme helpers are called **cofactors**. Many dietary vitamins and minerals are essential because they are cofactors or are precursors for them.

Some metal ions that act as cofactors stabilize the structure of an enzyme, in which case the enzyme denatures if the ions are removed. In other cases, metal cofactors play a functional role in a reaction by interacting with electrons in nearby atoms. Atoms of metal elements readily lose or gain electrons, so a metal cofactor can help bring on the transition state by donating electrons, accepting them, or simply tugging on them.

FIGURE 5.18 Example of a coenzyme. Coenzyme Q_{10} (above) is an essential part of the ATP-making machinery in your mitochondria. It carries electrons between enzymes of electron transfer chains during aerobic respiration. Your body makes it, but some foods—particularly red meats, soy oil, and peanuts—are rich dietary sources.

TABLE 5.1

Some Common Coenzymes

Coenzyme	Example of Function
ATP	Transfers energy with a phosphate group
NAD, NAD^+	Carries electrons during glycolysis
NADP, NADPH	Carries electrons, hydrogen atoms during photosynthesis
FAD, FADH, $FADH_2$	Carries electrons during aerobic respiration
CoA	Carries acetyl group ($COCH_3$) during glycolysis
Coenzyme Q_{10}	Carries electrons in electron transfer chains of aerobic respiration
Heme	Accepts and donates electrons
Ascorbic acid	Carries electrons during peroxide breakdown (in lysosomes)
Biotin (vitamin B_7)	Carries CO_2 during fatty acid synthesis

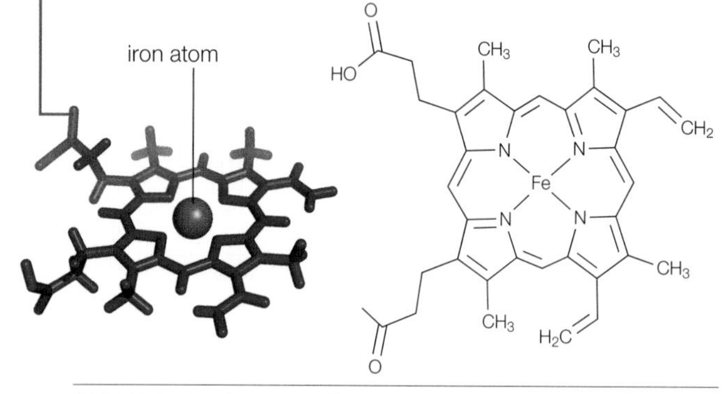

FIGURE 5.19 Heme. This organic molecule is part of the active site in many enzymes (such as catalase). In other contexts, it carries oxygen (e.g., in hemoglobin), or electrons (e.g., in molecules of electron transfer chains).

FIGURE IT OUT: Is heme a cofactor or a coenzyme?

Answer: It is both.

Organic cofactors are called **coenzymes** (**TABLE 5.1** and **FIGURE 5.18**). Coenzymes carry chemical groups, atoms, or electrons from one reaction to another, and often into or out of organelles. Unlike enzymes, many coenzymes are modified by taking part in a reaction. They are regenerated in separate reactions.

Consider NAD^+ (nicotinamide adenine dinucleotide), a coenzyme derived from niacin (vitamin B_3). NAD^+ can accept electrons and hydrogen atoms, thereby becoming reduced to NADH. When electrons and hydrogen atoms are removed from NADH (an oxidation reaction), NAD^+ forms again:

$$NAD^+ + \text{electrons} + H^+ \longrightarrow \textbf{NADH} \longrightarrow NAD^+ + \text{electrons} + H^+$$

In some reactions, cofactors participate as separate molecules. In others, they stay tightly bound to the enzyme. Catalase, an enzyme of peroxisomes, has four tightly bound cofactors called hemes. A heme is a small organic compound with an iron atom at its center (**FIGURE 5.19**). Catalase's substrate is hydrogen peroxide (H_2O_2), a highly reactive molecule that forms during some normal metabolic reactions. Hydrogen peroxide is dangerous because it can easily oxidize and destroy the organic molecules of life, or form free radicals that do. Catalase neutralizes this threat. When the enzyme binds to hydrogen peroxide, it holds the molecule close to a heme. Interacting with the iron atom in the heme causes peroxide molecules to break down to water.

Substances such as catalase that interfere with the oxidation of other molecules are called **antioxidants**. Antioxidants are essential to health because they reduce the amount of damage that cells sustain as a result of oxidation by free radicals or other molecules. Oxidative damage is associated with many diseases, including cancer, diabetes, atherosclerosis, stroke, and neurodegenerative problems such as Alzheimer's disease.

ATP—A SPECIAL COENZYME

In cells, the nucleotide ATP (adenosine triphosphate, Section 3.6) functions as a cofactor in many reactions. Bonds between phosphate groups hold a lot of energy compared to other bonds. ATP has two of of these bonds holding its three phosphate groups together (**FIGURE 5.20A**). When a phosphate group is transferred to or from a nucleotide, energy is transferred along with it. Thus, the nucleotide can receive energy from an exergonic reaction, and it can contribute energy to an endergonic one. ATP is such an important currency in a cell's energy economy that we use a cartoon coin to symbolize it.

A reaction in which a phosphate group is transferred from one molecule to another is called a **phosphorylation**. ADP (adenosine diphosphate) forms when an enzyme transfers a phosphate group from ATP to another molecule (**FIGURE 5.20B**). Cells constantly run this reaction in order to drive a variety of endergonic reactions. Thus, they must constantly replenish their stockpile of ATP—by running exergonic reactions that phosphorylate ADP. The cycle of using and replenishing ATP is called the **ATP/ADP cycle** (**FIGURE 5.20C**).

The ATP/ADP cycle couples endergonic reactions with exergonic ones (**FIGURE 5.21**). As you will see in Chapter 7, cells harvest energy from organic compounds by running metabolic pathways that break them down. Energy that cells harvest in these pathways is not released to the environment, but rather stored in the high-energy phosphate bonds of ATP molecules and in electrons carried by reduced coenzymes. Both the ATP and the reduced cofactors that form in these pathways can be used to drive many of the different kinds of endergonic reactions that a cell runs.

antioxidant Substance that prevents oxidation of other molecules.
ATP/ADP cycle Process by which cells regenerate ATP. ADP forms when a phosphate group is removed from ATP, then ATP forms again as ADP gains a phosphate group.
coenzyme An organic cofactor.
cofactor A metal ion or organic compound that associates with an enzyme and is necessary for its function.
phosphorylation A phosphate-group transfer.

TAKE-HOME MESSAGE 5.5

- Cofactors associate with enzymes and assist their function.
- Many coenzymes carry chemical groups, atoms, or electrons from one reaction to another.
- When a phosphate group is transferred from ATP to another molecule, energy is transferred along with it. This energy drives cellular work.

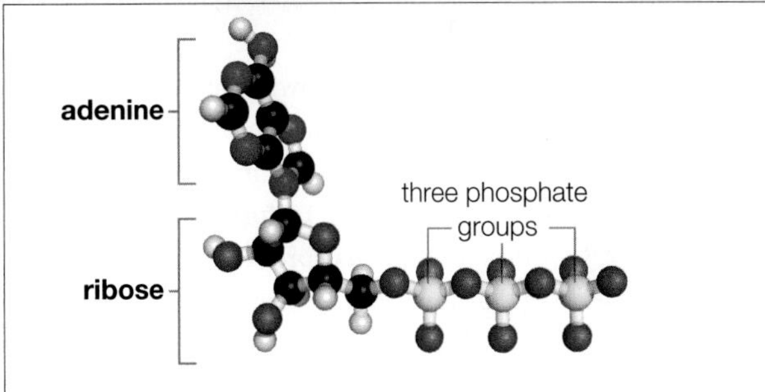

A ATP. Bonds between its phosphate groups hold a lot of energy.

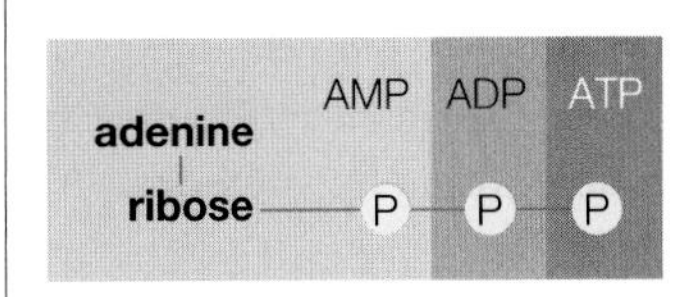

B After ATP loses one phosphate group, the nucleotide is ADP (adenosine diphosphate); after losing two, it is AMP (adenosine monophosphate).

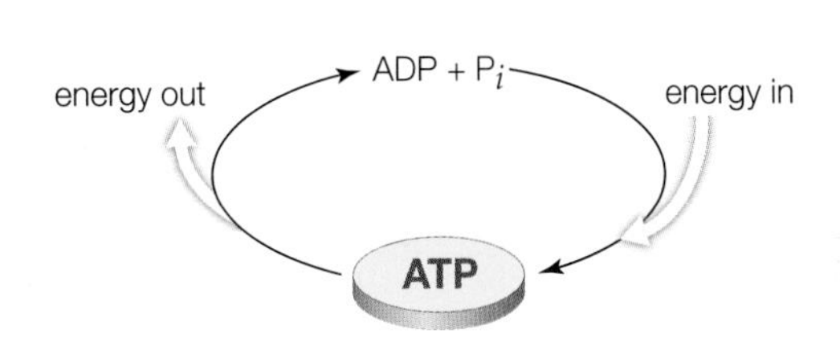

C The ATP/ADP cycle. ADP forms in a reaction that removes a phosphate group from ATP (P_i is an abbreviation for phosphate group). Energy released in this reaction drives other reactions that are the stuff of cellular work. ATP forms again in reactions that phosphorylate ADP.

FIGURE 5.20 ATP, an important energy currency in metabolism.

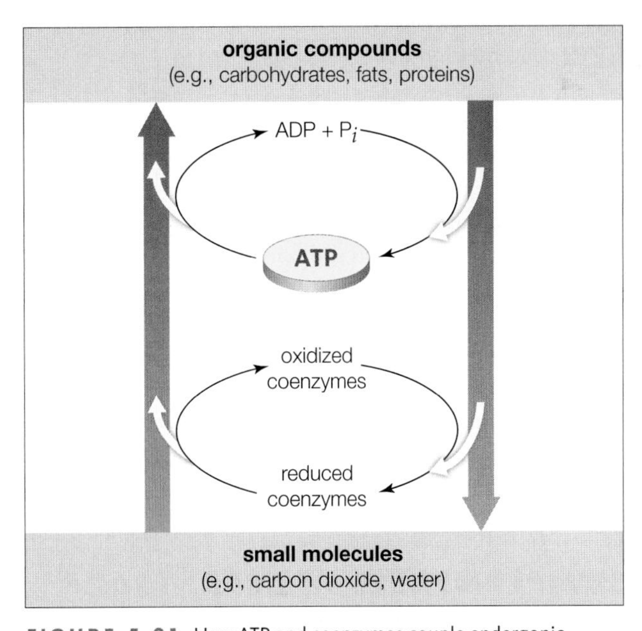

FIGURE 5.21 How ATP and coenzymes couple endergonic reactions with exergonic reactions. Yellow arrows indicate energy flow. Compare **FIGURES 5.8** and **5.20C**.

5.6 WHAT INFLUENCES THE MOVEMENT OF IONS AND MOLECULES?

Metabolic pathways require the participation of molecules that must move across membranes and through cells. **Diffusion** (*left*) is the spontaneous spreading of molecules or ions, and it is an essential way in which substances move into, through, and out of cells. An atom or molecule is always jiggling, and this internal movement causes it to randomly bounce off of nearby objects, including other atoms or molecules. Rebounds from such collisions propel solutes through a liquid or gas, with the result being a gradual and complete mixing. How fast this occurs depends on five factors:

Size It takes more energy to move a large object than it does to move a small one. Thus, smaller molecules diffuse more quickly than larger ones.

Temperature Atoms and molecules jiggle faster at higher temperature, so they collide more often. Thus, the higher the temperature, the faster the rate of diffusion.

Concentration A difference in solute concentration (Section 2.4) between adjacent regions of solution is called a concentration gradient. Solutes tend to diffuse "down" their concentration gradient, from a region of higher concentration to one of lower concentration. Why? Consider that moving objects (such as molecules) collide more often as they get more crowded. Thus, during a given interval, more molecules get bumped out of a region of higher concentration than get bumped into it.

Charge Each ion or charged molecule in a fluid contributes to the fluid's overall electric charge. A difference in charge between two regions of the fluid can affect the rate and direction of diffusion between them. For example, positively charged substances (such as sodium ions) will tend to diffuse toward a region with an overall negative charge.

Pressure Diffusion may be affected by a difference in pressure between two adjoining regions. Pressure squeezes objects—including atoms and molecules—closer together. Atoms and molecules that are more crowded collide and rebound more frequently. Thus, diffusion occurs faster at higher pressures.

SEMIPERMEABLE MEMBRANES

Remember from Section 4.3 that lipid bilayers are selectively permeable: Water can cross them, but ions and most polar molecules cannot (FIGURE 5.22). When two

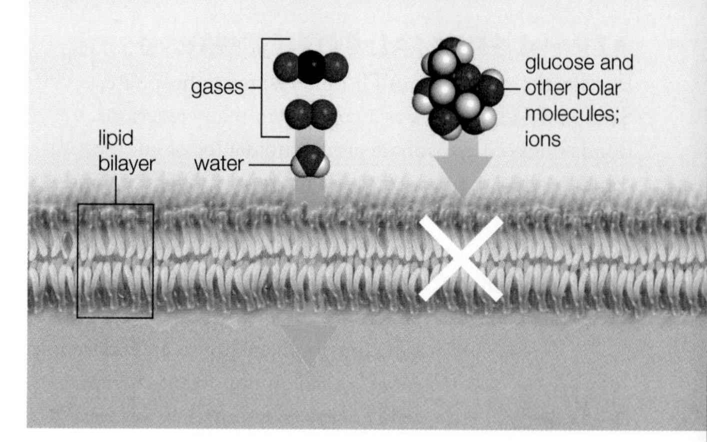

FIGURE 5.22 {Animated} Selective permeability of lipid bilayers. Hydrophobic molecules, gases, and water molecules can cross a lipid bilayer on their own. Ions in particular and most polar molecules such as glucose cannot.

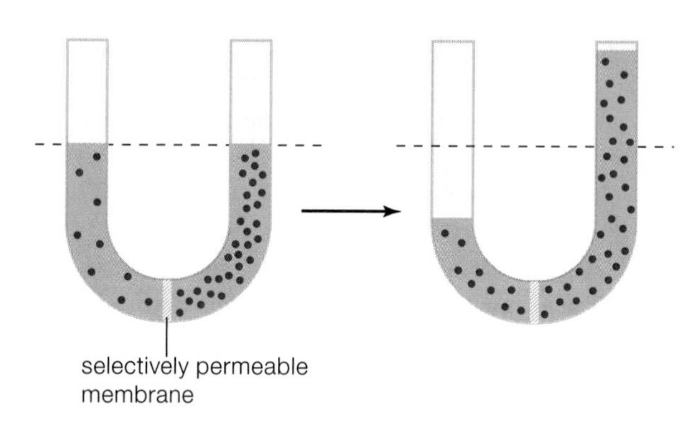

FIGURE 5.23 Osmosis. Water moves across a selectively permeable membrane that separates two fluids of differing solute concentration. The fluid volume changes in the two compartments as water diffuses across the membrane from the hypotonic solution to the hypertonic one.

fluids with different solute concentrations are separated by a selectively permeable membrane, water will diffuse across the membrane. The direction of water movement depends on the relative solute concentration of the two fluids. Tonicity refers to the solute concentration of one fluid relative to another that is separated by a selectively permeable membrane. Fluids that are **isotonic** have the same overall solute concentration. If the overall solute concentrations of the two fluids differ, the fluid with the lower concentration of solutes is said to be **hypotonic** (*hypo–*, under). The other one, with the higher solute concentration, is **hypertonic** (*hyper–*, over).

When a selectively permeable membrane separates two fluids that are not isotonic, water will move across the membrane from the hypotonic fluid into the hypertonic one (FIGURE 5.23). The diffusion will continue until the two fluids are isotonic, or until pressure against the hypertonic fluid counters it. The movement of water across membranes is so important in biology that it is given a special name: **osmosis**.

If a cell's cytoplasm is hypertonic with respect to the fluid outside of its plasma membrane, water diffuses into it. If the cytoplasm is hypotonic with respect to the fluid on the outside, water diffuses out. In either case, the solute concentration of the cytoplasm may change. If it changes enough, the cell's enzymes will stop working, with potentially lethal results. Many cells have built-in mechanisms that compensate for differences in solute concentration between cytoplasm and extracellular (external) fluid. In cells with no such mechanism, the volume—and solute concentration—of cytoplasm will change as water diffuses into or out of the cell (**FIGURE 5.24**).

TURGOR

The rigid cell walls of plants and many protists, fungi, and bacteria can resist an increase in the volume of cytoplasm even in hypotonic environments. In the case of plant cells, cytoplasm usually contains more solutes than soil water does. Thus, water usually diffuses from soil into a plant—but only up to a point. Stiff walls keep plant cells from expanding very much, so an inflow of water causes pressure to build up inside them. Pressure that a fluid exerts against a structure that contains it is called **turgor**. When enough pressure builds up inside a plant cell, water stops diffusing into its cytoplasm. The amount of turgor that is enough to stop osmosis is called **osmotic pressure**.

Osmotic pressure keeps walled cells plump, just as high air pressure inside a tire keeps it inflated. A young land plant can resist gravity to stay erect because its cells are plump with cytoplasm (**FIGURE 5.25A**). When soil dries out, it loses water but not solutes, so the concentration of solutes increases in soil water. If soil water becomes hypertonic with respect to cytoplasm, water will start diffusing out of the plant's cells, so their cytoplasm shrinks (**FIGURE 5.25B**). As turgor inside the cells decreases, the plant wilts.

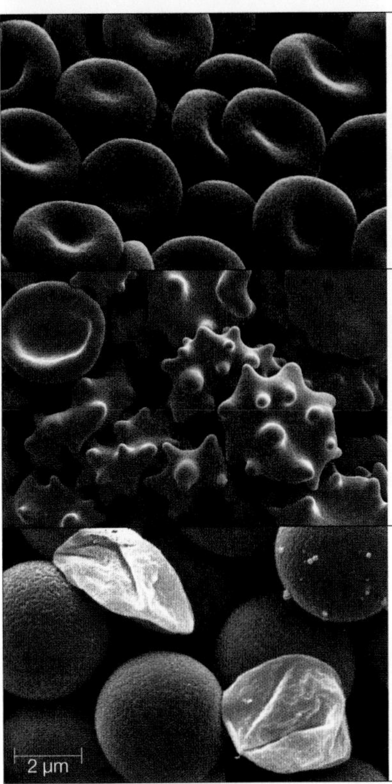

A Red blood cells in an isotonic solution (such as the fluid portion of blood) have a normal, indented disk shape.

B Water diffuses out of red blood cells immersed in a hypertonic solution, so they shrivel up.

C Water diffuses into red blood cells immersed in a hypotonic solution, so they swell up. Some of these have burst.

FIGURE 5.24 {Animated} Effects of tonicity in human red blood cells. These cells have no mechanism to compensate for differences in solute concentration between cytoplasm and extracellular fluid.

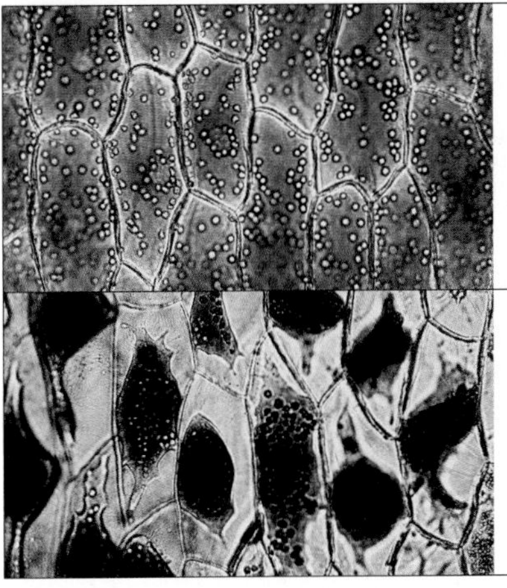

A Osmotic pressure keeps plant parts erect. These cells in an iris petal are plump with cytoplasm.

B Cells from a wilted iris petal. The cytoplasm shrank, and the plasma membrane moved away from the wall.

FIGURE 5.25 Turgor, as illustrated in cells of iris petals.

TAKE-HOME MESSAGE 5.6

Molecules or ions tend to diffuse into an adjoining region of fluid in which they are not as concentrated. The steepness of a concentration gradient as well as temperature, molecular size, charge, and pressure affect the rate of diffusion.

When two fluids of different solute concentration are separated by a selectively permeable membrane, water diffuses from the hypotonic to the hypertonic fluid. This movement, osmosis, is opposed by turgor.

diffusion Spontaneous spreading of molecules or ions.
hypertonic Describes a fluid that has a high solute concentration relative to another fluid separated by a semipermeable membrane.
hypotonic Describes a fluid that has a low solute concentration relative to another fluid separated by a semipermeable membrane.
isotonic Describes two fluids with identical solute concentrations and separated by a semipermeable membrane.
osmosis Diffusion of water across a selectively permeable membrane; occurs in response to a difference in solute concentration between the fluids on either side of the membrane.
osmotic pressure Amount of turgor that prevents osmosis into cytoplasm or other hypertonic fluid.
turgor Pressure that a fluid exerts against a structure that contains it.

CREDITS: (24A) Annie Cavanagh/Wellcome Images; (24B,C) CMSP/Getty Images; (25A,B) Claude Nuridsany & Marie Perennou/Science Source; (25 inset) © Evgenyi/Shutterstock.com.

5.7 HOW DO IONS AND CHARGED MOLECULES CROSS CELL MEMBRANES?

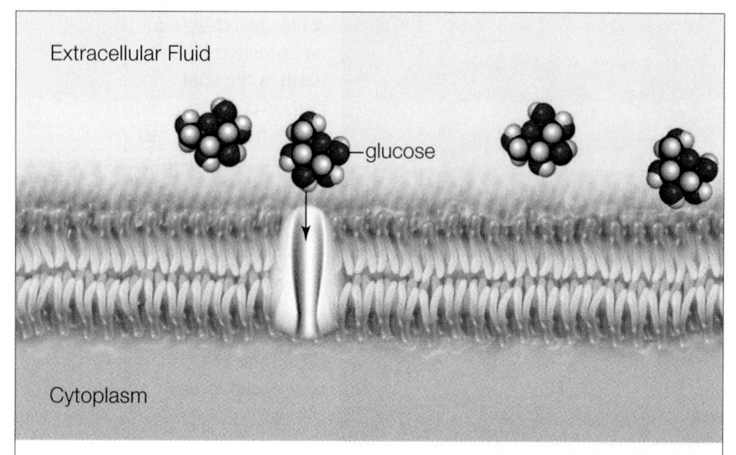

A A glucose molecule (here, in extracellular fluid) binds to a glucose transporter (gray) in the plasma membrane.

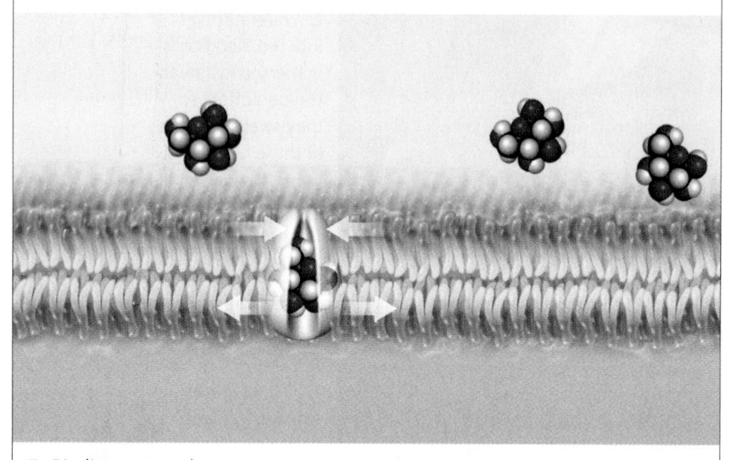

B Binding causes the transport protein to change shape.

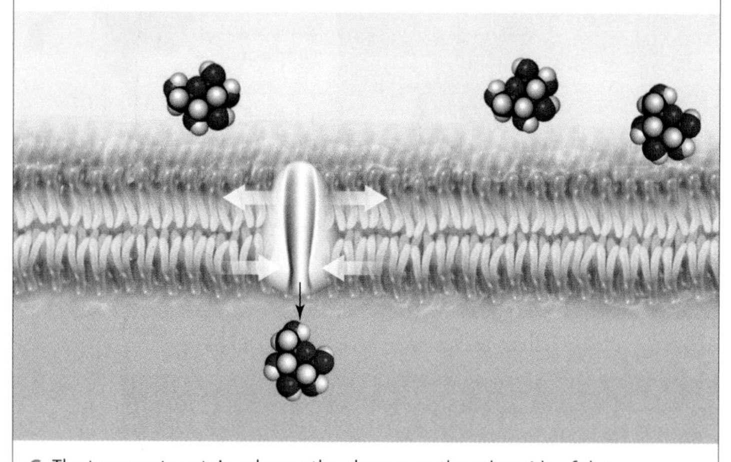

C The transport protein releases the glucose on the other side of the membrane (here, in cytoplasm) and resumes its original shape.

FIGURE 5.26 {Animated} Facilitated diffusion.

FIGURE IT OUT: In this example, which fluid is hypotonic: the extracellular fluid or cytoplasm?

Answer: Cytoplasm

TRANSPORT PROTEIN SPECIFICITY

Substances that cannot diffuse directly through lipid bilayers—ions in particular—cross cell membranes only with the help of transport proteins (Section 4.3). Each transport protein allows a specific substance to cross: Calcium pumps pump only calcium ions; glucose transporters transport only glucose; and so on. This specificity is an important part of homeostasis. For example, the composition of cytoplasm depends on the movement of particular solutes across the plasma membrane, which in turn depends on the transporters embedded in it. Glucose is an important source of energy for most cells, so they normally take up as much as they can from extracellular fluid. They do so with the help of glucose transporters in the plasma membrane. As soon as a molecule of glucose enters cytoplasm, an enzyme (hexokinase) phosphorylates it. Phosphorylation traps the molecule inside the cell because the transporters are specific for glucose, not phosphorylated glucose. Thus, phosphorylation prevents the molecule from moving back through the transporter and leaving the cell.

FACILITATED DIFFUSION

Osmosis is an example of **passive transport**, which is a membrane-crossing mechanism that requires no energy input. Diffusion of solutes through transport proteins is another example. In this case, the movement of the solute (and the direction of its movement) is driven entirely by the solute's concentration gradient. Some transport proteins form permanently open channels through a membrane. Others are gated, which means they open and close in response to a stimulus such as a shift in electric charge or binding to a particular signaling molecule.

With a passive transport mechanism called **facilitated diffusion**, a solute binds to a transport protein, which then changes shape so the solute is released to the other side of the membrane. A glucose transporter is an example of a transport protein that works in facilitated diffusion (**FIGURE 5.26**). This protein changes shape when it binds to a molecule of glucose. The shape change moves the solute to the opposite side of the membrane, where it detaches from the transport protein. Then, the transporter reverts to its original shape.

ACTIVE TRANSPORT

Maintaining a particular solute's concentration at a certain level often means transporting the solute against its gradient, to the side of the membrane where it is more concentrated. Pumping a solute against its gradient takes energy. In **active transport**, a transport protein uses energy to pump a solute against its gradient across a cell membrane.

After a solute binds to an active transport protein, an energy input (for example, in the form of a phosphate-group transfer from ATP) changes the shape of the protein. The change causes the transporter to release the solute to the other side of the membrane.

A calcium pump moves calcium ions across cell membranes by active transport (**FIGURE 5.27**). Calcium ions act as potent messengers inside cells, and they affect the activity of many enzymes, so their concentration in cytoplasm is very tightly regulated. Calcium pumps in the plasma membrane of all eukaryotic cells can keep the concentration of calcium ions in cytoplasm thousands of times lower than it is in extracellular fluid.

Another example of active transport involves sodium–potassium pumps (**FIGURE 5.28**). Nearly all of the cells in your body have these transport proteins. Sodium ions in cytoplasm diffuse into the pump's open channel and bind to its interior. A phosphate-group transfer from ATP causes the pump to change shape so that its channel opens to extracellular fluid, where it releases the sodium ions. Then, potassium ions from extracellular fluid diffuse into the channel and bind to its interior. The transporter releases the phosphate group and reverts to its original shape. The channel opens to the cytoplasm, where it releases the potassium ions.

Bear in mind that the membranes of all cells, not just those of animals, have active transport proteins. In plants, for example, active transport proteins in the plasma membranes of leaf cells pump sucrose into tubes that thread throughout the plant body.

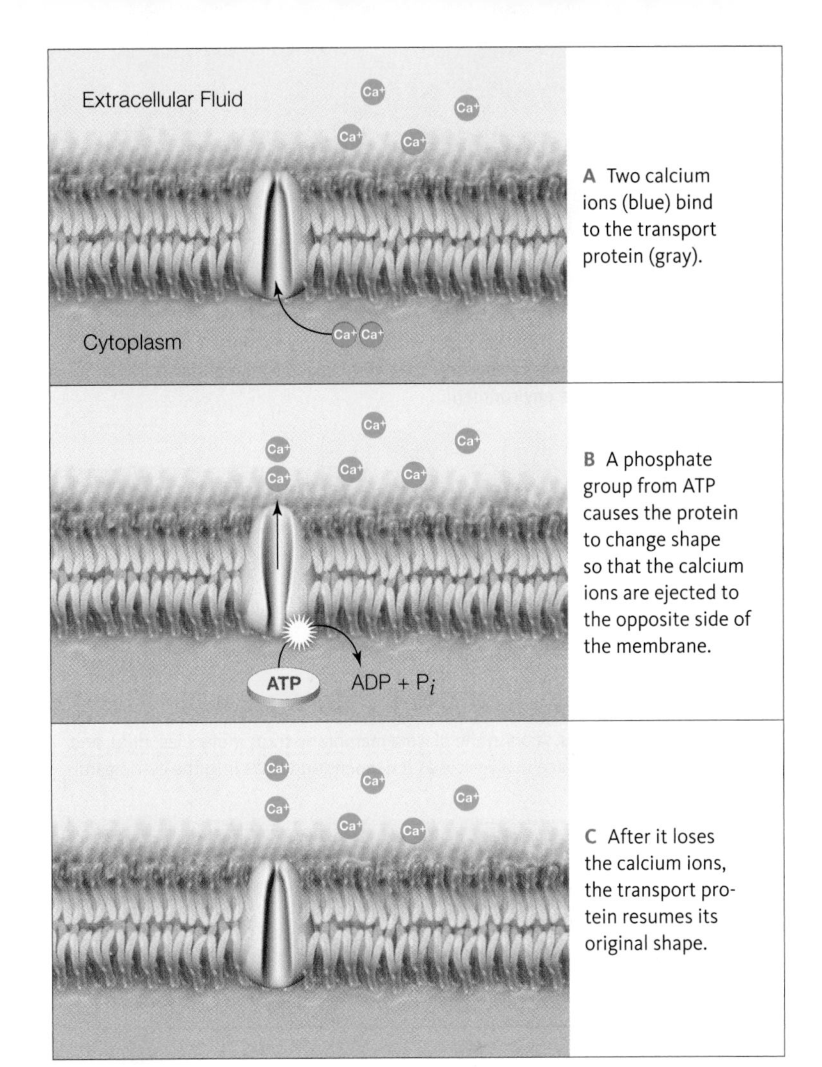

FIGURE 5.27 {Animated} Active transport of calcium ions.

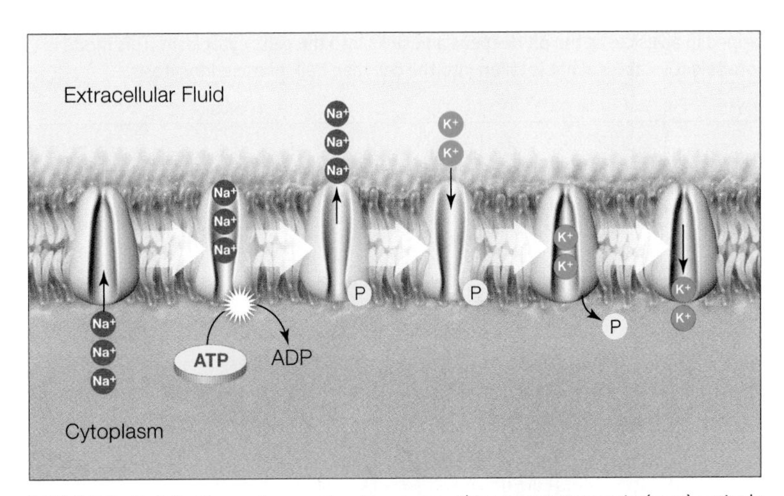

FIGURE 5.28 The sodium–potassium pump. This transport protein (gray) actively transports sodium ions (Na^+) from cytoplasm to extracellular fluid, and potassium ions (K^+) in the other direction. The transfer of a phosphate group (P) from ATP provides energy required for transporting the ions against their concentration gradient.

active transport Energy-requiring mechanism in which a transport protein pumps a solute across a cell membrane against its concentration gradient.
facilitated diffusion Passive transport mechanism in which a solute follows its concentration gradient across a membrane by moving through a transport protein.
passive transport Membrane-crossing mechanism that requires no energy input.

TAKE-HOME MESSAGE 5.7

Transport proteins move specific ions or molecules across a cell membrane. The amounts and types of these substances that cross a membrane depend on the transport proteins embedded in it.

In a type of passive transport called facilitated diffusion, a solute binds to a transport protein that releases it on the opposite side of the membrane. The movement is driven by the solute's concentration gradient.

In active transport, a transport protein pumps a solute across a membrane against its concentration gradient. The movement is driven by an energy input, as from ATP.

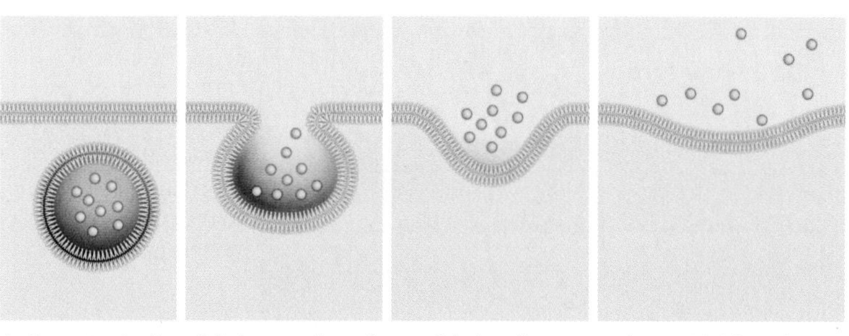

A Exocytosis. A vesicle in cytoplasm fuses with the plasma membrane. Lipids and proteins of the vesicle's membrane become part of the plasma membrane as its contents are expelled to the environment.

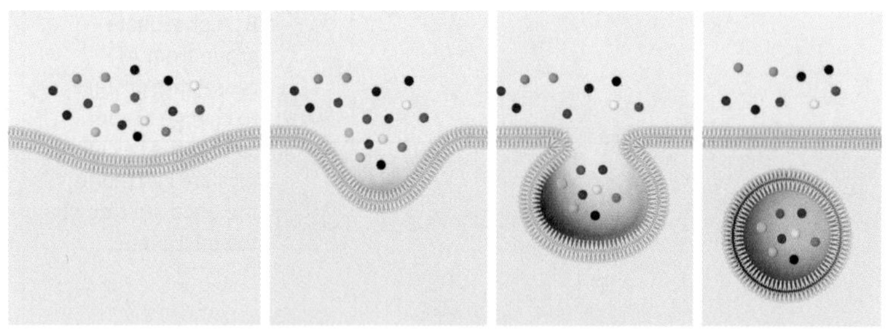

B Bulk-phase endocytosis. A pit in the plasma membrane traps molecules, fluid, and particles near the cell's surface in a vesicle as it deepens and sinks into the cytoplasm.

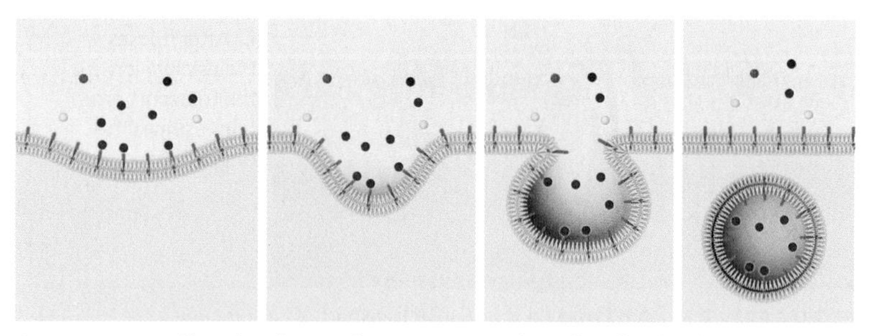

C Receptor-mediated endocytosis. Receptors on the cell surface bind a target molecule and trigger a pit to form in the plasma membrane. The target molecules are trapped in a vesicle as the pit deepens and sinks into the cell's cytoplasm. This mode is more selective about what is taken into the cell than bulk-phase endocytosis.

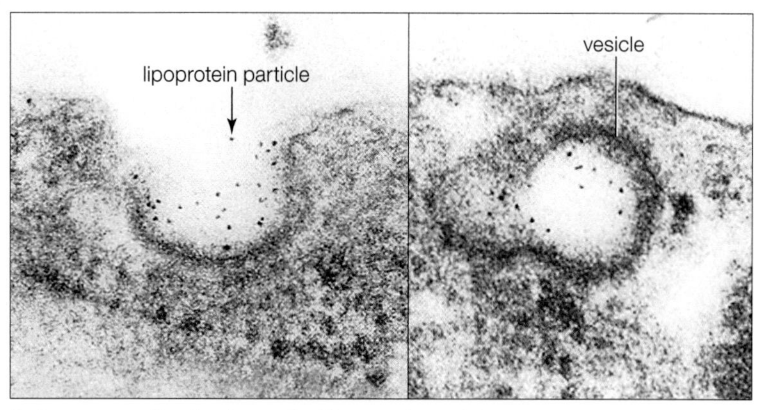

D Receptor-mediated endocytosis of lipoprotein particles.

FIGURE 5.29 {Animated} Exocytosis and endocytosis.

VESICLE MOVEMENT

Think back on the structure of a lipid bilayer (Section 4.3). When a bilayer is disrupted, it seals itself. Why? The disruption exposes the fatty acid tails of the phospholipids to their watery surroundings. Remember, in water, phospholipids spontaneously rearrange themselves so that their nonpolar tails stay together. A vesicle forms when a patch of membrane bulges into the cytoplasm because the hydrophobic tails of the lipids in the bilayer are repelled by the watery fluid on both sides. The fluid "pushes" the phospholipid tails together, which helps round off the bud as a vesicle, and also seals the rupture in the membrane.

Vesicles are constantly carrying materials to and from a cell's plasma membrane. This movement typically requires ATP because it involves motor proteins that drag the vesicles along cytoskeletal elements. We describe the movement based on where and how the vesicle originates, and where it goes.

By **exocytosis**, a vesicle in the cytoplasm moves to the cell's surface and fuses with the plasma membrane. As the exocytic vesicle loses its identity, its contents are released to the surroundings (**FIGURE 5.29A**).

There are several pathways of **endocytosis**, but all take up substances in bulk near the cell's surface (as opposed to one molecule or ion at a time via transport proteins). In bulk-phase endocytosis, a small patch of plasma membrane balloons inward, and then it pinches off after sinking farther into the cytoplasm. The membrane patch becomes the outer boundary of a vesicle (**FIGURE 5.29B**).

With receptor-mediated endocytosis, molecules of a hormone, vitamin, mineral, or another substance bind to receptors on the plasma membrane. The binding triggers a shallow pit to form in the membrane patch under the receptors. The pit sinks into the cytoplasm and traps the target substance in a vesicle as it closes back on itself (**FIGURE 5.29C,D**). LDL and other lipoproteins (Section 3.4) enter cells this way.

Phagocytosis (which literally means "cell eating") is a type of receptor-mediated endocytosis in which motile cells engulf microorganisms, cellular debris, or other large particles (**FIGURE 5.30**). Many single-celled protists such as amoebas feed by phagocytosis. Some of your white blood cells use phagocytosis to engulf viruses and bacteria, cancerous body cells, and other threats.

Phagocytosis begins when receptor proteins bind to a particular target. The binding causes microfilaments to assemble in a mesh under the plasma membrane. The microfilaments contract, forcing a lobe of membrane-enclosed cytoplasm to bulge outward as a pseudopod (Section 4.9). Pseudopods that merge around a target

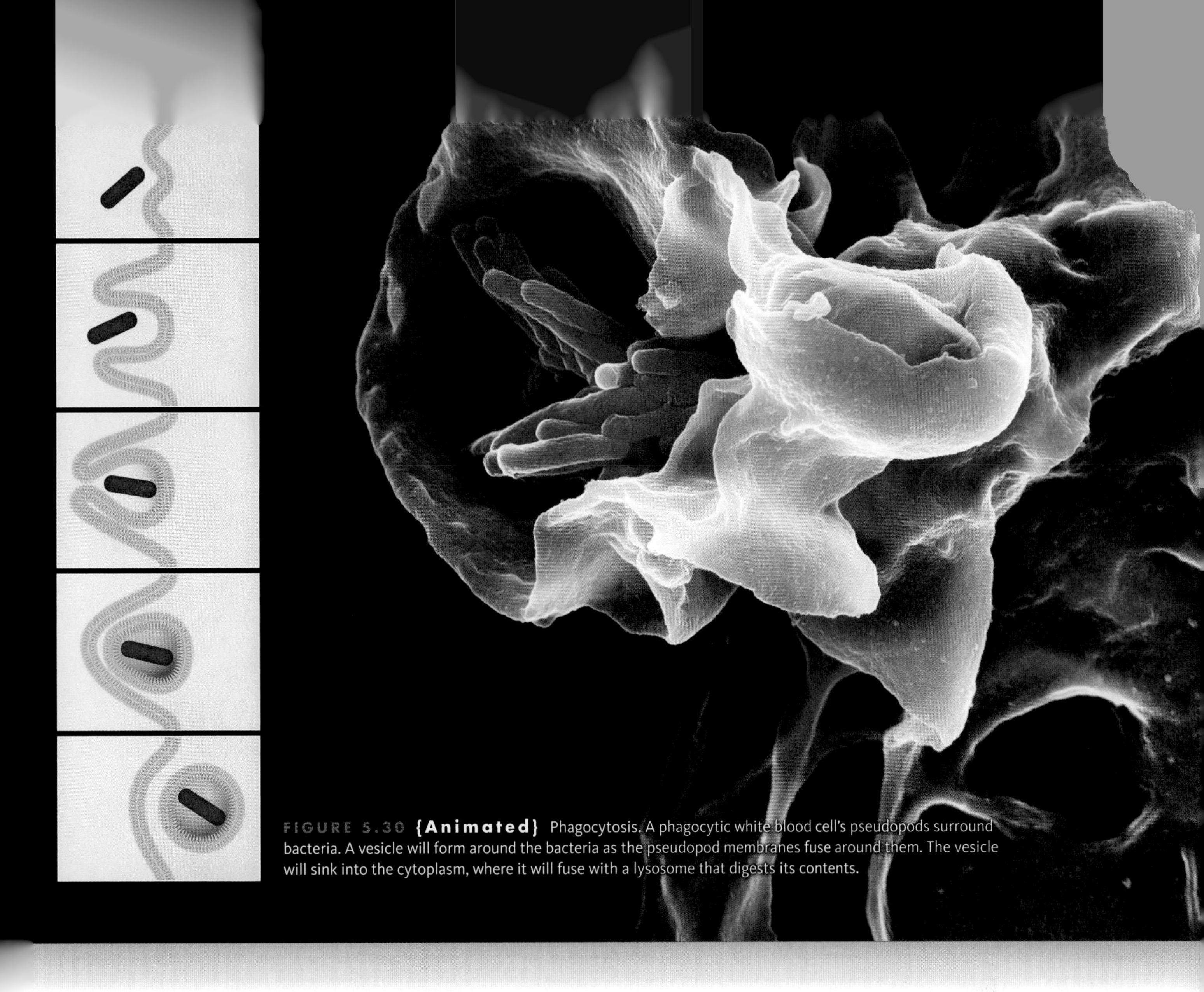

FIGURE 5.30 **{Animated}** Phagocytosis. A phagocytic white blood cell's pseudopods surround bacteria. A vesicle will form around the bacteria as the pseudopod membranes fuse around them. The vesicle will sink into the cytoplasm, where it will fuse with a lysosome that digests its contents.

trap it inside a vesicle that sinks into the cytoplasm. Material taken in by phagocytosis is typically digested by lysosomes, and the resulting molecular bits may be recycled by the cell, or expelled by exocytosis.

MEMBRANE TRAFFICKING

The composition of a plasma membrane begins in the ER. There, membrane proteins and lipids are made and modified, and both become part of vesicles that transport them to Golgi bodies for final modification. New plasma membrane forms when the finished proteins and lipids are repackaged as vesicles that travel to the plasma membrane and fuse with it.

As long as a cell is alive, exocytosis and endocytosis continually replace and withdraw patches of its plasma membrane. If the cell is not enlarging, the total area of the plasma membrane remains more or less constant. Membrane lost as a result of endocytosis is replaced by membrane arriving as exocytic vesicles.

endocytosis Process by which a cell takes in a small amount of extracellular fluid (and its contents) by the ballooning inward of the plasma membrane.
exocytosis Process by which a cell expels a vesicle's contents to extracellular fluid.
phagocytosis "Cell eating"; an endocytic pathway by which a cell engulfs particles such as microbes or cellular debris.

TAKE-HOME MESSAGE 5.8

Exocytosis and endocytosis move materials in bulk across plasma membranes.

In exocytosis, a cytoplasmic vesicle fuses with the plasma membrane and releases its contents to the outside of the cell.

In endocytosis, a patch of plasma membrane sinks inward and forms a vesicle in the cytoplasm.

Some cells can engulf large particles by phagocytosis.

5.9 Application: A TOAST TO ALCOHOL DEHYDROGENASE

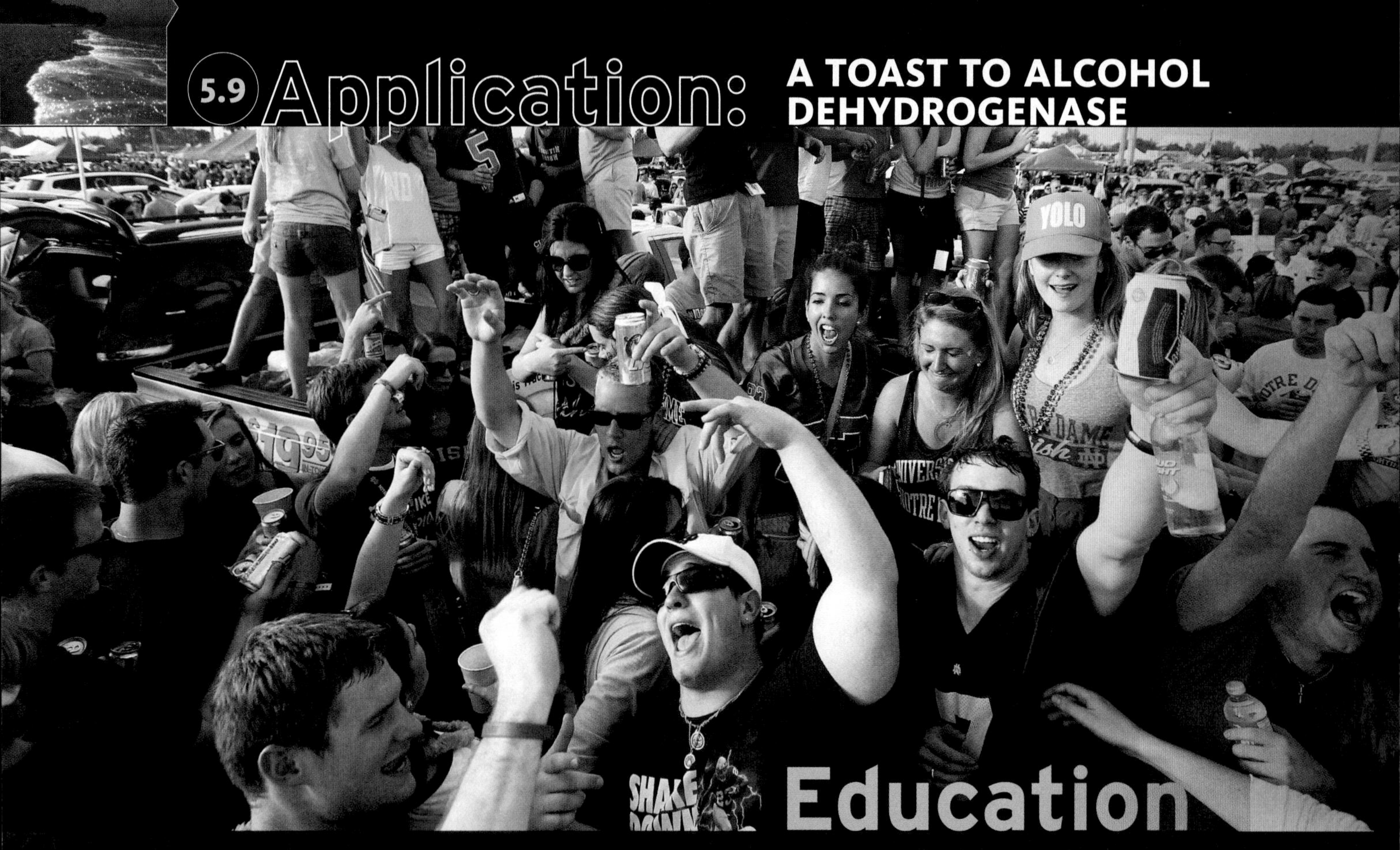

FIGURE 5.31 A tailgate party at a Notre Dame–Alabama football game. During 2012 alone, Indiana State police arrested 138 Notre Dame students for underage drinking at tailgate parties.

MOST COLLEGE STUDENTS ARE UNDER THE LEGAL DRINKING AGE, but alcohol abuse continues to be the most serious drug problem on college campuses throughout the United States (FIGURE 5.31). Every year, drinking kills more than 1,700 students and injures about 500,000 more; it is also a factor in 600,000 assaults and 100,000 rapes on college campuses.

Tens of thousands of undergraduate students have been polled about their drinking habits in recent surveys. More than half of them reported that they regularly drink five or more alcoholic beverages within a two-hour period.

Each alcoholic drink—a bottle of beer, glass of wine, shot of vodka, and so on—contains the same amount of alcohol or, more precisely, ethanol. Ethanol molecules move quickly from the stomach and small intestine into the bloodstream. Almost all of the ethanol ends up in the liver, a large organ in the abdomen. Liver cells have impressive numbers of enzymes. One of them, ADH (alcohol dehydrogenase), helps break down ethanol and other toxic compounds.

ADH converts ethanol to acetaldehyde, an organic molecule even more toxic than ethanol and the most likely source of various hangover symptoms. A different enzyme, ALDH, very quickly converts acetaldehyde to nontoxic acetate. Both ADH and ALDH use the coenzyme NAD$^+$ to accept electrons and hydrogen atoms. Thus, the overall pathway of ethanol metabolism in humans is:

$$\text{ethanol} \xrightarrow[\text{NAD}^+ \to \text{NADH}]{\text{ADH}} \text{acetaldehyde} \xrightarrow[\text{NAD}^+ \to \text{NADH}]{\text{ALDH}} \text{acetate}$$

In the average healthy adult, this metabolic pathway can detoxify between 7 and 14 grams of ethanol per hour. The typical alcoholic beverage contains between 10 and 20 grams of ethanol.

If you put more ethanol into your body than your enzymes can deal with, then you will damage it. Ethanol and acetaldehyde kill liver cells, so the more

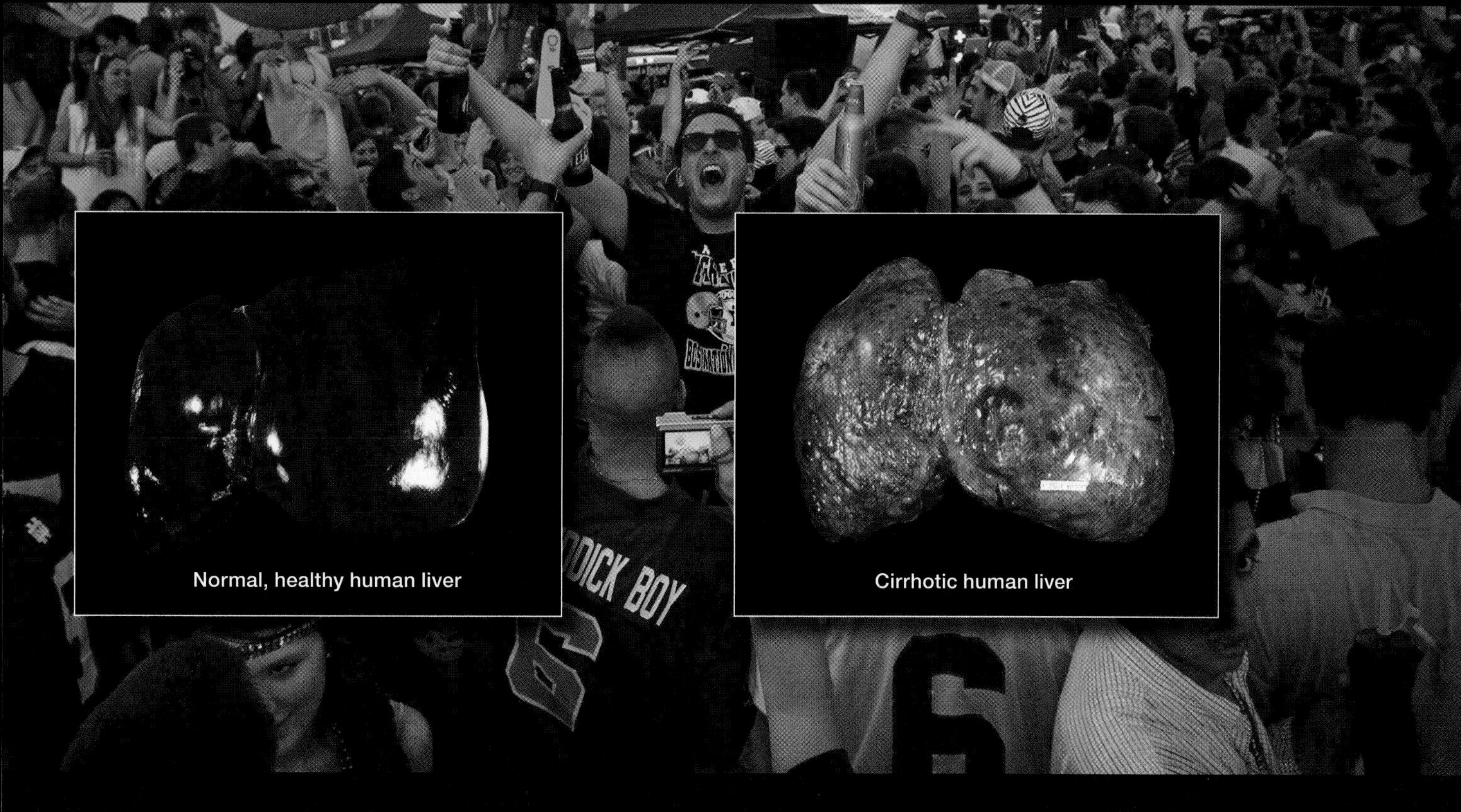

a person drinks, the fewer liver cells are left for detoxification . Ethanol also interferes with normal metabolic processes. For example, in the presence of ethanol, oxygen that would ordinarily take part in breaking down fatty acids is diverted to breaking down the ethanol. This is why fats accumulate as large globules in the tissues of heavy drinkers.

Long-term heavy drinking causes alcoholic hepatitis, a disease characterized by inflammation and destruction of liver tissue; and cirrhosis, a condition in which the liver becomes so scarred, hardened, and filled with fat that it loses its function. (The term cirrhosis is from the Greek *kirros*, meaning orange-colored, after the abnormal skin color of people with the disease.) The liver is the largest gland in the human body, and it has many important functions. A cirrhotic liver stops making the protein albumin, so the solute balance of body fluids is disrupted, and the legs and abdomen swell with watery fluid. It can no longer remove drugs and other toxins from the blood, so they accumulate in the brain—which impairs mental functioning and alters personality. Restricted blood flow through the liver causes veins to enlarge and rupture, so internal bleeding is a risk. The damage to the body results in a heightened susceptibility to diabetes and liver cancer. Once cirrhosis has been diagnosed, a person has about a 50 percent chance of dying within 10 years (FIGURE 5.32).

FIGURE 5.32 Gary Reinbach, who died at the age of 22 from alcoholic liver disease shortly after this photograph was taken, in 2009. The odd color of his skin is a symptom of cirrhosis.

Transplantation is a last-resort treatment for a failed liver, but there are not nearly enough liver donors for everyone who needs a transplant. Reinbach was refused a transplant that would have saved his life because he had not abstained from drinking for the prior 6 months.

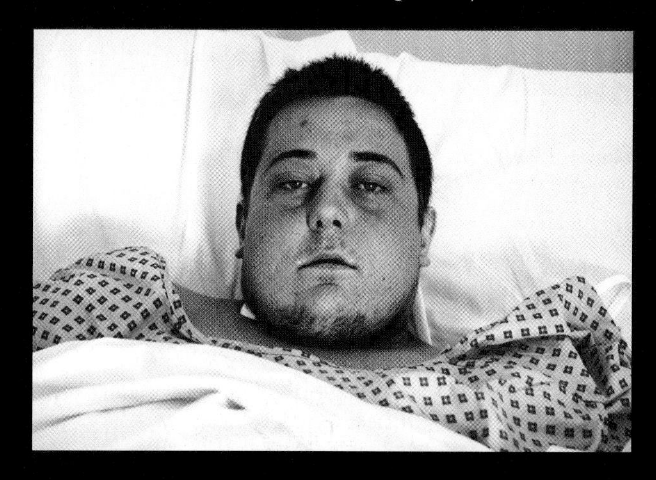

CREDITS: (31) left inset, Southern Illinois University/Science Source; right inset, Martin M. Rotker/Science Source; (32) Stuart Clark/The Sunday Times/nisyndication.

Summary

SECTION 5.1 **Kinetic energy** and **potential energy** are different forms of **energy**, the capacity to do work. Energy, which cannot be created or destroyed (**first law of thermodynamics**), tends to disperse spontaneously (**second law of thermodynamics**). **Entropy** is a measure of how much the energy of a system is dispersed. A bit disperses at each energy transfer, usually in the form of heat.

SECTION 5.2 Cells store and retrieve energy by making and breaking chemical bonds in reactions that convert **reactants** to **products**. **Endergonic** reactions require a net input of energy to proceed. **Exergonic** reactions end with a net release of energy. **Activation energy** is the minimum energy required to start a reaction.

SECTION 5.3 Enzymes greatly enhance the rate of reactions without being changed by them, a process called **catalysis**. They lower a reaction's activation energy, for example by boosting local concentrations of **substrates** or improving the fit between a substrate and the enzyme's **active site** (**induced-fit model**). Each enzyme works best within a characteristic range of conditions that reflect its evolutionary context.

SECTION 5.4 Cells build, convert, and dispose of substances in enzyme-mediated reaction sequences called **metabolic pathways**. Regulating these pathways allows a cell to conserve energy and resources. With **allosteric regulation**, a regulatory molecule or ion alters the activity of an enzyme by binding to it in a region other than the active site. Some products of metabolic pathways inhibit their own production, a regulatory mechanism called **feedback inhibition**. **Redox** (oxidation–reduction) **reactions** in **electron transfer chains** allow cells to harvest energy in small, manageable steps.

SECTION 5.5 Most enzymes require **cofactors**, which are metal ions or organic **coenzymes**. Some enzymes that act as **antioxidants** have cofactors that help them prevent dangerous oxidation reactions.

When a phosphate group is transferred from ATP to another molecule, energy is transferred along with it. Phosphate-group transfers (**phosphorylations**) to and from ATP couple exergonic with endergonic reactions. Cells regenerate ATP in the **ATP/ADP cycle**.

SECTION 5.6 The steepness of a concentration gradient, temperature, solute size, charge, and pressure influence the rate of **diffusion**. **Osmosis** is the diffusion of water across a selectively permeable membrane, from a **hypotonic** fluid toward a **hypertonic** fluid. There is no net movement of water between **isotonic** solutions. **Osmotic pressure** is the amount of **turgor** (fluid pressure against a cell membrane or wall) sufficient to halt osmosis.

SECTION 5.7 The types of transport proteins in a membrane determine which substances cross it. Ions and most polar molecules cross membranes with the help of a transport protein. With **facilitated diffusion**, a solute follows its concentration gradient across a membrane through a transport protein. Facilitated diffusion is a type of **passive transport** (no energy input is required). With **active transport**, a transport protein uses energy to pump a solute across a membrane against its concentration gradient. A phosphate-group transfer from ATP often supplies the necessary energy for active transport.

SECTION 5.8 Substances can be moved across plasma membranes in bulk by two ATP-requiring processes. With **exocytosis**, a cytoplasmic vesicle fuses with the plasma membrane, and its contents are released to the outside of the cell. With **endocytosis**, a patch of plasma membrane balloons into the cell, and forms a vesicle that sinks into the cytoplasm. Some cells engulf large particles such as prey or cell debris by the endocytic pathway of **phagocytosis**.

SECTION 5.9 Currently the most serious drug problem on college campuses is binge drinking, which is often a symptom of alcoholism. Drinking more alcohol than the body's enzymes can detoxify can be lethal in both the short term and the long term.

Self-Quiz Answers in Appendix VII

1. _______ is life's primary source of energy.
   a. Food b. Water c. Sunlight d. ATP

2. Which of the following statements is not correct?
   a. Energy cannot be created or destroyed.
   b. Energy cannot change from one form to another.
   c. Energy tends to disperse spontaneously.

3. Entropy _______ .
   a. disperses
   b. is a measure of disorder
   c. always increases, overall
   d. b and c

4. If we liken a chemical reaction to an energy hill, then a(n) _______ reaction is an uphill run.
   a. endergonic
   b. exergonic
   c. catalytic
   d. both a and c

5. If we liken a chemical reaction to an energy hill, then activation energy is like _______ .
   a. a burst of speed
   b. coasting downhill
   c. a bump at the top of the hill

6. _______ are always changed by participating in a reaction. (Choose all that are correct.)
   a. Enzymes b. Cofactors c. Reactants d. Coenzymes

7. Name one environmental factor that typically influences enzyme function.

8. A metabolic pathway may _______ .
   a. build or break down molecules
   b. include an electron transfer chain
   c. generate heat
   d. all of the above

Data Analysis Activities

Deep inside one of the most toxic sites in the United States: Iron Mountain Mine, in California. The water in this stream, which is about 1 meter (3 feet) wide in this view, is hot (around 40°C, or 104°F), heavily laden with arsenic and other toxic metals, and has a pH of zero. The slime streamers growing in it are a biofilm dominated by a species of archaea, *Ferroplasma acidarmanus*.

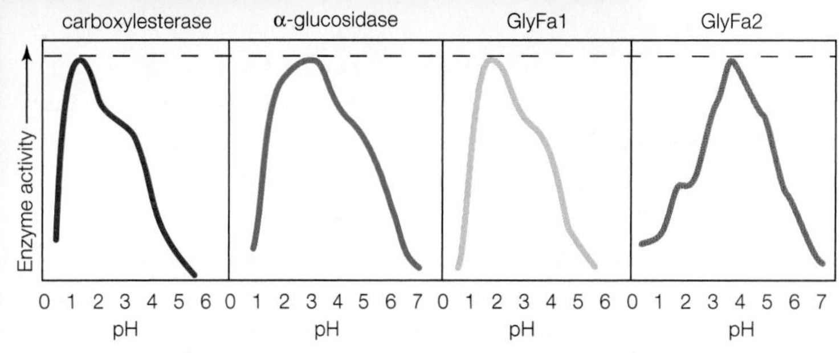

FIGURE 5.33 pH anomaly of *Ferroplasma* enzymes. Above are pH activity profiles of four enzymes isolated from *Ferroplasma*. Researchers had expected these enzymes to function best at the cells' cytoplasmic pH (5.0).

One Tough Bug *Ferroplasma acidarmanus* is a species of archaea discovered in an abandoned California copper mine. These cells use an energy-harvesting pathway that combines oxygen with iron–sulfur compounds in minerals such as pyrite. The reaction dissolves the minerals, so groundwater that seeps into the mine ends up with high concentrations of metal ions such as copper, zinc, cadmium, and arsenic. The reaction also produces sulfuric acid, which lowers the pH of the water around the cells to zero. *F. acidarmanus* cells maintain their internal pH at a cozy 5.0 despite living in an environment similar to hot battery acid. Thus, researchers investigating *Ferroplasma* were surprised to discover that most of the cells' enzymes function best at very low pH (**FIGURE 5.33**).

1. What does the dashed line in the graph signify?
2. Of the four enzymes, how many function optimally at a pH lower than 5?
3. What is the optimal pH for the carboxylesterase?

9. All antioxidants _______ .
   a. prevent other molecules from being oxidized
   b. are necessary in the human diet
   c. balance charge
   d. deoxidize free radicals

10. Solutes tend to diffuse from a region where they are _______ (more/less) concentrated to an adjacent region where they are _______ (more/less) concentrated.

11. _______ cannot easily diffuse across a lipid bilayer.
    a. Water c. Ions
    b. Gases d. all of the above

12. A transport protein requires ATP to pump sodium ions across a membrane. This is a case of _______ .
    a. passive transport c. facilitated diffusion
    b. active transport d. a and c

13. Immerse a human blood cell in a hypotonic solution, and water _______ .
    a. diffuses into the cell c. shows no net movement
    b. diffuses out of the cell d. moves in by endocytosis

14. Vesicles form during _______ .
    a. endocytosis c. phagocytosis
    b. exocytosis d. a and c

CENGAGE brain .com To access course materials, please visit www.cengagebrain.com.

15. Match each term with its most suitable description.

___ reactant	a. assists enzymes
___ phagocytosis	b. forms at reaction's end
___ first law of thermodynamics	c. enters a reaction
___ product	d. requires energy input
___ cofactor	e. one cell engulfs another
___ concentration gradient	f. energy cannot be created or destroyed
___ passive transport	g. basis of diffusion
___ active transport	h. no energy input required

Critical Thinking

1. Often, beginning physics students are taught the basic concepts of thermodynamics with two phrases: First, you can never win. Second, you can never break even. Explain.

2. How do you think a cell regulates the amount of glucose it brings into its cytoplasm from the extracellular environment?

3. The enzyme trypsin is sold as a dietary enzyme supplement. Explain what happens to trypsin that is taken with food.

4. Catalase combines two hydrogen peroxide molecules (H_2O_2 + H_2O_2) to make two molecules of water. A gas also forms. What is the gas?

Raindrops can separate light from the sun into its different component wavelengths, which we see as different colors in a rainbow.

6 WHERE IT STARTS—PHOTOSYNTHESIS

Links to Earlier Concepts

This chapter explores the main metabolic pathways (Section 5.4) by which organisms harvest energy from the sun (5.1). We revisit experimental design (1.6), electrons and energy levels (2.2), bonds (2.3), carbohydrates (3.2), membrane proteins (4.3), plastids (4.8), antioxidants (5.5), and concentration gradients (5.6).

KEY CONCEPTS

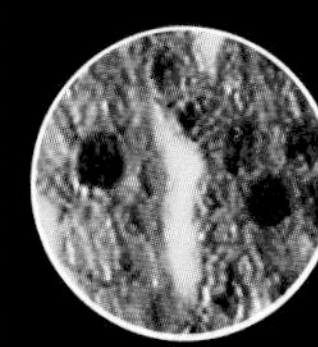

THE RAINBOW CATCHERS

The main flow of energy through the biosphere starts when photosynthetic pigments absorb light. In plants and other eukaryotes, these pigments occur in chloroplasts.

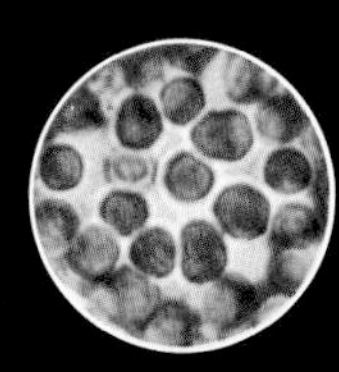

WHAT IS PHOTOSYNTHESIS?

Photosynthesis is a metabolic pathway that occurs in two stages. Light energy harvested in the first stage is used to make molecules that power sugar formation in the second.

MAKING ATP AND NADPH

In the first stage of photosynthesis, light drives ATP synthesis in one of two pathways. A cyclic pathway makes ATP alone. A noncyclic pathway makes ATP and NADPH, and it releases oxygen.

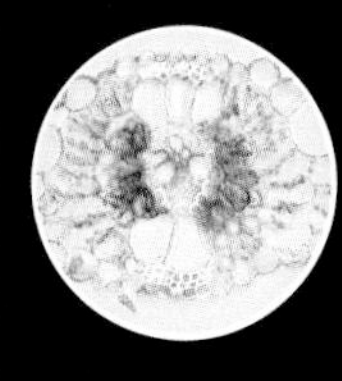

MAKING SUGARS

Sugars are assembled from carbon dioxide (CO_2) during the second stage of photosynthesis. In plants, the reactions run on ATP and NADPH—molecules that formed in the first stage.

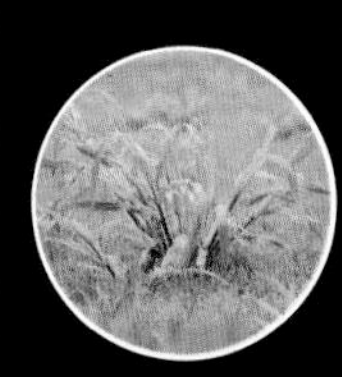

ALTERNATE PATHWAYS

Metabolic pathways are shaped by evolution. Variations in photosynthetic pathways are evolutionary adaptations that allow plants to thrive in a variety of environments.

Photograph by Robbie George, National Geographic Creative.

6.1 HOW DO PHOTOSYNTHESIZERS ABSORB LIGHT?

Energy flow through nearly all ecosystems on Earth begins when plants and other photosynthesizers intercept sunlight. These organisms are called producers because they produce the food that sustains ecosystems. Producers are **autotrophs,** organisms that make their own food—sugars—by harvesting energy directly from the environment (*auto–* means self; *–troph* refers to nourishment). All organisms need carbon; autotrophs obtain it from inorganic molecules such as carbon dioxide (CO_2). By contrast, **heterotrophs** get their carbon by breaking down organic molecules assembled by other organisms (*hetero–* means other). Heterotrophs are an ecosystem's consumers.

PROPERTIES OF LIGHT

Photosynthesizers make their own food by converting light energy to chemical energy. In order to understand how that happens, you have to know a little about the nature of light. Light is electromagnetic radiation, a type of energy that moves through space in waves, a bit like waves moving across an ocean. The distance between the crests of two successive waves is a **wavelength**, measured in nanometers (nm). Light that humans can see is a small part of the spectrum of electromagnetic radiation emitted by the sun (**FIGURE 6.1A**). Visible light travels in wavelengths between 380 and 750 nm, and this is the main form of energy that drives photosynthesis. Our eyes perceive all of these wavelengths combined as white light, and particular wavelengths in this range as different colors. White light separates into its component colors when it passes through a prism, or raindrops that act as tiny prisms. A prism bends longer wavelengths more than it bends shorter ones, so a rainbow of colors forms.

Light travels in waves, but it is also organized in packets of energy called photons. A photon's energy and its wavelength are related, so all photons traveling at the same wavelength carry the same amount of energy. Photons that carry the least amount of energy travel in longer wavelengths; those that carry the most energy travel in shorter wavelengths (**FIGURE 6.1B**).

CAPTURING A RAINBOW

Photosynthesizers use pigments to capture light. A **pigment** is an organic molecule that selectively absorbs light of specific wavelengths. Wavelengths of light that are not absorbed are reflected, and that reflected light gives each pigment its characteristic color.

Chlorophyll *a* is the most common photosynthetic pigment in plants and photosynthetic protists. It also occurs in some bacteria. Chlorophyll *a* absorbs violet, red, and orange light, and it reflects green light, so it appears green to us. Accessory pigments, including other chlorophylls, collectively harvest a wide range of additional light wavelengths for photosynthesis (**FIGURE 6.2**).

A pigment molecule is a bit like an antenna specialized for receiving light. Each has a light-trapping part in which single bonds alternate with double bonds; electrons

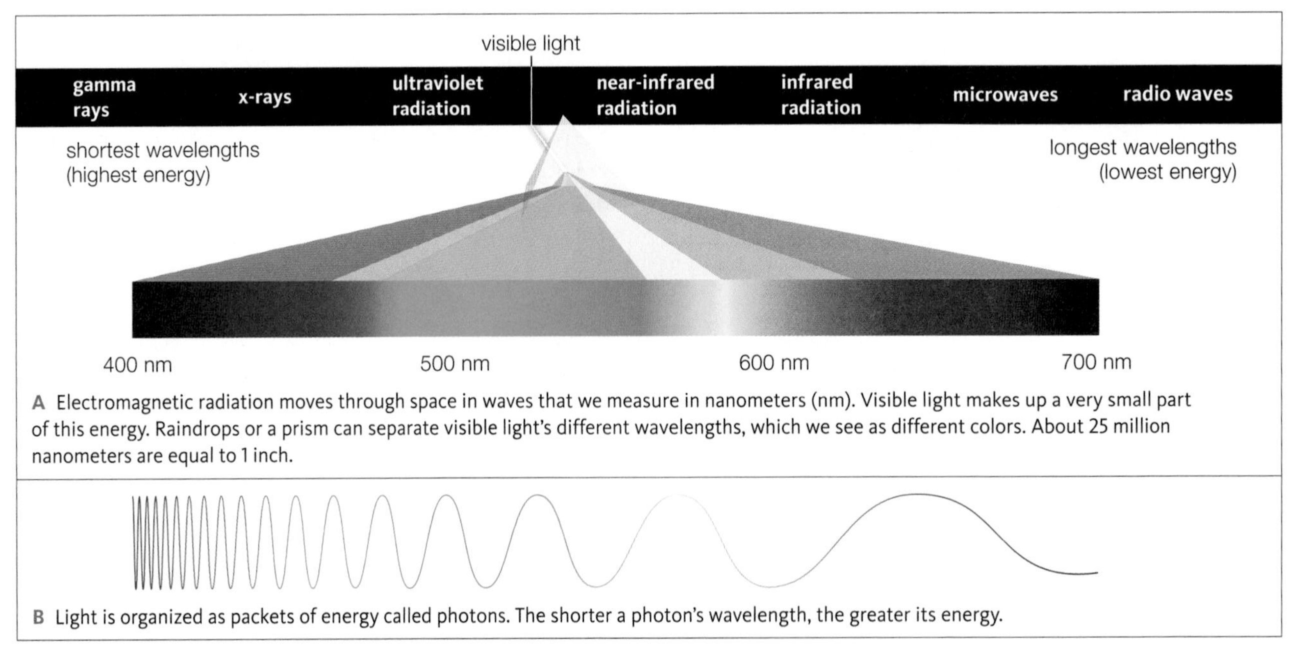

A Electromagnetic radiation moves through space in waves that we measure in nanometers (nm). Visible light makes up a very small part of this energy. Raindrops or a prism can separate visible light's different wavelengths, which we see as different colors. About 25 million nanometers are equal to 1 inch.

B Light is organized as packets of energy called photons. The shorter a photon's wavelength, the greater its energy.

FIGURE 6.1 Properties of light.

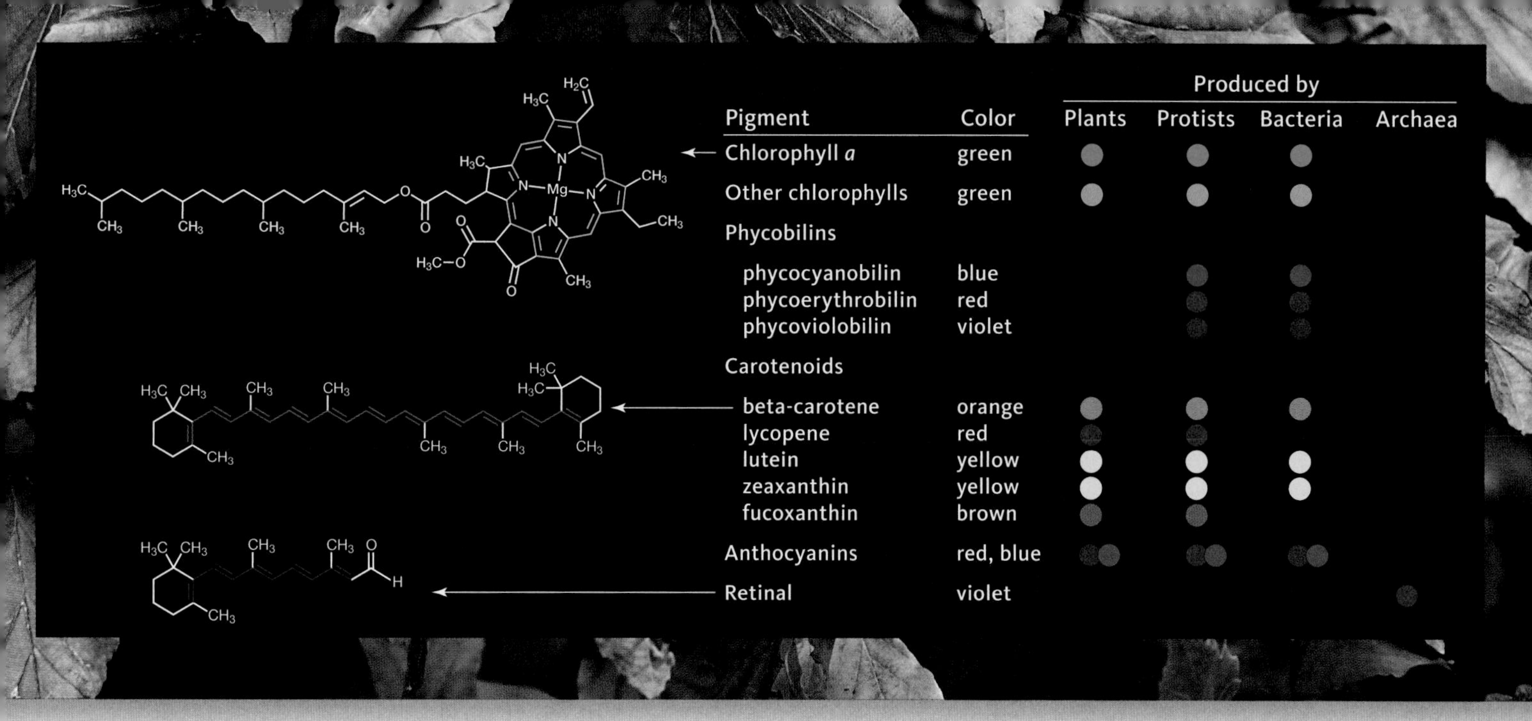

Pigment	Color	Plants	Protists	Bacteria	Archaea
Chlorophyll *a*	green	●	●	●	
Other chlorophylls	green	●	●	●	
Phycobilins					
phycocyanobilin	blue		●	●	
phycoerythrobilin	red		●	●	
phycoviolobilin	violet		●	●	
Carotenoids					
beta-carotene	orange	●	●	●	
lycopene	red	●	●		
lutein	yellow	●	●	●	
zeaxanthin	yellow	●	●	●	
fucoxanthin	brown	●	●		
Anthocyanins	red, blue	●	●	●	
Retinal	violet				●

FIGURE 6.2 Examples of photosynthetic pigments. Photosynthetic pigments can collectively absorb almost all visible light wavelengths. *Left*, the light-catching part of a pigment (shown in color) is the region in which single bonds alternate with double bonds. These and many other pigments (including heme, Section 5.5) are derived from evolutionary remodeling of the same compound. Animals convert dietary beta-carotene into a similar pigment (retinal) that is the basis of vision.

populating the atoms in such arrays easily absorb a photon—but not just any photon. Only a photon with exactly enough energy to boost an electron to a higher energy level is absorbed (Section 2.2). This is why a pigment absorbs light of only certain wavelengths.

An excited electron (one that has been boosted to a higher energy level) quickly emits its extra energy and returns to a lower energy level. As you will see in Section 6.4, photosynthetic cells capture energy emitted from an electron returning to a lower energy level.

Most photosynthetic organisms use a combination of pigments to capture light for photosynthesis—and often for additional purposes. Many accessory pigments are antioxidants that protect cells from the damaging effects of UV light in the sun's rays (Section 5.5). Appealing colors attract animals to ripening fruit or pollinators to flowers. You may already be familiar with some of these molecules. Carrots, for example, are orange because they contain beta-carotene (β-carotene); roses are red and violets are blue because of their anthocyanin content.

In green plants, chlorophylls are usually so abundant that they mask the colors of the other pigments. Plants that change color during autumn are preparing for a period of dormancy; they conserve resources by moving nutrients from tender parts that would be damaged by winter cold (such as leaves) to protected parts (such as roots). Chlorophylls are not needed during dormancy, so they are disassembled and their components recycled. Yellow and orange accessory pigments are also recycled, but not as quickly as chlorophylls. Their colors begin to show as the chlorophyll content declines in leaves. Anthocyanin synthesis also increases in some plants, adding red and purple tones to turning leaf colors. (Chapter 27 returns to the topic of dormancy in plants.)

autotroph Organism that makes its own food using energy from the environment and carbon from inorganic molecules such as CO_2.
chlorophyll *a* Main photosynthetic pigment in plants.
heterotroph Organism that obtains carbon from organic compounds assembled by other organisms.
pigment An organic molecule that can absorb light of certain wavelengths.
wavelength Distance between the crests of two successive waves.

TAKE-HOME MESSAGE 6.1

The sun emits electromagnetic radiation (light). Visible light is the main form of energy that drives photosynthesis.

Light travels in waves and is organized as photons. We see different wavelengths of visible light as different colors.

Pigments absorb light at specific wavelengths. Photosynthetic species use pigments such as chlorophyll *a* to harvest the energy of light for photosynthesis.

6.2 WHY DO CELLS USE MORE THAN ONE PHOTOSYNTHETIC PIGMENT?

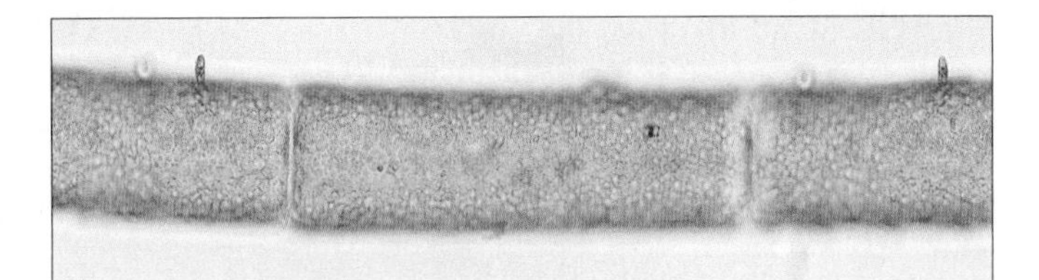

A Light micrograph of a filament of green algae. Each strand is a stack of individual photosynthetic cells. Theodor Engelmann used several species of algae in a series of experiments to determine whether some colors of light are better for photosynthesis than others.

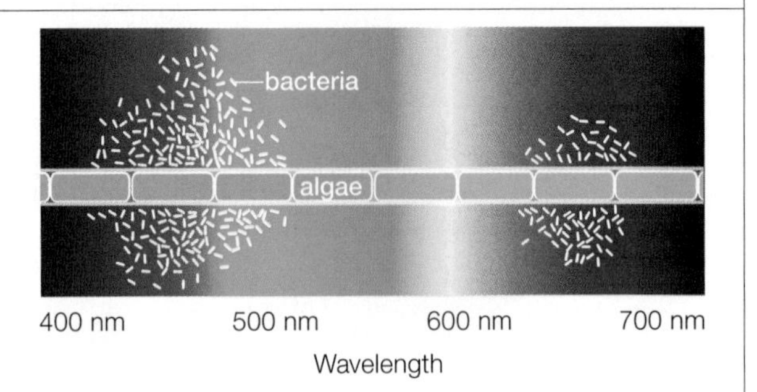

B Engelmann directed light through a prism so that bands of colors crossed a water droplet on a microscope slide. The water held a strand of photosynthetic algae, and also oxygen-requiring bacteria. The bacteria clustered around the algal cells that were releasing the most oxygen—the ones most actively engaged in photosynthesis. Those cells were under blue and red light.

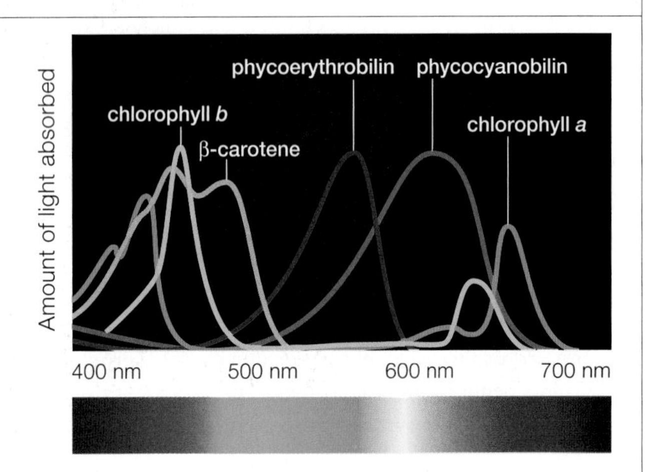

C Absorption spectra of chlorophylls *a* and *b*, β-carotene, and two phycobilins reveal the efficiency with which these pigments absorb different wavelengths of visible light. Line color indicates the characteristic color of each pigment.

FIGURE 6.3 {Animated} Discovery that photosynthesis is driven best by particular wavelengths of visible light.

FIGURE IT OUT: Which three pigments in **C** would you conclude are the main ones in the green algae tested in **B**?

Answer: Chlorophyll *a*, chlorophyll *b*, and β-carotene

In 1882, botanist Theodor Engelmann designed a set of experiments to test his hypothesis that the color of light affects the rate of photosynthesis. It had long been known that photosynthesis releases oxygen, so Engelmann used oxygen emission as an indirect measurement of photosynthetic activity. He directed a spectrum of light across individual strands of green algae suspended in water (**FIGURE 6.3A**). Oxygen-sensing equipment had not yet been invented, so Engelmann used motile, oxygen-requiring bacteria to show him where the oxygen concentration in the water was highest. The bacteria moved through the water and gathered mainly where blue and red light fell across the algal cells (**FIGURE 6.3B**). Engelmann concluded that photosynthetic cells illuminated by light of these colors were releasing the most oxygen—a sign that blue and red light are the best for driving photosynthesis in these algal cells.

Today we have equipment that can directly measure how efficiently a photosynthetic pigment absorbs different wavelengths of light. A graph that shows this efficiency is called an absorption spectrum. Peaks in the graph indicate wavelengths absorbed best (**FIGURE 6.3C**). Engelmann's results—where the bacteria clustered around the algal cells—represent the combined spectra of all the photosynthetic pigments present in the tested algae.

The combination of pigments used for photosynthesis differs among species. Why? Photosynthetic species are adapted to the environment in which they evolved, and light that reaches different environments varies in its proportions of wavelengths. Consider that seawater absorbs green and blue-green light less efficiently than other colors. Thus, more green and blue-green light penetrates deep ocean water. Algae that live in in this environment tend to have pigments—mainly phycobilins—that absorb green and blue-green light (*below*).

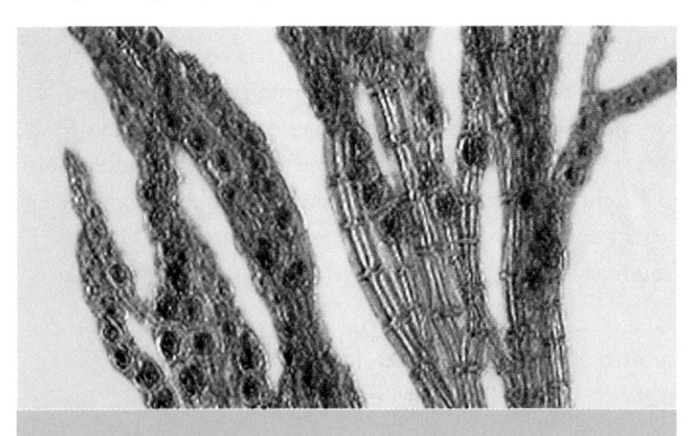

TAKE-HOME MESSAGE 6.2

A combination of pigments allows a photosynthetic organism to most efficiently capture the particular range of light wavelengths that reaches the habitat in which it evolved.

CREDITS: (3A) Jason Sonneman; (3B,C) © Cengage Learning; (in text) © Michael Davidson/The Florida State University.

6.3 WHAT HAPPENS DURING PHOTOSYNTHESIS?

All life is sustained by inputs of energy, but not all forms of energy can sustain life. Sunlight, for example, is abundant here on Earth, but it cannot be used to directly power protein synthesis or other energy-requiring reactions that keep organisms alive. Photosynthesis converts the energy of light into the energy of chemical bonds. Unlike light, chemical energy can power the reactions of life, and it can be stored for use at a later time.

In eukaryotes, photosynthesis takes place in chloroplasts (Section 4.8). Plant chloroplasts have two outer membranes, and they are filled with a thick, cytoplasm-like fluid called **stroma** (**FIGURE 6.4**). Suspended in the stroma are the chloroplast's own DNA, some ribosomes, and an inner, much-folded **thylakoid membrane**. The folds of a thylakoid membrane typically form stacks of interconnected disks called thylakoids. The space enclosed by the thylakoid membrane is a single, continuous compartment.

Photosynthesis is often summarized by this equation:

$$CO_2 + \text{water} \xrightarrow{\textit{light energy}} \text{sugars} + O_2$$

CO_2 is carbon dioxide, and O_2 is oxygen; both are gases abundant in the atmosphere. Keep in mind that photosynthesis is not a single reaction. Rather, it is a metabolic pathway (Section 5.4), a series of many reactions that occur in two stages. Molecules embedded in the thylakoid membrane carry out the reactions of the first stage, which are driven by light and thus called the **light-dependent reactions**. The "photo" in photosynthesis means light, and it refers to the conversion of light energy to the chemical bond energy of ATP during this stage. In addition to making ATP, the main light-dependent pathway in chloroplasts splits water molecules and releases O_2. Hydrogen ions and electrons from the water molecules end up in the coenzyme NADPH:

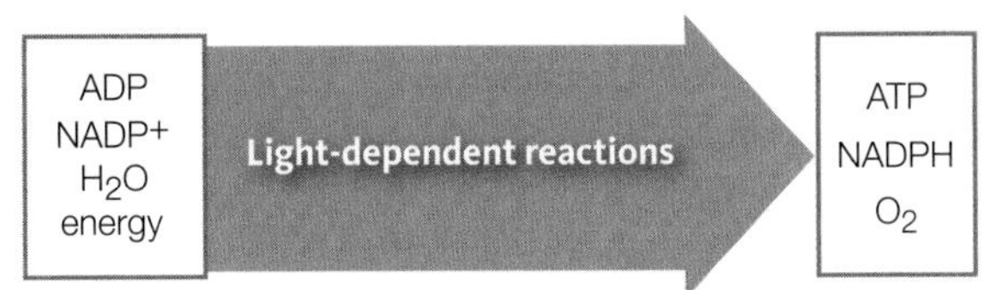

The "synthesis" part of photosynthesis refers to the reactions of the second stage, which build sugars from CO_2 and water. These sugar-building reactions run in the stroma. They are collectively called the **light-independent reactions** because light energy does not power them. Instead, they run on energy delivered by NADPH and ATP that formed during the first stage:

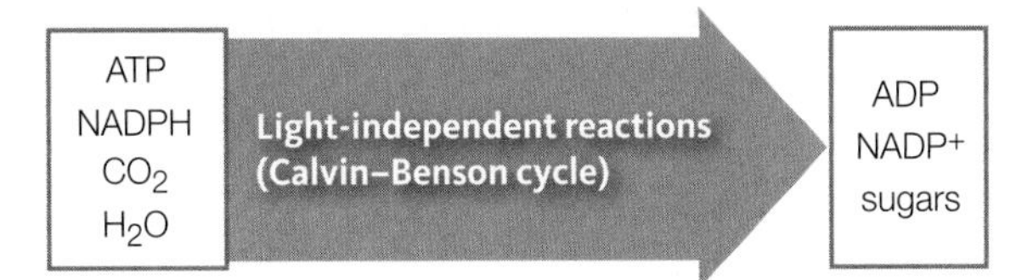

light-dependent reactions First stage of photosynthesis; convert light energy to chemical energy.
light-independent reactions Second stage of photosynthesis; use ATP and NADPH to assemble sugars from water and CO_2.
stroma The cytoplasm-like fluid between the thylakoid membrane and the two outer membranes of a chloroplast.
thylakoid membrane A chloroplast's highly folded inner membrane system; forms a continuous compartment in the stroma.

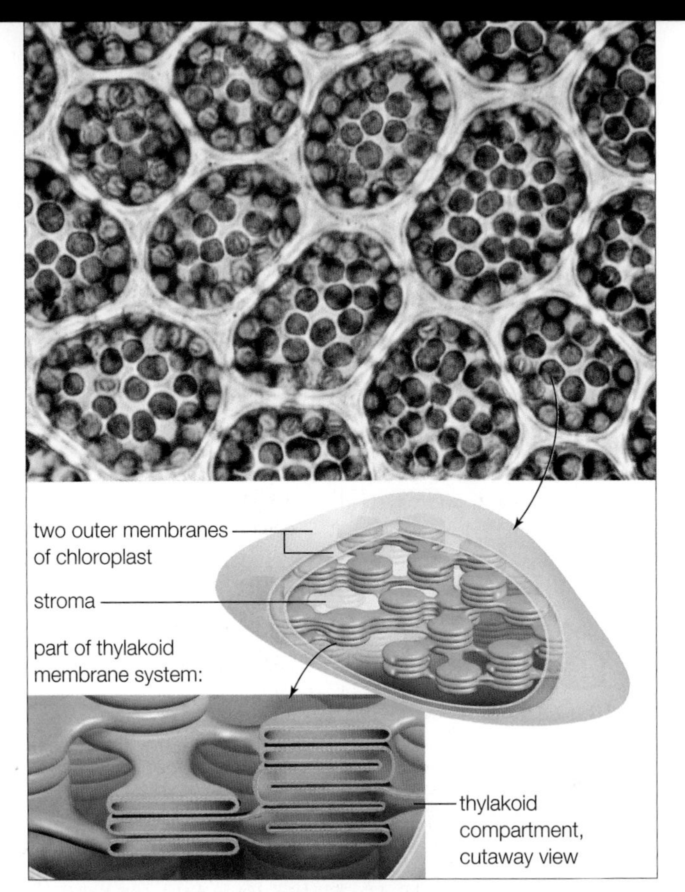

FIGURE 6.4 **{Animated}** Zooming in on the site of photosynthesis in a plant cell. The micrograph shows chloroplasts in cells of a moss leaf.

TAKE-HOME MESSAGE 6.3

In eukaryotic cells, the first stage of photosynthesis occurs at the thylakoid membrane of chloroplasts. During these light-dependent reactions, light energy drives the formation of ATP and NADPH.

In eukaryotic cells, the second stage of photosynthesis occurs in the stroma of chloroplasts. During these light-independent reactions, ATP and NADPH drive the synthesis of sugars from water and carbon dioxide.

6.4 HOW DO THE LIGHT-DEPENDENT REACTIONS WORK?

A chloroplast's thylakoid membrane contains millions of light-harvesting complexes, which are circular arrays of chlorophylls, various accessory pigments, and proteins (**FIGURE 6.5**). When a chlorophyll or accessory pigment in a light-harvesting complex absorbs light, one of its electrons jumps to a higher energy level, or shell (Section 2.2). The electron quickly drops back down to a lower shell by emitting its extra energy. Light-harvesting complexes hold on to that emitted energy by passing it back and forth, a bit like volleyball players pass a ball among team members. The reactions of photosynthesis begin when energy being passed around the thylakoid membrane reaches a photosystem. A **photosystem** is a group of hundreds of chlorophylls, accessory pigments, and other molecules that work as a unit to begin the reactions of photosynthesis.

FIGURE 6.5 A view of some components of the thylakoid membrane as seen from the stroma. Molecules of electron transfer chains and ATP synthases are also present, but not shown for clarity.

FIGURE 6.6 Summary of the inputs and outputs of the two pathways of light-dependent reactions.

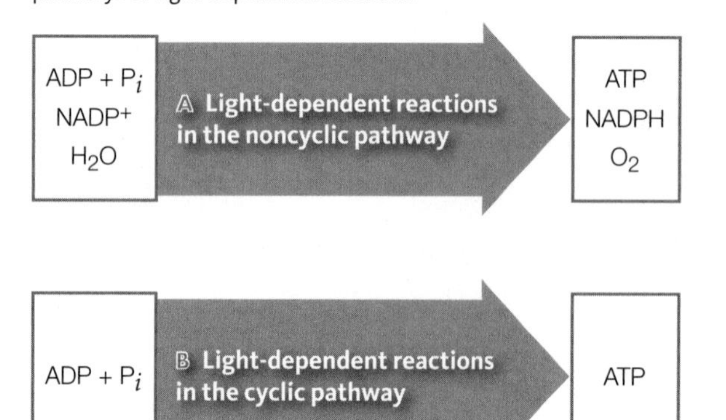

THE NONCYCLIC PATHWAY

Thylakoid membranes contain two kinds of photosystems, type I and type II, that were named in the order of their discovery. In cyanobacteria, plants, and all photosynthetic protists, both photosystem types work together in the noncyclic pathway of photosynthesis (**FIGURE 6.6A**). The pathway begins when energy being passed among light-harvesting complexes reaches a photosystem II (**FIGURE 6.7**). At the center of each photosystem are two very closely associated chlorophyll *a* molecules (a "special pair"). When a photosystem absorbs energy, electrons are ejected from its special pair ❶.

A photosystem that loses electrons must be restocked with more. Photosystem II restocks itself with electrons by pulling them off of water molecules in the thylakoid compartment. Pulling electrons off of water molecules causes them to split into hydrogen ions and oxygen atoms ❷. The oxygen atoms combine and diffuse out of the cell as oxygen gas (O_2). This and any other process by which a molecule is broken apart by light energy is called **photolysis**.

The actual conversion of light energy to chemical energy occurs when electrons ejected from photosystem II enter an electron transfer chain in the thylakoid membrane ❸. Remember that electron transfer chains can harvest the energy of electrons in a series of redox reactions, releasing a bit of their extra energy with each step (Section 5.4). In this case, molecules of the electron transfer chain use the released energy to actively transport hydrogen ions (H^+) across the membrane, from the stroma to the thylakoid compartment ❹. Thus, the flow of electrons through electron transfer chains sets up and maintains a hydrogen ion gradient across the thylakoid membrane.

The hydrogen ion gradient is a type of potential energy (Section 5.1) that can be tapped to make ATP. Hydrogen ions in the thylakoid compartment want to follow their concentration gradient by moving back into the stroma. However, ions cannot diffuse through lipid bilayers (Section 5.6). H^+ leaves the thylakoid compartment only by flowing through proteins called ATP synthases embedded in the thylakoid membrane ❼. An ATP synthase is both a transport protein and an enzyme. When hydrogen ions flow through its interior, the protein phosphorylates ADP, so ATP forms in the stroma ❽. The process by which the flow of electrons through electron transfer chains drives ATP formation is called **electron transfer phosphorylation**.

After the electrons have moved through the first electron transfer chain, they are accepted by a photosystem I. When this photosystem absorbs light energy, its special pair of chlorophylls emits electrons ❺. These electrons enter a second, different electron transfer chain. At the end of this

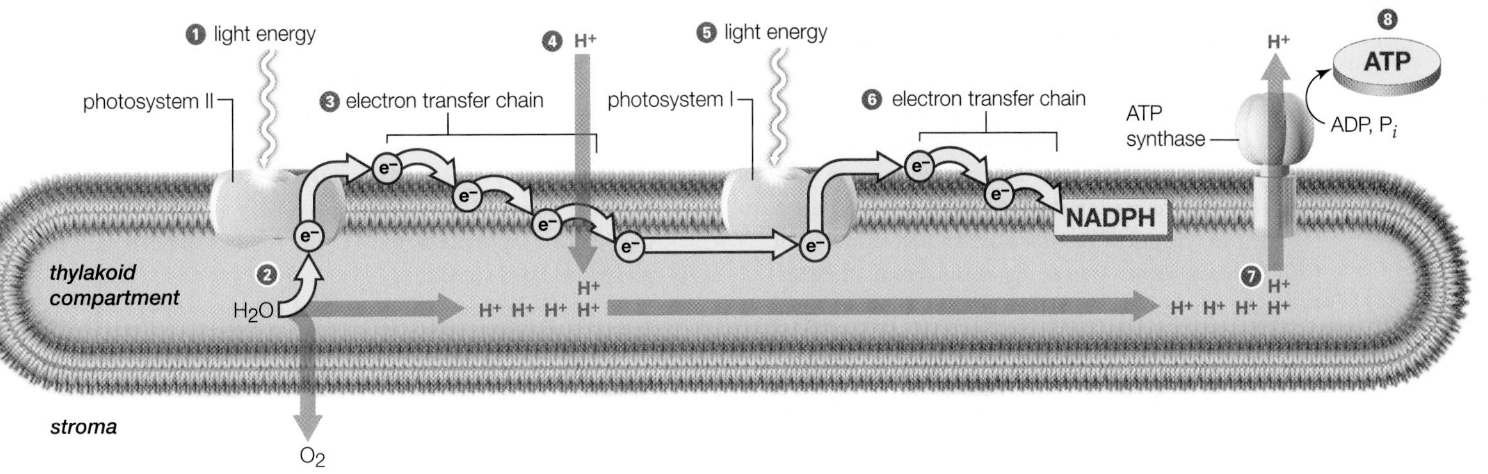

1 Light energy ejects electrons from a photosystem II.

2 The photosystem pulls replacement electrons from water molecules, which then break apart into oxygen and hydrogen ions. The oxygen leaves the cell as O_2.

3 The electrons enter an electron transfer chain in the thylakoid membrane.

4 Energy lost by the electrons as they move through the chain is used to actively transport hydrogen ions from the stroma into the thylakoid compartment. A hydrogen ion gradient forms across the thylakoid membrane.

5 Light energy ejects electrons from a photosystem I. Replacement electrons come from an electron transfer chain.

6 The ejected electrons move through a second electron transfer chain, then combine with $NADP^+$ and H^+, so NADPH forms.

7 Hydrogen ions in the thylakoid compartment are propelled through the interior of ATP synthases by their gradient across the thylakoid membrane.

8 Hydrogen ion flow causes ATP synthases to phosphorylate ADP, so ATP forms in the stroma.

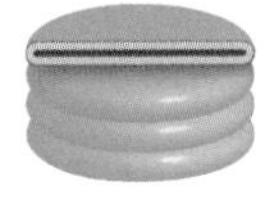

FIGURE 6.7 {Animated} Light-dependent reactions, noncyclic pathway. ATP and oxygen gas are produced in this pathway. Electrons that travel through two different electron transfer chains end up in NADPH.

chain, the coenzyme $NADP^+$ accepts the electrons along with H^+, so NADPH forms 6:

$$NADP^+ + 2e^- + H^+ \longrightarrow \textbf{NADPH}$$

NADPH is a powerful reducing agent (electron donor).

THE CYCLIC PATHWAY

As you will see in the next section, ATP and NADPH produced in the light-dependent reactions are used to make sugars. On its own, the noncyclic pathway does not yield enough ATP to balance NADPH use in sugar production pathways. The cyclic pathway produces additional ATP for this purpose (**FIGURE 6.6B**).

In the cyclic pathway, electrons that are ejected from photosystem I enter an electron transfer chain, and then return to photosystem I. As in the noncyclic pathway, the electron transfer chain uses electron energy to move hydrogen ions into the thylakoid compartment, and the resulting hydrogen ion gradient drives ATP formation. However, the cyclic pathway does not produce NADPH or oxygen gas.

The cyclic pathway also allows light-dependent reactions to continue when the noncyclic pathway stops, for example under intense illumination. Light energy in excess of what can be used for photosynthesis can result in the formation of dangerous free radicals (Section 2.2). A light-induced structural change in photosystem II prevents this from happening. The photosystem stops initiating the noncyclic pathway, and traps excess energy instead. At such times, the cyclic pathway predominates.

electron transfer phosphorylation Process in which electron flow through electron transfer chains sets up a hydrogen ion gradient that drives ATP formation.
photolysis Process by which light energy breaks down a molecule.
photosystem Cluster of pigments and proteins that converts light energy to chemical energy in photosynthesis.

TAKE-HOME MESSAGE 6.4

Photosynthetic pigments in the thylakoid membrane transfer the energy of light to photosystems, which eject electrons that enter electron transfer chains.

In both noncyclic and cyclic pathways, the flow of electrons through the transfer chains sets up hydrogen ion gradients that drive ATP formation.

In the noncyclic pathway, water molecules are split, oxygen is released, and electrons end up in NADPH.

In the cyclic pathway, no NADPH forms, and no oxygen is released.

6.5 HOW DO THE LIGHT-INDEPENDENT REACTIONS WORK?

THE CALVIN–BENSON CYCLE

The enzyme-mediated reactions of the **Calvin–Benson cycle** build sugars in the stroma of chloroplasts (**FIGURE 6.8**). These reactions are light-independent because light energy does not power them. Instead, they run on ATP and NADPH that formed in the light-dependent reactions.

Light-independent reactions use carbon atoms from CO_2 to make sugars. Extracting carbon atoms from an inorganic source and incorporating them into an organic molecule is a process called **carbon fixation**. In most plants, photosynthetic protists, and some bacteria, the enzyme **rubisco** fixes carbon by attaching CO_2 to RuBP (ribulose bisphosphate), a five-carbon molecule ❶.

FIGURE 6.8 {Animated} The Calvin–Benson cycle. This sketch shows a cross-section of a chloroplast with the reactions cycling in the stroma. The steps shown are a summary of six cycles of reactions (see Appendix III for details). Black balls are carbon atoms.

❶ Six CO_2 diffuse into a photosynthetic cell, and then into a chloroplast. Rubisco attaches each to a RuBP molecule. The resulting intermediates split, so twelve molecules of PGA form.

❷ Each PGA molecule gets a phosphate group from ATP, plus hydrogen and electrons from NADPH. Twelve PGAL form.

❸ Two PGAL may combine to form one six-carbon sugar (such as glucose).

❹ The remaining ten PGAL receive phosphate groups from ATP. The transfer primes them for endergonic reactions that regenerate the 6 RuBP.

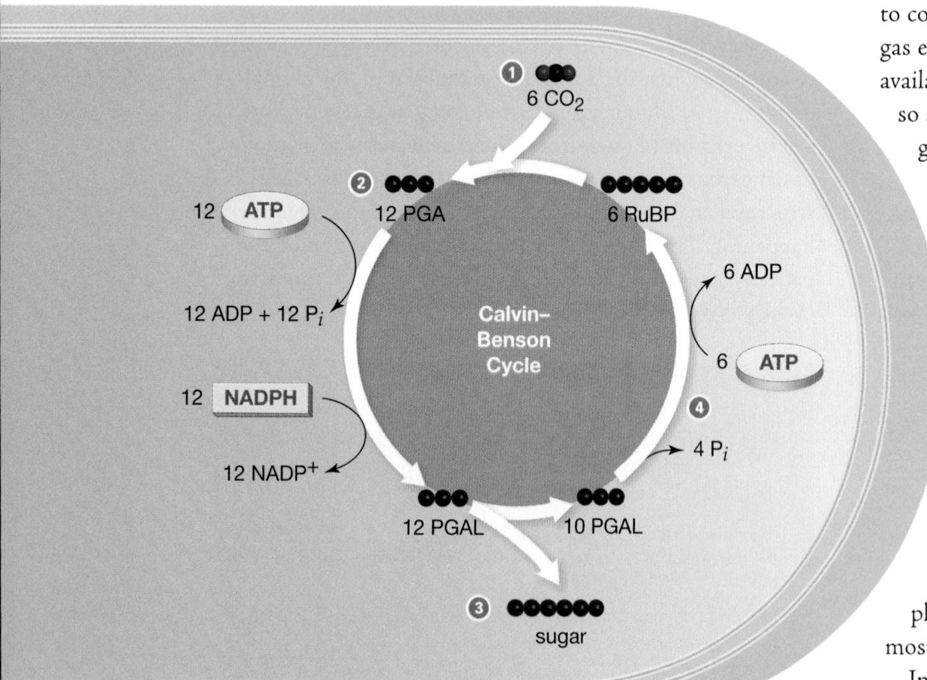

The six-carbon intermediate that forms by this reaction is unstable, so it splits right away into two three-carbon molecules of PGA (phosphoglycerate). Each PGA receives a phosphate group from ATP, and hydrogen and electrons from NADPH ❷. Thus, ATP energy and the reducing power of NADPH convert each molecule of PGA into a molecule of PGAL (phosphoglyceraldehyde), a phosphorylated sugar.

In later reactions, two or more of the three-carbon PGAL molecules can be combined and rearranged to form larger carbohydrates. Glucose, remember, has six carbon atoms. To make one glucose molecule, six CO_2 must be attached to six RuBP molecules, so twelve PGAL form. Two PGAL may combine to form one six-carbon glucose ❸. The ten remaining PGAL regenerate the starting compound of the cycle, RuBP ❹. Most of the glucose that a plant makes is converted at once to sucrose or starch by other pathways.

ADAPTATIONS TO CLIMATE

Mechanisms that help a plant prevent water loss also limit the gas exchange needed for photosynthesis. Most plants have a thin, waterproof cuticle that limits evaporation of water from their aboveground parts. Gases cannot diffuse across the cuticle, so the surfaces of leaves and stems are studded with tiny, closable gaps called stomata. Stomata are tiny gateways for gases. When they are open, CO_2 can diffuse from the air into photosynthetic tissues, and O_2 can diffuse out of the tissues into the air. Stomata close to conserve water on hot, dry days. When that happens, gas exchange comes to a halt. Closed stomata limit the availability of CO_2 for the light-independent reactions, so sugar synthesis slows. This detrimental effect is greatest in **C3 plants**, which fix carbon only by the Calvin–Benson cycle. They are called C3 plants because a three-carbon molecule (PGA) is the first stable intermediate in their light-independent reactions. When the CO_2 concentration inside a C3 plant declines, rubisco uses oxygen as a substrate in **photorespiration**, an alternate pathway that produces carbon dioxide. Thus, when a C3 plant closes stomata during the day, it loses carbon instead of fixing it (**FIGURE 6.9A**). In addition, ATP and NADPH are used to convert the pathway's intermediates to a molecule that can enter the Calvin–Benson cycle, so extra energy is required to make sugars. C3 plants compensate for photorespiration by making a lot of rubisco: It is the most abundant protein on Earth.

In many plant lineages, an additional set of reactions compensates for rubisco's inefficiency. Plants that use

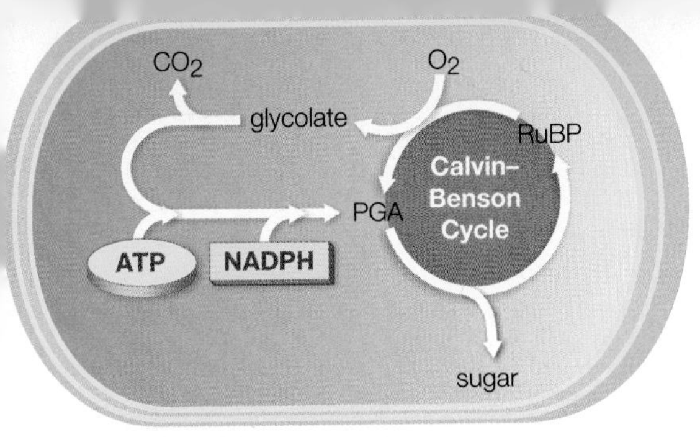

A When the CO_2 level declines in leaves of a C3 plant, rubisco uses oxygen as a substrate in photorespiration, and sugar production becomes inefficient.

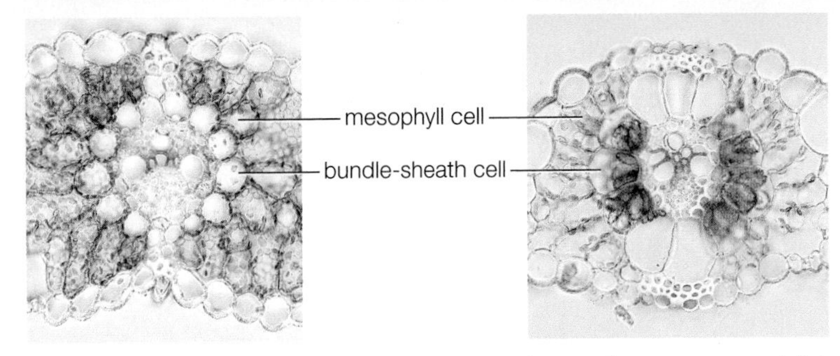

B In a C3 plant (barley, *left*), chloroplasts—the sites of carbon fixation—occur mainly in mesophyll cells. In a C4 plant (millet, *right*), carbon fixation occurs first in mesophyll cells, then in bundle-sheath cells. C4 adaptations maintain a high CO_2/O_2 ratio near rubisco.

FIGURE 6.9 {Animated} Anatomical and biochemical specializations minimize photorespiration in C4 plants. Micrographs show leaf cross sections.

these reactions also close stomata on dry days, but their sugar production does not decline. Examples include corn, switchgrass, and bamboo. We call these **C4 plants** because the first stable intermediate to form in their light-independent reactions is a four-carbon compound. C4 plants fix carbon twice, in two kinds of cells (**FIGURE 6.9B**). The first set of reactions occurs in mesophyll cells, where carbon is fixed by an enzyme that does not use oxygen even when the carbon dioxide level is low. The resulting intermediate is transported to bundle-sheath cells, where it is converted to CO_2. Rubisco fixes carbon for the second time as the CO_2 enters the Calvin–Benson cycle.

Bundle-sheath cells of C4 plants have chloroplasts that carry out light-dependent reactions, but only in the cyclic pathway. No oxygen is released, so the O_2 level near rubisco stays low. This, along with the high CO_2 level provided by the C4 reactions, minimizes photorespiration, so sugar production stays efficient in these plants even in hot, dry weather (**FIGURE 6.10**).

Succulents, cacti, and other **CAM plants** use a carbon-fixing pathway that allows them to conserve water even in desert regions with extremely high daytime temperatures. CAM stands for crassulacean acid metabolism, after the Crassulaceae family of plants in which this pathway was first studied. Like C4 plants, CAM plants fix carbon twice, but the reactions occur at different times rather than in different cells. Stomata on a CAM plant open at night, when typically lower temperatures minimize evaporative water loss. The plants fix carbon from CO_2 in the air at this time. The product of the cycle, a four-carbon acid, is stored in the cell's central vacuole. When the stomata close the next day, the acid moves out of the vacuole and becomes broken down to CO_2, which is fixed for the second time when it enters the Calvin–Benson cycle.

FIGURE 6.10 Crabgrass "weeds" overgrowing a lawn. Crabgrasses, which are C4 plants, thrive in hot, dry summers, when they easily outcompete Kentucky bluegrass and other fine-leaved C3 grasses commonly planted in residential lawns.

C3 plant Type of plant that uses only the Calvin–Benson cycle to fix carbon.
C4 plant Type of plant that minimizes photorespiration by fixing carbon twice, in two cell types.
Calvin–Benson cycle Cyclic carbon-fixing pathway that builds sugars from CO_2; the light-independent reactions of photosynthesis.
CAM plant Type of plant that conserves water by fixing carbon twice, at different times of day.
carbon fixation Process by which carbon from an inorganic source such as carbon dioxide gets incorporated into an organic molecule.
photorespiration Reaction in which rubisco attaches oxygen instead of carbon dioxide to ribulose bisphosphate.
rubisco Ribulose bisphosphate carboxylase. Carbon-fixing enzyme of the Calvin–Benson cycle.

TAKE-HOME MESSAGE 6.5

The light-independent reactions build sugars using carbon fixed from CO_2. They run on ATP and electrons from NADPH.

C3 plants use only the Calvin–Benson cycle. In these plants, photorespiration on hot, dry days reduces the efficiency of sugar production, so it limits growth.

Plants adapted to hot, dry conditions minimize photorespiration by fixing carbon twice. C4 plants separate the two sets of reactions in space; CAM plants separate them in time.

6.6 Application: GREEN ENERGY

FIGURE 6.11 Switchgrass growing wild in a North American prairie.

TODAY, THE EXPRESSION "FOOD IS FUEL" IS NOT JUST ABOUT EATING. With fossil fuel prices soaring, there is an increasing demand for biofuels, which are oils, gases, or alcohols made from organic matter that is not fossilized. Most materials we use for biofuel production today consist of food crops—mainly corn, soybeans, and sugarcane. Growing these crops in large quantities is typically damaging to the environment, and using them to make biofuel competes with our food supply.

How did we end up competing with our vehicles for food? We both run on the same fuel: energy that plants have stored in chemical bonds. Fossil fuels such as petroleum, coal, and natural gas formed from the remains of ancient swamp forests that decayed and compacted over millions of years. These fuels consist of molecules originally assembled by ancient plants. Biofuels—and foods—consist mainly of molecules originally assembled by modern plants.

A lot of energy is locked up in the chemical bonds of molecules made by plants. That energy can fuel heterotrophs, as when an animal cell powers ATP synthesis by breaking the bonds of sugars (a topic detailed in the next chapter). It can also fuel our cars, which run on energy released by burning biofuels or fossil fuels. Both processes are fundamentally the same: They release energy by breaking the bonds of organic molecules. Both use oxygen to break those bonds, and both produce carbon dioxide.

Photosynthesis removes carbon dioxide from the atmosphere, and fixes its carbon atoms in organic compounds. When we burn fossil fuels, the carbon that has been sequestered in them for hundreds of millions of years is released back into the atmosphere, mainly as CO_2 that reenters the atmosphere. Our extensive use of fossil fuels has put Earth's atmospheric cycle of carbon dioxide out of balance: We are adding far more CO_2 to the atmosphere than photosynthetic organisms are removing from it. Atmospheric carbon dioxide affects Earth's climate, so this increase in CO_2 is contributing to global climate change.

CREDIT: (11) Photo by Peggy Greb/USDA.

Ratna Sharma and Mari Chinn researching ways to reduce the cost of producing biofuel from renewable sources such as wild grasses and agricultural waste.

Unlike fossil fuels, biofuels are a renewable source of energy: We can always make more of them simply by growing more plants. Also unlike fossil fuels, biofuels do not contribute to global climate change, because growing plant matter for fuel recycles carbon that is already in the atmosphere.

Corn and other food crops are rich in oils, starches, and sugars that can be easily converted to biofuels. The starch in corn kernels, for example, can be enzymatically broken down to glucose, which is converted to ethanol by heterotrophic bacteria or yeast. Making biofuels from other types of plant matter requires additional steps, because these materials contain a higher proportion of cellulose. Breaking down this tough, insoluble carbohydrate to its glucose monomers adds a lot of cost to the biofuel product. Researchers are currently working on cost-effective ways to break down the abundant cellulose in fast-growing weeds such as switchgrass (FIGURE 6.11), and agricultural wastes such as wood chips, wheat straw, cotton stalks, and rice hulls.

PEOPLE MATTER

National Geographic Grantee
WILLIE SMITS

One of Indonesia's most ardent rain forest protection activists is in what may seem an unlikely position: spearheading a project to produce biofuel from trees. But tropical forest scientist Willie Smits, after 30 years of studying fragile ecosystems in these Southeast Asian islands, wants to draw world attention to a powerhouse of a tree—the Arenga sugar palm, *Arenga pinnata*. Smits says that this deep-rooted palm could serve as the core of a waste-free system that produces a premium organic sugar as well as ethanol for fuel, providing food products and jobs to villagers while it helps preserve the existing native rain forest. Scientists who have studied the unique harvesting and production process developed by Smits agree the system would protect the atmosphere rather than add to Earth's growing carbon dioxide burden. "The palm juice chiefly consists of water and sugar—made from rain, sunshine, carbon dioxide, and nothing else," says Smits. "You are basically only harvesting sunshine." The project, being funded in part by a grant from National Geographic's Great Energy Challenge initiative, has the potential to disrupt a cycle of poverty and environmental devastation that has gripped one of the most vulnerable and remote areas of the planet, while providing a new source of sustainable fuel.

Summary

SECTIONS 6.1, 6.2 Plants and other **autotrophs** make their own food using energy from the environment and carbon from inorganic sources such as CO_2. **Heterotrophs** get carbon from molecules that other organisms have already assembled. Visible light drives photosynthesis, which begins when light is absorbed by photosynthetic pigments. **Pigments** absorb light of particular **wavelengths** only; wavelengths not captured are reflected as its characteristic color. The main photosynthetic pigment, **chlorophyll *a***, absorbs violet and red light, so it appears green. Accessory pigments absorb additional wavelengths.

SECTION 6.3 In chloroplasts, the **light-dependent reactions** of photosynthesis occur at a much-folded **thylakoid membrane**. The membrane forms a compartment in the chloroplast's interior (**stroma**), in which the **light-independent reactions** occur:

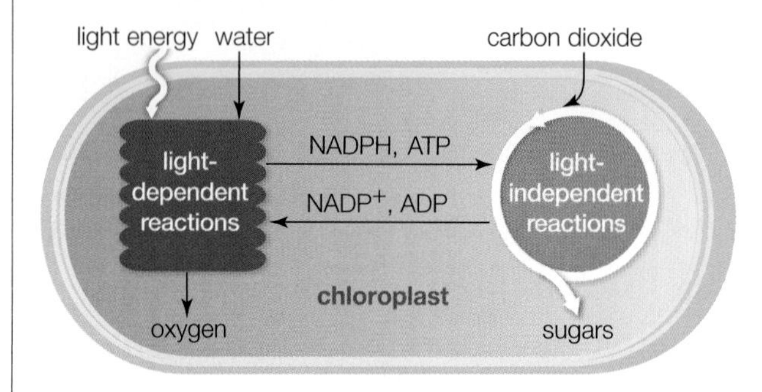

SECTION 6.4 In the light-dependent reactions, light-harvesting complexes in the thylakoid membrane absorb photons and pass the energy to **photosystems**. Receiving energy causes photosystems to release electrons.

In the noncyclic pathway, electrons released from photosystem II flow through an electron transfer chain, then to photosystem I. Photosystem II replaces lost electrons by pulling them from water, which then splits into H^+ and O_2 (an example of **photolysis**). Electrons released from photosystem I end up in NADPH.

In the cyclic pathway, electrons released from photosystem I enter an electron transfer chain, then cycle back to photosystem I. NADPH does not form and O_2 is not released.

ATP forms by **electron transfer phosphorylation** in both pathways. Electrons flowing through electron transfer chains cause H^+ to accumulate in the thylakoid compartment. The H^+ follows its gradient back across the membrane through ATP synthases, driving ATP synthesis.

SECTION 6.5 **Carbon fixation** occurs in light-independent reactions. Inside the stroma, the enzyme **rubisco** attaches a carbon from CO_2 to RuBP to start the **Calvin–Benson cycle**. This cyclic pathway uses energy from ATP, carbon and oxygen from CO_2, and hydrogen and electrons from NADPH to make sugars.

On hot, dry days, plants conserve water by closing stomata, so CO_2 for light-independent reactions cannot enter their tissues. In **C3 plants**, the resulting low CO_2 level causes **photorespiration**, which reduces the efficiency of sugar production. Other types of plants minimize photorespiration by fixing carbon twice, thus keeping the CO_2 level high and the O_2 level low near rubisco. **C4 plants** carry out the two sets of reactions in different cell types; **CAM plants** carry them out at different times.

SECTION 6.6 Autotrophs remove CO_2 from the atmosphere; the metabolic activity of most organisms puts it back. Humans disrupt this cycle by burning fossil fuels, which adds extra CO_2 to the atmosphere. The resulting imbalance is contributing to global warming.

Self-Quiz Answers in Appendix VII

1. A cat eats a bird, which ate a caterpillar that chewed on a weed. Which of these organisms are autotrophs? Which ones are heterotrophs?

2. Which of the following statements is incorrect?
   a. Pigments absorb light of certain wavelengths only.
   b. Many accessory pigments are multipurpose molecules.
   c. Chlorophyll is green because it absorbs green light.

3. In plants, the light-dependent reactions proceed at/in the ______ .
   a. thylakoid membrane
   b. plasma membrane
   c. stroma
   d. cytoplasm

4. When a photosystem absorbs light, ______ .
   a. sugar phosphates are produced
   b. electrons are transferred to ATP
   c. RuBP accepts electrons
   d. electrons are ejected from its special pair

5. In the light-dependent reactions, ______ .
   a. carbon dioxide is fixed
   b. ATP forms
   c. sugars form
   d. CO_2 accepts electrons
   e. b and c
   f. a and c

6. What accumulates inside the thylakoid compartment during the light-dependent reactions?
   a. sugars b. hydrogen ions c. O_2 d. CO_2

7. The atoms in the molecular oxygen released during photosynthesis come from split ______ molecules.
   a. sugar
   b. CO_2
   c. water
   d. O_2

8. Light-independent reactions in plants proceed at/in the ______ of chloroplasts.
   a. thylakoid membrane
   b. plasma membrane
   c. stroma
   d. cytoplasm

Data Analysis Activities

Energy Efficiency of Biofuel Production Most material currently used for biofuel production in the United States consists of food crops—mainly corn, soybeans, and sugarcane. In 2006, David Tilman and his colleagues published the results of a 10-year study comparing the net energy output of various biofuels. The researchers grew a mixture of native perennial grasses without irrigation, fertilizer, pesticides, or herbicides, in sandy soil that was so depleted by intensive agriculture that it had been abandoned. They measured the usable energy in biofuels made from the grasses, from corn, and from soy. They also measured the energy it took to grow and produce each kind of biofuel (**FIGURE 6.12**).

1. About how much energy did ethanol produced from one hectare of corn yield? How much energy did it take to grow the corn to make that ethanol?
2. Which of the biofuels tested had the highest ratio of energy output to energy input?
3. Which of the three crops would require the least amount of land to produce a given amount of biofuel energy?

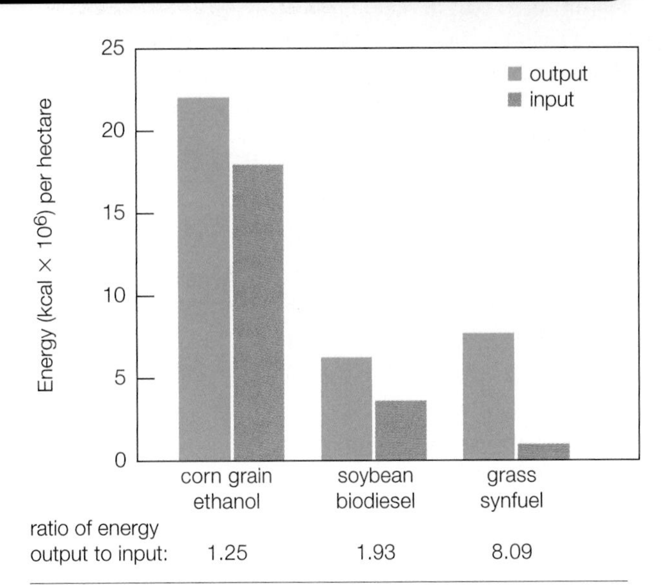

FIGURE 6.12 Energy inputs and outputs of biofuels from different sources. One hectare is about 2.5 acres.

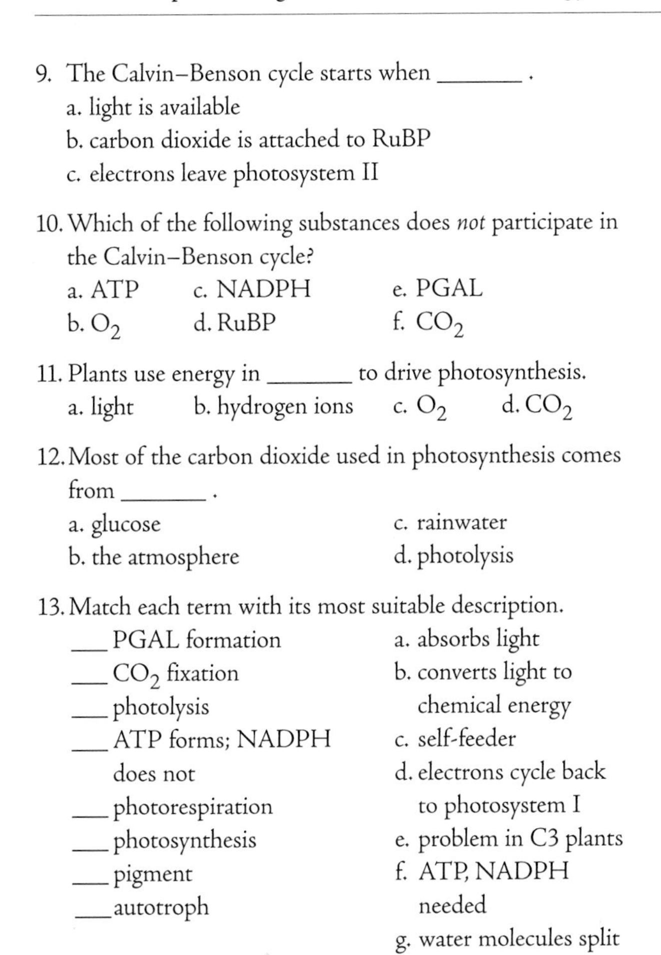

9. The Calvin–Benson cycle starts when ________ .
   a. light is available
   b. carbon dioxide is attached to RuBP
   c. electrons leave photosystem II

10. Which of the following substances does *not* participate in the Calvin–Benson cycle?
    a. ATP c. NADPH e. PGAL
    b. O_2 d. RuBP f. CO_2

11. Plants use energy in ________ to drive photosynthesis.
    a. light b. hydrogen ions c. O_2 d. CO_2

12. Most of the carbon dioxide used in photosynthesis comes from ________ .
    a. glucose c. rainwater
    b. the atmosphere d. photolysis

13. Match each term with its most suitable description.

___ PGAL formation	a. absorbs light
___ CO_2 fixation	b. converts light to chemical energy
___ photolysis	c. self-feeder
___ ATP forms; NADPH does not	d. electrons cycle back to photosystem I
___ photorespiration	e. problem in C3 plants
___ photosynthesis	f. ATP, NADPH needed
___ pigment	g. water molecules split
___ autotroph	h. rubisco function

Critical Thinking

1. About 200 years ago, Jan Baptista van Helmont wanted to know where growing plants get the materials necessary for increases in size. He planted a tree seedling weighing 5 pounds in a barrel filled with 200 pounds of soil and then watered the tree regularly. After five years, the tree weighed 169 pounds, 3 ounces, and the soil weighed 199 pounds, 14 ounces. Because the tree had gained so much weight and the soil had lost so little, he concluded that the tree had gained all of its additional weight by absorbing the water he had added to the barrel, but of course he was incorrect. What really happened?

2. While gazing into an aquarium, you observe bubbles coming from an aquatic plant (*left*). What are the bubbles and where do they come from?

3. A C3 plant absorbs a carbon radioisotope (as part of $^{14}CO_2$). In which stable, organic compound does the labeled carbon appear first?

4. As you learned in this chapter, cell membranes are required for electron transfer phosphorylation. Thylakoid membranes in chloroplasts serve this purpose in photosynthetic eukaryotes. Prokaryotic cells do not have this organelle, but many are photosynthesizers. How do you think they carry out the light-dependent reactions, given that they have no chloroplasts?

Like you, a whale breathes air to provide its cells with a fresh supply of oxygen for aerobic respiration. Carbon dioxide released from aerobically respiring cells leaves the body in each exhalation.

7 HOW CELLS RELEASE CHEMICAL ENERGY

Links to Earlier Concepts

This chapter focuses on metabolic pathways (Section 5.4) that harvest energy (5.1) stored in the chemical bonds of sugars (3.2). Some reactions (3.1) of these pathways occur in mitochondria (4.7). You will revisit free radicals (2.2), lipids (3.3), proteins (3.4), electron transfer chains (5.4), coenzymes (5.5), membrane transport (5.6, 5.7), and photosynthesis (6.4).

KEY CONCEPTS

ENERGY FROM SUGARS

Most cells can make ATP by breaking down sugars in either aerobic respiration or anaerobic fermentation pathways. Aerobic respiration yields the most ATP.

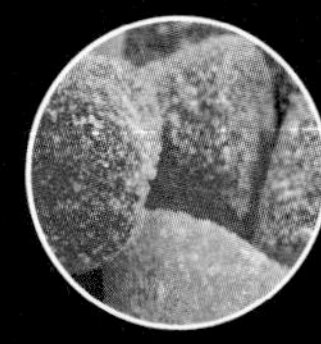

GLYCOLYSIS

Aerobic respiration and fermentation start in the cytoplasm with glycolysis, a pathway that splits glucose into two pyruvate molecules and yields two ATP.

AEROBIC RESPIRATION

Eukaryotes break down pyruvate to CO_2 in mitochondria. Many coenzymes are reduced; these deliver electrons and hydrogen ions to electron transfer chains that drive ATP formation.

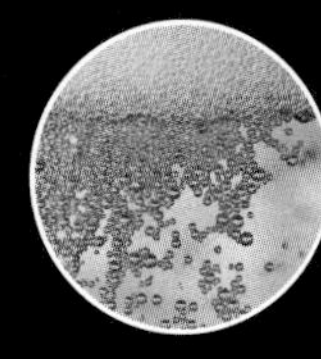

FERMENTATION

Fermentation ends in the cytoplasm, where organic molecules accept electrons from pyruvate. The net yield of ATP is small compared with that from aerobic respiration.

OTHER METABOLIC PATHWAYS

Cells also make ATP by breaking down molecules other than sugars. Dietary lipids and proteins are converted to molecules that enter glycolysis or another step in the aerobic respiration pathway.

Photograph by Ralph Lee Hopkins, National Geographic Creative.

7.1 HOW DO CELLS ACCESS THE CHEMICAL ENERGY IN SUGARS?

Photosynthetic organisms capture energy from the sun and store it in the form of sugars. They and most other organisms use energy stored in sugars to run the diverse reactions that sustain life. However, sugars rarely participate in such reactions, so how do cells harness their energy? In order to use the energy stored in sugars, cells must first transfer it to molecules—ATP in particular—that do participate in energy-requiring reactions. The energy transfer occurs when cells break the bonds of a sugar's carbon backbone. Energy released as those bonds are broken drives ATP synthesis. The two main mechanisms by which organisms break down sugars to make ATP are aerobic respiration and fermentation.

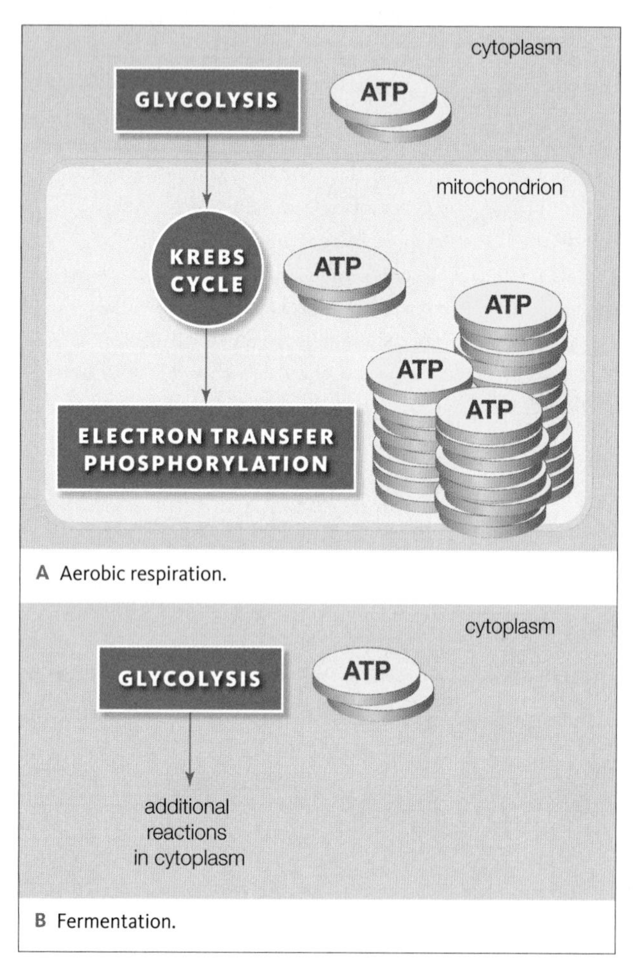

FIGURE 7.1 {Animated} Comparison of aerobic respiration and fermentation.

FIGURE IT OUT: Which pathway produces more ATP?

Answer: Aerobic respiration

aerobic respiration Oxygen-requiring metabolic pathway that breaks down sugars to produce ATP.

fermentation A metabolic pathway that breaks down sugars to produce ATP and does not require oxygen.

AEROBIC RESPIRATION AND FERMENTATION COMPARED

Aerobic respiration is an oxygen-requiring metabolic pathway that breaks down sugars to make ATP. It is the main energy-releasing pathway in nearly all eukaryotes and some bacteria. Aerobic respiration occurs in three stages (**FIGURE 7.1A**). The first stage, glycolysis, is a linear pathway that occurs in cytoplasm. Glycolysis begins the breakdown of one sugar molecule for a net yield of 2 ATP. In eukaryotes, the next two stages occur in mitochondria. The second stage, the Krebs cycle, completes the breakdown of the sugar molecule to CO_2. This cyclic pathway produces 2 ATP and reduces many coenzymes. In the third stage, electron transfer phosphorylation (Section 6.4), coenzymes reduced during glycolysis and the Krebs cycle deliver electrons and hydrogen ions to electron transfer chains. Energy released by electrons as they move through the chains drives the formation of as many as 32 ATP. Water forms when oxygen accepts hydrogen ions and electrons at the end of the electron transfer chains. Aerobic respiration, which means "taking a breath of air," refers to this pathway's requirement for oxygen as the final acceptor of electrons.

Fermentation refers to sugar breakdown pathways that do not require oxygen to make ATP (**FIGURE 7.1B**). Like aerobic respiration, fermentation begins with glycolysis in cytoplasm. Unlike aerobic respiration, fermentation occurs entirely in cytoplasm, and does not include electron transfer chains. The reactions that conclude fermentation produce no additional ATP, and the final acceptor of electrons is an organic molecule (not oxygen).

Aerobic respiration produces about 36 ATP per sugar molecule; fermentation produces only 2. Fermentation provides enough ATP to sustain many single-celled species. It also helps cells of multicelled species produce ATP under anaerobic conditions, but aerobic respiration is a much more efficient way of harvesting energy from sugars. You and other large, multicelled organisms could not live without its higher yield.

TAKE-HOME MESSAGE 7.1

- Most cells can make ATP by breaking down sugars, either in aerobic respiration or fermentation.
- Aerobic respiration and fermentation begin in cytoplasm.
- Fermentation does not require oxygen and ends in cytoplasm.
- Aerobic respiration requires oxygen and, in eukaryotes, it ends in mitochondria. This pathway yields much more ATP than fermentation.

7.2 HOW DID ENERGY-RELEASING PATHWAYS EVOLVE?

PEOPLE MATTER

National Geographic Grantee
DR. J. WILLIAM SCHOPF

For the first 85 percent of its history, Earth was populated by what was essentially pond scum. In 1987, National Geographic grantee J. William Schopf discovered what may be evidence of cells 3.465 billion years old, opening the floodgates to decades of controversy—and research that is filling in the holes in our understanding of how and when life evolved.

The first cells we know of appeared on Earth about 3.4 billion years ago. Like some modern prokaryotes, these ancient organisms did not tap into sunlight: They extracted the energy they needed from simple molecules such as methane and hydrogen sulfide. When the cyclic pathway of photosynthesis first evolved (Section 6.4), sunlight offered cells that used it an essentially unlimited supply of energy. Shortly afterward, this pathway became modified. The new noncyclic pathway split water molecules into hydrogen and oxygen. Cells that used the pathway were very successful. Oxygen gas (O_2) released from uncountable numbers of water molecules began seeping out of photosynthetic prokaryotes. The gas reacts easily with metals, so at first, most of it combined with metal atoms in exposed rocks. After the exposed minerals became saturated with oxygen, the gas began to accumulate in the ocean and the atmosphere. From that time on, the world of life would never be the same.

Before photosynthesis evolved, molecular oxygen had been a very small component of Earth's early atmosphere. In what may have been the earliest case of catastrophic pollution, the new abundance of this gas exerted tremendous pressure on all life at the time. Why? Then, like now, enzymes that require metal cofactors were a critical part of metabolism. Oxygen reacts with metal cofactors, and free radicals (Section 2.2) form during those reactions. Free radicals damage DNA and other biological molecules, so they are dangerous to life. Cells with no way to cope with them quickly died out. Only a few lineages persisted in deep water, muddy sediments, and other **anaerobic** (oxygen-free) habitats.

Antioxidants evolved in the survivors. Cells that made these molecules were the first **aerobic** organisms—they could live in the presence of oxygen. As they evolved, their antioxidant molecules became incorporated into new metabolic pathways that put oxygen's reactive properties to use. One of the new pathways was aerobic respiration. This pathway requires oxygen, and it produces carbon dioxide—the raw materials from which photosynthetic organisms make sugars. It also combines molecular oxygen with hydrogen ions and electrons—exactly the reverse of the reaction that splits water during photosynthesis:

$$O_2 + 4e^- + 4H^+ \longrightarrow 2H_2O$$

With this connection, the cycling of carbon, hydrogen, and oxygen through living things came full circle (*left*).

energy
PHOTOSYNTHESIS
CO_2
H_2O
sugar
O_2
AEROBIC RESPIRATION
energy

aerobic Involving or occurring in the presence of oxygen.
anaerobic Occurring in the absence of oxygen.

TAKE-HOME MESSAGE 7.2

Molecular oxygen produced by early photosynthetic prokaryotes put severe pressure on all life.

The evolution of antioxidants allowed the reactive properties of oxygen to be put to use in new metabolic pathways.

Today, carbon dioxide, water, sugar, and oxygen cycle through the world of life via photosynthesis (energy capture) and aerobic respiration (energy release).

7.3 WHAT IS GLYCOLYSIS?

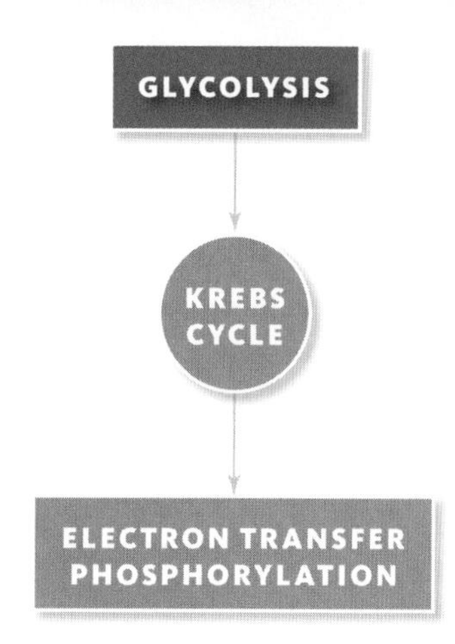

Glycolysis is a series of reactions that begin the sugar breakdown pathways of aerobic respiration (*above*) and fermentation. The reactions of glycolysis, which occur with some variation in the cytoplasm of almost all cells, convert one six-carbon molecule of sugar (such as glucose) into two molecules of **pyruvate**, an organic compound with a three-carbon backbone:

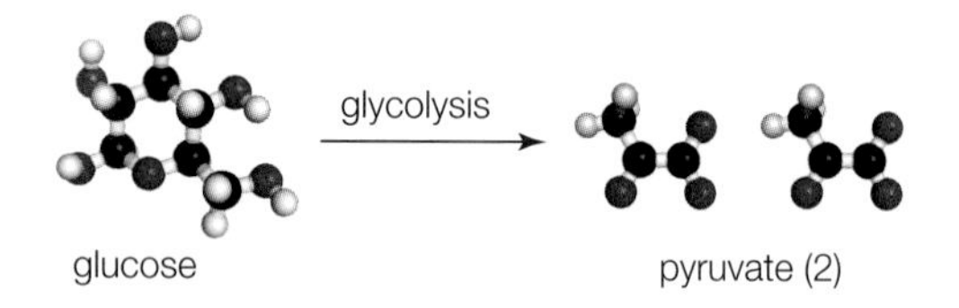

The word glycolysis comes from two Greek words: *glyk*–, sweet, and –*lysis*, loosening; it refers to the release of chemical energy from sugars. Other six-carbon sugars such as fructose and galactose can enter glycolysis, but for clarity we focus here on glucose.

Glycolysis begins when a molecule of glucose enters a cell through a glucose transporter, a passive transport protein you encountered in Section 5.7. The cell invests two ATP in the endergonic reactions that begin the pathway (**FIGURE 7.2**). In the first reaction, a phosphate group is transferred from ATP to the glucose, thus forming glucose-6-phosphate ❶. A model of hexokinase, the enzyme that catalyzes this reaction, is pictured in Section 5.3.

Glycolysis continues as the glucose-6-phosphate accepts a phosphate group from another ATP, then splits ❷ to form two PGAL (phosphoglyceraldehyde). Remember from Section 6.5 that this phosphorylated sugar also forms during the Calvin–Benson cycle.

glycolysis Set of reactions in which a six-carbon sugar (such as glucose) is converted to two pyruvate for a net yield of two ATP.
pyruvate Three-carbon end product of glycolysis.
substrate-level phosphorylation The formation of ATP by the direct transfer of a phosphate group from a substrate to ADP.

FIGURE 7.2 {Animated} Glycolysis (*opposite*).

For clarity, we track only the six carbon atoms (black balls) that enter the reactions as part of glucose. Cells invest two ATP to start glycolysis, so the net yield from one glucose molecule is two ATP. Two NADH also form, and two pyruvate molecules are the end products. Appendix III has more details for interested students.

In the next reaction, each PGAL receives a second phosphate group, and each gives up two electrons and a hydrogen ion. Two molecules of PGA (phosphoglycerate) form as products of this reaction ❸. The electrons and hydrogen ions are accepted by two NAD^+, which thereby become reduced to NADH. Aerobic respiration's third stage requires this NADH, as does fermentation.

Next, a phosphate group is transferred from each PGA to ADP, so two ATP form ❹. The direct transfer of a phosphate group from a substrate to ADP is called **substrate-level phosphorylation**. Substrate-level phosphorylation is a completely different process from the way ATP forms during electron transfer phosphorylation (Section 6.4).

Glycolysis ends with the formation of two more ATP by substrate-level phosphorylation ❺. Remember, two ATP were invested to begin the reactions of glycolysis. A total of four ATP form, so the net yield is two ATP per molecule of glucose ❻. The pathway also produces two three-carbon pyruvate molecules. Pyruvate is a substrate for the second-stage reactions of aerobic respiration, and also for fermentation reactions.

TAKE-HOME MESSAGE 7.3

Glycolysis is the first stage of sugar breakdown in both aerobic respiration and fermentation.

The reactions of glycolysis occur in the cytoplasm.

Glycolysis converts one molecule of glucose to two molecules of pyruvate, with a net yield of two ATP. Two NADH also form.

GLYCOLYSIS

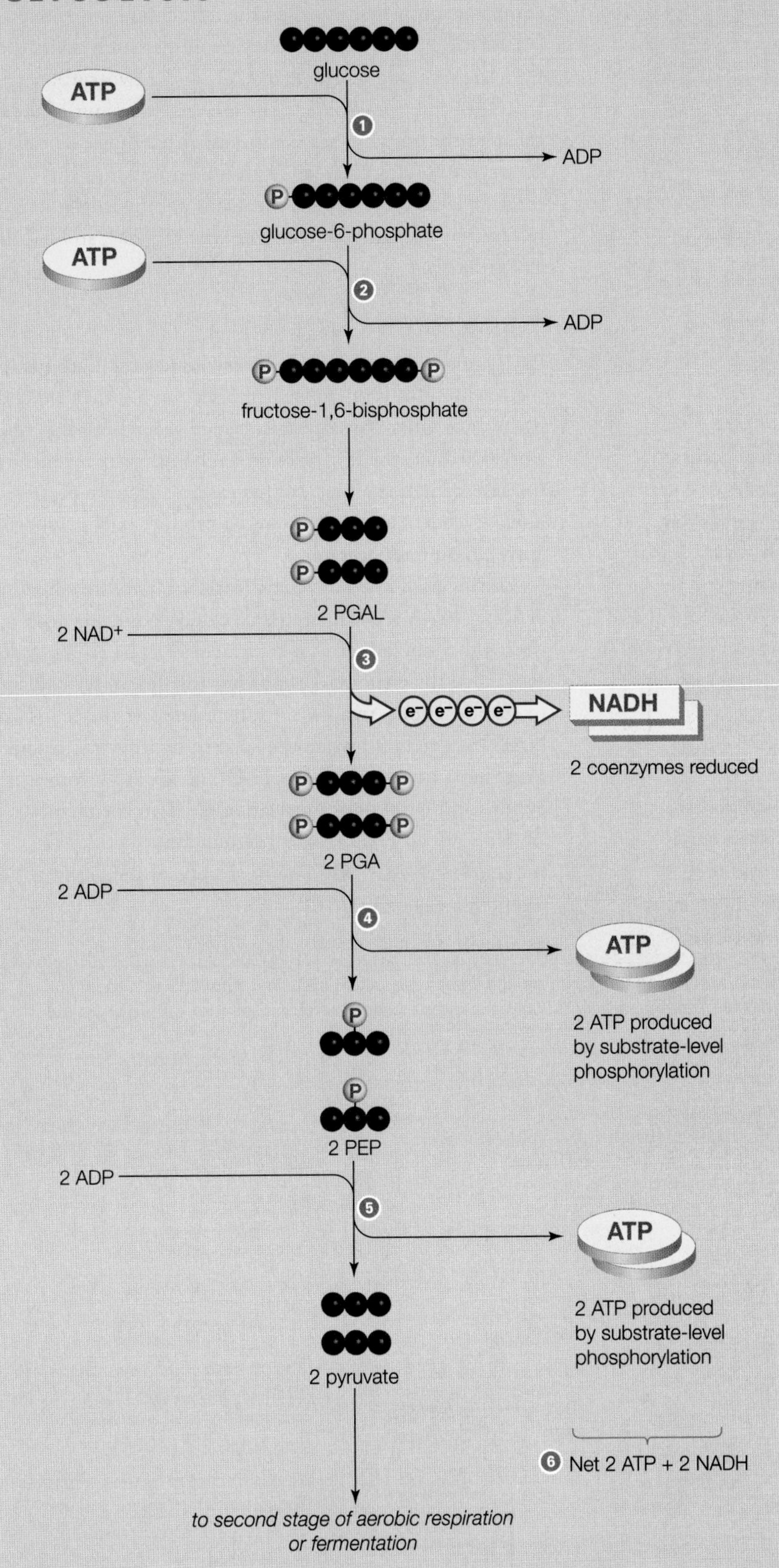

ATP-Requiring Steps

1. A phosphate group is transferred from ATP to glucose, forming glucose-6-phosphate. (You learned about the enzyme that catalyzes this reaction, hexokinase, in **FIGURE 5.10** and Section 5.7).

2. A phosphate group from a second ATP is transferred to the glucose-6-phosphate. The resulting molecule is unstable, and it splits into two three-carbon molecules. The molecules are interconvertible, so we will call them both PGAL (phosphoglyceraldehyde).

Two ATP have now been invested in the reactions.

ATP-Generating Steps

3. An enzyme attaches a phosphate to the two PGAL, so two PGA (phosphoglycerate) form. Two electrons and a hydrogen ion (not shown) from each PGAL are accepted by NAD^+, so two NADH form.

4. An enzyme transfers a phosphate group from each PGA to ADP, forming two ATP and two intermediate molecules (PEP).

The original energy investment of two ATP has now been recovered.

5. An enzyme transfers a phosphate group from each PEP to ADP, forming two more ATP and two molecules of pyruvate.

6. Summing up, glycolysis yields two NADH, two ATP (net), and two pyruvate for each glucose molecule.

Depending on the type of cell and environmental conditions, the pyruvate may enter the second stage of aerobic respiration or it may be used in other ways, such as in fermentation.

7.4 WHAT HAPPENS DURING THE SECOND STAGE OF AEROBIC RESPIRATION?

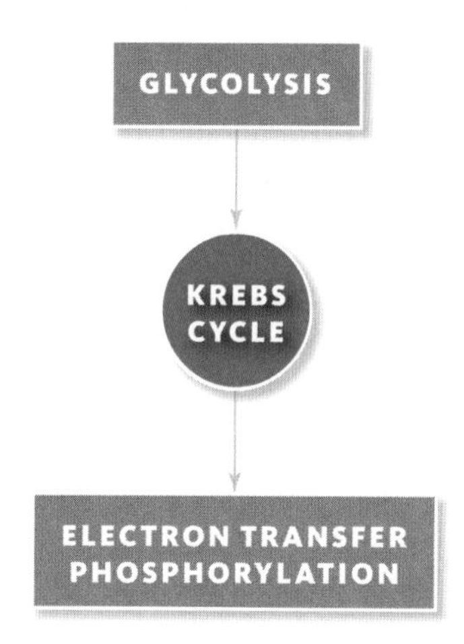

The second stage of aerobic respiration (*above*) occurs inside mitochondria (**FIGURE 7.3**). It includes two sets of reactions, acetyl–CoA formation and the **Krebs cycle**, that break down pyruvate, the product of glycolysis. All of the carbon atoms that were once part of glucose end up in CO_2, which departs the cell. Only two ATP form, but the reactions reduce many coenzymes. The energy of electrons carried by these coenzymes will drive the reactions of the third stage of aerobic respiration.

ACETYL–CoA FORMATION

Aerobic respiration's second stage begins when the two pyruvate molecules that formed during glycolysis enter a mitochondrion. Pyruvate is transported across the mitochondrion's two membranes and into the inner compartment, which is called the mitochondrial matrix (**FIGURE 7.4**). There, an enzyme immediately splits each pyruvate into one molecule of CO_2 and a two-carbon acetyl group ($—COCH_3$). The CO_2 diffuses out of the cell, and the acetyl group combines with a molecule called coenzyme A (abbreviated CoA). The product of this reaction is acetyl–CoA ❶. Electrons and hydrogen ions released by the reaction combine with NAD^+, so NADH also forms.

THE KREBS CYCLE

Each molecule of acetyl–CoA now carries two carbons into the Krebs cycle. Remember from Section 5.4 that a cyclic pathway is not a physical object, such as a wheel. It is called a cycle because the last reaction in the pathway regenerates the substrate of the first. In this case, a substrate of the Krebs cycle's first reaction—and a product of the last—is four-carbon oxaloacetate.

During each cycle of Krebs reactions, two carbon atoms of acetyl–CoA are transferred to oxaloacetate, forming citrate, the ionized form of citric acid ❷. The Krebs cycle is also called the citric acid cycle after this first intermediate. In later reactions, two CO_2 form and depart the cell. Two NAD^+ are reduced when they accept hydrogen ions and electrons, so two NADH form ❸ and ❹. ATP forms by substrate-level phosphorylation ❺. Two coenzymes are reduced: an FAD (flavin adenine dinucleotide) ❻, and another NAD^+ ❼. The final steps of the pathway regenerate oxaloacetate ❽.

FIGURE 7.3 The second stage of aerobic respiration, acetyl–CoA formation and the Krebs cycle, occurs inside mitochondria. *Left*, an inner membrane divides a mitochondrion's interior into two fluid-filled compartments. *Right*, the second stage of aerobic respiration takes place in the mitochondrion's innermost compartment, or matrix.

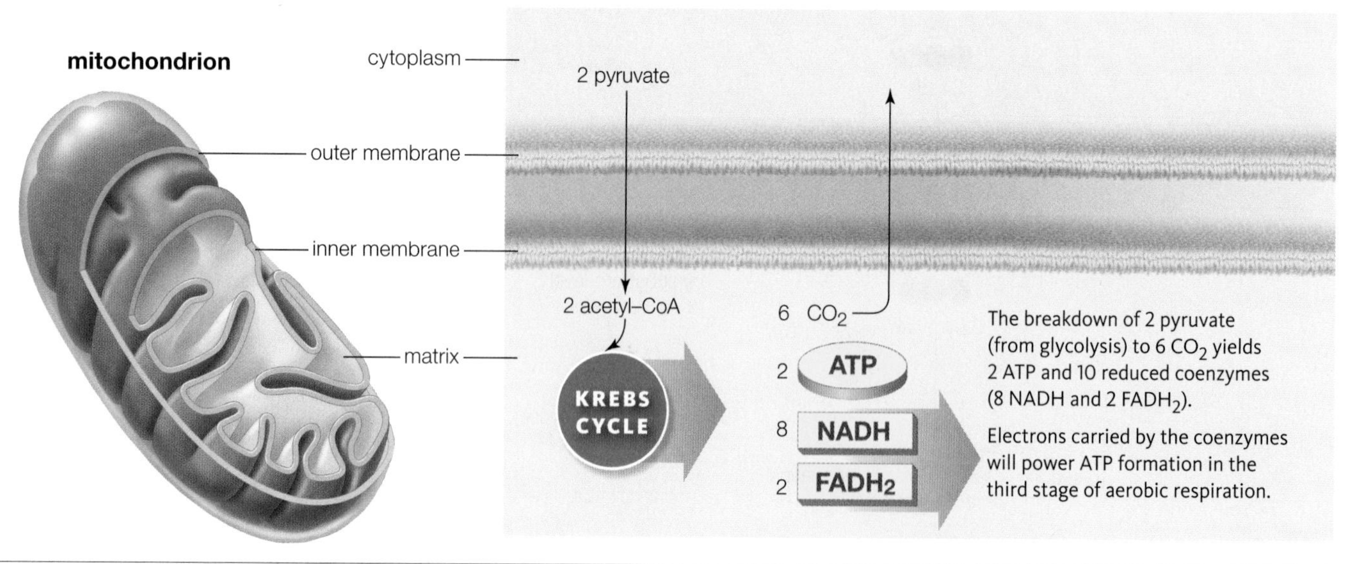

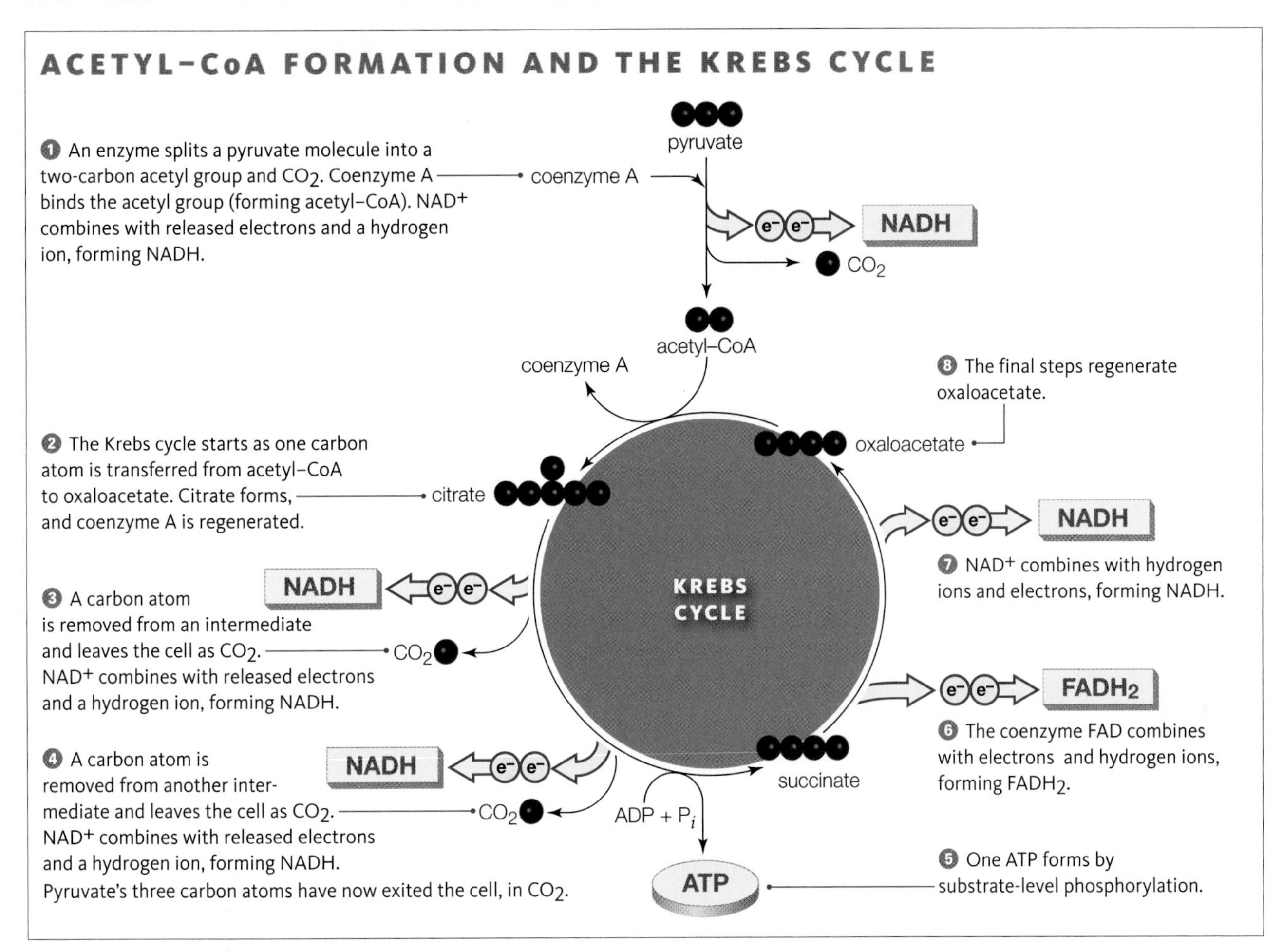

FIGURE 7.4 {Animated} Acetyl–CoA formation and the Krebs cycle. It takes two cycles of Krebs reactions to break down two pyruvate molecules that formed during glycolysis of one glucose molecule. After two cycles, all six carbons that entered glycolysis in one glucose molecule have left the cell, in six CO_2. Electrons and hydrogen ions are released as each carbon is removed from the backbone of intermediate molecules; ten coenzymes will carry them to the third and final stage of aerobic respiration. Not all reactions are shown; see Appendix III for details.

After two cycles of Krebs reactions, the two carbon atoms carried by each acetyl–CoA end up in CO_2. Thus, the combined second-stage reactions of aerobic respiration break down two pyruvate to six CO_2:

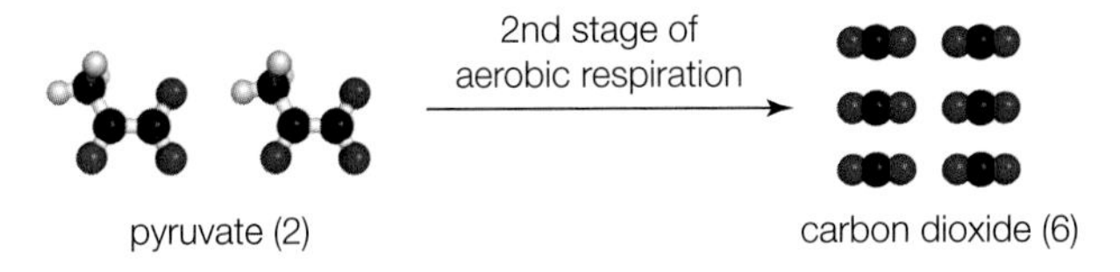

Remember, the two pyruvate were a product of glycolysis. So, at this point in aerobic respiration, the carbon backbone of one glucose molecule has been broken down completely, its six carbon atoms having exited the cell in CO_2.

Krebs cycle Cyclic pathway that, along with acetyl–CoA formation, breaks down pyruvate to carbon dioxide during aerobic respiration.

Two ATP that form during the second stage add to the small net yield of 2 ATP from glycolysis. However, ten coenzymes (eight NAD^+ and two FAD) are reduced during this stage. Add in the two NAD^+ that were reduced in glycolysis, and the full breakdown of each glucose molecule has a big potential payoff. Twelve reduced coenzymes will deliver electrons—and the energy they carry—to the third stage of aerobic respiration.

TAKE-HOME MESSAGE 7.4

The second stage of aerobic respiration, acetyl–CoA formation and the Krebs cycle, occurs in the inner compartment (matrix) of mitochondria.

The second-stage reactions convert the two pyruvate that formed in glycolysis to six CO_2. Two ATP form, and ten coenzymes (eight NAD^+ and two FAD) are reduced.

7.5 WHAT HAPPENS DURING THE THIRD STAGE OF AEROBIC RESPIRATION?

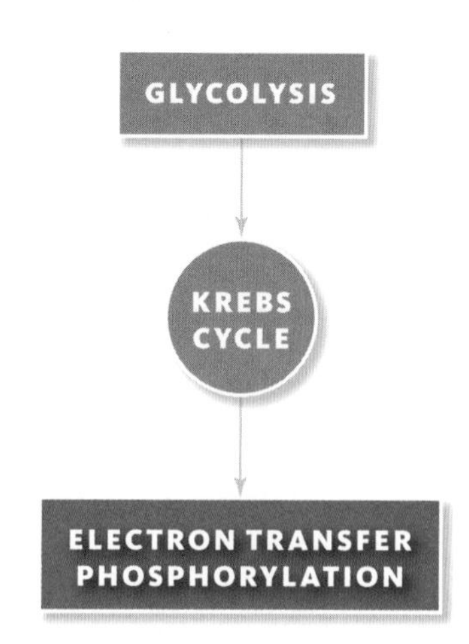

FIGURE 7.5 {Animated} The third and final stage of aerobic respiration, electron transfer phosphorylation.

❶ NADH and $FADH_2$ deliver their cargo of electrons and hydrogen ions to electron transfer chains in the inner mitochondrial membrane.

❷ Electron flow through the chains causes the hydrogen ions (H^+) to be pumped from the matrix to the intermembrane space. A hydrogen ion gradient forms across the inner mitochondrial membrane.

❸ Hydrogen ion flow back to the matrix through ATP synthases drives the formation of ATP from ADP and phosphate (P_i).

❹ Oxygen combines with electrons and hydrogen ions at the end of the electron transfer chains, so water forms.

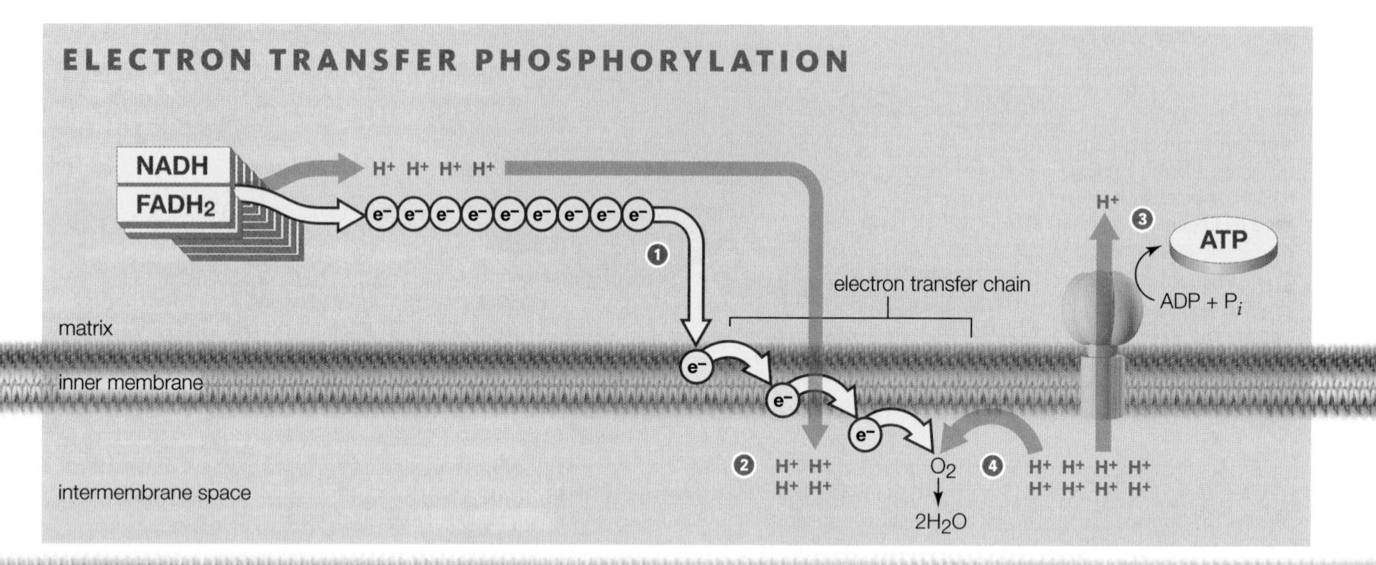

The third stage of aerobic respiration occurs at the inner mitochondrial membrane (**FIGURE 7.5**). Electron transfer phosphorylation reactions begin with the coenzymes NADH and $FADH_2$, which became reduced during the first two stages of aerobic respiration. These coenzymes now deliver their cargo of electrons and hydrogen ions to electron transfer chains embedded in the inner mitochondrial membrane ❶.

As the electrons move through the chains, they give up energy little by little (Section 5.4). Some molecules of the transfer chains harness that energy to actively transport the hydrogen ions across the inner membrane, from the matrix to the intermembrane space ❷. The accumulating ions form a hydrogen ion gradient across the inner mitochondrial membrane. This gradient attracts the ions back toward the matrix. However, ions cannot diffuse through a lipid bilayer on their own (Section 5.7). Hydrogen ions cross the inner mitochondrial membrane only by flowing through ATP synthases embedded in the membrane. The flow of hydrogen ions through ATP synthases causes these proteins to attach phosphate groups to ADP, so ATP forms ❸.

Oxygen accepts electrons at the end of mitochondrial electron transfer chains ❹. When oxygen accepts electrons, it combines with hydrogen ions to form water, which is a product of the third-stage reactions.

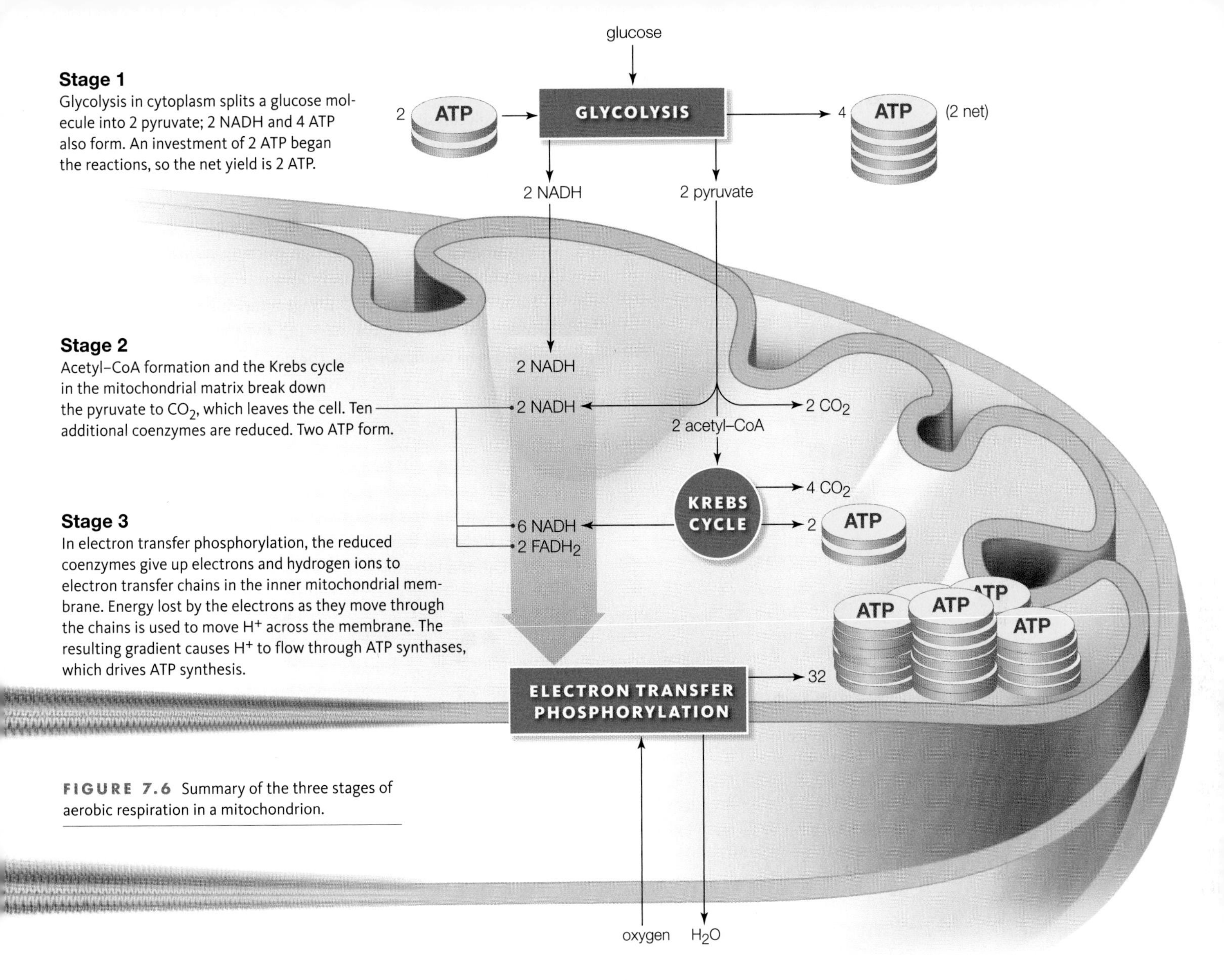

FIGURE 7.6 Summary of the three stages of aerobic respiration in a mitochondrion.

For each glucose molecule that enters aerobic respiration, four ATP form in the first- and second-stage reactions. The twelve coenzymes reduced in these two stages deliver enough electrons to fuel synthesis of about thirty-two additional ATP during the third stage. Thus, the breakdown of one glucose molecule yields about thirty-six ATP (**FIGURE 7.6**). The ATP yield varies depending on cell type. For example, the typical yield of aerobic respiration in brain and skeletal muscle cells is thirty-eight ATP, not thirty-six.

Remember that some energy dissipates with every transfer (Section 5.2). Even though aerobic respiration is a very efficient way of retrieving energy from sugars, about 60 percent of the energy harvested in this pathway disperses as metabolic heat.

TAKE-HOME MESSAGE 7.5

In aerobic respiration's third stage, electron transfer phosphorylation, energy released by electrons moving through electron transfer chains is ultimately captured in the attachment of phosphate to ADP.

The third-stage reactions begin when coenzymes that were reduced in the first and second stages deliver electrons and hydrogen ions to electron transfer chains in the inner mitochondrial membrane.

Energy released by electrons as they pass through electron transfer chains is used to pump H^+ from the mitochondrial matrix to the intermembrane space. The H^+ gradient that forms across the inner mitochondrial membrane drives the flow of hydrogen ions through ATP synthases, which results in ATP formation.

About thirty-two ATP form during the third-stage reactions, so a typical net yield of all three stages of aerobic respiration is thirty-six ATP per glucose.

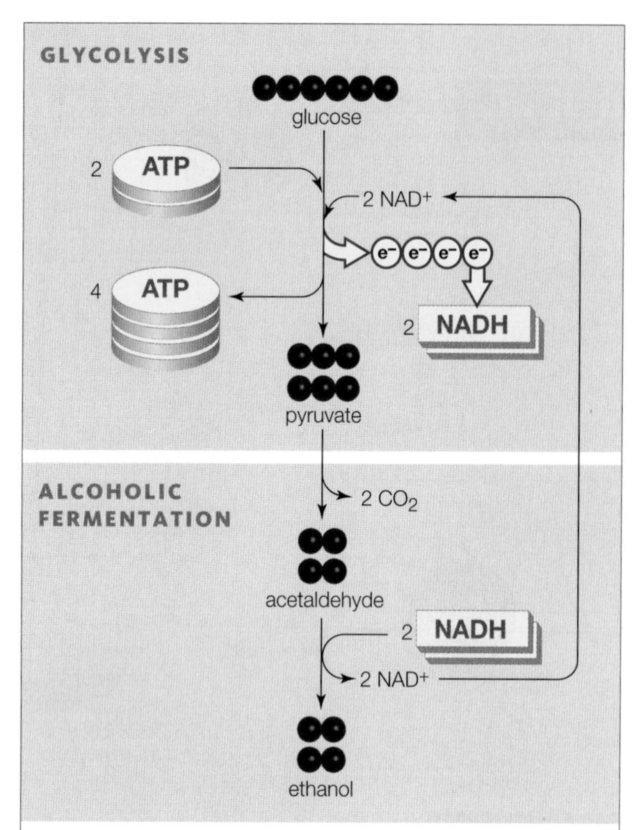

A Alcoholic fermentation begins with glycolysis, and the final steps regenerate NAD^+. The net yield of these reactions is two ATP per molecule of glucose (from glycolysis).

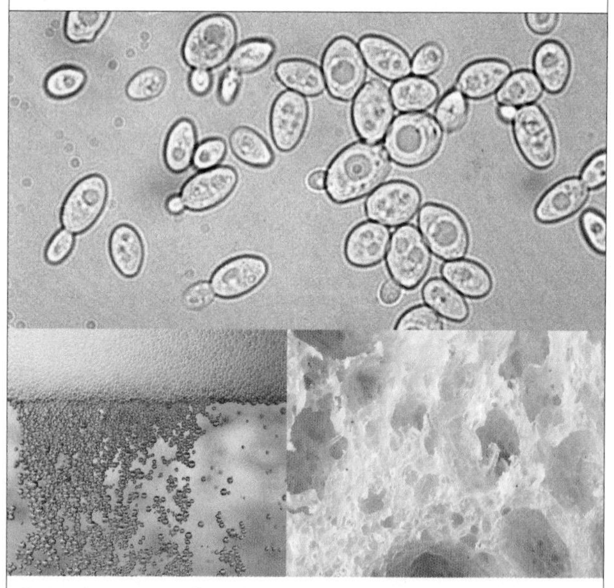

B *Saccharomyces* cells (*top*). One product of alcoholic fermentation in these cells (ethanol) makes beer alcoholic; another (CO_2) makes it bubbly. Holes in bread are pockets where CO_2 released by fermenting *Saccharomyces* cells accumulated in the dough.

FIGURE 7.7 {Animated} Alcoholic fermentation.

TWO FERMENTATION PATHWAYS

Aerobic respiration and fermentation begin with the same set of glycolysis reactions in cytoplasm. After glycolysis, the two pathways differ. The final steps of fermentation occur in the cytoplasm and do not require oxygen. In these reactions, pyruvate is converted to other molecules, but it is not fully broken down to CO_2 (as occurs in aerobic respiration). Electrons do not move through electron transfer chains, so no additional ATP forms. However, electrons are removed from NADH, so NAD^+ is regenerated. Regenerating this coenzyme allows glycolysis—and the small ATP yield it offers—to continue. Thus, the net ATP yield of fermentation consists of the two ATP that form in glycolysis.

Alcoholic Fermentation In **alcoholic fermentation**, the pyruvate from glycolysis is converted to ethanol (**FIGURE 7.7A**). First, 3-carbon pyruvate is split into carbon dioxide and 2-carbon acetaldehyde. Then, electrons and hydrogen are transferred from NADH to the acetaldehyde, forming NAD^+ and ethanol:

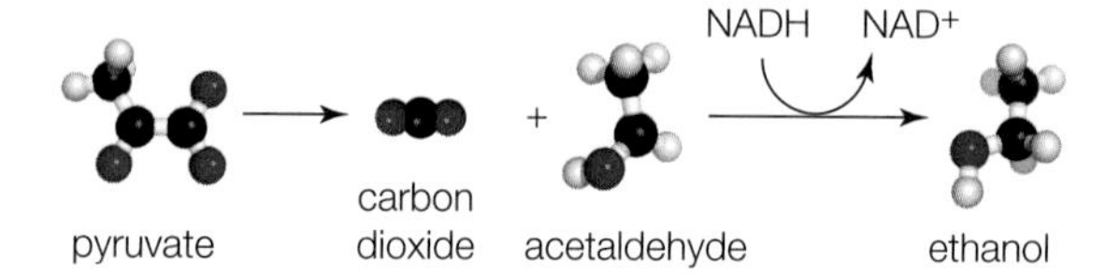

Alcoholic fermentation in a fungus, *Saccharomyces cerevisiae*, sustains these yeast cells as they grow and reproduce. It also helps us produce beer, wine, and bread (**FIGURE 7.7B**). Beer brewers typically use germinated, roasted, and crushed barley as a sugar source for *Saccharomyces* fermentation. Ethanol produced by the fermenting yeast cells makes the beer alcoholic, and CO_2 makes it bubbly. Flowers of the hop plant add flavor and help preserve the finished product. Winemakers start with crushed grapes for *Saccharomyces* fermentation. The yeast cells convert sugars in the grape juice to ethanol.

Bakers take advantage of alcoholic fermentation by *Saccharomyces* cells to make bread from flour, which contains starches and a protein called gluten. When flour is kneaded with water, the gluten forms polymers in long, interconnected strands that make the resulting dough stretchy and resilient. Yeast cells in the dough produce CO_2 as they ferment the starches. The gas accumulates in bubbles that are trapped by the mesh of gluten strands. As the bubbles expand, they cause the dough to rise. Ethanol produced by the fermentation reactions evaporates during baking.

Lactate Fermentation In **lactate fermentation**, the electrons and hydrogen ions carried by NADH are transferred directly to pyruvate (**FIGURE 7.8A**). This reaction converts pyruvate to 3-carbon lactate (the ionized form of lactic acid), and also converts NADH to NAD^+:

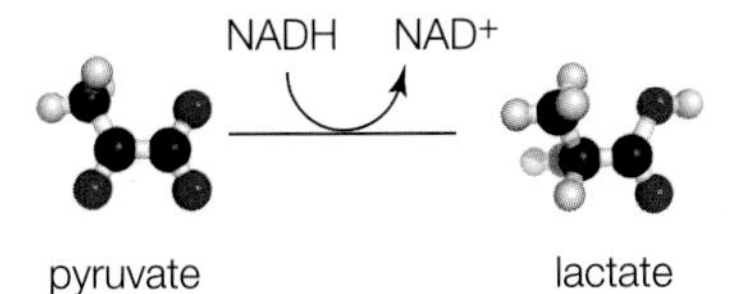

Some lactate fermenters spoil food, but we use others to preserve it. For instance, *Lactobacillus* bacteria break down lactose in milk by fermentation. We use this bacteria to produce dairy products such as buttermilk, cheese, and yogurt, and also to pickle vegetables and other foods.

Cells in animal skeletal muscles are fused as long fibers that carry out aerobic respiration, lactate fermentation, or both. Red fibers have many mitochondria and produce ATP mainly by aerobic respiration. These fibers sustain prolonged activity such as marathon runs. They are red because they have an abundance of myoglobin, a protein that stores oxygen for aerobic respiration (**FIGURE 7.8B**). White muscle fibers contain few mitochondria and no myoglobin, so they do not carry out a lot of aerobic respiration. Instead, they make most of their ATP by lactate fermentation. This pathway makes ATP quickly but not for long, so it is useful for quick, strenuous activities such as weight lifting or sprinting (**FIGURE 7.8C**). The low ATP yield does not support prolonged activity.

Most human muscles are a mixture of white and red fibers, but the proportions vary among muscles and among individuals. Great sprinters tend to have more white fibers. Great marathon runners tend to have more red fibers. Chickens cannot fly very far because their flight muscles consist mostly of white fibers (thus, the "white" breast meat). A chicken most often walks or runs. Its leg muscles consist mostly of red muscle fibers, the "dark meat." Section 32.5 returns to skeletal muscle fibers and how they work.

alcoholic fermentation Anaerobic sugar breakdown pathway that produces ATP, CO_2, and ethanol.
lactate fermentation Anaerobic sugar breakdown pathway that produces ATP and lactate.

TAKE-HOME MESSAGE 7.6

ATP can form by sugar breakdown in fermentation pathways, which are anaerobic.

The end product of lactate fermentation is lactate. The end product of alcoholic fermentation is ethanol.

Both pathways have a net yield of two ATP per glucose molecule. The ATP forms during glycolysis.

Fermentation reactions regenerate the coenzyme NAD^+, without which glycolysis (and ATP production) would stop.

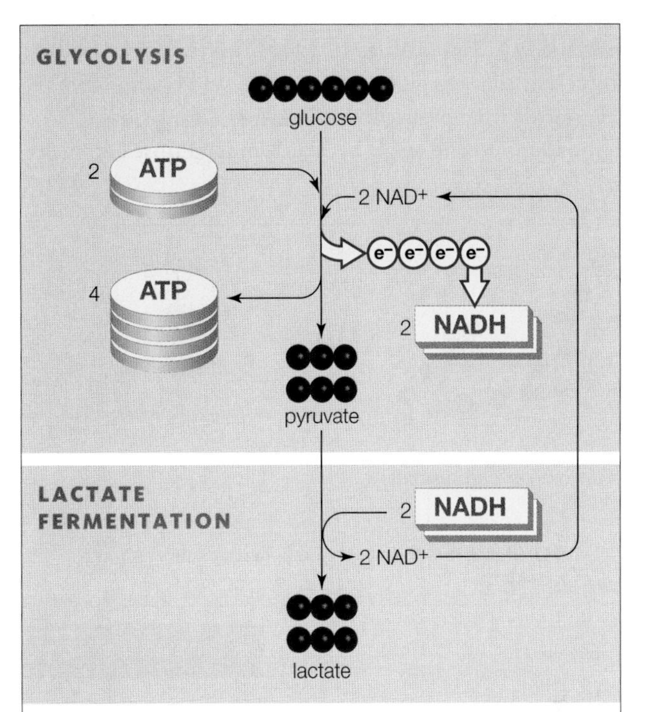

A Lactate fermentation begins with glycolysis, and the final steps regenerate NAD^+. The net yield of these reactions is two ATP per molecule of glucose (from glycolysis).

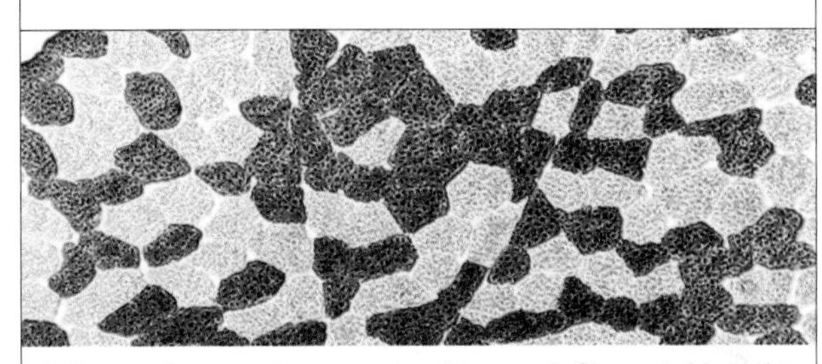

B Lactate fermentation occurs in white muscle fibers, visible in this cross-section of human thigh muscle. The red fibers, which make ATP by aerobic respiration, sustain endurance activities.

C Intense activity such as sprinting quickly depletes oxygen in muscles. Under anaerobic conditions, ATP is produced mainly by lactate fermentation in white muscle fibers. Fermentation does not make enough ATP to sustain this type of activity for long.

FIGURE 7.8 Lactate fermentation.

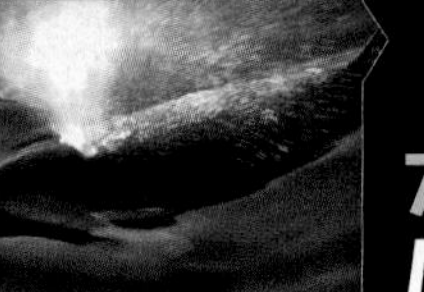

7.7 CAN THE BODY USE ANY ORGANIC MOLECULE FOR ENERGY?

ENERGY FROM DIETARY MOLECULES

Glycolysis converts glucose to pyruvate, and electrons are transferred from pyruvate to coenzymes during aerobic respiration's second stage. In other words, glucose becomes oxidized (it gives up electrons) and coenzymes become reduced (they accept electrons). Oxidizing an organic molecule can break the covalent bonds of its carbon backbone. Aerobic respiration generates a lot of ATP by fully oxidizing glucose, completely dismantling it carbon by carbon.

Cells also dismantle other organic molecules by oxidizing them. Complex carbohydrates, fats, and proteins in food can be converted to molecules that enter glycolysis or the Krebs cycle (**FIGURE 7.9**). As in glucose metabolism, many coenzymes are reduced, and the energy of the electrons they carry ultimately drives the synthesis of ATP in electron transfer phosphorylation.

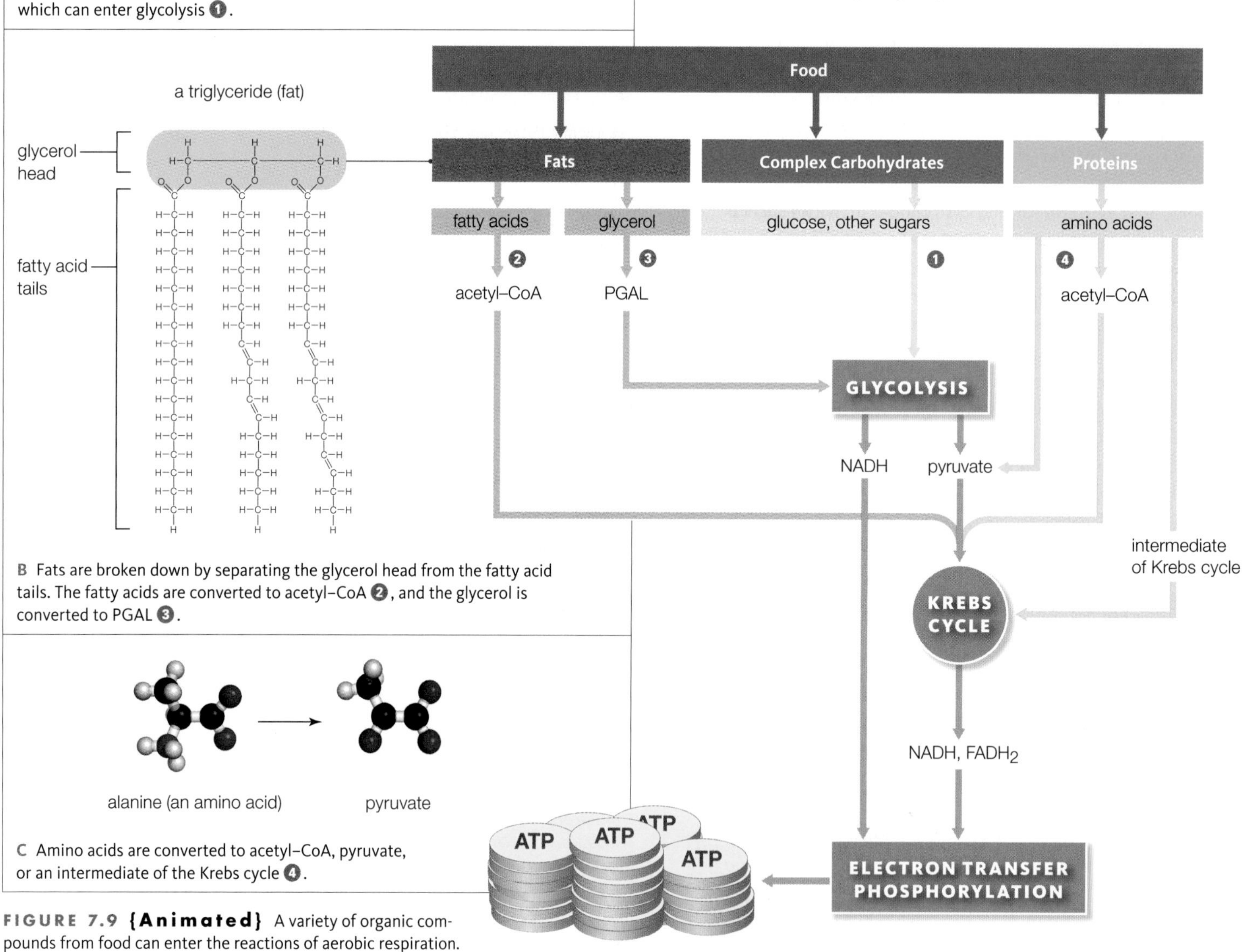

FIGURE 7.9 **{Animated}** A variety of organic compounds from food can enter the reactions of aerobic respiration.

Complex Carbohydrates In humans and other mammals, the digestive system breaks down starch and other complex carbohydrates to monosaccharides (**FIGURE 7.9A**). These sugars are quickly taken up by cells and converted to glucose-6-phosphate for glycolysis ❶. When a cell produces more ATP than it uses, the concentration of ATP rises in the cytoplasm. A high concentration of ATP causes glucose-6-phosphate to be diverted away from glycolysis and into a pathway that forms glycogen. Liver and muscle cells especially favor the conversion of glucose to glycogen, and these cells contain the body's largest stores of it. Between meals, the liver maintains the glucose level in blood by converting the stored glycogen to glucose.

Fats A fat molecule has a glycerol head and one, two, or three fatty acid tails (Section 3.3). Cells dismantle these molecules by first breaking the bonds that connect the fatty acid tails to the glycerol head (**FIGURE 7.9B**). Nearly all cells in the body can oxidize free fatty acids by splitting their long backbones into two-carbon fragments. These fragments are converted to acetyl–CoA, which can enter the Krebs cycle ❷. Enzymes in liver cells convert the glycerol to PGAL, an intermediate of glycolysis ❸.

On a per carbon basis, fats are a richer source of energy than carbohydrates. Carbohydrate backbones have many oxygen atoms, so they are partially oxidized. A fat's long fatty acid tails are hydrocarbon chains that typically have no oxygen atoms bonded to them, so they have a longer way to go to become oxidized—more reactions are required to fully break them down. Coenzymes accept electrons in these oxidation reactions. The more reduced coenzymes that form, the more electrons can be delivered to the ATP-forming machinery of electron transfer phosphorylation.

What happens if you eat too many carbohydrates? When the blood level of glucose gets too high, acetyl–CoA is diverted away from the Krebs cycle and into a pathway that makes fatty acids. That is why excess dietary carbohydrate ends up as fat.

Proteins Enzymes in the digestive system split dietary proteins into their amino acid subunits, which are absorbed into the bloodstream. Cells use the amino acids to build proteins or other molecules. When you eat more protein than your body needs for this purpose, the amino acids are broken down. The amino (NH_3^+) group is removed, and it becomes ammonia (NH_3), a waste product that is eliminated in urine. The carbon backbone is split, and acetyl–CoA, pyruvate, or an intermediate of the Krebs cycle forms, depending on the amino acid (**FIGURE 7.9C**). These molecules enter aerobic respiration's second stage ❹.

PEOPLE MATTER

DR. BENJAMIN RAPOPORT

MIT engineers Benjamin Rapoport, Jakub Kedzierski, and Rahul Sarpeshkar have developed a tiny fuel cell that runs on the same sugar that powers human cells: glucose.

The fuel cell strips electrons from glucose molecules to create a small electric current. In this way, it mimics the activity of cellular enzymes that break down glucose to generate ATP.

The researchers fabricated the fuel cell on a silicon chip, so it can be integrated with other implantable circuits that could, for example, be implanted in the spinal cord to help paralyzed patients move their arms and legs again. Current devices can do this too, but they require an external power source. The new fuel cell can use glucose in the fluid that bathes the brain. Glucose is normally the brain's only fuel, and this fluid contains a lot of it. The fuel cell uses only a tiny amount of glucose, so it would have a minimal impact on brain function.

"It will be a few more years into the future before you see people with spinal-cord injuries receive such implantable systems in the context of standard medical care, but those are the sorts of devices you could envision powering from a glucose-based fuel cell," says Rapoport.

TAKE-HOME MESSAGE 7.7

Oxidizing organic molecules can break their carbon backbones, releasing electrons whose energy can be harnessed to drive ATP formation in aerobic respiration.

Fats, complex carbohydrates, and proteins can be oxidized in aerobic respiration to yield ATP. First the digestive system and then individual cells convert molecules in food into substrates of glycolysis or aerobic respiration's second-stage reactions.

7.8 Application: MITOCHONDRIAL MALFUNCTION

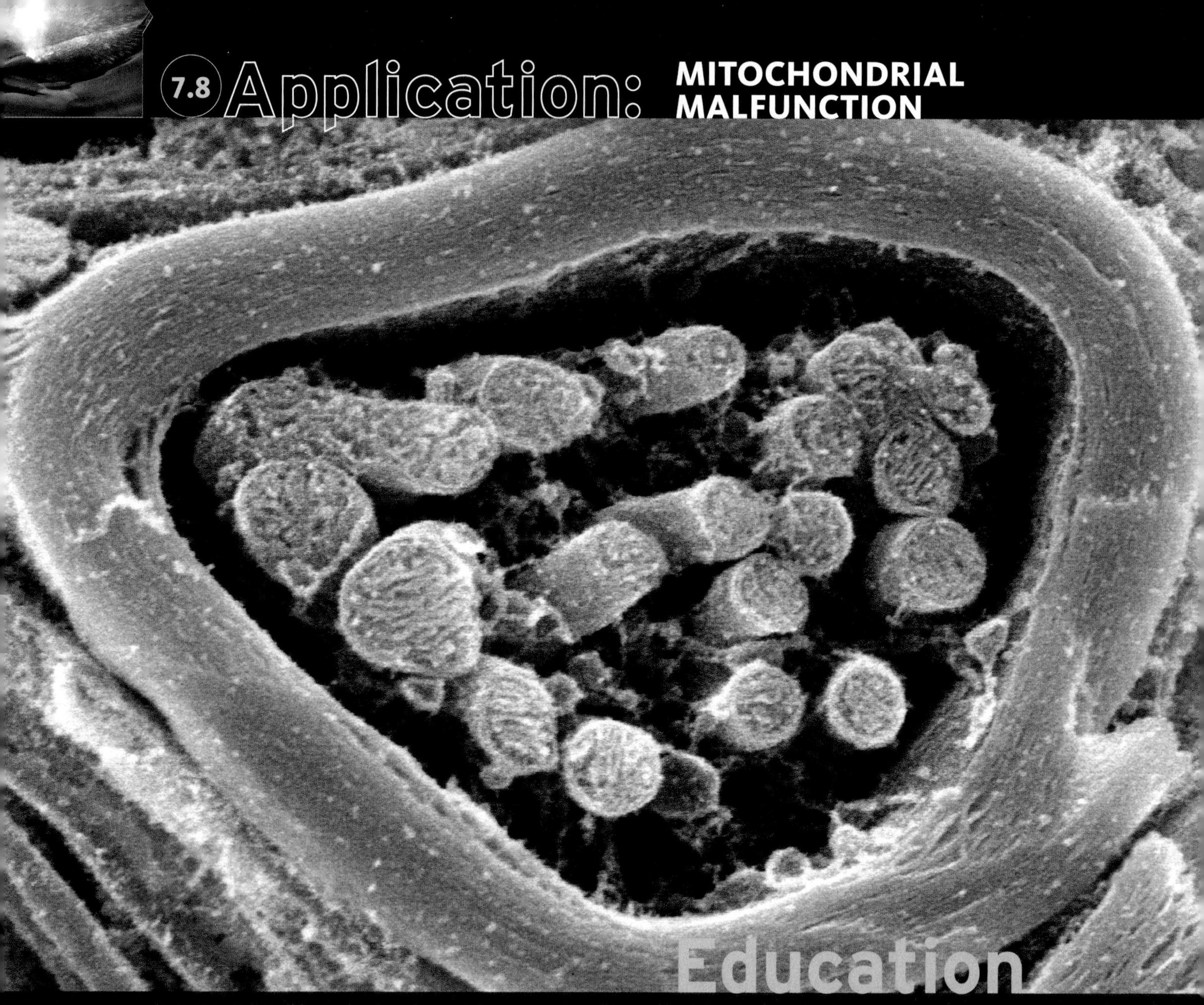

FIGURE 7.10 This cross-section of a nerve shows how these cells are packed with mitochondria. Mitochondria are powerhouses of all eukaryotic cells. When they malfunction, the lights go off in cellular businesses.

AEROBIC RESPIRATION IS A DANGEROUS OCCUPATION. When an oxygen molecule accepts electrons from an electron transfer chain in a mitochondrion, it dissociates into oxygen atoms. These atoms immediately combine with hydrogen ions and end up in water molecules. Occasionally, however, an oxygen atom escapes this final reaction. The atom has an unpaired electron, so it is a free radical. Free radicals can easily strip electrons from (oxidize) biological molecules and break their carbon backbones.

Mitochondria cannot detoxify free radicals, so they rely on antioxidant enzymes and vitamins in the cell's cytoplasm to do it for them. The system works well, at least most of the time. However, a genetic disorder or an unfortunate encounter with a toxin or pathogen can result in a missing antioxidant, or a defective component of the mitochondrial electron transfer chain. In either case, the normal cellular balance of aerobic respiration and free radical formation is tipped. Free radicals accumulate and destroy first the function of mitochondria, then the cell. The resulting tissue damage is called oxidative stress.

At least 83 proteins are directly involved in mitochondrial electron transfer chains. A defect in any one of them—or in any of the thousands of other

"Tom does not look sick, but inside his organs are all getting badly damaged," said Martine Martin, pictured here with her eight-year-old son. Tom was born with a mitochondrial disease. He eats with the help of a machine, suffers intense pain, and will soon be blind. Despite intensive medical intervention, he is not expected to reach his teens.

proteins used by mitochondria—can wreak havoc in the body. Hundreds of incurable disorders are associated with such defects (FIGURE 7.10), and more are being discovered all the time. Nerve and brain cells, which require a lot of ATP, are particularly affected. Symptoms range from mild to major progressive neurological deficits, blindness, deafness, diabetes, strokes, seizures, gastrointestinal malfunction, and disabling muscle weakness. New research is showing that mitochondrial malfunction is also involved in many other illnesses, including cancer, hypertension, and Alzheimer's and Parkinson's diseases.

Summary

SECTIONS 7.1, 7.2 Most organisms can make ATP by breaking down sugars in fermentation or aerobic respiration. Both pathways begin in cytoplasm. **Aerobic respiration** requires oxygen and, in eukaryotes, ends in mitochondria. It includes electron transfer chains, and ATP forms by electron transfer phosphorylation. **Fermentation** pathways end in cytoplasm and do not require oxygen. Aerobic respiration yields much more ATP per glucose molecule than fermentation.

Photosynthesis by early prokaryotes changed the composition of Earth's atmosphere, with profound effects on life's evolution. Organisms that could not tolerate the increased atmospheric oxygen persisted only in **anaerobic** habitats. The evolution of antioxidants allowed organisms to tolerate the increase in the atmospheric content of oxygen, and to thrive under **aerobic** conditions. Over time, the antioxidants became incorporated into aerobic respiration and other pathways that harnessed the reactive properties of oxygen.

SECTION 7.3 **Glycolysis**, the first stage of aerobic respiration and fermentation, occurs in cytoplasm. In the reactions, enzymes use two ATP to convert one molecule of glucose or another six-carbon sugar to two molecules of 3-carbon **pyruvate**. Electrons and hydrogen ions are transferred to two NAD^+, which are thereby reduced to NADH. Four ATP also form by **substrate-level phosphorylation**.

SECTION 7.4 In eukaryotes, aerobic respiration continues in mitochondria. The second stage of aerobic respiration, acetyl–CoA formation and the **Krebs cycle**, takes place in the inner compartment (matrix) of the mitochondrion. The first steps convert the two pyruvate from glycolysis to two acetyl–CoA and two CO_2. The acetyl–CoA delivers carbon atoms to the Krebs cycle. Electrons and hydrogen ions are transferred to NAD^+ and FAD, which are thereby reduced to NADH and $FADH_2$. ATP forms by substrate-level phosphorylation. Two cycles of Krebs reactions break down the two pyruvate from glycolysis. At this stage of aerobic respiration, the glucose molecule that entered glycolysis has been dismantled completely: All of its carbon atoms have exited the cell in CO_2.

SECTION 7.5 In the third and final stage of aerobic respiration, electron transfer phosphorylation, the many coenzymes that were reduced in the first two stages now deliver their cargo of electrons and hydrogen ions to electron transfer chains in the inner mitochondrial membrane. The electrons move through the chains, releasing energy bit by bit; molecules of the chain use that energy to move H^+ from the matrix to the intermembrane space. Hydrogen ions accumulate in the intermembrane space, forming a gradient across the inner

Summary continued

membrane. The ions follow the gradient back to the matrix through ATP synthases. H^+ flow through these transport proteins drives ATP synthesis.

Oxygen accepts electrons at the end of the chains and combines with hydrogen ions, so water forms.

The ATP yield of aerobic respiration varies, but typically it is about thirty-six ATP for each glucose molecule that enters glycolysis.

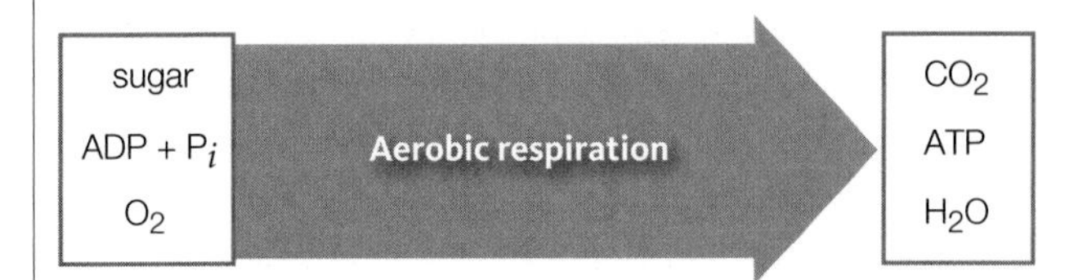

SECTION 7.6 Anaerobic fermentation pathways begin with glycolysis, and they run entirely in the cytoplasm. An organic molecule, rather than oxygen, accepts electrons at the end of these reactions. The end product of **alcoholic fermentation** is ethyl alcohol, or ethanol. The end product of **lactate fermentation** is lactate.

The final steps of fermentation regenerate NAD^+, which is required for glycolysis to continue, but they produce no ATP. Thus, the breakdown of one glucose molecule in either alcoholic or lactate fermentation yields only the two ATP from glycolysis.

Skeletal muscle consists of two types of fiber. ATP produced primarily by aerobic respiration in red fibers sustains activities that require endurance. Lactate fermentation in white fibers supports activities that occur in short, intense bursts.

SECTION 7.7 Oxidizing an organic molecule can break its carbon backbone. Aerobic respiration fully oxidizes glucose, dismantling its backbone carbon by carbon. Each carbon removed releases electrons that drive ATP formation in electron transfer phosphorylation. Organic molecules other than sugars are also broken down (oxidized) for energy. In humans and other mammals, first the digestive system and then individual cells convert fats, proteins, and complex carbohydrates in food to molecules that are substrates of glycolysis or the second-stage reactions of aerobic respiration.

SECTION 7.8 Free radicals that form during aerobic respiration are detoxified by antioxidant molecules in the cell's cytoplasm. Missing antioxidant molecules or heritable defects in mitochondrial electron transfer chain components can cause a buildup of free radicals that damage the cell—and ultimately, the individual. Symptoms can be lethal. Oxidative stress due to mitochondrial malfunction plays a role in many illnesses such as cancer, and Alzheimer's and Parkinson's diseases.

Self-Quiz

Answers in Appendix VII

1. Is the following statement true or false? Unlike animals, which make many ATP by aerobic respiration, plants make all of their ATP by photosynthesis.

2. Glycolysis starts and ends in the _______ .
   a. nucleus　　c. plasma membrane
   b. mitochondrion　　d. cytoplasm

3. Which of the following metabolic pathways require(s) molecular oxygen (O_2)?
   a. aerobic respiration
   b. lactate fermentation
   c. alcoholic fermentation
   d. all of the above

4. Which molecule does not form during glycolysis?
   a. NADH　　c. oxygen (O_2)
   b. pyruvate　　d. ATP

5. In eukaryotes, aerobic respiration is completed in the _______ .
   a. nucleus　　c. plasma membrane
   b. mitochondrion　　d. cytoplasm

6. In eukaryotes, fermentation is completed in the _______ .
   a. nucleus　　c. plasma membrane
   b. mitochondrion　　d. cytoplasm

7. Which of the following reaction pathways is not part of the second stage of aerobic respiration?
   a. electron transfer phosphorylation　　c. Krebs cycle
   b. acetyl–CoA formation　　d. glycolysis
   e. a and d

8. After the Krebs reactions run through _______ cycle(s), one glucose molecule has been completely broken down to CO_2.
   a. one　　b. two　　c. three　　d. six

9. In the third stage of aerobic respiration, _______ is the final acceptor of electrons.
   a. water　　c. oxygen (O_2)
   b. hydrogen　　d. NADH

10. Most of the energy that is released by the full breakdown of glucose to CO_2 and water ends up in _______ .
    a. NADH　　c. heat
    b. ATP　　d. electrons

11. _______ accepts electrons in alcoholic fermentation.
    a. Oxygen　　c. Acetaldehyde
    b. Pyruvate　　d. Ethanol

12. Your body cells can break down _______ as a source of energy to fuel ATP production.
    a. fatty acids　　c. amino acids
    b. glycerol　　d. all of the above

13. Which of the following is *not* produced by an animal muscle cell operating under anaerobic conditions?
a. heat
b. lactate
c. ATP
d. NAD^+
e. pyruvate
f. all are produced

14. Hydrogen ion flow drives ATP synthesis during ______ .
a. glycolysis
b. the Krebs cycle
c. aerobic respiration
d. fermentation
e. a and c

15. Match the term with the best description.

___ mitochondrial matrix	a. needed for glycolysis
___ pyruvate	b. inner space
___ NAD^+	c. makes many ATP
___ mitochondrion	d. product of glycolysis
___ NADH	e. reduced coenzyme
___ anaerobic	f. no oxygen required

Critical Thinking

1. The higher the altitude, the lower the oxygen level in air. Climbers of very tall mountains risk altitude sickness, which is characterized by shortness of breath, weakness, dizziness, and confusion.

The early symptoms of cyanide poisoning are the same as those for altitude sickness. Cyanide binds tightly to cytochrome *c* oxidase, the protein that reduces oxygen molecules in the final step of mitochondrial electron transfer chains. Cytochrome *c* oxidase with bound cyanide can no longer transfer electrons. Explain why cyanide poisoning starts with the same symptoms as altitude sickness.

2. As you learned, membranes impermeable to hydrogen ions are required for electron transfer phosphorylation. Membranes in mitochondria serve this purpose in eukaryotes. Bacteria do not have this organelle, but they do make ATP by electron transfer phosphorylation. How do you think they do it, given that they have no mitochondria?

3. The bar-tailed godwit is a type of shorebird that makes an annual migration from Alaska to New Zealand and back. The birds make each 11,500-kilometer (7,145-mile) trip by flying over the Pacific Ocean in about nine days, depending on weather, wind speed, and direction of travel. One bird was observed to make the entire journey uninterrupted, a feat that is comparable to a human running a nonstop seven-day marathon at 70 kilometers per hour (43.5 miles per hour). Would you expect the flight (breast) muscles of bar-tailed godwits to be light or dark colored? Explain your answer.

Data Analysis Activities

Mitochondrial Abnormalities in Tetralogy of Fallot

Tetralogy of Fallot (TF) is a genetic disorder characterized by four major malformations of the heart. The circulation of blood is abnormal, so TF patients have too little oxygen in their blood. Inadequate oxygen levels result in damaged mitochondrial membranes, which in turn cause cells to self-destruct.

In 2004, Sarah Kuruvilla and her colleagues looked at abnormalities in the mitochondria of heart muscle in TF patients. Some of their results are shown in **FIGURE 7.11**.

1. Which abnormality was most strongly associated with tetralogy of Fallot?
2. Can you make any correlations between blood oxygen content and mitochondrial abnormalities in these TF patients?

FIGURE 7.11 Mitochondrial changes in tetralogy of Fallot (TF).

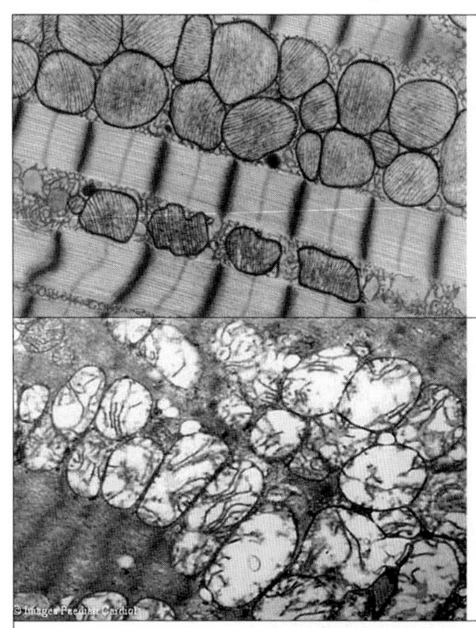

A Normal heart muscle. Many mitochondria between the fibers provide muscle cells with ATP for contraction.

B Heart muscle from a person with TF has swollen, broken mitochondria.

Patient (age)	SPO_2 (%)	Mitochondrial Abnormalities in TF: Number	Shape	Size	Broken
1 (5)	55	+	+	–	–
2 (3)	69	+	+	–	–
3 (22)	72	+	+	–	–
4 (2)	74	+	+	–	–
5 (3)	76	+	+	–	+
6 (2.5)	78	+	+	–	+
7 (1)	79	+	+	–	–
8 (12)	80	+	–	+	–
9 (4)	80	+	+	–	–
10 (8)	83	+	–	+	–
11 (20)	85	+	+	–	–
12 (2.5)	89	+	–	+	–

C Types of mitochondrial abnormalities in TF patients. SPO_2 is oxygen saturation of the blood. A normal value of SPO_2 is 96%. Abnormalities are marked +.

CREDITS: (11A) Steve Gschmeissner/Science Source.; (11B) © Images Paediatr Cardiol; (11C) © Cengage Learning 2014.

Information encoded in DNA is the basis of visible traits that define species and distinguish individuals. Identical twins are identical because they inherited copies of the same DNA.

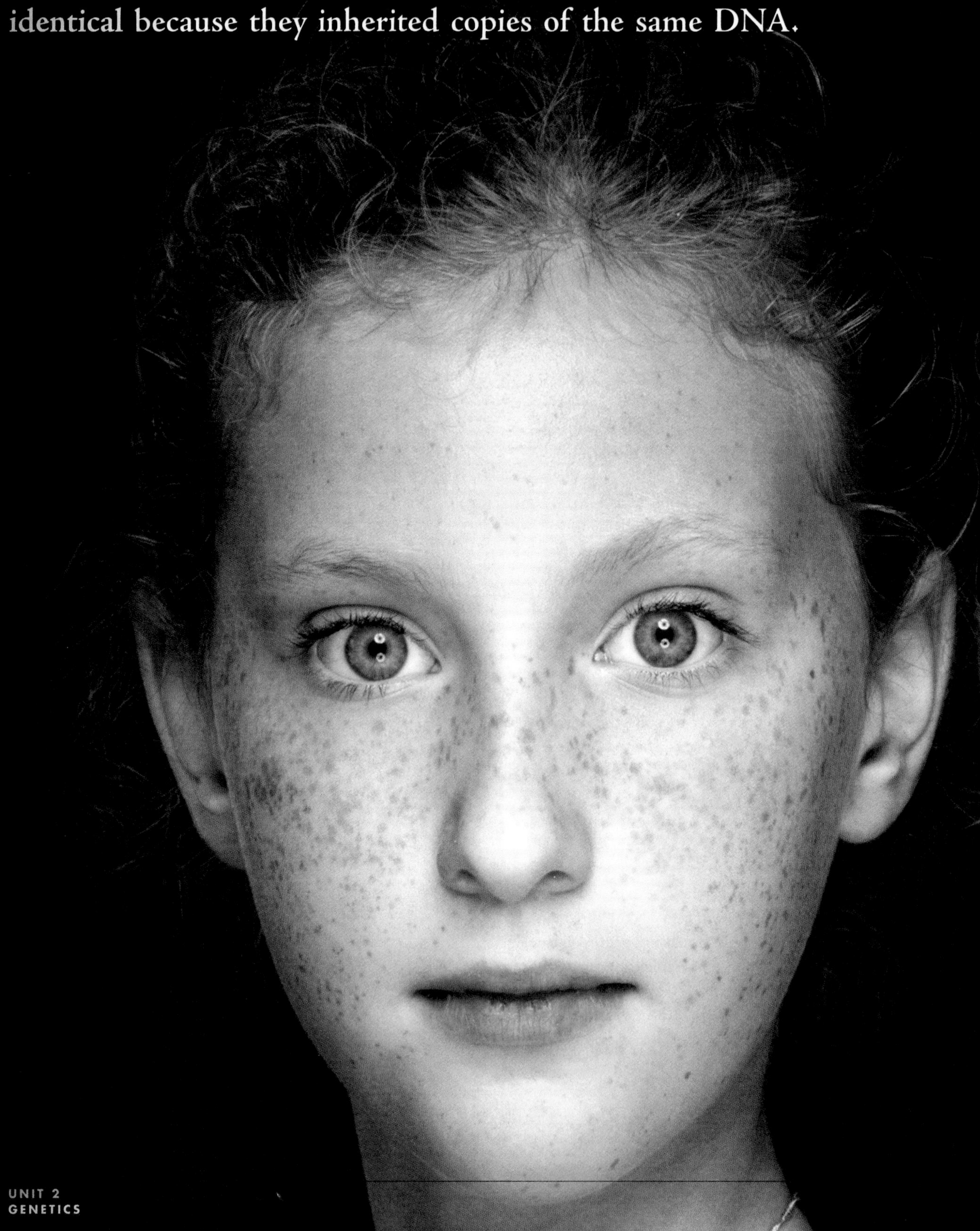

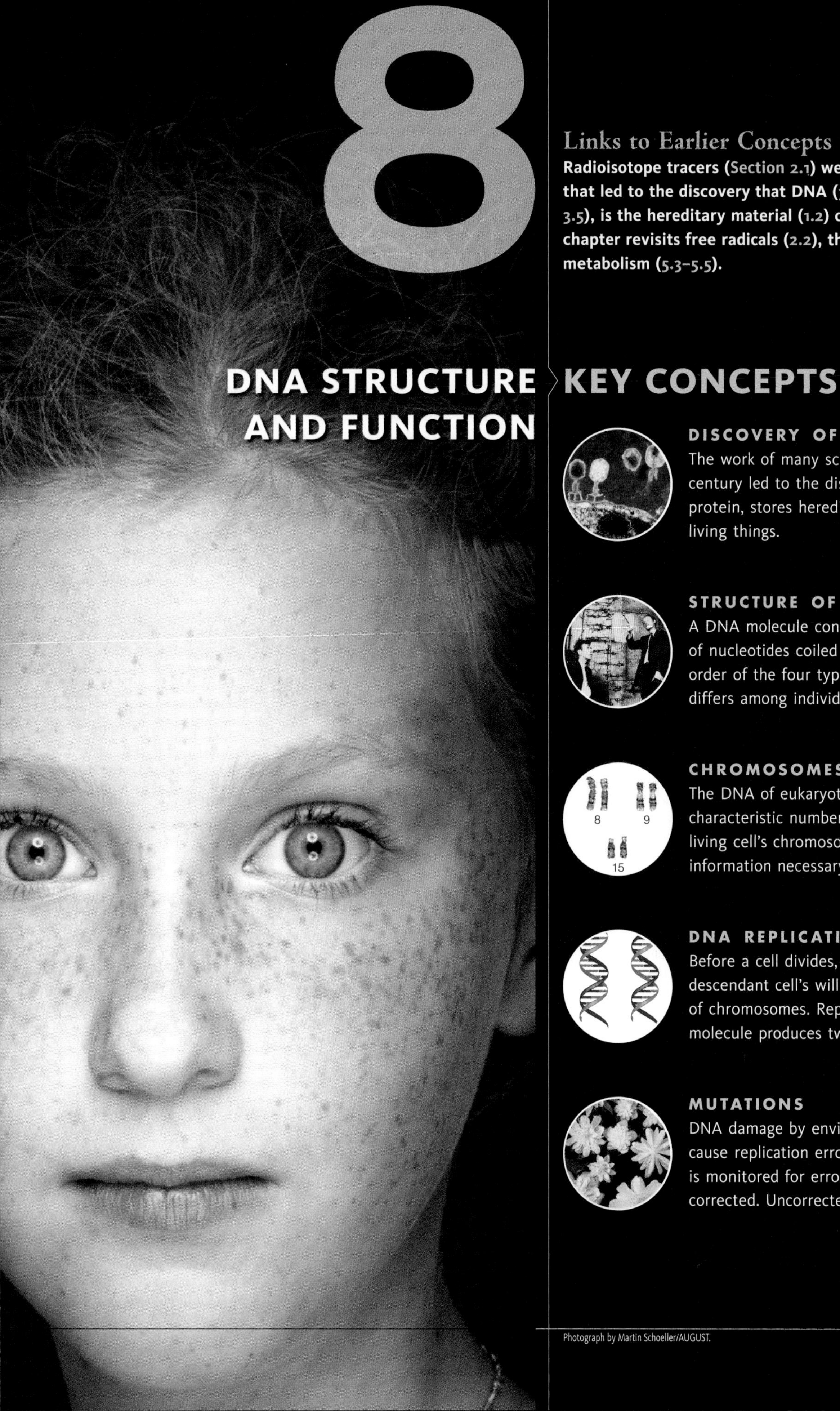

8

DNA STRUCTURE AND FUNCTION

Links to Earlier Concepts

Radioisotope tracers (Section 2.1) were used in research that led to the discovery that DNA (3.6), not protein (3.4, 3.5), is the hereditary material (1.2) of all organisms. This chapter revisits free radicals (2.2), the cell nucleus (4.5), and metabolism (5.3–5.5).

KEY CONCEPTS

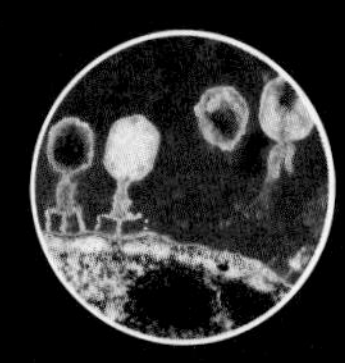

DISCOVERY OF DNA'S FUNCTION

The work of many scientists over nearly a century led to the discovery that DNA, not protein, stores hereditary information in all living things.

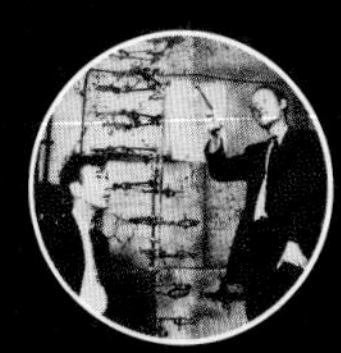

STRUCTURE OF DNA MOLECULES

A DNA molecule consists of two long chains of nucleotides coiled into a double helix. The order of the four types of nucleotides in a chain differs among individuals and species.

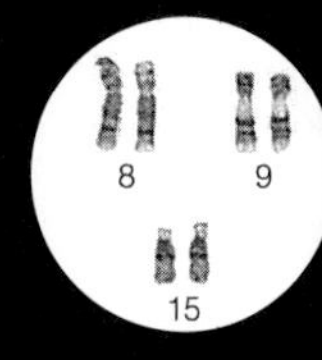

CHROMOSOMES

The DNA of eukaryotes is divided among a characteristic number of chromosomes. A living cell's chromosomes contain all of the information necessary to build a new individual.

DNA REPLICATION

Before a cell divides, it copies its DNA so both descendant cell's will inherit a full complement of chromosomes. Replication of each DNA molecule produces two duplicates.

MUTATIONS

DNA damage by environmental agents can cause replication errors. Newly forming DNA is monitored for errors, most of which are corrected. Uncorrected errors become mutations.

Photograph by Martin Schoeller/AUGUST.

8.1 HOW WAS DNA'S FUNCTION DISCOVERED?

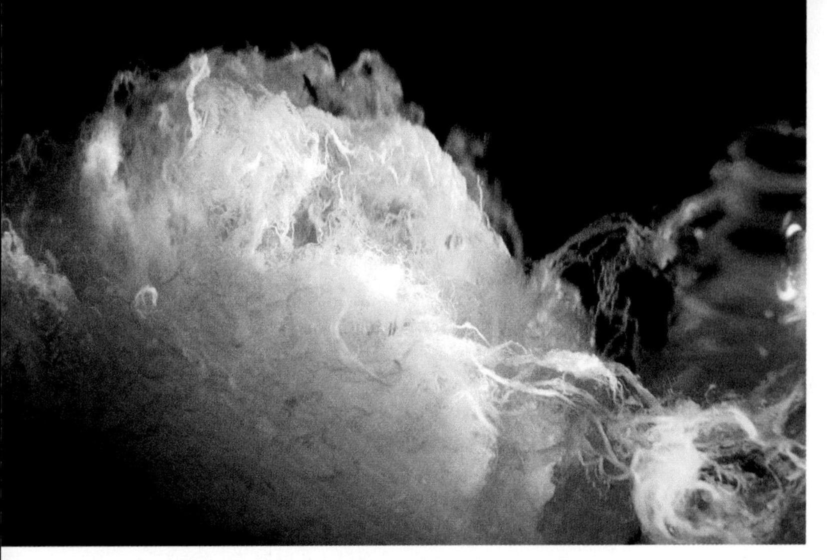

FIGURE 8.1 DNA extracted from human cells.

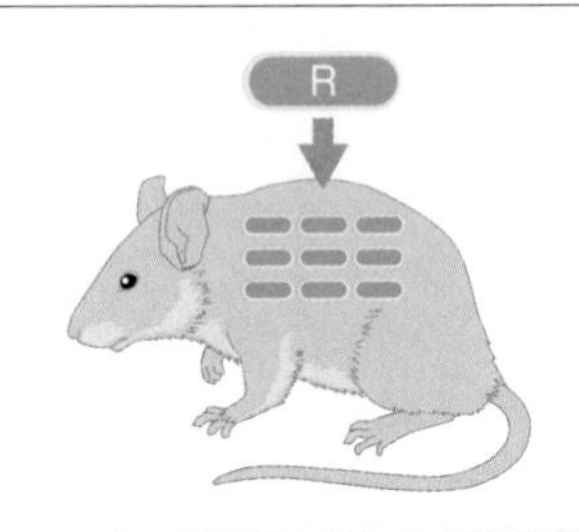

A Griffith's first experiment showed that R cells were harmless. When injected into mice, the bacteria multiplied, but the mice remained healthy.

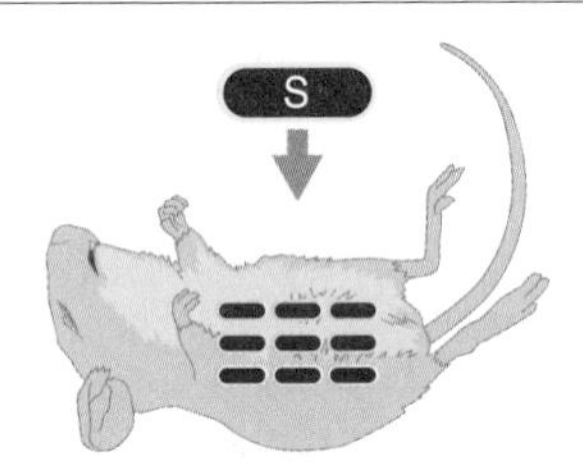

B The second experiment showed that an injection of S cells caused mice to develop fatal pneumonia. Their blood contained live S cells.

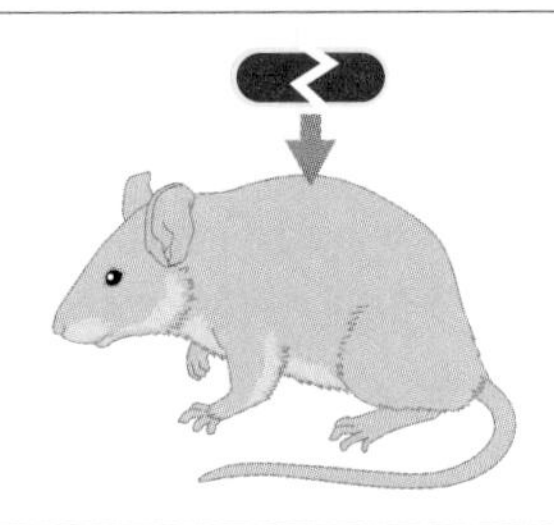

C For a third experiment, Griffith killed S cells with heat before injecting them into mice. The mice remained healthy, indicating that the heat-killed S cells were harmless.

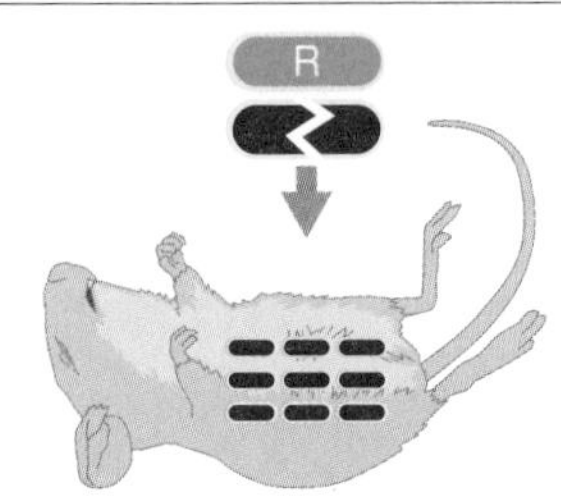

D In his fourth experiment, Griffith injected a mixture of heat-killed S cells and live R cells. To his surprise, the mice became fatally ill, and their blood contained live S cells.

FIGURE 8.2 {Animated} Fred Griffith's experiments with two strains (R and S) of *Streptococcus pneumoniae* bacteria.

The substance we now call DNA (**FIGURE 8.1**) was first described in 1869 by Johannes Miescher, a chemist who extracted it from cell nuclei. Miescher determined that DNA is not a protein, and that it is rich in nitrogen and phosphorus, but he never learned its function. That would take many more years and experiments by many scientists.

Sixty years after Miescher's work, Frederick Griffith unexpectedly uncovered a clue about DNA's function. Griffith was studying pneumonia-causing bacteria in the hope of creating a vaccine. He isolated two strains (types) of the bacteria, one harmless (R), the other lethal (S). Griffith used R and S cells in a series of experiments testing their ability to cause pneumonia in mice (**FIGURE 8.2**). He discovered that heat destroyed the ability of lethal S bacteria to cause pneumonia, but it did not destroy their hereditary material, including whatever specified "kill mice." That material could be transferred from the dead S cells to the live R cells, which put it to use. The transformation was permanent and heritable: Even after hundreds of generations, descendants of transformed R cells retained the ability to kill mice.

What substance had caused the transformation? In 1940, Oswald Avery and Maclyn McCarty set out to identify that substance, which they termed the "transforming principle," by a process of elimination. The researchers made an extract of S cells that contained only lipid, protein, and nucleic acids. The S cell extract could still transform R cells after it had been treated with lipid- and protein-destroying enzymes. Thus, the transforming principle could not be lipid or protein, and Avery and McCarty realized that the substance they were seeking must be nucleic acid—DNA or RNA. DNA-degrading enzymes destroyed the extract's ability to transform cells, but RNA-degrading enzymes did not. Thus, DNA had to be the transforming principle.

The result surprised Avery and McCarty, who, along with most other scientists, had assumed that proteins were the material of heredity. After all, traits are diverse, and proteins are the most diverse of all biological molecules. The two scientists were so skeptical that they published their results only after they had convinced themselves, by years of painstaking experimentation, that DNA was indeed hereditary material. They were also careful to point out that they had not proven DNA was the *only* hereditary material.

Avery and McCarty's tantalizing results prompted a stampede of other scientists into the field of DNA research. The resulting explosion of discovery confirmed the molecule's role as carrier of hereditary information. Key in this advance was the realization that any molecule—DNA or otherwise—had to have certain properties in order to function as the sole repository of hereditary material.

CREDITS: (1) Patrick Landmann/Science Source; (2) © Cengage Learning.

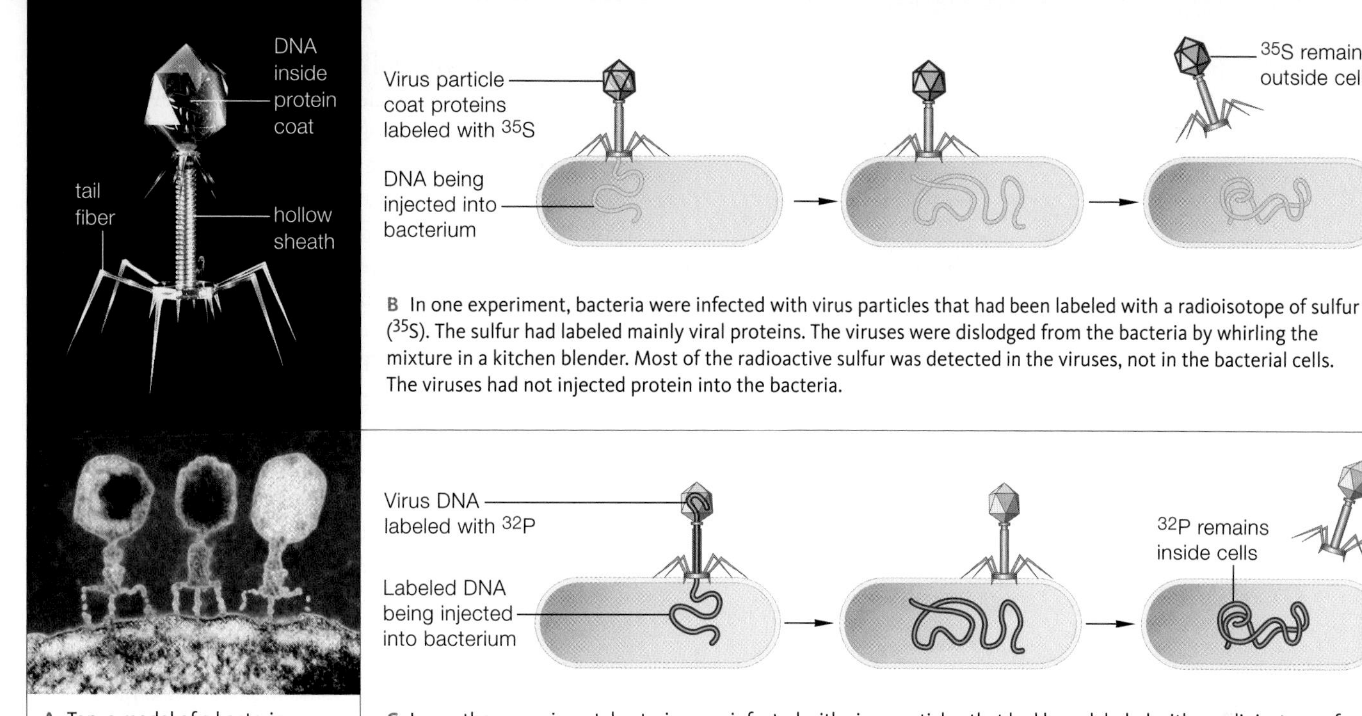

A Top, a model of a bacteriophage. Bottom, micrograph of three viruses injecting DNA into an *E. coli* cell.

B In one experiment, bacteria were infected with virus particles that had been labeled with a radioisotope of sulfur (^{35}S). The sulfur had labeled mainly viral proteins. The viruses were dislodged from the bacteria by whirling the mixture in a kitchen blender. Most of the radioactive sulfur was detected in the viruses, not in the bacterial cells. The viruses had not injected protein into the bacteria.

C In another experiment, bacteria were infected with virus particles that had been labeled with a radioisotope of phosphorus (^{32}P). The phosphorus had labeled mainly viral DNA. When the viruses were dislodged from the bacteria, the radioactive phosphorus was detected mainly inside the bacterial cells. The viruses had injected DNA into the cells—evidence that DNA is the genetic material of this virus.

FIGURE 8.3 {Animated} The Hershey–Chase experiments. Alfred Hershey and Martha Chase carried out experiments to determine the composition of the hereditary material that bacteriophage inject into bacteria. The experiments were based on the knowledge that proteins contain more sulfur (S) than phosphorus (P), and DNA contains more phosphorus than sulfur.

First, a full complement of hereditary information must be transmitted along with the molecule; second, each cell of a given species should contain the same amount of it; third, because the molecule functions as a genetic bridge between generations, it has to be exempt from change; and fourth, it must be capable of encoding the almost unimaginably huge amount of information required to build a new individual.

In the late 1940s, Alfred Hershey and Martha Chase proved that DNA, and not protein, satisfies the first property of a hereditary molecule: It transmits a full complement of hereditary information. Hershey and Chase specialized in working with **bacteriophage**, a type of virus that infects bacteria (**FIGURE 8.3**). Like all viruses, these infectious particles carry information about how to make new viruses in their hereditary material. After one injects a cell with this material, the cell starts making new virus particles. Hershey and Chase carried out an elegant series of experiments proving that the material bacteriophage injects into bacteria is DNA, not protein (**FIGURE 8.3B,C**).

The second property expected of a hereditary molecule was pinned on DNA by André Boivin and Roger Vendrely, who meticulously measured the amount of DNA in cell nuclei from a number of species. In 1948, they proved that body cells of any individual of a species contain precisely the same amount of DNA. Daniel Mazia's laboratory discovered that the protein and RNA content of cells varies over time, but not the DNA content, demonstrating that DNA is not involved in metabolism (and proving DNA has the third property expected of a hereditary molecule). The fourth property—that a hereditary molecule must somehow encode a huge amount of information—would be proven along with the elucidation of DNA's structure, a topic we continue in the next section.

TAKE-HOME MESSAGE 8.1

DNA, the molecule of inheritance, was first discovered in the late 1800s. Its role as the carrier of hereditary information was uncovered over many years, as scientists built upon one another's discoveries.

bacteriophage Virus that infects bacteria.

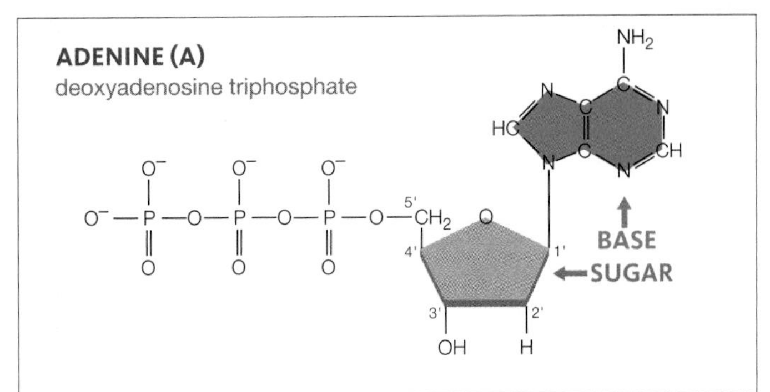

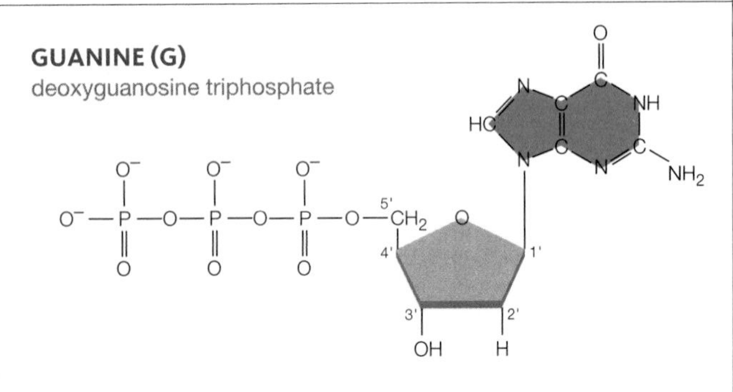

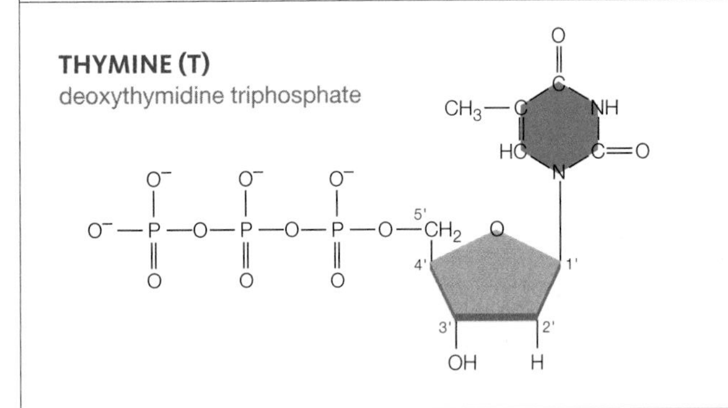

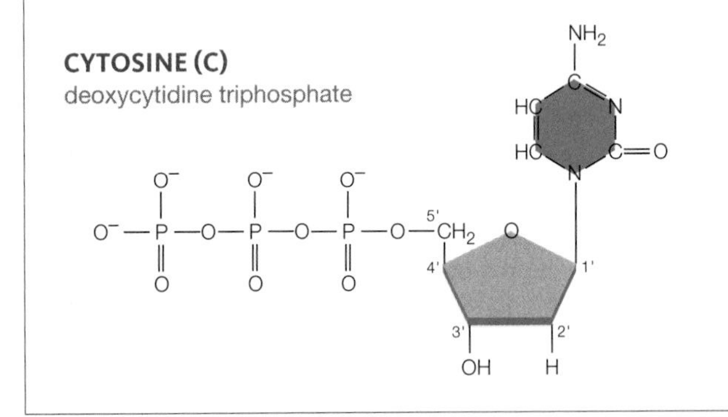

FIGURE 8.4 {Animated} The four nucleotides in DNA. All four have three phosphate groups, a deoxyribose sugar (orange), and a nitrogen-containing base (blue) after which it is named. Biochemist Phoebus Levene identified the structure of these bases and how they are connected in nucleotides in the early 1900s. Levene worked with DNA for almost 40 years.

Adenine and guanine bases are purines; thymine and cytosine, pyrimidines. Numbering the carbons in the sugars allows us to keep track of the orientation of nucleotide chains (compare **FIGURE 8.6**).

BUILDING BLOCKS OF DNA

DNA is a polymer of nucleotides, each with a five-carbon sugar, three phosphate groups, and one of four nitrogen-containing bases (**FIGURE 8.4**). Just how those four nucleotides—adenine (A), guanine (G), thymine (T), and cytosine (C)—are arranged in a DNA molecule was a puzzle that took over 50 years to solve.

Clues about DNA's structure started coming together around 1950, when Erwin Chargaff, one of many researchers investigating DNA's function, made two important discoveries about the molecule. First, the amounts of thymine and adenine are identical, as are the amounts of cytosine and guanine (A = T and G = C). We call this discovery Chargaff's first rule. Chargaff's second discovery, or rule, is that the DNA of different species differs in its proportions of adenine and guanine.

Meanwhile, American biologist James Watson and British biophysicist Francis Crick had been sharing ideas about the structure of DNA. The helical (coiled) pattern of secondary structure that occurs in many proteins (Section 3.4) had just been discovered, and Watson and Crick suspected that the DNA molecule was also a helix. The two spent many hours arguing about the size, shape, and bonding requirements of the four kinds of nucleotides that make up DNA. They pestered chemists to help them identify bonds they might have overlooked, fiddled with cardboard cutouts, and made models from scraps of metal connected by suitably angled "bonds" of wire.

Biochemist Rosalind Franklin had also been working on the structure of DNA. Like Crick, Franklin specialized in x-ray crystallography, a technique in which x-rays are directed through a purified and crystallized substance. Atoms in the substance's molecules scatter the x-rays in a pattern that can be captured as an image. Researchers can use the pattern to calculate the size, shape, and spacing between any repeating elements of the molecules—all of which are details of molecular structure.

As molecules go, DNA is gigantic, and it was difficult to crystallize given the techniques of the time. Franklin made the first clear x-ray diffraction image (*left*) of DNA as it occurs in cells. From the information in this image, she calculated that DNA is very long compared to its 2-nanometer diameter. She also identified a repeating pattern every 0.34 nanometer along its length, and another every 3.4 nanometers.

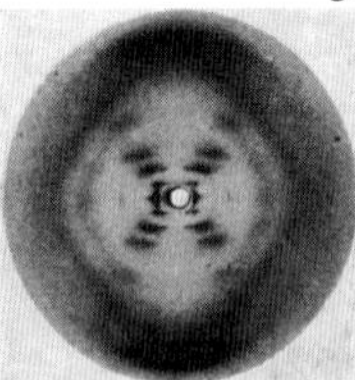

Franklin's image and data came to the attention of Watson and Crick, who now had all the information they needed to build a model of the DNA helix (**FIGURE 8.5**), one with two sugar–phosphate chains running in opposite directions, and paired bases inside (**FIGURE 8.6**). Bonds between the sugar of one nucleotide and the phosphate of the next form the backbone of each chain (or strand). Hydrogen bonds between the internally positioned bases hold the two strands together. Only two kinds of base pairings form: A to T, and G to C, which explains the first of Chargaff's rules. Most scientists had assumed (incorrectly) that the bases had to be on the outside of the helix, because they would be more accessible to DNA-copying enzymes that way. You will see in Section 8.4 how DNA replication enzymes access the bases on the inside of the double helix.

DNA'S BASE SEQUENCE

A small piece of DNA from a tulip, a human, or any other organism might be:

T G A G G A C T C C T C
A C T C C T G A G G A G
one base pair

Notice how the two strands of DNA match. They are complementary—the base of each nucleotide on one strand pairs with a suitable partner base on the other. This base-pairing pattern (A to T, G to C) is the same in all molecules of DNA. How can just two kinds of base pairings give rise to the incredible diversity of traits we see among living things? Even though DNA is composed of only four nucleotides, the *order* in which one nucleotide follows the next along a strand—the **DNA sequence**—varies tremendously among species (which explains Chargaff's second rule). DNA molecules can be hundreds of millions of nucleotides long, so their sequence can encode a massive amount of information (we return to the nature of that information in the next chapter). DNA sequence variation is the basis of traits that define species and distinguish individuals. Thus DNA, the molecule of inheritance in every cell, is the basis of life's unity. Variations in its nucleotide sequence are the foundation of life's diversity.

DNA sequence Order of nucleotides in a strand of DNA.

TAKE-HOME MESSAGE 8.2

A DNA molecule consists of two nucleotide chains (strands) running in opposite directions and coiled into a double helix. Internally positioned nucleotide bases hydrogen-bond between the two strands. A pairs with T, and G with C.

The sequence of bases along a DNA strand varies among species and among individuals. This variation is the basis of life's diversity.

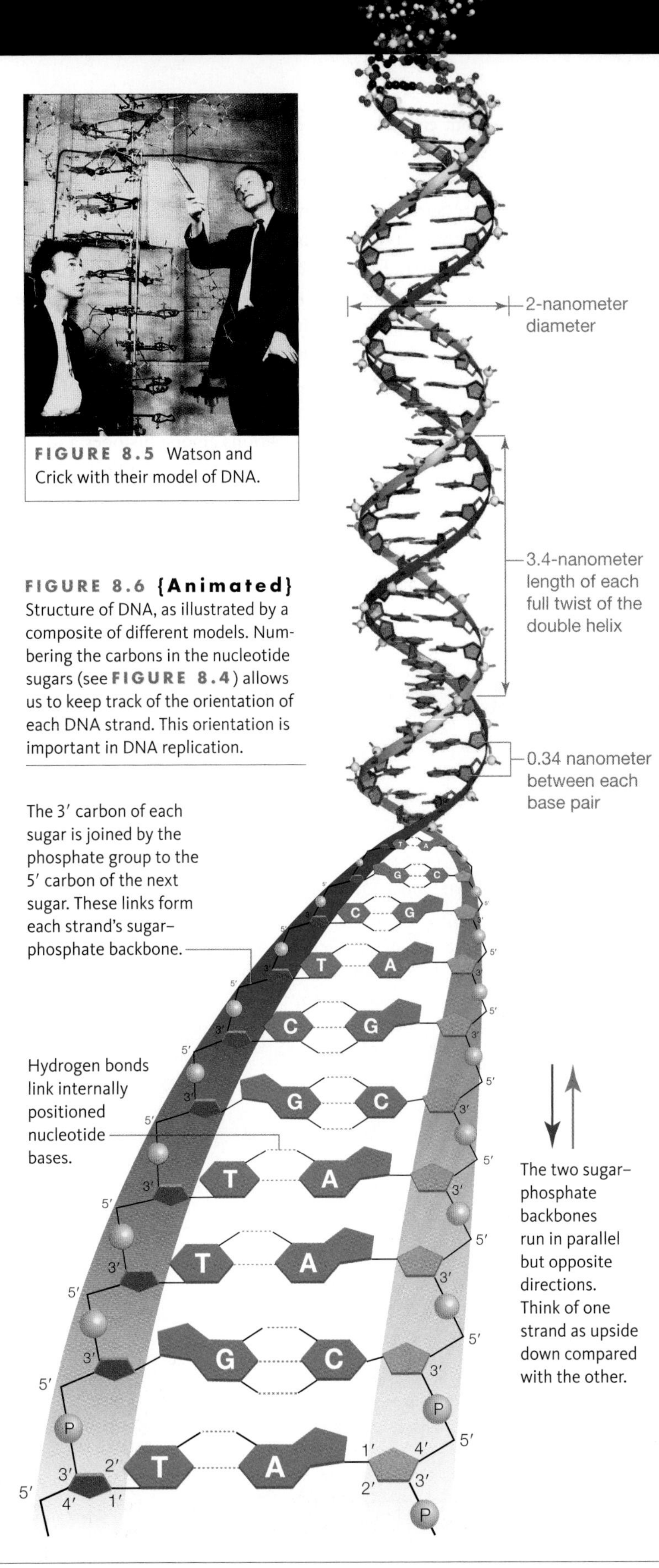

FIGURE 8.5 Watson and Crick with their model of DNA.

FIGURE 8.6 {Animated} Structure of DNA, as illustrated by a composite of different models. Numbering the carbons in the nucleotide sugars (see **FIGURE 8.4**) allows us to keep track of the orientation of each DNA strand. This orientation is important in DNA replication.

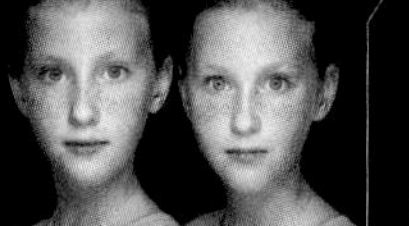

8.3 WHAT IS A CHROMOSOME?

Stretched out end to end, the DNA in a single human cell would be about 2 meters (6.5 feet) long. How can that much DNA pack into a nucleus that is less than 10 micrometers in diameter? Such tight packing is possible because proteins associate with the DNA and help keep it organized. In cells, DNA molecules and their associated proteins form structures

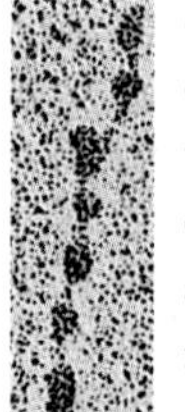

called **chromosomes** (FIGURE 8.7). In a chromosome of a eukaryotic cell, the DNA double helix ❶ wraps twice at regular intervals around "spools" of proteins called **histones** ❷. These DNA–histone spools, which are called **nucleosomes**, look like beads on a string in micrographs (*left*). Interactions among histones and other proteins twist the spooled DNA into a tight fiber ❸. This fiber coils, and then it coils again into a hollow cylinder, a bit like an old-style telephone cord ❹.

During most of the cell's life, each chromosome consists of one DNA molecule. When the cell prepares to divide, it duplicates its chromosomes by DNA replication (more about this process in the next section). After replication, each chromosome consists of two DNA molecules, or **sister chromatids**, attached to one another at a constricted region called the **centromere**:

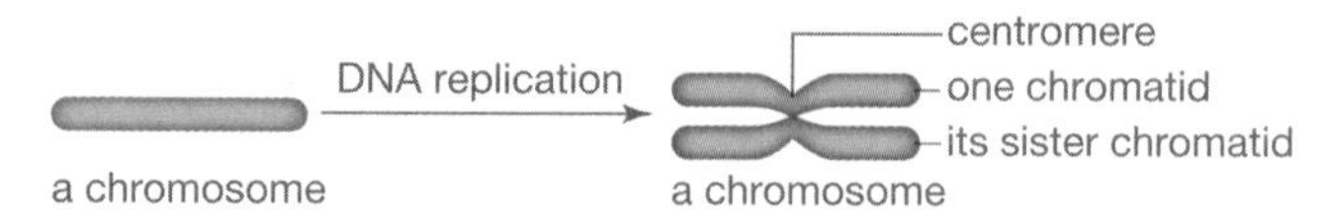

The chromosomes condense into their familiar "X" shapes ❺ just before the cell divides.

CHROMOSOME NUMBER AND TYPE

The DNA of a eukaryotic cell is divided among several chromosomes that differ in length and shape ❻. Each species has a characteristic **chromosome number**—the number of chromosomes in its cells. For example, the chromosome

FIGURE 8.7
{Animated}
Chromosome structure.

① Two strands of DNA twist into a double helix.
② At regular intervals, the DNA (blue) wraps around a core of histone proteins (purple).
③ The DNA and proteins associated with it twist tightly into a fiber.
④ The fiber coils and then coils again to form a hollow cylinder.
⑤ At its most condensed, a duplicated chromosome has an X shape.
⑥ The DNA in the nucleus of a eukaryotic cell is typically divided into a number of chromosomes.

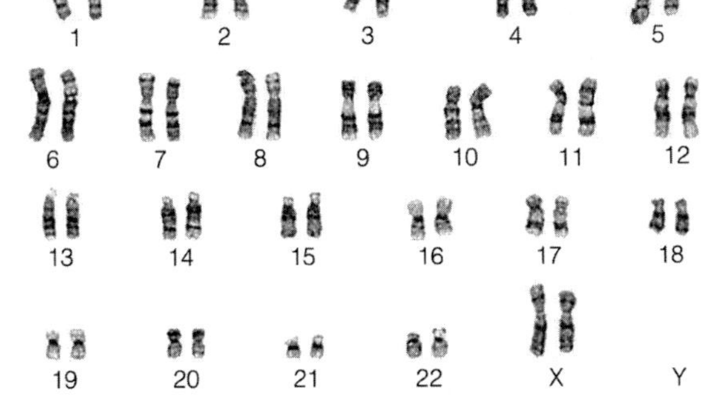

FIGURE 8.8 {Animated} A karyotype of a human female, showing 22 pairs of autosomes and a pair of X chromosomes (XX).

number of oak trees is 12, so the nucleus of a cell from an oak tree contains 12 chromosomes. The chromosome number of humans is 46, so our cells have 46 chromosomes.

Actually, human body cells have two sets of 23 chromosomes—two of each type. Having two sets of chromosomes means these cells are **diploid**, or 2*n*. A **karyotype** is an image of an individual's diploid set of chromosomes (**FIGURE 8.8**). To create a karyotype, cells taken from an individual are treated to make the chromosomes condense, and then stained so the chromosomes can be distinguished under a microscope. A micrograph of a single cell is digitally rearranged so the images of the chromosomes are lined up by centromere location, and arranged according to size, shape, and length.

In a human body cell, all but one pair of chromosomes are **autosomes**, which are the same in both females and males. The two autosomes of a pair have the same length, shape, and centromere location. They also hold information about the same traits. Think of them as two sets of books on how to build a house. Your father gave you one set. Your mother had her own ideas about wiring, plumbing, and so on. She gave you an alternate set that says slightly different things about many of those tasks.

Members of a pair of **sex chromosomes** differ between females and males, and the differences determine an individual's sex. The sex chromosomes of humans are called X and Y. The body cells of typical human females have two X chromosomes (XX); those of typical human males have one X and one Y chromosome (XY). XX females and XY males are the rule among fruit flies, mammals, and many other animals, but there are other patterns. In butterflies, moths, birds, and certain fishes, males are the ones with identical sex chromosomes. Environmental factors (not sex chromosomes) determine sex in some invertebrates and reptiles (**FIGURE 8.9**).

FIGURE 8.9 National Geographic Explorer Mariana Fuentes studies sea turtle populations. Climate change is an immediate, serious threat to these reptiles, in part because the temperature of the sand in which their eggs are buried—not sex chromosomes—determines the sex of the hatchlings. Fuentes predicts that rising global temperatures will soon skew the gender ratio of hatchlings toward all female, with disastrous results for sea turtle populations.

TAKE-HOME MESSAGE 8.3

In cells, DNA and associated proteins are organized as chromosomes.

A eukaryotic cell's DNA is divided among some characteristic number of chromosomes, which differ in length and shape.

Members of a pair of sex chromosomes differ between males and females. Chromosomes that are the same in males and females are called autosomes.

autosome A chromosome that is the same in males and females.
centromere Of a duplicated eukaryotic chromosome, constricted region where sister chromatids attach to each other.
chromosome A structure that consists of DNA and associated proteins; carries part or all of a cell's genetic information.
chromosome number The total number of chromosomes in a cell of a given species.
diploid Having two of each type of chromosome characteristic of the species (2*n*).
histone Type of protein that structurally organizes eukaryotic chromosomes.
karyotype Image of an individual's set of chromosomes arranged by size, length, shape, and centromere location.
nucleosome A length of DNA wound twice around a spool of histone proteins.
sex chromosome Member of a pair of chromosomes that differs between males and females.
sister chromatids The two attached DNA molecules of a duplicated eukaryotic chromosome.

During most of its life, a typical cell contains one set of chromosomes. When the cell reproduces, it divides. The two descendant cells must inherit a full complement of chromosomes—a complete copy of genetic information—or they will not function properly. Thus, in preparation for division, the cell copies its chromosomes so that it contains two sets: one for each of its future offspring.

The process by which a cell copies its DNA is called **DNA replication**. During this energy-intensive metabolic pathway, enzymes and other molecules open the double helix of a DNA molecule to expose the internally positioned bases, then link nucleotides into new strands of DNA according to the sequence of those bases.

Each chromosome is replicated in its entirety. Two identical molecules of DNA are the result. In eukaryotes, these molecules are sister chromatids that remain attached at the centromere until cell division occurs.

① As replication begins, enzymes begin to unwind and separate the two strands of DNA.

② Primers base-pair with the exposed single DNA strands.

③ Starting at primers, DNA polymerases (green boxes) assemble new strands of DNA from nucleotides, using the parent strands as templates.

④ DNA ligase seals any gaps that remain between bases of the "new" DNA, so a continuous strand forms.

⑤ Each parental DNA strand serves as a template for assembly of a new strand of DNA. Both strands of the double helix serve as templates, so two double-stranded DNA molecules result. One strand of each is parental (old), and the other is new, so DNA replication is said to be semiconservative.

enzymes

primer

DNA polymerase

nucleotide

DNA ligase

FIGURE 8.10 {Animated} DNA replication, in which a double-stranded molecule of DNA is copied in entirety. The Y-shaped structure of a DNA molecule undergoing replication is called a replication fork.

SEMICONSERVATIVE REPLICATION

Before DNA replication, a chromosome consists of one molecule of DNA—one double helix (**FIGURE 8.10**). As replication begins, enzymes break the hydrogen bonds that hold the double helix together, so the two DNA strands unwind and separate ①. Another enzyme constructs **primers**—short, single strands of nucleotides that serve as attachment points for **DNA polymerase**, the enzyme that assembles new strands of DNA. The nucleotide bases of a primer can form hydrogen bonds with exposed bases of a single strand of DNA ②. Thus, a primer can base-pair with a complementary strand of DNA:

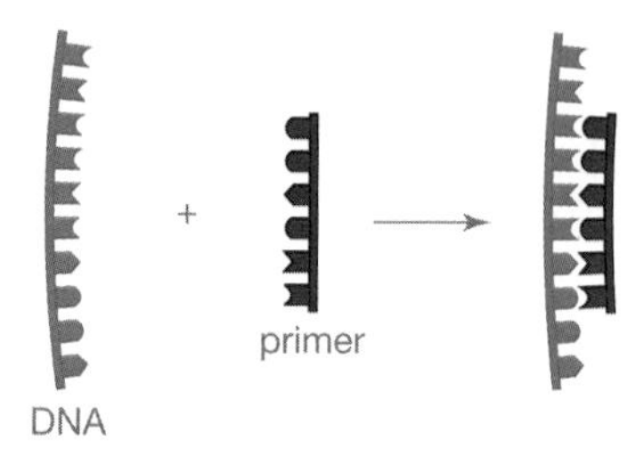

The establishment of base-pairing between two strands of DNA (or DNA and RNA) is called **nucleic acid hybridization**. Hybridization is spontaneous, driven by hydrogen bonding between bases of complementary strands.

DNA polymerases attach to the hybridized primers and begin DNA synthesis. As a DNA polymerase moves along a strand, it uses the sequence of exposed nucleotide bases as a template, or guide, to assemble a new strand of DNA from free nucleotides ③.

Each nucleotide provides energy for its own attachment to the end of a growing strand of DNA. Remember from Section 5.5 that the bonds between a nucleotide's phosphate groups hold a lot of free energy. Two of the three phosphate groups are removed when the nucleotide is added to a DNA strand. Breaking those bonds releases enough free energy to drive the attachment.

A DNA polymerase follows base-pairing rules: It adds a T to the end of the new DNA strand when it reaches an A in the template strand; it adds a G when it reaches a C; and so on. Thus, the nucleotide sequence of each new strand of DNA is complementary to its template (parental) strand. The enzyme **DNA ligase** seals any gaps, so the new DNA strands are continuous ④.

Both of the two strands of the parent molecule are copied at the same time. As each new DNA strand lengthens, it winds up with its template strand into a double helix. So, after replication, two double-stranded molecules of DNA have formed ❺. One strand of each molecule is parental (old), and the other is new; hence the name of the process, **semiconservative replication.** Each new strand of DNA is complementary in sequence to one of the two parent strands, so both double-stranded molecules produced by DNA replication are duplicates of the parent molecule.

DIRECTIONAL SYNTHESIS

Numbering the carbons of the sugars in nucleotides allows us to keep track of the orientation of DNA strands in a double helix (see **FIGURES 8.4** and **8.6**). Each strand has two ends. The last carbon atom on one end of the strand is a 5′ (5 prime) carbon of a sugar; the last carbon atom on the other end is a 3′ (three prime) carbon of a sugar:

5′ ——— 3′
3′ ——— 5′

DNA polymerase can attach a nucleotide only to a 3′ end. Thus, during DNA replication, only one of two new strands of DNA can be constructed in a single piece (**FIGURE 8.11**). Synthesis of the other strand occurs in segments that must be joined by DNA ligase where they meet up. This is why we say that DNA synthesis proceeds only in the 5′ to 3′ direction.

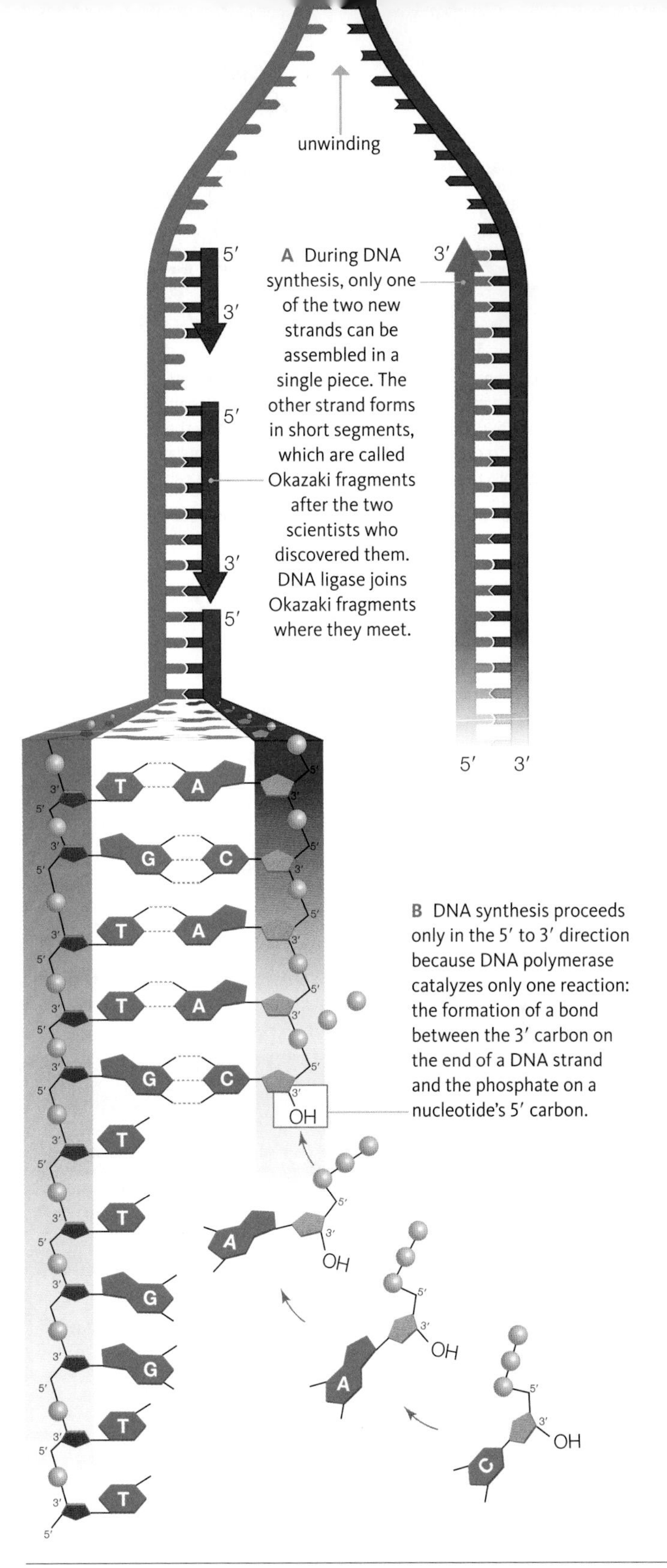

FIGURE 8.11 Discontinuous synthesis of DNA. This close-up of a replication fork shows that only one of the two new DNA strands is assembled in one piece.

DNA ligase Enzyme that seals gaps in double-stranded DNA.
DNA polymerase DNA replication enzyme. Uses one strand of DNA as a template to assemble a complementary strand of DNA from nucleotides.
DNA replication Process by which a cell duplicates its DNA before it divides.
nucleic acid hybridization Convergence of complementary nucleic acid strands. Arises because of base-pairing interactions.
primer Short, single strand of DNA that base-pairs with a targeted DNA sequence.
semiconservative replication Describes the process of DNA replication, which produces two copies of a DNA molecule: one strand of each copy is new, and the other is parental.

TAKE-HOME MESSAGE 8.4

- DNA replication is an energy-intensive metabolic pathway by which a cell copies its chromosomes.
- During replication of a molecule of DNA, each strand of its double helix serves as a template for synthesis of a new, complementary strand of DNA.
- Replication of a molecule of DNA produces two double helices that are duplicates of the parent molecule. One strand of each is parental; the other is new.

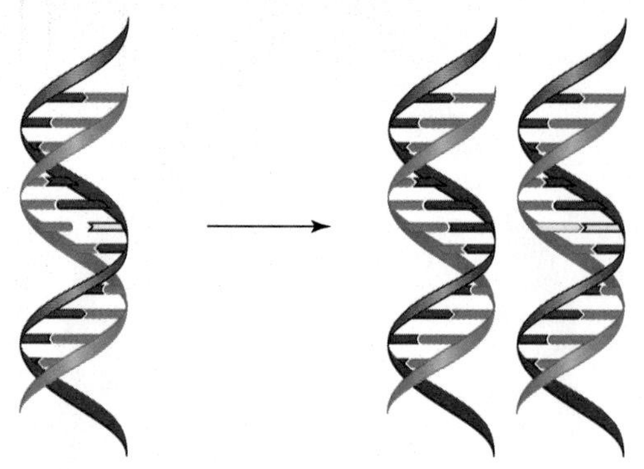

A Repair enzymes can recognize a mismatched base (yellow), but sometimes fail to correct it before DNA replication.

B After replication, both strands base-pair properly. Repair enzymes can no longer recognize the error, which has now become a mutation that will be passed on to the cell's descendants.

FIGURE 8.12 How a replication error can become a mutation.

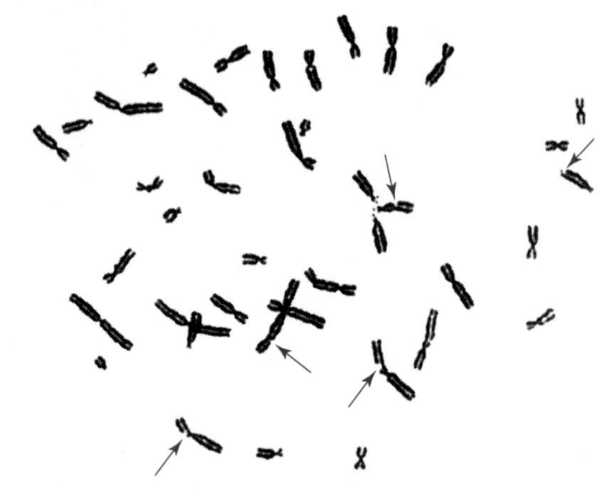

A Major breaks (red arrows) in chromosomes of a human white blood cell after exposure to ionizing radiation. Pieces of broken chromosomes often become lost during DNA replication.

B These *Ranunculus* flowers were grown from plants harvested around Chernobyl, Ukraine, where in 1986 an accident at a nuclear power plant released huge amounts of radiation. A normal flower is shown for comparison, in the inset.

FIGURE 8.13 Exposure to ionizing radiation causes mutations.

REPLICATION ERRORS

Mistakes can and do occur during DNA replication. Sometimes, the wrong base is added to a growing DNA strand; at other times, a nucleotide gets lost, or an extra one slips in. Either way, the newly synthesized DNA strand will no longer be complementary to its parent strand. Most of these replication errors occur simply because DNA polymerases work very fast, copying about 50 nucleotides per second in eukaryotes, and up to 1,000 per second in bacteria. Mistakes are inevitable, and some types of DNA polymerases make a lot of them. Luckily, most DNA polymerases also proofread their work. They can correct a mismatch by reversing the synthesis reaction to remove the mispaired nucleotide, then resuming synthesis in the forward direction.

Replication errors also occur after the cell's DNA gets broken or otherwise damaged, because DNA polymerases do not copy damaged DNA very well. In most cases, repair enzymes and other proteins remove and replace damaged or mismatched bases in DNA before replication begins.

When proofreading and repair mechanisms fail, an error becomes a **mutation**, a permanent change in the DNA sequence of a cell's chromosome(s). Repair enzymes cannot fix a mutation after DNA replication has occurred, because they do not recognize correctly paired bases (**FIGURE 8.12**). Thus, a mutation is passed to the cell's descendants, their descendants, and so on.

Mutations can form in any type of cell. Those that occur during egg or sperm formation can be passed to offspring, and in fact each human child is born with an average of 36 new ones. Mutations that alter DNA's instructions may have a harmful or lethal outcome; most cancers begin with them (we return to this topic in Section 11.5). However, not all mutations are dangerous: As you will see in Chapter 17, they give rise to the variation in traits that is the raw material of evolution.

AGENTS OF DNA DAMAGE

Electromagnetic energy with a wavelength shorter than 320 nanometers, including x-rays, most ultraviolet (UV) light, and gamma rays, can knock electrons out of atoms. Such ionizing radiation damages DNA, breaking it into pieces that get lost during replication (**FIGURE 8.13A**). Ionizing radiation can also cause covalent bonds to form between bases on opposite strands of the double helix, an outcome that permanently blocks replication. (We consider cancer-causing effects of such cell cycle interruptions in Chapter 11.) High-energy radiation also fatally alters nucleotide bases. Repair enzymes can remove bases damaged in this way, but they leave an empty space in the double helix or even a strand break. Sometimes the enzymes cut out the entire nucleotide from the strand, leaving an unpaired nucleotide on the opposite strand. Any of these events can result in mutations (**FIGURE 8.13B**).

UV light in the range of 320–380 nanometers can boost electrons to a higher energy level, but not enough to knock them out of atoms. UV light in this range is still dangerous, because it has enough energy to open up the double bond in the ring of a cytidine or thymine base. The open ring can form a covalent bond with the ring of an adjacent cytidine or thymine (*left*). The resulting dimer kinks the DNA strand. DNA polymerase tends to copy the kinked part incorrectly during replication, and mutations are the outcome. Mutations that arise as a result of nucleotide dimers are the primary cause of skin cancer. Exposing unprotected skin to sunlight increases the risk of cancer because its UV wavelengths cause dimers to form. For every second a skin cell spends in the sun, 50–100 of these dimers form in its DNA.

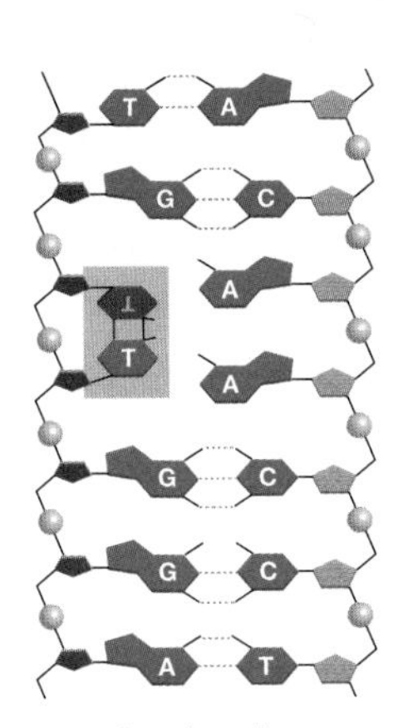

a thymine dimer

Exposure to some natural or synthetic chemicals also causes mutations. For instance, several of the fifty-five or more cancer-causing chemicals in tobacco smoke transfer methyl groups ($-CH_3$) to the nucleotide bases in DNA. Nucleotides altered in this way do not base-pair correctly. Other chemicals in the smoke are converted by the body to compounds that are easier to excrete, and the breakdown products bind irreversibly to DNA. Replication errors that can lead to mutation may be the outcome in both cases. Cigarette smoke also contains free radicals, which inflict the same damage on DNA as ionizing radiation.

mutation Permanent change in the nucleotide sequence of DNA.

TAKE-HOME MESSAGE 8.5

Proofreading and repair mechanisms usually maintain the integrity of a cell's genetic information by correcting mispaired bases and fixing damaged DNA before replication.

Mismatched or damaged nucleotides that are not repaired can become mutations—permanent changes in the DNA sequence of a chromosome.

DNA damage by environmental agents such as UV light and chemicals can result in mutations, because damaged DNA is not replicated very well.

PEOPLE MATTER

DR. ROSALIND FRANKLIN

Rosalind Franklin had been told she would be the only one in her department working on the structure of DNA, so she did not know that Maurice Wilkins was already doing the same thing just down the hall. Franklin's meticulous work yielded the first clear x-ray diffraction image of DNA as it occurs inside cells, and she gave a presentation on this work in 1952. DNA, she said, had two chains twisted into a double helix, with a backbone of phosphate groups on the outside, and bases arranged in an as-yet unknown way on the inside. She had calculated DNA's diameter, the distance between its chains and between its bases, the angle of the helix, and the number of bases in each coil. Francis Crick, with his crystallography background, would have recognized the significance of the work—if he had been there. James Watson was in the audience but he did not fully understand the implications of Franklin's x-ray diffraction image or her calculations.

Franklin started to write a research paper on her findings. Meanwhile, and perhaps without her knowledge, Watson reviewed Franklin's x-ray diffraction image with Wilkins, and Watson and Crick read Franklin's unpublished data. That data provided Watson and Crick with the last piece of the DNA puzzle. In 1953, they put together all of the clues that had been accumulating for fifty years and built the first accurate model of DNA structure. On April 25, 1953, Rosalind Franklin's work appeared third in a series of articles about the structure of DNA in the journal *Nature*. Wilkins's research paper was the second article in the series. The work of Franklin and Wilkins supported with experimental evidence Watson and Crick's theoretical model, which was presented in the first article.

Rosalind Franklin died in 1958 at the age of 37, of ovarian cancer probably caused by extensive exposure to x-rays during her work. At the time, the link between x-rays, mutations, and cancer was not understood. Because the Nobel Prize is not given posthumously, Franklin did not share in the 1962 honor that went to Watson, Crick, and Wilkins for the discovery of the structure of DNA.

The word "cloning" means making an identical copy of something, and it can refer to deliberate interventions in reproduction that produce an exact genetic copy of an organism. Genetically identical organisms occur all the time in nature, arising most often by the process of asexual reproduction (which we discuss in Chapter 11). Embryo splitting, another natural process, results in identical twins. The first few divisions of a fertilized egg form a ball of cells that sometimes splits spontaneously. If both halves of the ball continue to develop independently, identical twins result.

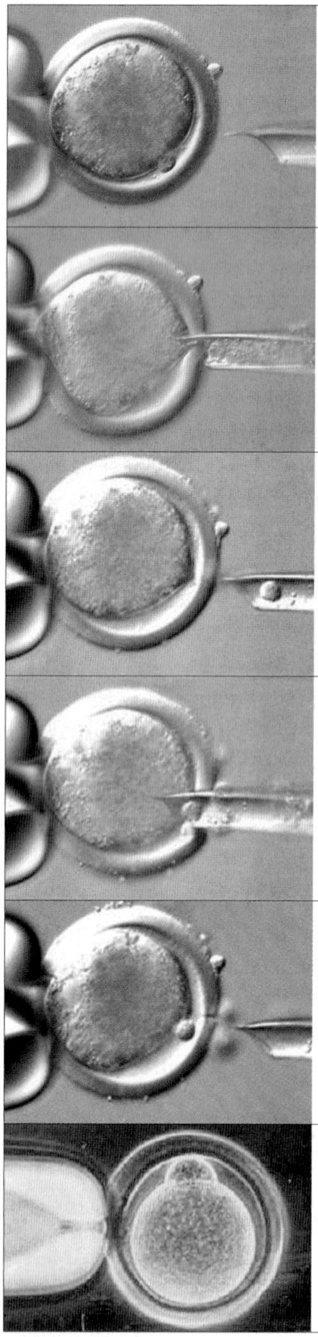

Artificial embryo splitting has been used in research and animal husbandry for decades. With this technique, a ball of cells is grown from a fertilized egg in a laboratory. The tiny ball is teased apart into two halves, each of which goes on to develop as a separate embryo. The embryos are implanted in surrogate mothers, who give birth to identical twins. Artificial twinning and any other technology that yields genetically identical individuals is called **reproductive cloning**.

Twins get their DNA from two parents that typically differ in their DNA sequence. Thus, although twins produced by embryo splitting are identical to one another, they are not identical to either parent. Animal breeders who want an exact copy of a specific individual may turn to a cloning method that starts with a somatic cell taken from an adult organism (a somatic cell is a body cell, as opposed to a reproductive cell; *soma* is a Greek word for body). All cells descended from a fertilized egg inherit the same DNA. Thus, the DNA in each living cell of an individual is like a master blueprint that contains enough information to build an entirely new individual. However, a somatic cell taken from an adult will not automatically start dividing to produce an embryo. It must first be tricked into rewinding its developmental clock. During development, cells in an embryo start using different subsets of their DNA. As they do, the cells become different in form and function, a process called **differentiation**. Differentiation is usually a one-way path in animal cells. Once a cell has become specialized, all of its descendant cells will be specialized the same way. By the time a liver cell, muscle cell, or other differentiated cell forms, most of its DNA has been turned off, and is no longer used. To clone an adult, scientists transform one of its differentiated cells into an undifferentiated cell by turning its unused DNA back on. One way to do this is **somatic cell nuclear transfer (SCNT)**, a laboratory procedure in which an unfertilized egg's nucleus is replaced with the nucleus of a donor's somatic cell (**FIGURE 8.14**). If all goes well, the egg's cytoplasm reprograms the transplanted DNA to direct the development of an embryo, which is then implanted into a surrogate mother. The animal that is born to the surrogate is genetically identical with the donor of the nucleus—a clone.

SCNT is now a common practice among people who breed prized livestock. Among other benefits, many more offspring can be produced in a given time frame by cloning than by traditional breeding methods. Cloned animals have the same championship features as their DNA donors

A A cow's egg is held in place by suction through a hollow glass tube called a micropipette. DNA is identified by a purple stain.

B Another micropipette punctures the egg and sucks out the DNA. All that remains inside the egg's plasma membrane is cytoplasm.

C A new micropipette prepares to enter the egg at the puncture site. The pipette contains a cell grown from the skin of a donor animal.

D The micropipette enters the egg and delivers the skin cell to a region between the cytoplasm and the plasma membrane.

E After the pipette is withdrawn, the donor's skin cell is visible next to the cytoplasm of the egg. The transfer is now complete.

F An electric current causes the foreign cell to fuse with and deposit its nucleus into the cytoplasm of the egg. The egg begins to divide, and an embryo forms.

FIGURE 8.14 {Animated} An example of somatic cell nuclear transfer, using cattle cells. This series of micrographs was taken at a company that specializes in cloning livestock.

FIGURE 8.15 Champion Holstein dairy cow (*right*) and her clone (*left*), who was produced by somatic cell nuclear transfer in 2003.

(**FIGURE 8.15**). Offspring can also be produced from a donor animal that is castrated or even dead.

As the techniques become routine, cloning humans is no longer only within the realm of science fiction. SCNT is already being used to produce human embryos for medical purposes, a practice called **therapeutic cloning**. Undifferentiated (stem) cells taken from the cloned human embryos are used to treat human patients and to study human diseases. For example, embryos created using cells from people with genetic heart defects are allowing researchers to study how the defect causes developing heart cells to malfunction. Such research may ultimately lead to treatments for people who suffer from fatal diseases. (We return to the topic of stem cells and their potential medical benefits in Chapter 28.) Human cloning is not the intent of such research, but if it were, SCNT would indeed be the first step toward that end.

differentiation Process by which cells become specialized during development; occurs as different cells in an embryo begin to use different subsets of their DNA.
reproductive cloning Technology that produces genetically identical individuals.
somatic cell nuclear transfer (SCNT) Reproductive cloning method in which the DNA of an adult donor's body cell is transferred into an unfertilized egg.
therapeutic cloning The use of SCNT to produce human embryos for research purposes.

TAKE-HOME MESSAGE 8.6

Reproductive cloning technologies produce genetically identical individuals.

The DNA inside a living cell contains all the information necessary to build a new individual.

In somatic cell nuclear transfer (SCNT), the DNA of an adult donor's body cell is transferred to an egg with no nucleus. The hybrid cell may develop into an embryo that is genetically identical to the donor's.

CREDITS: (15) Courtesy of Cyagra, Inc., www.cyagra.com; (16) © James Symington.

8.7 A HERO DOG'S GOLDEN CLONES

FIGURE 8.16 James Symington and his dog Trakr at Ground Zero, 9/11/2001.

WHY CLONE ANIMALS? Consider the story of Canadian police officer James Symington and his search dog Trakr. On September 11, 2001, Symington drove Trakr from Nova Scotia to Manhattan. Within hours of arriving, the dog led rescuers to the area where the final survivor of the World Trade Center attacks was buried. She had been clinging to life, pinned under rubble from the building where she had worked. Symington and Trakr helped with the search and rescue efforts for three days nonstop, until Trakr collapsed from smoke and chemical inhalation, burns, and exhaustion (**FIGURE 8.16**).

Trakr survived the ordeal, but later lost the use of his limbs, probably because of toxic smoke exposure at Ground Zero. The hero dog died in April 2009, but his DNA lives on—in his clones. Symington's essay about Trakr's superior nature and abilities as a search and rescue dog won the Golden Clone Giveaway, a contest to find the world's most clone-worthy dog. Trakr's DNA was inserted into donor dog eggs, which were then implanted into surrogate mother dogs. Five puppies, all clones of Trakr, were delivered to Symington in July 2009. Today, the clones are search and rescue dogs for Team Trakr Foundation, Symington's international humanitarian organization.

Cloning animals raises uncomfortable ethical questions about cloning humans. For example, if cloning a lost animal for a grieving owner is acceptable, why would it not be acceptable to clone a lost child for a grieving parent? Different people have very different answers to such questions, so controversy over cloning continues to rage even as techniques improve.

Summary

SECTION 8.1 Eighty years of experimentation with cells and **bacteriophage** offered solid evidence that deoxyribonucleic acid (DNA), not protein, is the hereditary material of all life.

SECTION 8.2 A DNA nucleotide has a five-carbon sugar (deoxyribose), three phosphate groups, and one of four nitrogen-containing bases after which the nucleotide is named: adenine, thymine, guanine, or cytosine. DNA is a polymer that consists of two strands of these nucleotides coiled into a double helix. Hydrogen bonding between the internally positioned bases holds the strands together. The bases pair in a consistent way: adenine with thymine (A–T), and guanine with cytosine (G–C). The order of bases along a strand of DNA—the **DNA sequence**—varies among species and among individuals, and this variation is the basis of life's diversity.

SECTION 8.3 The DNA of eukaryotes is typically divided among a number of **chromosomes** that differ in length and shape. In eukaryotic chromosomes, the DNA wraps around **histone** proteins to form **nucleosomes**. When duplicated, a eukaryotic chromosome consists of two **sister chromatids** attached at a **centromere**. **Diploid** cells have two of each type of chromosome. **Chromosome number** is the total number of chromosomes in a cell of a given species. A human body cell has twenty-three pairs of chromosomes. Members of a pair of **sex chromosomes** differ among males and females. Chromosomes that are the same in males and females are **autosomes**. Autosomes of a pair have the same length, shape, and centromere location. A **karyotype** is an individual's complete set of chromosomes.

SECTION 8.4 A cell copies its chromosomes before it divides so each of its offspring will inherit a complete set of genetic information. **DNA replication** is the energy-intensive metabolic pathway in which a cell copies its chromosomes. For each double-stranded molecule of DNA that is copied, two double-stranded DNA molecules that are duplicates of the parent are produced. One strand of each molecule is new, and the other is parental; thus the name **semiconservative replication**. During DNA replication, enzymes unwind the double helix. **Primers** base-pair with the exposed single strands of DNA, a process called **nucleic acid hybridization**. Starting at the primers, **DNA polymerase** enzymes use each strand as a template to assemble new, complementary strands of DNA from free nucleotides. Synthesis of one strand necessarily occurs discontinuously. **DNA ligase** seals any gaps to form continuous strands.

SECTION 8.5 Proofreading by DNA polymerases corrects most DNA replication errors as they occur. DNA damage by environmental agents, including ionizing and nonionizing radiation, free radicals, and some other natural and synthetic chemicals, can lead to replication errors because DNA polymerase does not copy damaged DNA very well. Most types of DNA damage can be repaired before replication begins. Uncorrected replication errors become **mutations**, which are permanent changes in the nucleotide sequence of a cell's DNA. Cancer begins with mutations, but not all mutations are harmful.

SECTIONS 8.6, 8.7 **Somatic cell nuclear transfer (SCNT)** and other types of **reproductive cloning** technologies produce genetically identical individuals (clones). SCNT using human cells is called **therapeutic cloning**. The DNA in each living cell contains all the information necessary to build a new individual. During development, cells of an embryo become specialized as they begin to use different subsets of their DNA (a process called **differentiation**).

Self-Quiz

Answers in Appendix VII

1. Which is not a nucleotide base in DNA?
   a. adenine c. glutamine e. cytosine
   b. guanine d. thymine f. All are in DNA.
2. What are the base-pairing rules for DNA?
   a. A–G, T–C b. A–C, T–G c. A–T, G–C
3. Variation in _______ is the basis of variation in traits.
   a. karyotype c. the double helix
   b. the DNA sequence d. chromosome number
4. One species' DNA differs from others in its _______ .
   a. nucleotides c. sugar–phosphate backbone
   b. DNA sequence d. all of the above
5. In eukaryotic chromosomes, DNA wraps around _______ .
   a. histone proteins c. centromeres
   b. nucleosomes d. none of the above
6. Chromosome number _______ .
   a. refers to a particular chromosome in a cell
   b. is an identifiable feature of a species
   c. is the number of autosomes in cells of a given type
7. Human body cells are diploid, which means _______ .
   a. they are complete
   b. they have two sets of chromosomes
   c. they contain sex chromosomes
8. When DNA replication begins, _______ .
   a. the two DNA strands unwind from each other
   b. the two DNA strands condense for base transfers
   c. old strands move to find new strands
9. DNA replication requires _______ .
   a. DNA polymerase c. primers
   b. nucleotides d. all are required
10. Energy that drives DNA synthesis comes from _______ .
   a. ATP only c. DNA nucleotides
   b. DNA polymerase d. a and c

Data Analysis Activities

Hershey-Chase Experiments The graph in **FIGURE 8.17** is reproduced from Hershey and Chase's original publication. The data are from the two experiments described in Section 8.1, in which bacteriophage DNA and protein were labeled with radioactive tracers and allowed to infect bacteria. The virus–bacteria mixtures were whirled in a blender to dislodge the viruses, and the tracers were tracked inside and outside of the bacteria.

1. Before blending, what percentage of each isotope, ^{35}S and ^{32}P, was outside the bacteria?
2. After 4 minutes in the blender, what percentage of each isotope was outside the bacteria?
3. How did the researchers know that the radioisotopes in the fluid came from outside of the bacterial cells (extracellular) and not from bacteria that had been broken apart by whirling in the blender?
4. The extracellular concentration of which isotope increased the most with blending?
5. Do these results imply that viruses inject DNA or protein into bacteria? Why or why not?

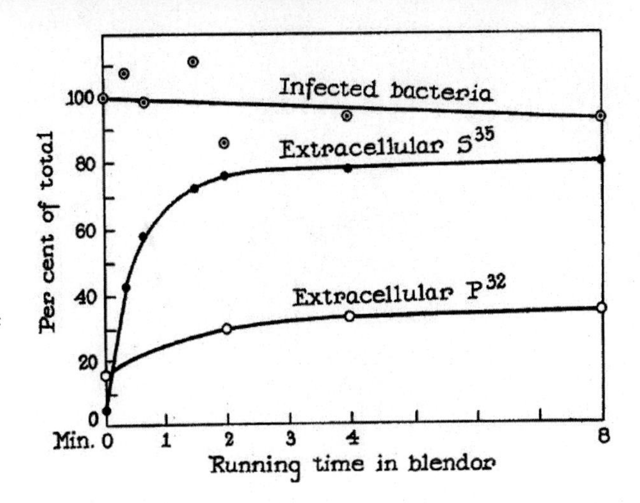

FIGURE 8.17 Detail of Alfred Hershey and Martha Chase's 1952 publication describing their experiments with bacteriophage. "Infected bacteria" refers to the percentage of bacteria that survived the blender.

11. The phrase "5′ to 3′" refers to the ______ .
    a. timing of DNA replication
    b. directionality of DNA synthesis
    c. number of phosphate groups

12. After DNA replication, a eukaryotic chromosome ______ .
    a. consists of two sister chromatids
    b has a characteristic X shape
    c. is constricted at the centromere
    d. all of the above

13. All mutations ______ .
    a. cause cancer
    b. lead to evolution
    c. are caused by radiation
    d. change the DNA sequence

14. ______ is an example of reproductive cloning.
    a. Somatic cell nuclear transfer (SCNT)
    b. Multiple offspring from the same pregnancy
    c. Artificial embryo splitting
    d. a and c

15. Match the terms appropriately.

___ bacteriophage	a. nitrogen-containing base, sugar, phosphate group(s)
___ clone	b. copy of an organism
___ nucleotide	c. does not determine sex
___ diploid	d. only DNA and protein
___ DNA ligase	e. seals breaks in a DNA strand
___ DNA polymerase	f. can cause cancer
___ autosome	g. two chromosomes of each type
___ mutation	h. adds nucleotides to a growing DNA strand

Critical Thinking

1. Show the complementary strand of DNA that forms on this template DNA fragment during replication:

   5′—GGTTTCTTCAAGAGA—3′

2. Woolly mammoths have been extinct for about 10,000 years, but we often find their well-preserved remains in Siberian permafrost. Research groups are now planning to use SCNT to resurrect these huge elephant-like mammals. No mammoth eggs have been recovered so far, so elephant eggs would be used instead. An elephant would also be the surrogate mother for the resulting embryo. The researchers may try a modified SCNT technique used to clone a mouse that had been dead and frozen for sixteen years. Ice crystals that form during freezing break up cell membranes, so cells from the frozen mouse were in bad shape. Their DNA was transferred into donor mouse eggs, and cells from the resulting embryos were fused with mouse stem cells. Four healthy clones were born from the hybrid embryos. What are some of the pros and cons of cloning an extinct animal?

3. Xeroderma pigmentosum is an inherited disorder characterized by rapid formation of skin sores that develop into cancers. All forms of radiation trigger these symptoms, including fluorescent light, which contains UV light in the range of 320–400 nm. What normal function has been compromised in affected individuals?

CREDIT: (17) *Journal of General Physiology*, 36(1), Sept. 20, 1952.

The hairless appearance of a sphynx cat arises from a single base-pair mutation in its DNA. The change results in an altered form of the keratin protein that makes up cat fur.

FROM DNA TO PROTEIN

Links to Earlier Concepts

Your knowledge of base pairing (Section 8.2) and chromosomes (8.3) will help you understand how cells use nucleic acids (3.6) to build proteins (3.4). You will revisit cell structure, including membrane proteins (4.3), the nucleus (4.5) and endomembrane system (4.6); as well as concepts of hydrophobicity (2.4), pathogenic bacteria (4.12), cofactors (5.5), enzyme function (5.3), DNA replication (8.4), and mutation (8.5).

KEY CONCEPTS

GENE EXPRESSION

The information encoded in DNA occurs in subsets called genes. The conversion of genetic information to a protein product occurs in two steps: transcription and translation.

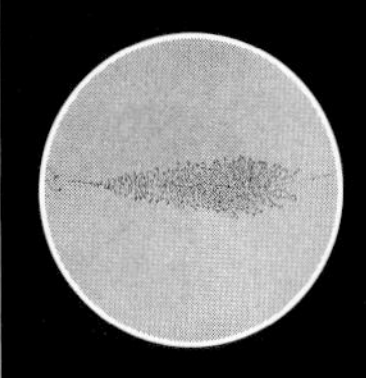

DNA TO RNA: TRANSCRIPTION

During transcription, a gene region in one strand of DNA serves as a template for assembling a strand of RNA. In eukaryotes, a new RNA is modified before leaving the nucleus.

RNA

A messenger RNA carries a gene's protein-building instructions as a string of three-nucleotide codons. Transfer RNA and ribosomal RNA translate those instructions into a protein.

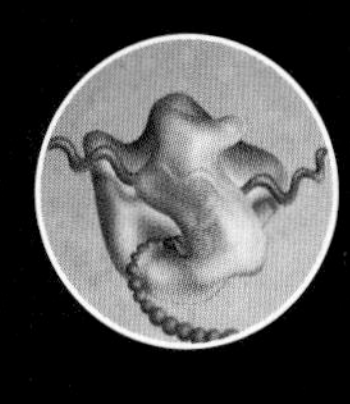

RNA TO PROTEIN: TRANSLATION

During translation, amino acids are assembled into a polypeptide in the order determined by the sequence of codons in an mRNA.

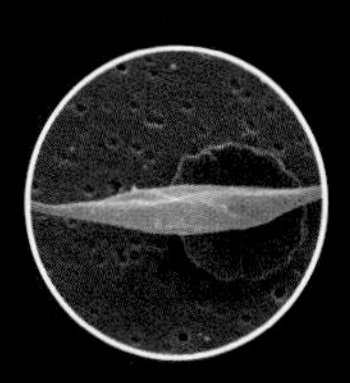

ALTERED PROTEINS

Mutations that change a gene's DNA sequence alter the instructions it encodes. A protein built using altered instructions may function improperly or not at all.

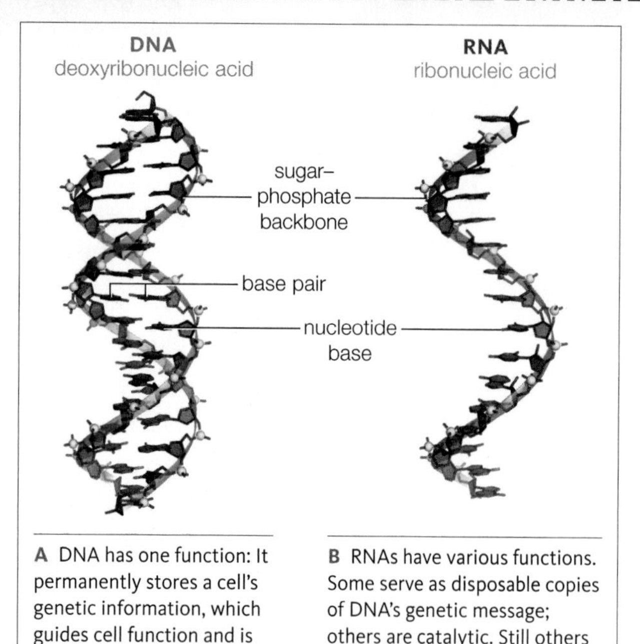

A DNA has one function: It permanently stores a cell's genetic information, which guides cell function and is passed to offspring.

B RNAs have various functions. Some serve as disposable copies of DNA's genetic message; others are catalytic. Still others have roles in gene control.

FIGURE 9.1 Comparing structure and function of DNA and RNA.

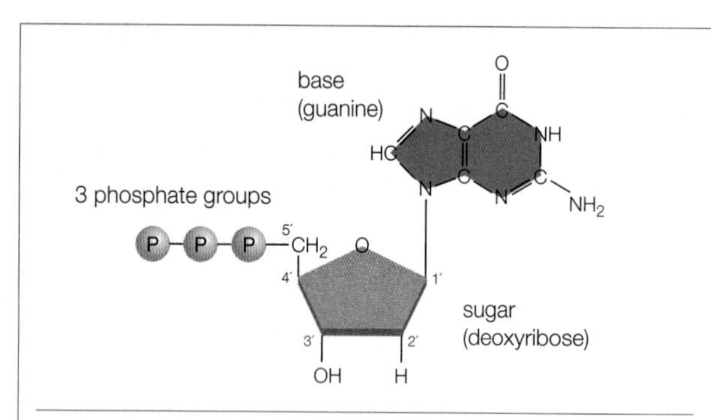

A **The DNA nucleotide guanine (G)**, or deoxyguanosine triphosphate, one of the four nucleotides in DNA. The other nucleotides—adenine, uracil, and cytosine—differ only in their component bases (blue). Three of the four bases in RNA nucleotides are identical to the bases in DNA nucleotides.

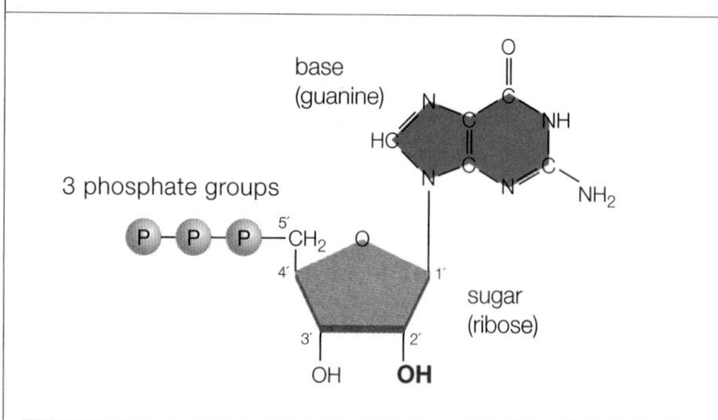

B **The RNA nucleotide guanine (G)**, or guanosine triphosphate. The only difference between the DNA and RNA versions of guanine (or adenine, or cytosine) is that RNA has a hydroxyl group (shown in red) at the 2′ carbon of the sugar.

FIGURE 9.2 Comparing nucleotides of DNA and RNA.

You learned in Chapter 8 that an individual's chromosomes are like a set of books that provide building and operating instructions. You already know the alphabet used to write that book: the four letters A, T, G, and C, for the four nucleotides in DNA: adenine, thymine, guanine, and cytosine. In this chapter, we investigate the nature of information represented by the sequence of nucleotides in DNA, and how a cell uses that information.

DNA TO RNA

Information encoded within a chromosome's DNA sequence occurs in hundreds or thousands of units called genes. The DNA sequence of a **gene** encodes (contains instructions for building) an RNA or protein product. Converting the information encoded by a gene into a product starts with RNA synthesis, or transcription. During **transcription**, enzymes use the gene's DNA sequence as a template to assemble a strand of RNA:

$$\text{DNA} \xrightarrow{\textit{transcription}} \text{RNA}$$

Most of the RNA inside cells occurs as a single strand that is similar in structure to a single strand of DNA (**FIGURE 9.1**). Both RNA and DNA are chains of nucleotides. Like a DNA nucleotide, an RNA nucleotide has three phosphate groups, a sugar, and one of four bases. However, the sugar in an RNA nucleotide is a ribose, which differs just a bit from deoxyribose, the sugar in a DNA nucleotide (**FIGURE 9.2**). Three bases (adenine, cytosine, and guanine) occur in both RNA and DNA nucleotides, but the fourth base differs. In DNA, the fourth base is thymine (T); in RNA, it is uracil (U).

DNA's important but only role is to store a cell's genetic information. By contrast, a cell makes several kinds of RNAs, each with a different function. Three types of RNA have roles in protein synthesis. **Ribosomal RNA (rRNA)** is the main component of ribosomes (Section 4.4), which assemble amino acids into polypeptide chains (Section 3.4). **Transfer RNA (tRNA)** delivers the amino acids to ribosomes, one by one, in the order specified by a **messenger RNA (mRNA)**.

RNA TO PROTEIN

Messenger RNA was named for its function as the "messenger" between DNA and protein. An mRNA's protein-building message is encoded by sets of three nucleotides, "genetic words" that occur one after another along its length. Like the words of a sentence, a series of these genetic words can form a meaningful parcel of information—in this case, the sequence of amino acids of a protein.

Excavating a *Tyrannosaurus rex* proved even more exciting than legendary paleontologist Jack Horner and his colleagues had anticipated when they discovered branching blood vessels and bone matrix inside its thigh bone. The team had never expected to find unfossilized tissues in the 68-million-year-old remains, because the molecules of life tend to break down relatively quickly. The tightly wound, durable structure of collagen (the main protein component of bone) may hold the key to the seemingly inexplicable preservation of the ancient tissue.

Fragments of collagen protein isolated from the tissue have a primary structure very similar to that of chicken bone collagen, providing the first molecular support for the hypothesis that modern birds are descended from dinosaurs. Until this discovery, the dinosaur–bird connection had been entirely based on physical similarities in fossils' body structures.

Researchers often compare DNA sequences to investigate evolutionary relationships, but no one has found DNA in such an ancient fossil. The sequence of amino acids in a protein is encoded by a gene, so protein similarities can also be used as evidence of hereditary connection. "If we spend time getting as deep into the sediment as we can, I think we're going to find that many specimens are like this," Horner said.

National Geographic Grantee
DR. JOHN "JACK" HORNER

By the process of **translation**, the protein-building information in an mRNA is decoded (translated) into a sequence of amino acids. The result is a polypeptide chain that twists and folds into a protein:

mRNA —translation→ protein

Transcription and translation are part of **gene expression**, the multistep process by which information encoded in a gene guides the assembly of an RNA or protein product. During gene expression, this information flows from DNA to RNA to protein:

DNA —transcription→ mRNA —translation→ protein

A cell's DNA sequence contains all the information it needs to make the molecules of life. Each gene encodes an RNA, and RNAs interact to assemble proteins from amino acids (Section 3.4). Proteins (enzymes, in particular) assemble lipids and carbohydrates, replicate DNA, make RNA, and perform many other functions that keep the cell alive.

gene A part of a chromosome that encodes an RNA or protein product in its DNA sequence.
gene expression Process by which the information in a gene guides assembly of an RNA or protein product.
messenger RNA (mRNA) RNA that has a protein-building message.
ribosomal RNA (rRNA) RNA that becomes part of ribosomes.
transcription Process by which enzymes assemble an RNA using the nucleotide sequence of a gene as a template.
transfer RNA (tRNA) RNA that delivers amino acids to a ribosome during translation.
translation Process by which a polypeptide chain is assembled from amino acids in the order specified by an mRNA.

TAKE-HOME MESSAGE 9.1

Information in a DNA sequence occurs in units called genes. A cell uses the information encoded in a gene to make an RNA or protein product, a process called gene expression.

The DNA sequence of a gene is transcribed into RNA.

Information carried by a messenger RNA (mRNA) is translated into a protein.

9.2 HOW IS RNA ASSEMBLED?

Remember that DNA replication begins with one DNA double helix and ends with two DNA double helices (Section 8.4). The two double helices are identical to the parent molecule because base-pairing rules are followed during DNA replication. A nucleotide can be added to a growing strand of DNA only if it base-pairs with the corresponding nucleotide of the parent strand: G pairs with C, and A pairs with T (Section 8.2):

The same base-pairing rules also govern RNA synthesis in transcription. An RNA strand is structurally so similar to a DNA strand that the two can base-pair if their nucleotide sequences are complementary. In such hybrid molecules, G pairs with C, and A pairs with U (uracil):

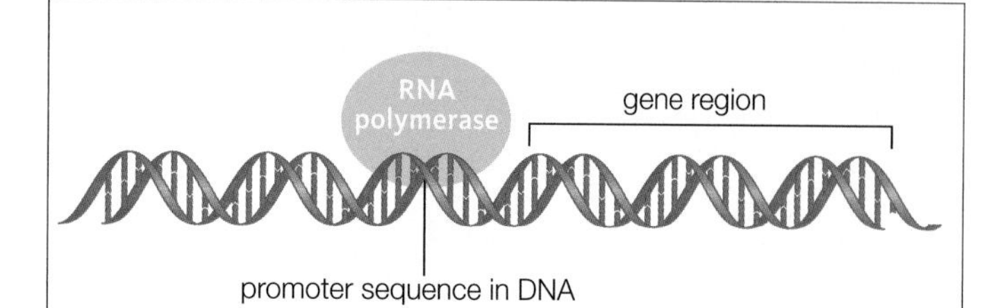

❶ The enzyme RNA polymerase binds to a promoter in the DNA. The binding positions the polymerase near a gene. Only the DNA strand complementary to the gene sequence will be translated into RNA.

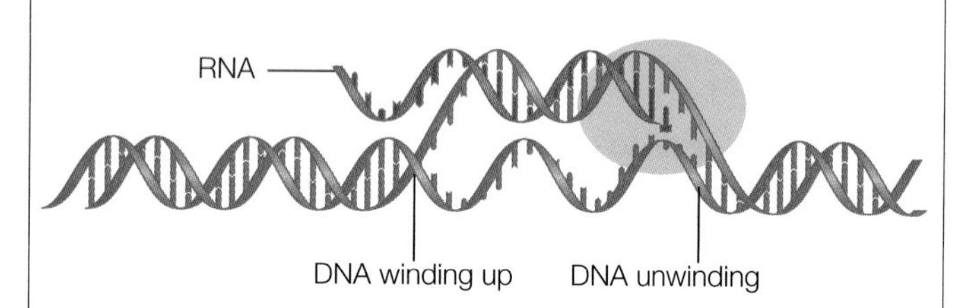

❷ RNA polymerase begins to move along the gene and unwind the DNA. As it does, it links RNA nucleotides in the order specified by the base sequence of the complementary (noncoding) DNA strand. The DNA winds up again after the polymerase passes.

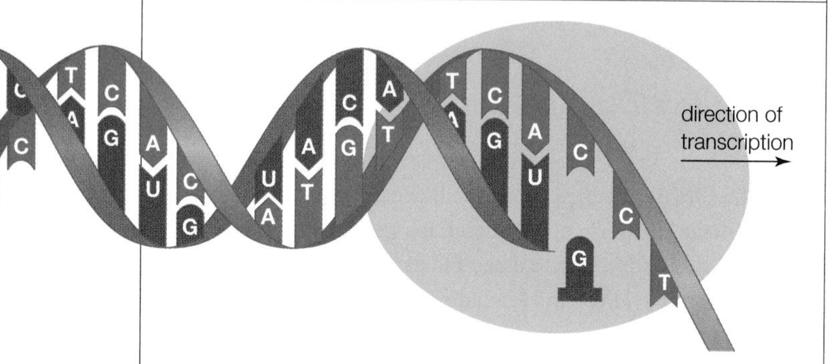

❸ Zooming in on the site of transcription, we can see that RNA polymerase covalently bonds successive nucleotides into an RNA strand. The base sequence of the new RNA strand is complementary to the base sequence of its DNA template strand, so it is an RNA copy of the gene.

FIGURE 9.3 {Animated} Transcription. By this process, a strand of RNA is assembled from nucleotides. A gene region in the DNA serves as the template for RNA synthesis.

FIGURE IT OUT: After the guanine (G), what nucleotide will be added to this growing strand of RNA?

Answer: Another guanine

During transcription, a strand of DNA acts as a template upon which a strand of RNA is assembled from nucleotides. A nucleotide can be added to a growing RNA only if it is complementary to the corresponding nucleotide of the parent strand of DNA. Thus, a new RNA is complementary in sequence to the DNA strand that served as its template. As in DNA replication, each nucleotide provides the energy for its own attachment to the end of a growing strand.

Transcription is similar to DNA replication in that one strand of a nucleic acid serves as a template for synthesis of another. However, in contrast with DNA replication, only part of one DNA strand, not the whole molecule, is used as a template for transcription. The enzyme **RNA polymerase**, not DNA polymerase, adds nucleotides to the end of a growing RNA. Also, transcription produces a single strand of RNA, not two DNA double helices.

In eukaryotic cells, transcription occurs in the nucleus; in prokaryotes, it occurs in cytoplasm. The process begins when an RNA polymerase and regulatory proteins attach to the DNA at a site called a **promoter** (FIGURE 9.3 ❶). Binding positions the polymerase close to the gene that will be transcribed. The strand that is complementary to the gene sequence (the noncoding strand) is the one that serves as the template for transcription.

Like DNA polymerase, RNA polymerase moves along DNA (Section 8.4). As the RNA polymerase moves over a gene region, it unwinds the double helix just a bit so it can "read" the base sequence of the DNA strand ❷. The polymerase joins free RNA nucleotides into a chain, in the order dictated by that DNA sequence. As in

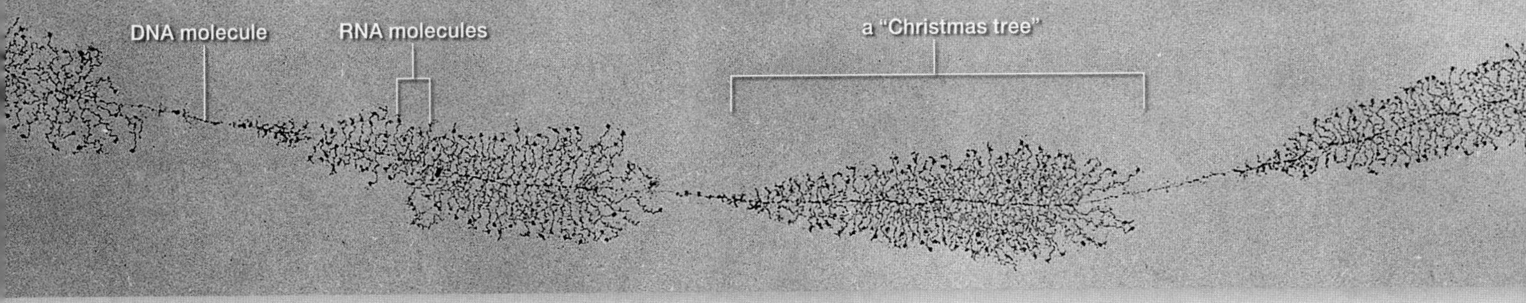

FIGURE 9.4 Typically, many RNA polymerases simultaneously transcribe the same gene, producing a structure often called a "Christmas tree" after its shape. Here, four genes next to one another on the same chromosome are being transcribed.

FIGURE IT OUT: Are the polymerases transcribing this DNA molecule moving from left to right or from right to left?

Answer: Left to right (the RNAs get longer as the polymerases move along the DNA)

DNA replication, the synthesis is directional: An RNA polymerase adds nucleotides only to the 3′ end of the growing strand of RNA.

When the polymerase reaches the end of the gene region, it releases the DNA and the new RNA. RNA polymerase follows base-pairing rules, so the new RNA strand is complementary in base sequence to the DNA strand from which it was transcribed ❸. It is an RNA copy of a gene, the same way that a paper transcript of a conversation carries the same information in a different format. Typically, many polymerases transcribe a particular gene region at the same time, so many new RNA strands can be produced very quickly (**FIGURE 9.4**).

FIGURE 9.5 {Animated} Post-transcriptional modification of RNA. Introns are removed and exons spliced together. Messenger RNAs also get a poly-A tail and modified guanine "cap."

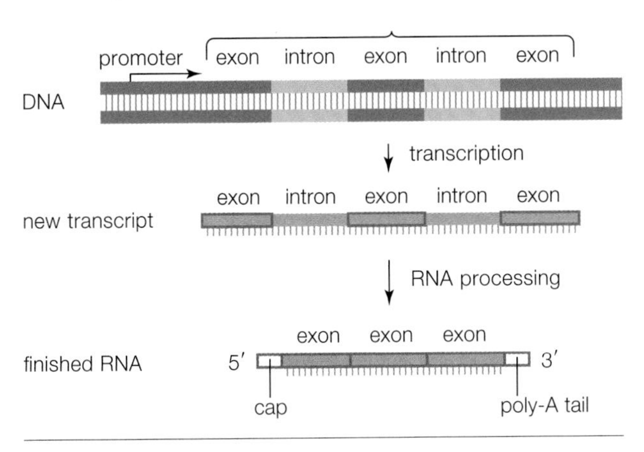

alternative splicing Post-translational RNA modification process in which some exons are removed or joined in various combinations.
exon Nucleotide sequence that remains in an RNA after post-transcriptional modification.
intron Nucleotide sequence that intervenes between exons and is removed during post-transcriptional modification.
promoter In DNA, a sequence to which RNA polymerase binds.
RNA polymerase Enzyme that carries out transcription.

POST-TRANSCRIPTIONAL MODIFICATIONS

Just as a dressmaker may snip off loose threads or add bows to a dress before it leaves the shop, so do eukaryotic cells tailor their RNA before it leaves the nucleus. Consider that most eukaryotic genes contain intervening sequences called **introns**. Introns are removed in chunks from a newly transcribed RNA before it leaves the nucleus. Sequences that stay in the RNA are called **exons** (**FIGURE 9.5**). Exons can be rearranged and spliced together in different combinations—a process called **alternative splicing**—so one gene may encode different proteins.

A newly transcribed RNA that will become an mRNA is further tailored after splicing. Enzymes attach a modified guanine "cap" to the 5′ end; later, this cap will help the finished mRNA bind to a ribosome. Between 50 and 300 adenines are also added to the 3′ end of a new mRNA. This poly-A tail is a signal that allows an mRNA to be exported from the nucleus, and as you will see in Chapter 10, it helps regulate the timing and duration of the mRNA's translation.

TAKE-HOME MESSAGE 9.2

Transcription is an energy-requiring process that uses the information in a gene to produce an RNA.

RNA polymerase uses a gene region in a chromosome as a template to assemble a strand of RNA. The new strand is an RNA copy of the gene from which it was transcribed.

Post-transcriptional modification of RNA occurs in the nucleus of eukaryotes.

9.3 WHAT ROLES DO mRNA, rRNA, AND tRNA PLAY DURING TRANSLATION?

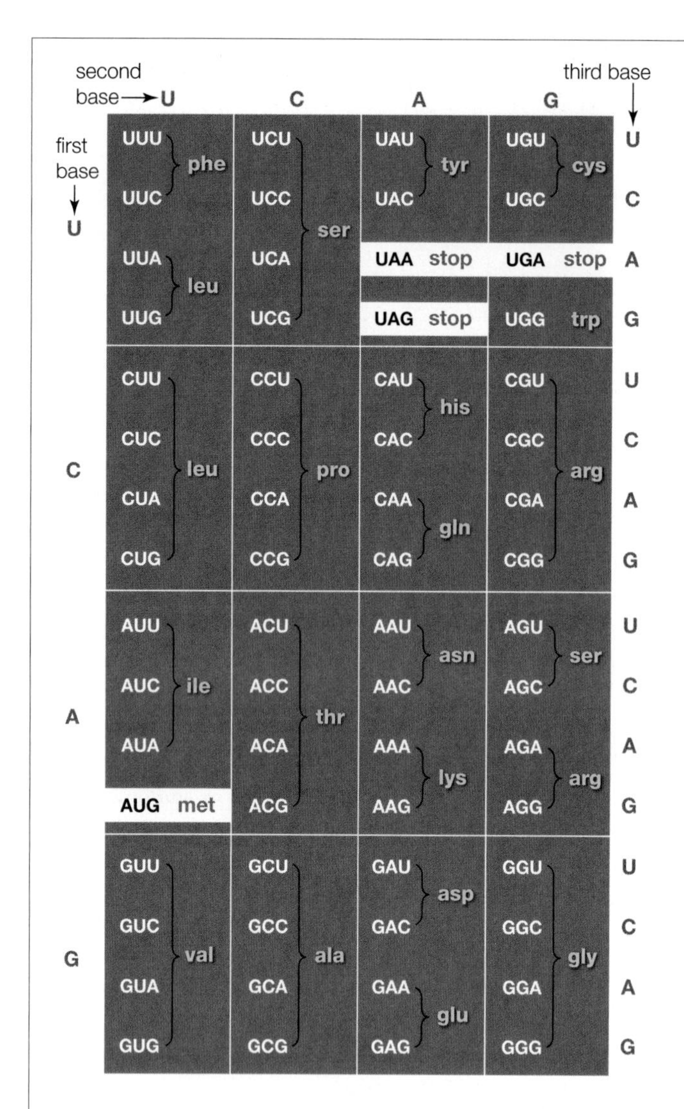

first base ↓ / second base →	U	C	A	G	third base ↓
U	UUU phe	UCU ser	UAU tyr	UGU cys	U
	UUC phe	UCC ser	UAC tyr	UGC cys	C
	UUA leu	UCA ser	UAA stop	UGA stop	A
	UUG leu	UCG ser	UAG stop	UGG trp	G
C	CUU leu	CCU pro	CAU his	CGU arg	U
	CUC leu	CCC pro	CAC his	CGC arg	C
	CUA leu	CCA pro	CAA gln	CGA arg	A
	CUG leu	CCG pro	CAG gln	CGG arg	G
A	AUU ile	ACU thr	AAU asn	AGU ser	U
	AUC ile	ACC thr	AAC asn	AGC ser	C
	AUA ile	ACA thr	AAA lys	AGA arg	A
	AUG met	ACG thr	AAG lys	AGG arg	G
G	GUU val	GCU ala	GAU asp	GGU gly	U
	GUC val	GCC ala	GAC asp	GGC gly	C
	GUA val	GCA ala	GAA glu	GGA gly	A
	GUG val	GCG ala	GAG glu	GGG gly	G

A codon table. Each codon in mRNA is a set of three nucleotide bases. Left column lists a codon's first base, the top row lists the second, and the right column lists the third. Sixty-one of the triplets encode amino acids; one of those, AUG, both codes for methionine and serves as a signal to start translation. Three codons are signals that stop translation.

ala alanine (A)	gly glycine (G)	pro proline (P)
arg arginine (R)	his histidine (H)	ser serine (S)
asn asparagine (N)	ile isoleucine (I)	thr threonine (T)
asp aspartic acid (D)	leu leucine (L)	trp tryptophan (W)
cys cysteine (C)	lys lysine (K)	tyr tyrosine (Y)
glu glutamic acid (E)	met methionine (M)	val valine (V)
gln glutamine (Q)	phe phenylalanine (F)	

Amino acid names and abbreviations.

FIGURE 9.6 The genetic code.

FIGURE IT OUT: Which codons specify the amino acid lysine (lys)?

Answer: AAA and AAG

DNA stores heritable information about proteins, but making those proteins requires messenger RNA (mRNA), transfer RNA (tRNA), and ribosomal RNA (rRNA). The three types of RNA interact to translate DNA's information into a protein.

THE MESSENGER: mRNA

An mRNA is essentially a disposable copy of a gene. Its job is to carry the gene's protein-building information to the other two types of RNA during translation. That protein-building information consists of a linear sequence of genetic "words" spelled with an alphabet of the four nucleotide bases A, C, G, and U. Each of the genetic "words" carried by an mRNA is three bases long, and each is a code—a **codon**—for a particular amino acid. With four possible nucleotides in each of the three positions of a codon, there are a total of sixty-four (or 4^3) mRNA codons. Collectively, the sixty-four codons constitute the **genetic code** (FIGURE 9.6). The sequence of bases in a triplet determines which amino acid the codon specifies. For instance, the codon UUU codes for the amino acid phenylalanine (phe), and UUA codes for leucine (leu).

Codons occur one after another along the length of an mRNA. When an mRNA is translated, the order of its codons determines the order of amino acids in the resulting polypeptide. Thus, the base sequence of a gene is transcribed into the base sequence of an mRNA, which is in turn translated into an amino acid sequence (FIGURE 9.7).

With a few exceptions, twenty naturally occurring amino acids are encoded by the sixty-four codons in the genetic

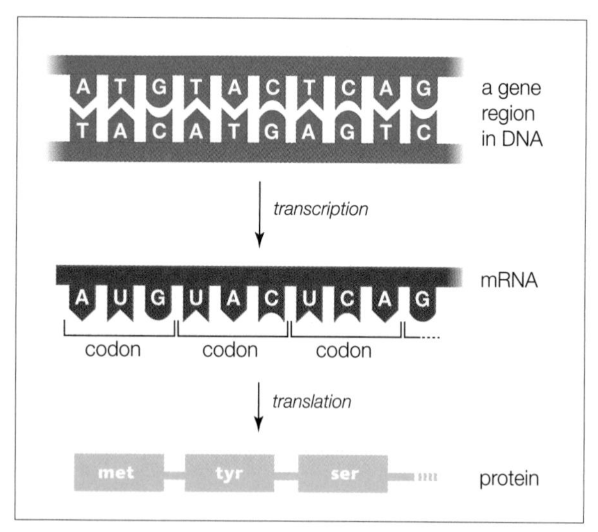

FIGURE 9.7 Example of the correspondence between DNA, RNA, and protein. A gene region in a strand of chromosomal DNA is transcribed into an mRNA, and the codons of the mRNA specify a chain of amino acids—a protein.

code. Sixty-four codons are more than are needed to specify twenty amino acids, so some amino acids are specified by more than one codon. For instance, the amino acid tyrosine (tyr) is specified by two codons: UAA and UAC.

Other codons signal the beginning and end of a protein-coding sequence. In most species, the first AUG in an mRNA serves as the signal to start translation. AUG is the codon for methionine, so methionine is always the first amino acid in new polypeptides of such organisms. The codons UAA, UAG, and UGA do not specify an amino acid. These are signals that stop translation, so they are called stop codons. A stop codon marks the end of the protein-coding sequence in an mRNA.

The genetic code is highly conserved, which means that most organisms use the same code and probably always have. Bacteria, archaea, and some protists have a few codons that differ from the eukaryotic code, as do mitochondria and chloroplasts—a clue that led to a theory of how these two organelles evolved (we return to this topic in Section 18.5).

THE TRANSLATORS: rRNA AND tRNA

Ribosomes interact with transfer RNAs (tRNAs) to translate the sequence of codons in an mRNA into a polypeptide. A ribosome has two subunits, one large and one small (**FIGURE 9.8**). Both subunits consist mainly of rRNA, with some associated structural proteins. During translation, a large and a small ribosomal subunit converge as an intact ribosome on an mRNA. Ribosomal RNA is one example of RNA with enzymatic activity: rRNA catalyzes formation of a peptide bond between amino acids as they are delivered to the ribosome.

Each tRNA has two attachment sites. The first is an **anticodon**, which is a triplet of nucleotides that base-pairs with an mRNA codon (**FIGURE 9.9A**). The other attachment site binds to an amino acid—the one specified by the codon. Transfer RNAs with different anticodons carry different amino acids.

During translation, tRNAs deliver amino acids to a ribosome, one after the next in the order specified by the codons in an mRNA (**FIGURE 9.9B**). As the amino acids are delivered, the ribosome joins them via peptide bonds into a new polypeptide (Section 3.4). Thus, the order of codons in an mRNA—DNA's protein-building message—becomes translated into a new protein.

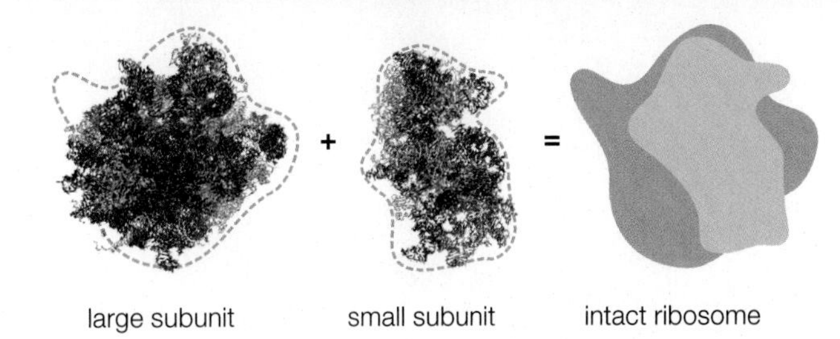

FIGURE 9.8 {Animated} Ribosome structure. Each intact ribosome consists of a large and a small subunit. The structural protein components of the two subunits are shown in green; the catalytic rRNA components, in brown.

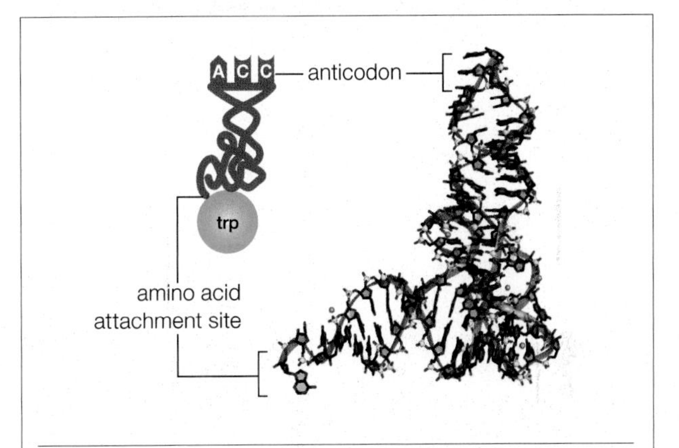

A Icon and model of the tRNA that carries the amino acid tryptophan. Each tRNA's anticodon is complementary to an mRNA codon. Each also carries the amino acid specified by that codon.

B During translation, tRNAs dock at an intact ribosome (for clarity, only the small subunit is shown, in tan). Here, the anticodons of two tRNAs have base-paired with complementary codons on an mRNA (red). The amino acids they carry are not shown, for clarity.

FIGURE 9.9 tRNA structure.

TAKE-HOME MESSAGE 9.3

The sequence of nucleotide triplets (codons) in an mRNA encode a gene's protein-building message.

The genetic code consists of sixty-four codons. Three are signals that stop translation; the remaining codons specify an amino acid. In most mRNAs, the first occurrence of the codon that specifies methionine is a signal to begin translation.

Ribosomes, which consist of two subunits of rRNA and proteins, assemble amino acids into polypeptide chains.

A tRNA has an anticodon complementary to an mRNA codon, and a binding site for the amino acid specified by that codon. During translation, tRNAs deliver amino acids to ribosomes.

anticodon In a tRNA, set of three nucleotides that base-pairs with an mRNA codon.
codon In an mRNA, a nucleotide base triplet that codes for an amino acid or stop signal during translation.
genetic code Complete set of sixty-four mRNA codons.

9.4 HOW IS mRNA TRANSLATED INTO PROTEIN?

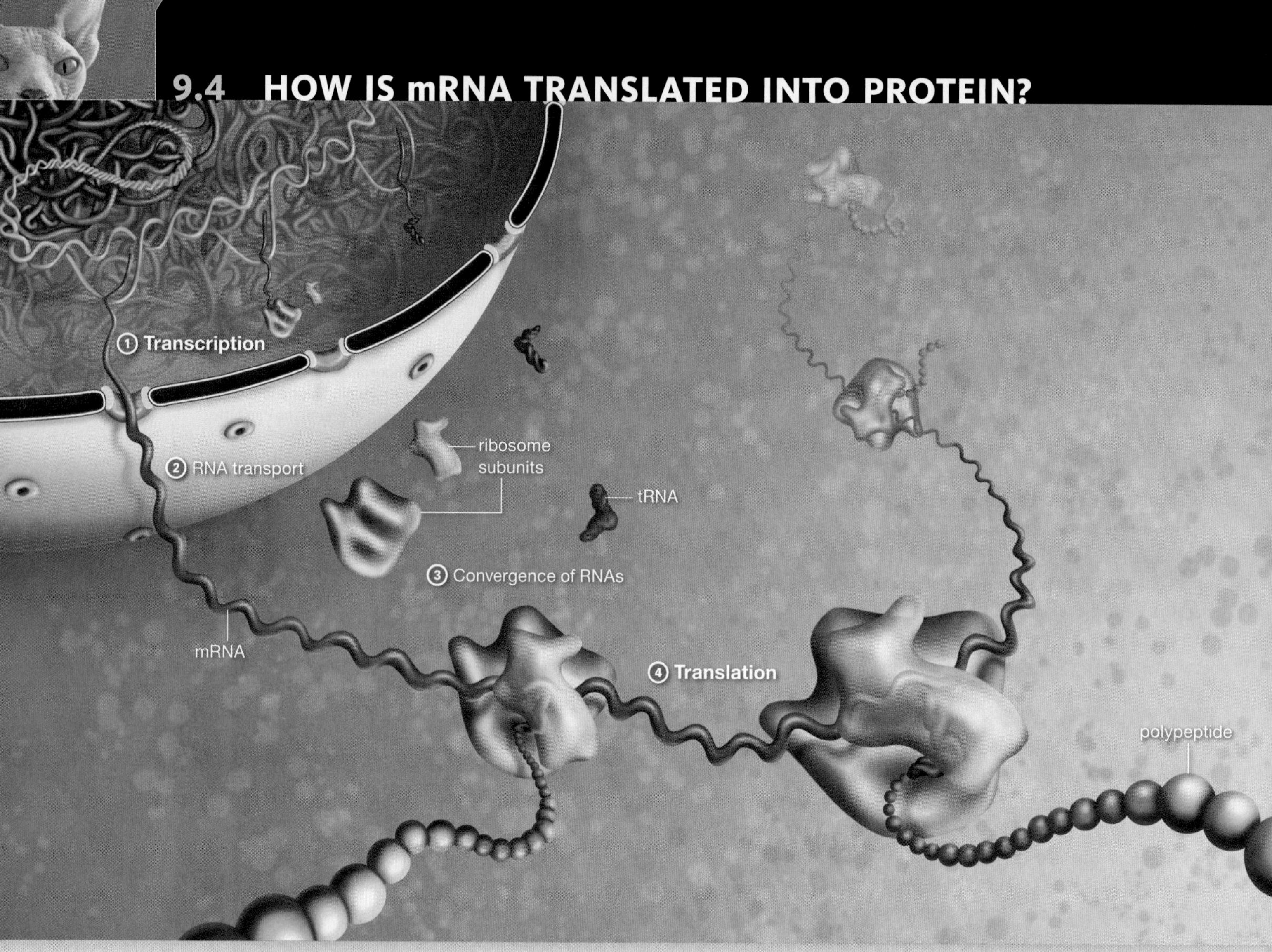

FIGURE 9.10 Overview of translation. In eukaryotes, RNA transcribed in the nucleus moves into the cytoplasm through nuclear pores. Translation occurs in the cytoplasm. Ribosomes simultaneously translating the same mRNA are called polysomes.

Translation, the second part of protein synthesis, occurs in the cytoplasm of all cells. Cytoplasm has many free amino acids, tRNAs, and ribosomal subunits available to participate in the process.

FIGURE 9.10 shows an overview of translation as it occurs in eukaryotes. An mRNA is transcribed in the nucleus ❶, and then transported through nuclear pores into the cytoplasm ❷. Translation begins when a small ribosomal subunit binds to the mRNA. Next, the anticodon of a special tRNA called an initiator base-pairs with the first AUG codon of the mRNA. Then, a large ribosomal subunit joins the small subunit ❸, and the intact ribosome begins to assemble a polypeptide chain as it moves along the mRNA ❹.

FIGURE 9.11 shows details of translation. Initiator tRNAs carry methionine, so the first amino acid of the new polypeptide chain is methionine. Another tRNA joins the complex when its anticodon base-pairs with the second codon in the mRNA ❺. This tRNA brings with it the second amino acid. The ribosome then catalyzes formation of a peptide bond between the first two amino acids ❻.

As the ribosome moves to the next codon, it releases the first tRNA. Another tRNA brings the third amino acid to the complex as its anticodon base-pairs with the third codon of the mRNA ❼. A peptide bond forms between the second and third amino acids ❽.

The second tRNA is released and the ribosome moves to the next codon. Another tRNA brings the fourth amino acid to the complex as its anticodon base-pairs with the fourth codon of the mRNA ❾. A peptide bond forms between the third and fourth amino acids. Elongation of the new polypeptide chain continues as the ribosome catalyzes peptide bonds between amino acids delivered by successive tRNAs.

Termination occurs when the ribosome reaches a stop codon in the mRNA ⑩. The mRNA and the polypeptide detach from the ribosome, and the ribosomal subunits separate from each other. Translation is now complete. The new polypeptide either joins the pool of proteins in the cytoplasm, or it enters rough ER of the endomembrane system (Section 4.6).

In cells that are making a lot of protein, many ribosomes may simultaneously translate the same mRNA, in which case they are called polysomes:

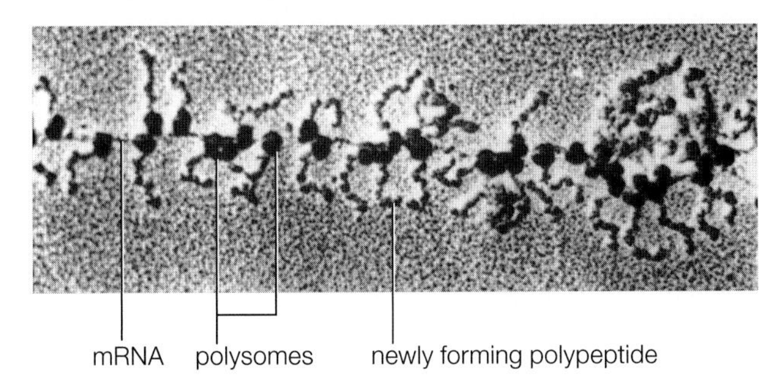

In bacteria and archaea, transcription and translation both occur in the cytoplasm, and these processes are closely linked in time and space. Translation begins before transcription ends, so a transcription "Christmas tree" is often decorated with polysome "balls."

Given that many polypeptides can be translated from one mRNA, why would any cell also make many copies of an mRNA? Compared with DNA, RNA is not very stable. An mRNA may last only a few minutes in cytoplasm before enzymes disassemble it. The fast turnover allows cells to adjust their protein synthesis quickly in response to changing needs.

Translation is energy intensive. That energy is provided mainly in the form of phosphate-group transfers from the RNA nucleotide GTP (shown in **FIGURE 9.2B**) to molecules involved in the process.

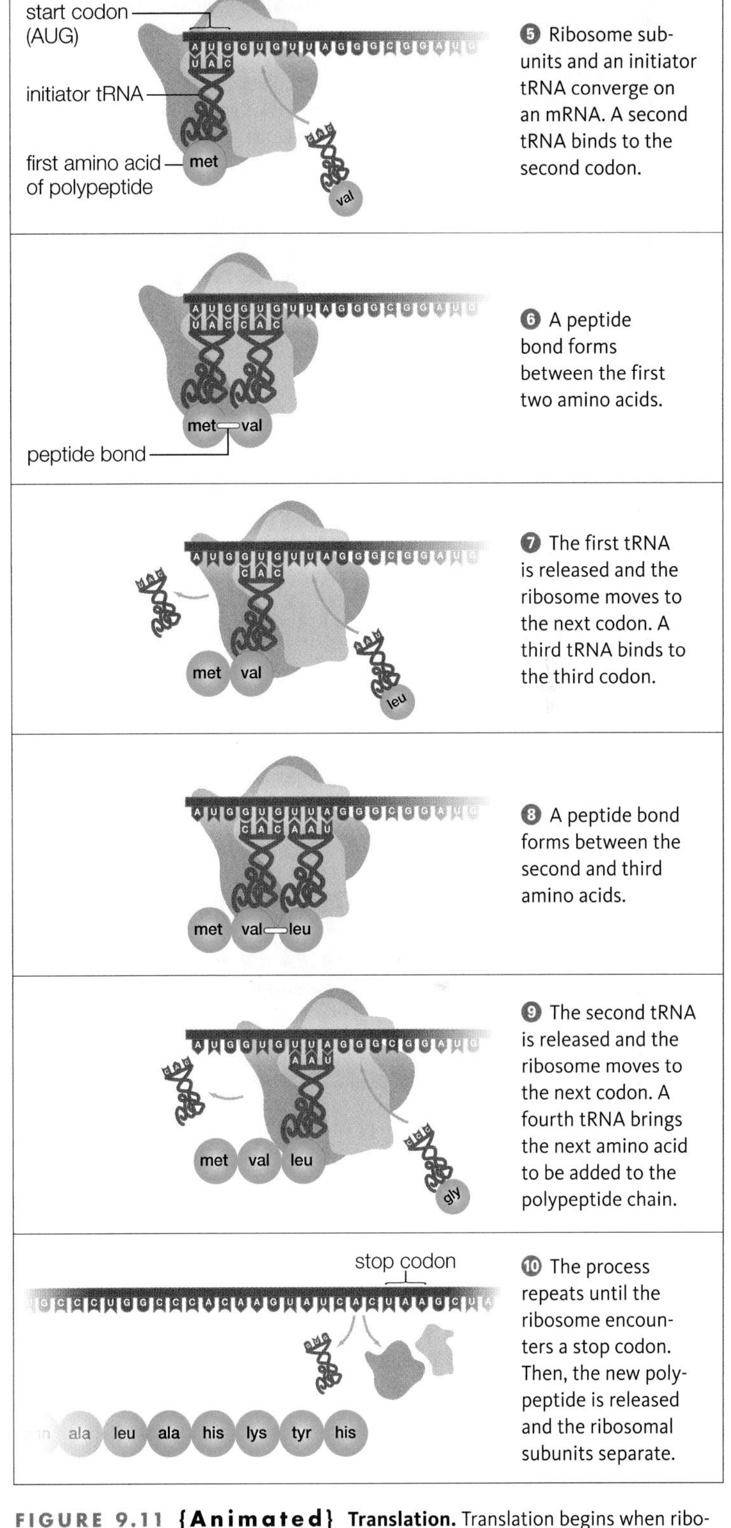

FIGURE 9.11 {Animated} Translation. Translation begins when ribosomal subunits and an initiator tRNA converge on an mRNA. tRNAs deliver amino acids in the order dictated by successive codons in the mRNA. The ribosome links the amino acids together as it moves along the mRNA, so a polypeptide forms and elongates. Translation ends when the ribosome reaches a stop codon.

TAKE-HOME MESSAGE 9.4

Translation is an energy-requiring process that converts the protein-building information carried by an mRNA into a polypeptide.

Translation begins when an mRNA joins with an initiator tRNA and two ribosomal subunits.

Amino acids are delivered to the complex by tRNAs in the order dictated by successive mRNA codons. As amino acids arrive, the ribosome joins each to the end of the polypeptide.

Translation ends when the ribosome encounters a stop codon in the mRNA. The mRNA and the polypeptide are released, and the ribosome disassembles.

9.5 WHAT HAPPENS AFTER A GENE BECOMES MUTATED?

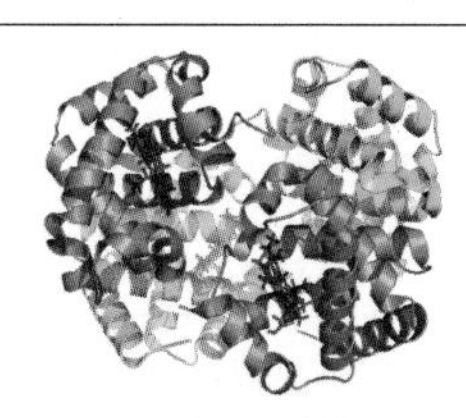

A Hemoglobin, an oxygen-binding protein in red blood cells. This protein consists of four polypeptides: two alpha globins (blue) and two beta globins (green). Each globin has a pocket that cradles a heme (red). Oxygen molecules bind to the iron atom at the center of each heme.

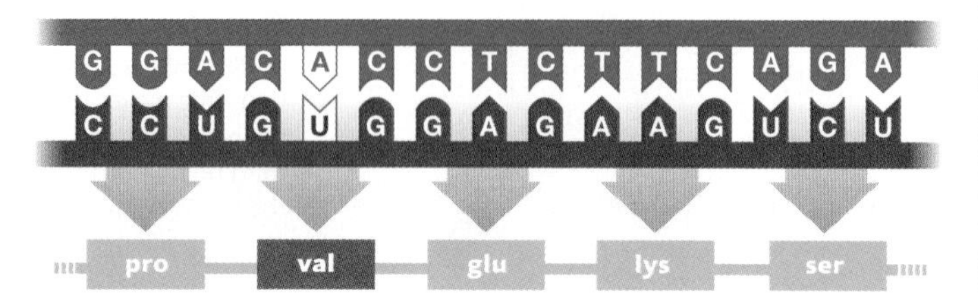

B Part of the DNA (blue), mRNA (brown), and amino acid sequence of human beta globin. Numbers indicate nucleotide position in the mRNA.

C A base-pair substitution replaces a thymine with an adenine. When the altered mRNA is translated, valine replaces glutamic acid as the sixth amino acid. Hemoglobin with this form of beta globin is called HbS, or sickle hemoglobin.

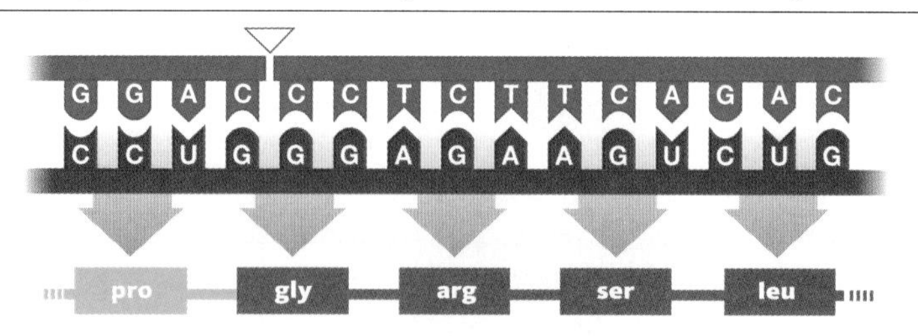

D A base-pair deletion shifts the reading frame for the rest of the mRNA, so a completely different protein product forms. The mutation shown results in a defective beta globin. The outcome is beta thalassemia, a genetic disorder in which a person has an abnormally low amount of hemoglobin.

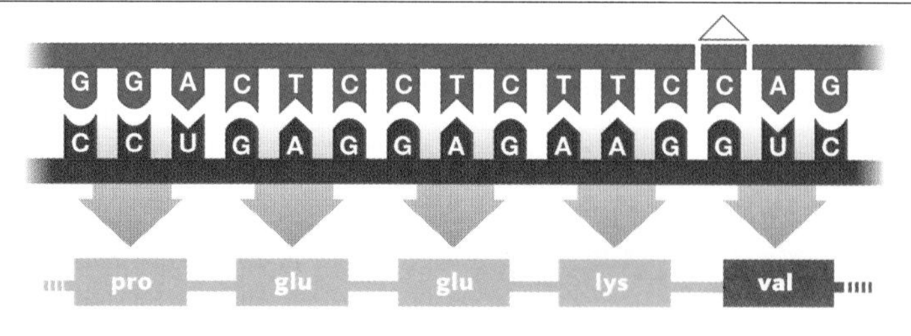

E An insertion of one nucleotide causes the reading frame for the rest of the mRNA to shift. The protein translated from this mRNA is too short and does not assemble correctly into hemoglobin molecules. As in **D**, the outcome is beta thalassemia.

FIGURE 9.12 {Animated} Examples of mutations.

Mutations, remember, are permanent changes in a DNA sequence (Section 8.5). A mutation in which one base pair is replaced by a different base pair is a **base-pair substitution**. Other mutations may involve the loss of one or more nucleotides (a **deletion**) or the addition of one or more extra nucleotides (an **insertion**).

Mutations are relatively uncommon events in a normal cell. Consider that the chromosomes in a diploid human cell collectively consist of about 6.5 billion nucleotides, any of which may become mutated each time that cell divides. On average, about 175 nucleotides do change during DNA replication. However, only about 3 percent of the cell's DNA encodes protein products, so there is a low probability that any of those mutations will be in a protein-coding region.

When a mutation does occur in a protein-coding region, the redundancy of the genetic code offers the cell a margin of safety. For example, a mutation that changes a CCC codon to CCG may not have further effects, because both of these codons specify serine. Other mutations may change an amino acid in a protein, or result in a premature stop codon that shortens it.

Mutations that alter a protein can have drastic effects on an organism. Consider the effects of mutations on hemoglobin, an oxygen-transporting protein in your red blood cells. Hemoglobin's structure allows it to bind and release oxygen. In adult humans, a hemoglobin molecule consists of four polypeptides called globins: two alpha globins and two beta globins (**FIGURE 9.12A**). Each globin folds around a heme, a cofactor with an iron atom at its center (Section 5.5). Oxygen molecules bind to hemoglobin at those iron atoms.

Mutations in the genes for alpha or beta globin cause a condition called anemia, in which a person's blood is deficient in red blood cells or in hemoglobin. Both outcomes limit the blood's ability to carry oxygen, and the resulting symptoms range from mild to life-threatening.

Sickle-cell anemia, a type of anemia that is most common in people of African ancestry, arises because of a base-pair substitution in the beta globin gene. The substitution causes the body to produce a version of beta globin in which the sixth amino acid is valine instead of glutamic acid (**FIGURE 9.12B,C**). Hemoglobin assembled with this altered beta globin chain is called sickle hemoglobin, or HbS.

Unlike glutamic acid, which carries a negative charge, valine carries no charge. As a result of that one base-pair substitution, a tiny patch of the beta globin polypeptide that is normally hydrophilic becomes hydrophobic. This change slightly alters the hemoglobin's behavior. Under certain

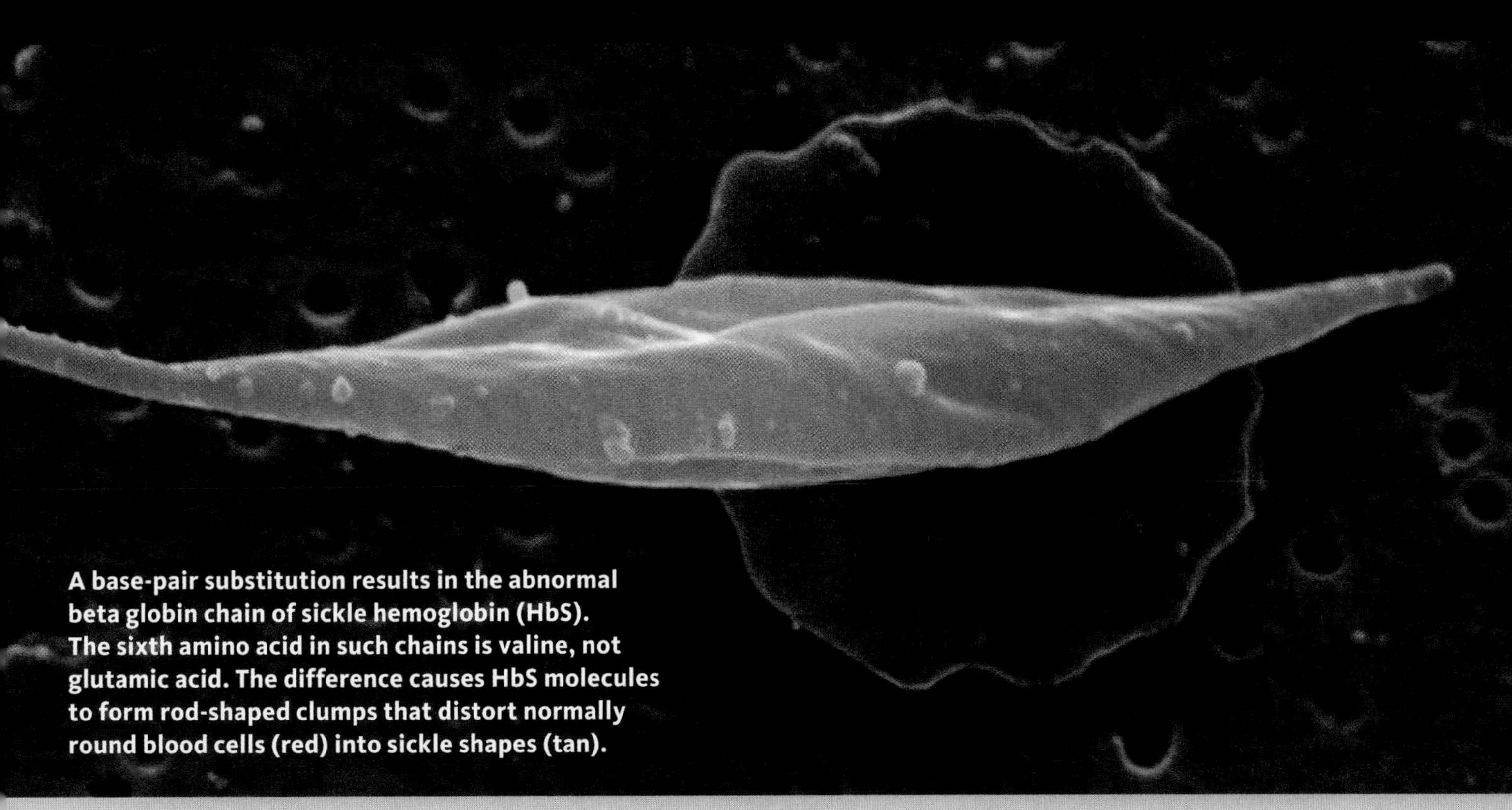

FIGURE 9.13 An amino acid substitution results in abnormally shaped red blood cells characteristic of sickle-cell anemia.

conditions, HbS molecules stick together and form large, rodlike clumps. Red blood cells that contain the clumps become distorted into a crescent (sickle) shape (**FIGURE 9.13**). Sickled cells clog tiny blood vessels, thus disrupting blood circulation throughout the body. Over time, repeated episodes of sickling can damage organs and cause death.

A different type of anemia, beta thalassemia, is caused by the deletion of the twentieth nucleotide in the coding region of the beta globin gene (**FIGURE 9.12D**). Like many other deletions, this one causes the reading frame of the mRNA codons to shift. A frameshift usually has drastic consequences because it garbles the genetic message, just as incorrectly grouping a series of letters garbles the meaning of a sentence:

The fat cat ate the sad rat
T hef atc ata tet hes adr at

The frameshift caused by the beta globin deletion results in a polypeptide that differs drastically from normal beta globin in amino acid sequence and length. This outcome is the source of the anemia. Beta thalassemia can also be caused by insertion mutations, which, like deletions, often result in frameshifts (**FIGURE 9.12E**).

Not all mutations that affect protein structure disrupt codons for amino acids. DNA also contains special nucleotide sequences that influence the expression of nearby genes (we return to this topic in the next chapter). A promoter is one example; an intron–exon splice site is another. Consider the mutation that causes hairlessness in cats (as shown in the chapter opening photo). In this case, a base-pair substitution disrupts an intron–exon splice site in the gene for keratin, a fibrous protein (Section 3.4). The intron, which is not correctly removed from the RNA, becomes an insertion in the mRNA. The altered protein translated from this mRNA cannot properly assemble into filaments that make up cat fur.

TAKE-HOME MESSAGE 9.5

Mutations that result in an altered protein can have drastic consequences.

A base-pair substitution may change an amino acid in a protein, or it may introduce a premature stop codon.

Frameshifts that occur after an insertion or deletion can change an mRNA's codon reading frame, thus garbling its protein-building instructions.

base-pair substitution Type of mutation in which a single base pair changes.
deletion Mutation in which one or more nucleotides are lost.
insertion Mutation in which one or more nucleotides become inserted into DNA.

9.6 Application: THE APTLY ACRONYMED RIPs

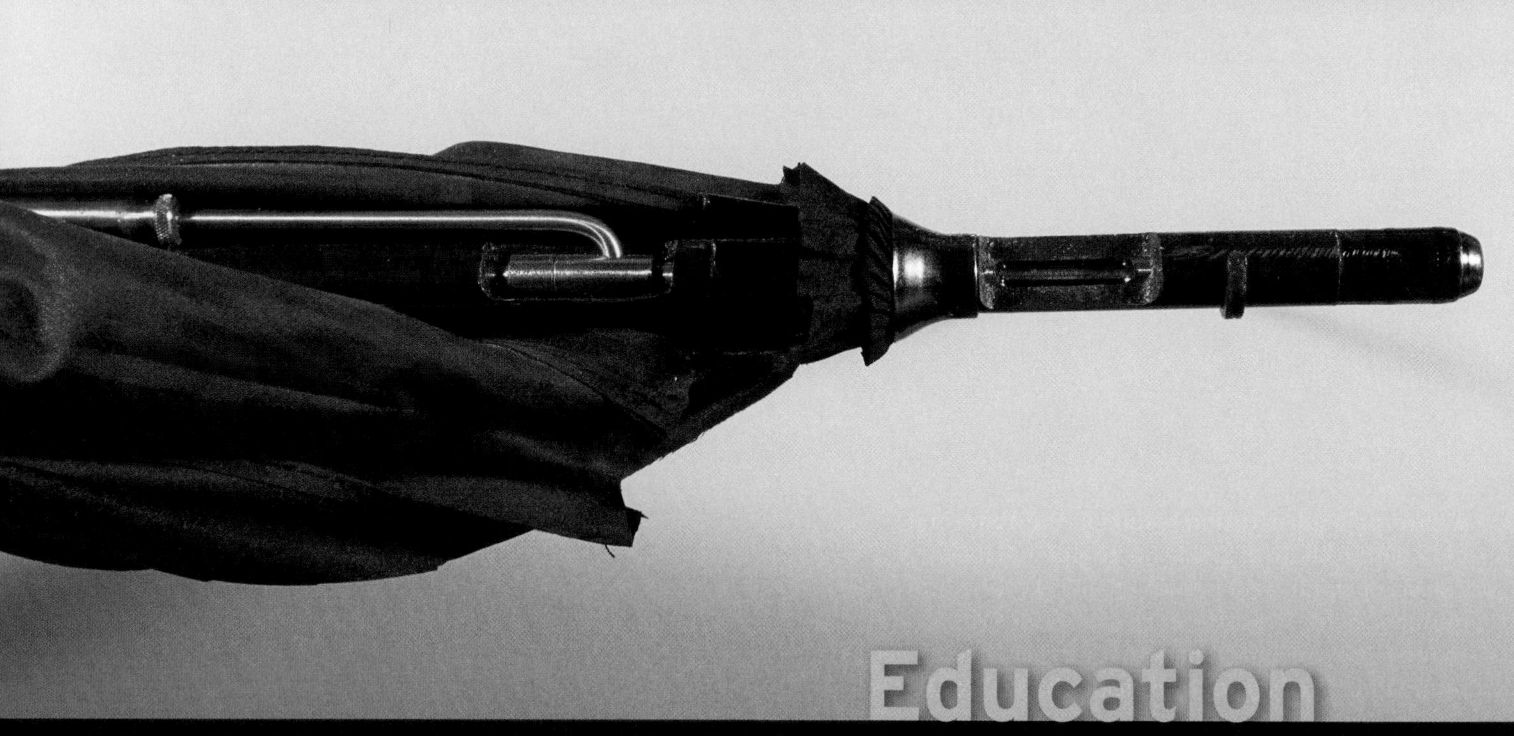

FIGURE 9.14 Bulgarian spy's weapon: an umbrella modified to fire a tiny pellet of ricin into a victim. An umbrella like this one was used to assassinate Georgi Markov on the streets of London in 1978.

A DOSE OF RICIN AS SMALL AS A FEW GRAINS OF SALT CAN KILL AN ADULT HUMAN, and there is no antidote. Ricin is a protein that deters beetles, birds, mammals, and other animals from eating seeds of the castor-oil plant (*Ricinus communis*), which grows wild in tropical regions worldwide and is widely cultivated. Castor-oil seeds are the source of castor oil, an ingredient in plastics, cosmetics, paints, soaps, polishes, and many other items. After the oil is extracted from the seeds, the ricin is typically discarded along with the leftover seed pulp.

Lethal effects of ricin were being exploited as long ago as 1888, but using ricin as a weapon is now banned by most countries under the Geneva Protocol. However, controlling its production is impossible, because it takes no special skills or equipment to manufacture the toxin from easily obtained raw materials. Thus, ricin appears periodically in the news as a tool of terrorists. For example, in June, 2013, a Texas actress was arrested for sending ricin-laced letters to President Obama and the mayor of New York City. Perhaps the most famous example occurred in 1978, at the height of the Cold War when defectors from countries under Russian control were targets for assassination. Bulgarian journalist Georgi Markov had defected to England and was working for the BBC. As he made his way to a bus stop on a London street, an assassin used a modified umbrella (FIGURE 9.14) to fire a tiny, ricin-laced ball into Markov's leg. Markov died in agony three days later.

Ricin is called a ribosome-inactivating protein (RIP) because it inactivates ribosomes. RIPs are enzymes that remove a particular adenine base from one of the rRNAs in the ribosome's heavy subunit. The adenine is part of a binding site for proteins involved in GTP-requiring steps of elongation. After the base has been removed, the ribosome can no longer bind to these proteins, and elongation stops. If enough ribosomes are affected, protein synthesis grinds to a halt. Proteins are critical to all life processes, so cells that cannot make them die very quickly. Someone who inhales ricin can die from low

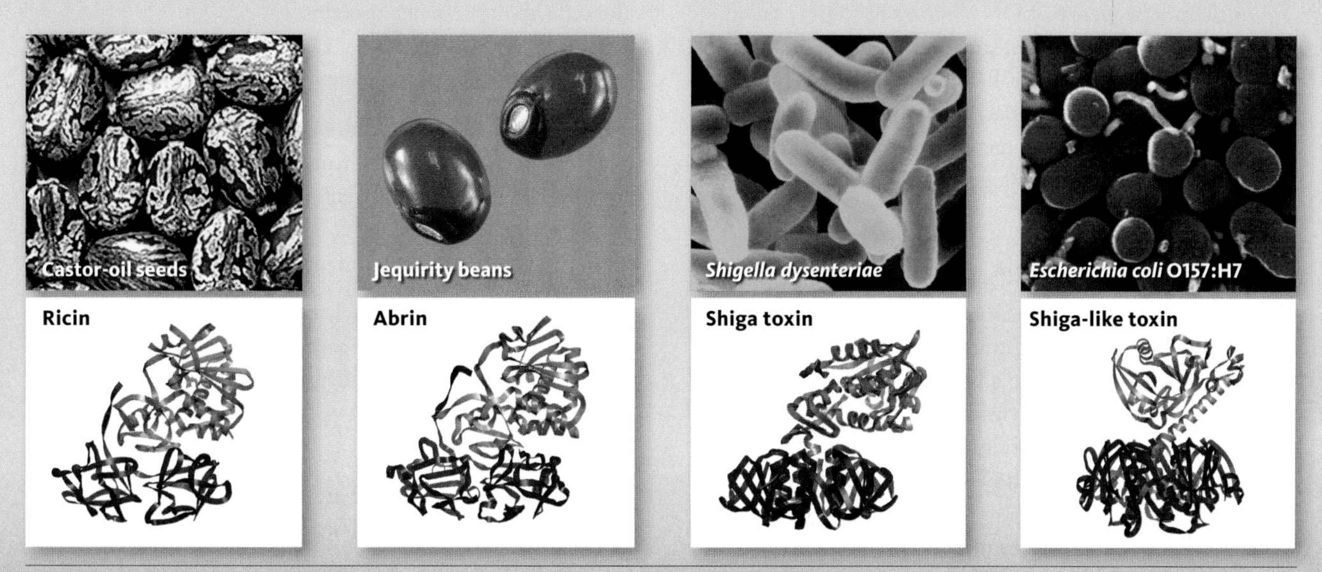

FIGURE 9.15 Lethal lineup: a few toxic ribosome-inactivating proteins (RIPs) and their sources. One of the two chains of a toxic RIP helps the molecule cross a cell's plasma membrane; the other is an enzyme that inactivates ribosomes.

blood pressure and respiratory failure within a few days of exposure.

Other RIPs are made by some bacteria, mushrooms, algae, and many plants (including food crops such as tomatoes, barley, and spinach). Most of these proteins are not particularly toxic in humans because they do not cross intact cell membranes very well. Those that do, including ricin, have a domain that binds tightly to glycolipids attached to proteins on our plasma membranes (FIGURE 9.15). Binding causes the cell to take up the RIP by endocytosis (Section 5.8). Once inside cytoplasm, the enzyme part of the molecule quickly goes to work: One molecule of ricin can inactivate more than 1,000 ribosomes per minute.

Fortunately, few people actually encounter ricin. Other RIPs are more prevalent. Bracelets made from beautiful seeds were recalled from stores in 2011 after a botanist recognized the seeds as jequirity beans. These beans contain abrin, an RIP even more toxic than ricin. Shiga toxin, an RIP made by *Shigella dysenteriae* bacteria, causes a severe bloody diarrhea (dysentery) that can be lethal. Some strains of *E. coli* bacteria make Shiga-like toxin, an RIP that is the source of intestinal illness (Section 4.12).

Despite their toxicity, the main function of RIPs may not be destroying ribosomes. Many are part of the immune system in plants, but it is their antiviral and anticancer activity that has researchers abuzz. Plants that make RIPs have been used as traditional medicines for many centuries. Western scientists are now investigating RIPs as potential weapons to combat HIV and cancer.

The unique properties of RIPs are proving particularly useful in drug design. For example, researchers who design drugs for cancer therapy have modified ricin's glycolipid-binding domain to recognize plasma membrane proteins (Section 4.3) especially abundant in cancer cells. The modified ricin preferentially enters—and kills—cancer cells. Ricin's toxic enzyme has also been attached to an antibody that can find cancer cells in a person's body. The intent of both strategies: to assassinate the cancer cells without harming normal ones.

CREDITS: (15) top from left, Vaughan Fleming/Science Source; Steve Hurst/USDA-NRCS PLANTS Database; Dr. Kari Lounatmaa/Science Source; Stephanie Schuller/Photo Researchers, Inc.; bottom art, From Starr/Taggart/Evers/Starr, Biology, 13E. © 2013 Cengage Learning.

Summary

SECTION 9.1 Information encoded within the nucleotide sequence of DNA occurs in subsets called **genes**. Converting the information in a gene to an RNA or protein product is called **gene expression**. RNA is produced during **transcription**. **Ribosomal RNA** (**rRNA**) and **transfer RNA** (**tRNA**) interact during **translation** of a **messenger RNA** (**mRNA**) into a protein product:

DNA —transcription→ mRNA —translation→ protein

SECTION 9.2 During transcription, the enzyme **RNA polymerase** binds to a **promoter** near a gene on a chromosome. The polymerase moves over the gene region, linking RNA nucleotides in the order dictated by the nucleotide sequence of the DNA. The new RNA strand is an RNA copy of the gene.

The RNA of eukaryotes is modified before it leaves the nucleus. **Introns** are removed, and the remaining **exons** may be rearranged and spliced in different combinations, a process called **alternative splicing**. A cap and poly-A tail are also added to a new mRNA.

SECTION 9.3 An mRNA carries DNA's protein-building information. The information consists of a series of **codons**, which are sets of three nucleotides. Sixty-four codons constitute the **genetic code**. Three codons function as signals that terminate translation. The remaining codons specify a particular amino acid. Some amino acids are specified by multiple codons.

Each tRNA has an **anticodon** that can base-pair with a codon, and it binds to the amino acid specified by that codon. Enzymatic rRNA and proteins make up the two subunits of ribosomes.

SECTION 9.4 During translation, protein-building information that is carried by an mRNA directs the synthesis of a polypeptide. First, an mRNA, an initiator tRNA, and two ribosomal subunits converge. Next, amino acids are delivered by tRNAs in the order specified by the codons in the mRNA. The intact ribosome catalyzes formation of a peptide bond between the successive amino acids, so a polypeptide forms. Translation ends when the ribosome encounters a stop codon in the mRNA.

SECTION 9.5 **Insertions**, **deletions**, and **base-pair substitutions** are mutations. A mutation that changes a gene's product may have harmful effects. Sickle-cell anemia, which is caused by a base-pair substitution in the gene for the beta globin chain of hemoglobin, is one example. Beta thalassemia is an outcome of frameshift mutations in the beta globin gene.

SECTION 9.6 The ability to make proteins is critical to all life processes. Ribosome-inactivating proteins (RIPs) have an enzyme domain that permanently disables ribosomes. Ricin and other toxic RIPs have an additional protein domain that triggers endocytosis. Once inside cytoplasm, the molecule's enzyme domain destroys the cell's ability to make proteins.

Self-Quiz

Answers in Appendix VII

1. A chromosome contains many different gene regions that are transcribed into different _______ .
   a. proteins
   b. polypeptides
   c. RNAs
   d. a and b

2. A binding site for RNA polymerase is called a _______ .
   a. gene
   b. promoter
   c. codon
   d. protein

3. An RNA molecule is typically _______ ; a DNA molecule is typically _______ .
   a. single-stranded; double-stranded
   b. double-stranded; single-stranded
   c. both are single-stranded
   d. both are double-stranded

4. RNAs form by _______ ; proteins form by _______ .
   a. replication; translation
   b. translation; transcription
   c. transcription; translation
   d. replication; transcription

5. The main function of a DNA molecule is to _______ .
   a. store heritable information
   b. carry RNA's message for translation
   c. form peptide bonds between amino acids
   d. carry amino acids to ribosomes

6. The main function of an mRNA molecule is to _______ .
   a. store heritable information
   b. carry DNA's genetic message for translation
   c. form peptide bonds between amino acids
   d. carry amino acids to ribosomes

7. Energy that drives transcription is provided mainly by _______ .
   a. ATP
   b. RNA nucleotides
   c. GTP
   d. all are correct

8. Most codons specify a(n) _______ .
   a. protein
   b. polypeptide
   c. amino acid
   d. mRNA

9. Anticodons pair with _______ .
   a. mRNA codons
   b. DNA codons
   c. RNA anticodons
   d. amino acids

10. Up to _______ amino acids can be encoded by an mRNA that consists of 45 nucleotides plus a stop codon.
    a. 15
    b. 45
    c. 90
    d. 135

Data Analysis Activities

RIPs as Cancer Drugs Researchers are taking a page from the structure–function relationship of RIPs in their quest for cancer treatments. The most toxic RIPs, remember, have one domain that interferes with ribosomes, and another that carries them into cells. Melissa Cheung and her colleagues incorporated a peptide that binds to skin cancer cells into the enzymatic part of an RIP, the *E. coli* Shiga-like toxin. The researchers created a new RIP that specifically kills skin cancer cells, which are notoriously resistant to established therapies. Some of their results are shown in **FIGURE 9.16**.

1. Which cells had the greatest response to an increase in concentration of the engineered RIP?
2. At what concentration of RIP did all of the different kinds of cells survive?
3. Which cells survived best at 10^{-6} grams per liter RIP?
4. On which type of cancer cells did the RIP have the least effect?

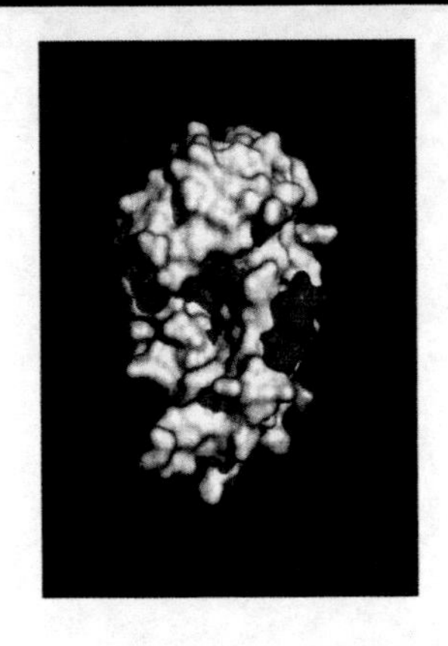

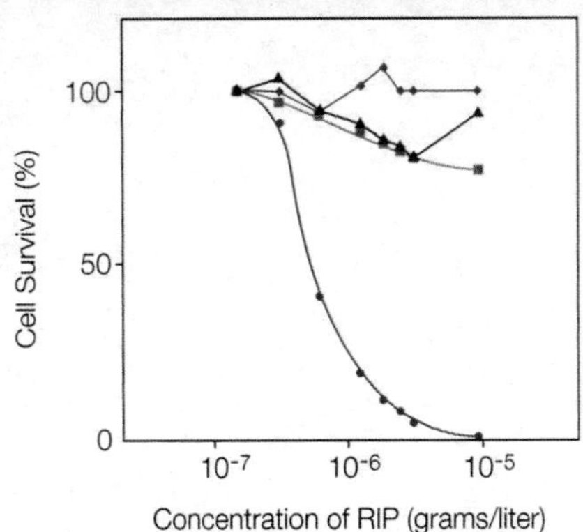

FIGURE 9.16 Effect of an engineered RIP on cancer cells.

The model on the *left* shows the enzyme portion of *E. coli* Shiga-like toxin that has been engineered to carry a small sequence of amino acids (in blue) that targets skin cancer cells. (Red indicates the active site.)

The graph on the *right* shows the effect of this engineered RIP on human cancer cells of the skin (●); breast (◆); liver (▲); and prostate (■).

11. _______ are removed from new mRNAs.
    a. Introns c. Telomeres
    b. Exons d. Amino acids

12. Where does transcription take place in a typical eukaryotic cell?
    a. the nucleus c. the cytoplasm
    b. ribosomes d. b and c are correct

13. Where does translation take place in a typical eukaryotic cell?
    a. the nucleus c. a and b
    b. the cytoplasm d. neither a nor b

14. Energy that drives translation is provided mainly by _______.
    a. ATP c. GTP
    b. amino acids d. all are correct

15. Match the terms with the best description.

___ genetic message	a. protein-coding segment
___ promoter	b. RNA polymerase binding site
___ polysome	c. read as base triplets
___ exon	d. removed before translation
___ genetic code	e. occurs only in groups
___ intron	f. complete set of 64 codons

Critical Thinking

1. Researchers are designing and testing antisense drugs as therapies for a variety of diseases, including cancer, AIDS, diabetes, and muscular dystrophy. The drugs are also being tested to fight infection by deadly viruses such as Ebola. Antisense drugs consist of short mRNA strands that are complementary in base sequence to mRNAs linked to the diseases. Speculate on how these drugs work.

2. An anticodon has the sequence GCG. What amino acid does this tRNA carry? What would be the effect of a mutation that changed the C of the anticodon to a G?

3. Each position of a codon can be occupied by one of four (4) nucleotides. What is the minimum number of nucleotides per codon necessary to specify all 20 of the amino acids that are found in proteins?

4. Refer to **FIGURE 9.6**, then translate the following mRNA nucleotide sequence into an amino acid sequence, starting at the first base:

(5′) UGUCAUGCUCGUCUUGAAUCUUGU GAUGCUCGUUGGAUUAAUUGU (3′)

5. Translate the sequence of bases in the previous question, starting at the second base.

6. Can you spell your name using the one-letter amino acid abbreviations shown in **FIGURE 9.6**? If so, construct an mRNA sequence that encodes your "protein" name.

CREDIT: Source: Cheung et al., *Molecular Cancer*, 9:28, 2010.

A multicelled eukaryote develops by repeated cell divisions. These are early frog embryos, each a product of three divisions of one fertilized egg.

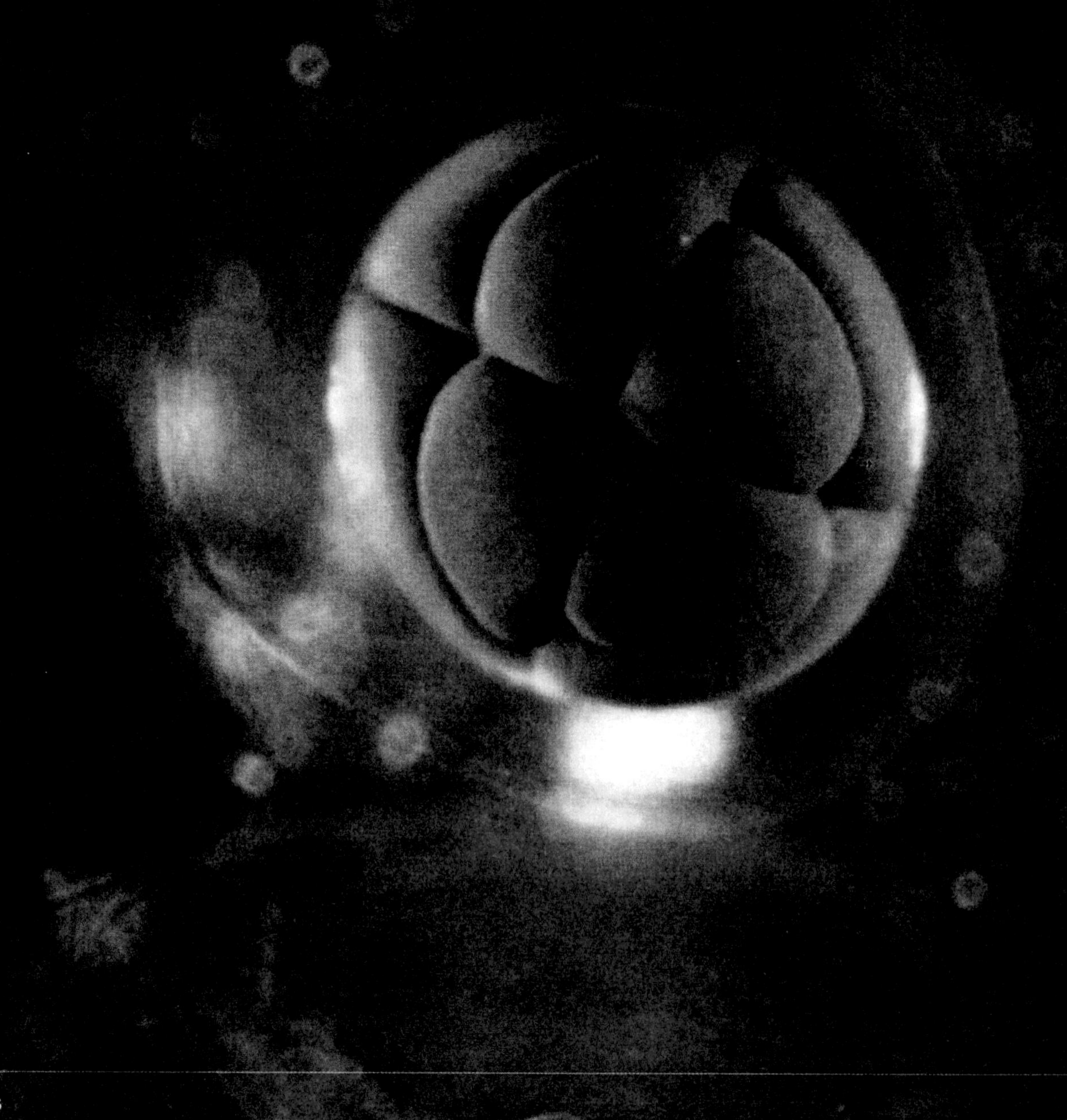

11 HOW CELLS REPRODUCE

Links to Earlier Concepts

Before beginning this chapter, be sure you understand cell structure (Sections 4.1, 4.5, 4.9, 4.10); chromosomes (8.3); and DNA replication (8.4) and repair (8.5). What you know about receptors and recognition proteins (4.3), free radicals (5.5) and mutations (9.5), fermentation (7.6), and eukaryotic gene control (10.1) will help you understand how cancer develops.

KEY CONCEPTS

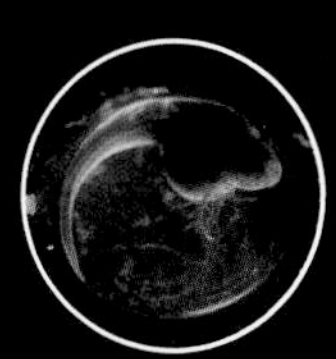

THE CELL CYCLE

A cell cycle starts when a new cell forms by division of a parent cell, and ends when the cel completes its own division. Built-in checkpoints control the timing and rate of the cycle.

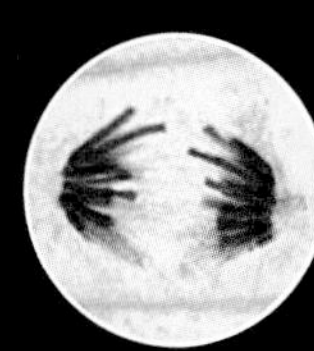

MITOSIS

Mitosis, a mechanism by which a cell's nucleus divides, maintains the chromosome number. Fou sequential stages parcel the cell's duplicated chromosomes into two new nuclei.

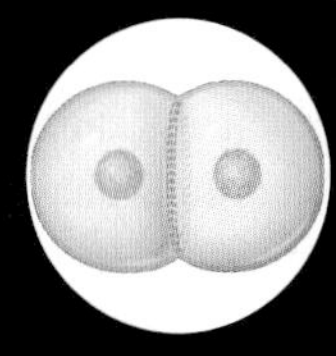

CYTOPLASMIC DIVISION

After nuclear division, the cytoplasm may divide, so one nucleus ends up in each of two new cells. The division proceeds by different mechanisms in animal and plant cells.

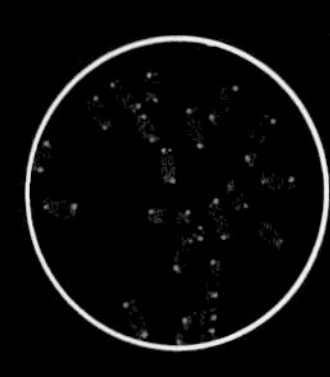

MITOTIC CLOCKS

Built into eukaryotic chromosomes are DNA sequences that protect the cell's genetic information. Degradation of these sequences is associated with cell death and aging.

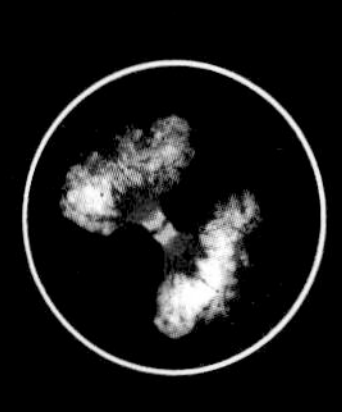

THE CELL CYCLE GONE AWRY

On rare occasions, checkpoint mechanisms fail, and cell division becomes uncontrollable. Tumou formation and cancer are outcomes.

Photograph by Carolina Biological Supply Company/Phototake.

11.1 HOW DO CELLS REPRODUCE?

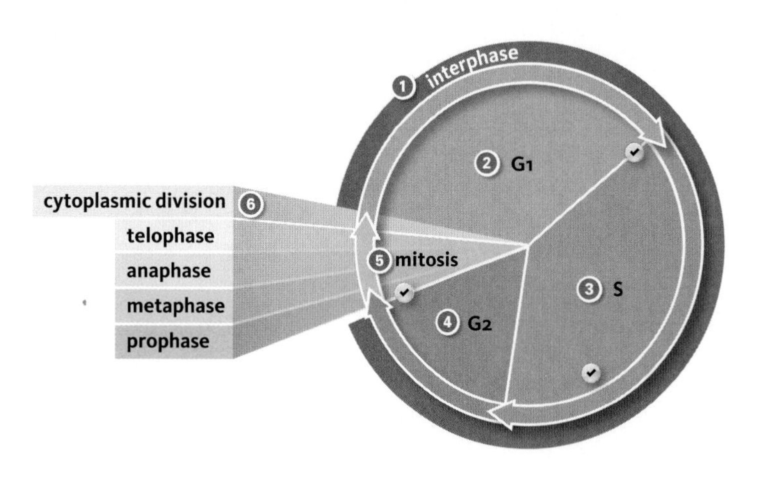

1 A cell spends most of its life in interphase, which includes three stages: G1, S, and G2.

2 G1 is the interval of growth before DNA replication. The cell's chromosomes are unduplicated.

3 S is the time of synthesis, during which the cell copies its DNA (duplicates its chromosomes).

4 G2 is the interval after DNA replication and before mitosis. The cell prepares to divide during this stage.

5 The nucleus divides during mitosis, the four stages of which are detailed in the next section. After mitosis, the cytoplasm may divide. Each descendant cell begins the cycle anew, in interphase.

Built-in checkpoints stop the cycle from proceeding until certain conditions are met.

FIGURE 11.1 {Animated} The eukaryotic cell cycle. The length of the intervals differs among cells. G1, S, and G2 are part of interphase.

MULTIPLICATION BY DIVISION

A life cycle is the collective series of events that an organism passes through during its lifetime. Multicelled organisms and free-living cells have life cycles, but what about cells that make up a multicelled body? Biologists consider such cells to be individually alive, each with its own life that passes through a series of recognizable stages. The events that occur from the time a cell forms until the time it divides are collectively called the **cell cycle** (**FIGURE 11.1**).

A typical cell spends most of its life in **interphase** 1. During this phase, the cell increases its mass, roughly doubles the number of its cytoplasmic components, and replicates its DNA in preparation for division. Interphase is typically the longest part of the cycle, and it consists of three stages: G1, S, and G2. G1 and G2 were named "Gap" intervals because outwardly they seem to be periods of inactivity, but they are not.

Most cells going about their metabolic business are in G1 2. Cells preparing to divide enter S 3, the time of DNA synthesis, when they duplicate their chromosomes (Section 8.4). During G2 4, the cell prepares to divide by making the proteins that will drive the process of division. Once S begins, DNA replication usually proceeds at a predictable rate until division begins.

The remainder of the cycle consists of the division process itself. When a cell divides, both of its cellular offspring end up with DNA and a blob of cytoplasm. Each of the offspring of a eukaryotic cell inherits its DNA packaged inside a nucleus. Thus, a eukaryotic cell's nucleus has to divide before its cytoplasm does.

Mitosis is a nuclear division mechanism that maintains the chromosome number 5. In multicelled organisms, mitosis and cytoplasmic division 6 are the basis of increases in body size and tissue remodeling during development (**FIGURE 11.2**), as well as ongoing replacements of damaged or dead cells. Mitosis and cytoplasmic division are also part of **asexual reproduction**, a reproductive mode by which offspring are produced by one parent only. This mode of reproduction is used by some multicelled eukaryotes and many single-celled ones. (Prokaryotes do not have a nucleus and do not undergo mitosis. We discuss their reproduction in Section 19.4.)

When a cell divides by mitosis, it produces two descendant cells, each with the chromosome number of the parent. However, if only the total number of chromosomes mattered, then one of the descendant cells might get, say, two pairs of chromosome 22 and no chromosome 9. A cell cannot function properly without a full complement of DNA, which means it needs to have *a copy of each*

FIGURE 11.2 A tadpole develops from repeated mitotic divisions of an egg. After it hatches, the individual will grow and undergo metamorphosis to develop into a frog.

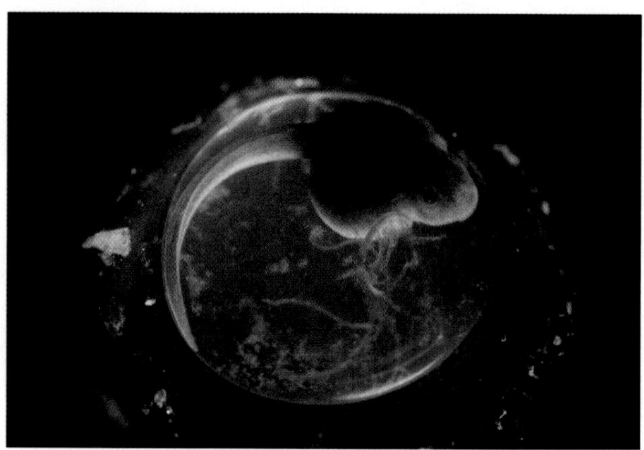

chromosome. Thus, the two cells produced by mitosis have the same number and types of chromosomes as the parent.

Remember from Section 8.3 that your body's cells are diploid, which means their nuclei contain pairs of chromosomes—two of each type. One chromosome of a pair was inherited from your father; the other, from your mother. Except for a pairing of nonidentical sex chromosomes (XY) in males, the chromosomes of each pair are homologous. **Homologous chromosomes** have the same length, shape, and genes (*hom–* means alike).

FIGURE 11.3 shows how homologous chromosomes are distributed to descendant cells when a diploid cell divides by mitosis. When a cell is in G1, each of its chromosomes consists of one double-stranded DNA molecule. The cell replicates its DNA in S, so by G2, each of its chromosomes consists of two double-stranded DNA molecules. These molecules stay attached to one another at the centromere as sister chromatids until mitosis is almost over, and then they are pulled apart and packaged into two separate nuclei. The next section details this process.

When sister chromatids are pulled apart, each becomes an individual chromosome that consists of one double-stranded DNA molecule. Thus, each of the two new nuclei that form in mitosis contains a full complement of (unduplicated) chromosomes. When the cytoplasm divides, these nuclei are packaged into separate cells. Each new cell starts the cell cycle over again in G1 of interphase.

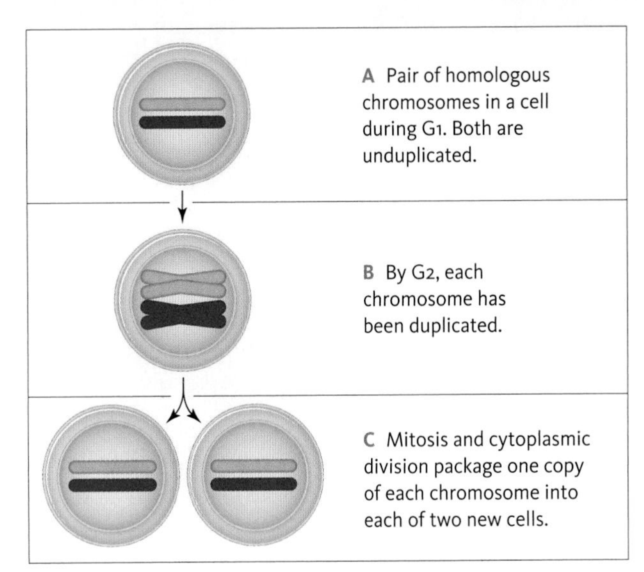

FIGURE 11.3 How mitosis maintains chromosome number in a diploid cell. For clarity, only one homologous pair is shown. The maternal chromosome is shown in pink, the paternal one in blue.

CONTROL OVER THE CELL CYCLE

When a cell divides—and when it does not—is determined by mechanisms of gene expression control (Section 10.1). Like the accelerator of a car, some of these mechanisms cause the cell cycle to advance. Others are like brakes, preventing the cycle from proceeding. In the adult body, brakes on the cell cycle normally keep the vast majority of cells in G1. Most of your nerve cells, skeletal muscle cells, heart muscle cells, and fat-storing cells have been in G1 since you were born, for example.

Control over the cell cycle also ensures that a dividing cell's descendants receive intact copies of its chromosomes. Built-in checkpoints monitor whether the cell's DNA has been copied completely, whether it is damaged, or even whether enough nutrients to support division are available. Protein products of "checkpoint genes" interact to carry out this type of control. For example, a checkpoint that operates in S puts the brakes on the cycle if the cell's chromosomes are damaged during DNA replication (Section 8.5). Checkpoint proteins that function as sensors recognize damaged DNA and bind to it. Upon binding, they trigger other events that stall the cell cycle, and also enhance expression of genes involved in DNA repair. After the problem has been corrected, the brakes are lifted and the cell cycle proceeds. If the problem remains uncorrected, other checkpoint proteins may initiate a series of events that eventually cause the cell to self-destruct.

asexual reproduction Reproductive mode of eukaryotes by which offspring arise from a single parent only.
cell cycle A series of events from the time a cell forms until its cytoplasm divides.
homologous chromosomes Chromosomes with the same length, shape, and genes.
interphase In a eukaryotic cell cycle, the interval between mitotic divisions when a cell grows, roughly doubles the number of its cytoplasmic components, and replicates its DNA.
mitosis Nuclear division mechanism that maintains the chromosome number.

TAKE-HOME MESSAGE 11.1

A cell cycle is the sequence of stages through which a cell passes during its lifetime (interphase, mitosis, and cytoplasmic division).

A eukaryotic cell reproduces by division: nucleus first, then cytoplasm. Each descendant cell receives a set of chromosomes and some cytoplasm.

When a nucleus divides by mitosis, each new nucleus has the same chromosome number as the parent cell.

Mechanisms of gene expression control can advance, delay, or block the cell cycle in response to internal and external conditions. Checkpoints built into the cycle allow problems to be corrected before the cycle proceeds.

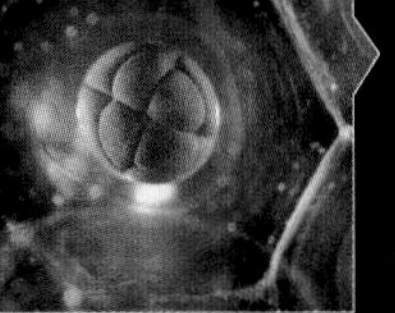

11.2 WHAT IS THE SEQUENCE OF EVENTS DURING MITOSIS?

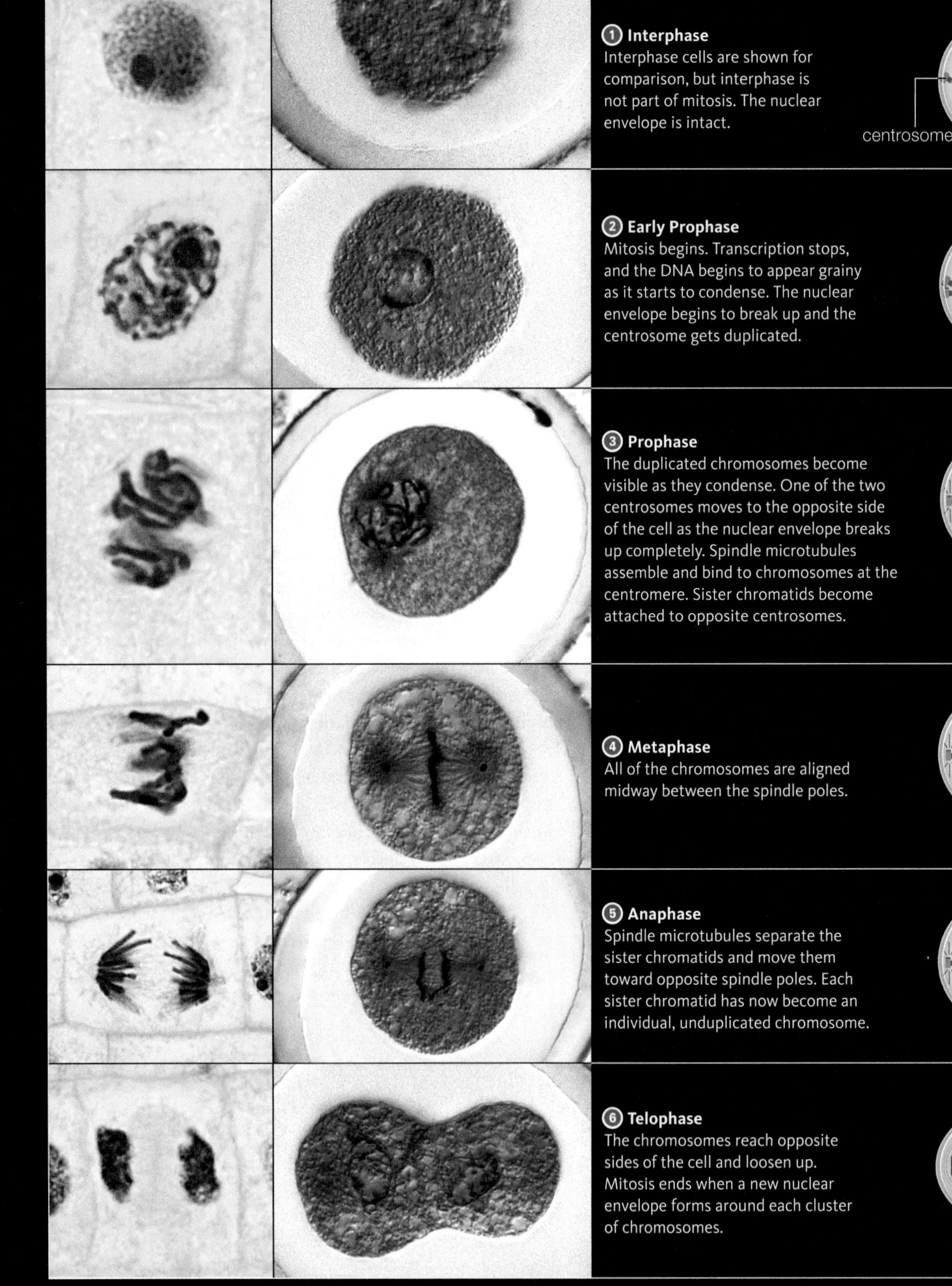

FIGURE 11.4 **{Animated}** Mitosis. Micrographs show nuclei of plant cells (onion root, *left*), and animal cells (fertilized eggs of a roundworm, *right*). A diploid ($2n$) animal cell with two chromosome pairs is illustrated.

During interphase, a cell's chromosomes are loosened to allow transcription and DNA replication. Loosened chromosomes are spread out, so they are not easily visible under a light microscope (**FIGURE 11.4** ❶). In preparation for nuclear division, the chromosomes begin to pack tightly ❷. Transcription and DNA replication stop as the chromosomes condense into their most compact "X" forms (Section 8.3). Tight condensation keeps the chromosomes from getting tangled and breaking during nuclear division.

Just before prophase, the centrosome becomes duplicated. Most animal cells have these structures, which typically consist of a pair of centrioles surrounded by a region of dense cytoplasm. (Remember from Section 4.9 that barrel-shaped centrioles help microtubules assemble.)

A cell reaches **prophase**, the first stage of mitosis, when its chromosomes have condensed so much that they are visible under a light microscope ❸. "Mitosis" is from *mitos*, the Greek word for thread, after the threadlike appearance of the chromosomes during nuclear division. If the cell has centrosomes, one of them now moves to the opposite side of the cell. Microtubules begin to assemble and lengthen from the centrosomes (or from other structures in cells with no centrosomes). The lengthening microtubules form a **spindle**, which is a temporary structure for moving chromosomes during nuclear division (**FIGURE 11.5**). The general area from which the spindle forms on each side of the cell is now called a spindle pole.

Spindle microtubules penetrate the nuclear region as the nuclear envelope breaks up. Some of the microtubules stop lengthening when they reach the middle of the cell. Others lengthen until they reach a chromosome and attach to it at the centromere. By the end of prophase, one sister chromatid of each chromosome has become attached to microtubules extending from one spindle pole, and the other sister has become attached to microtubules extending from the other spindle pole.

The opposing sets of microtubules then begin a tug-of-war by adding and losing tubulin subunits. As the microtubules lengthen and shorten, they push and pull

FIGURE 11.5 The spindle in a dividing cell of an amphibian. Microtubules (green) have extended from two centrosomes to form the spindle, which has attached to and aligned the chromosomes (blue) midway between its two poles. Red shows actin microfilaments.

the chromosomes. When all the microtubules are the same length, the chromosomes are aligned midway between spindle poles ❹. The alignment marks **metaphase** (from *meta*, the ancient Greek word for between).

During **anaphase**, the spindle pulls the sister chromatids of each duplicated chromosome apart and moves them toward opposite spindle poles ❺. Each DNA molecule has now become a separate chromosome.

Telophase begins when two clusters of chromosomes reach the spindle poles ❻. Each cluster has the same number and kinds of chromosomes as the parent cell nucleus had: two of each type of chromosome, if the parent cell was diploid. A new nuclear envelope forms around each set of chromosomes as they loosen up again. At this point, telophase—and mitosis—are finished.

TAKE-HOME MESSAGE 11.2

Chromosomes are duplicated before mitosis begins. Each now consists of two DNA molecules attached as sister chromatids.

In prophase, the chromosomes condense and a spindle forms. Spindle microtubules attach to the chromosomes as the nuclear envelope breaks up.

At metaphase, the spindle has aligned all of the (still duplicated) chromosomes in the middle of the cell.

In anaphase, sister chromatids separate and move toward opposite spindle poles. Each DNA molecule is now an individual chromosome.

In telophase, two clusters of chromosomes reach opposite spindle poles. A new nuclear envelope forms around each cluster, so two new nuclei form.

At the end of mitosis, each new nucleus has the same number and types of chromosomes as the parent cell's nucleus.

anaphase Stage of mitosis during which sister chromatids separate and move toward opposite spindle poles.
metaphase Stage of mitosis at which all chromosomes are aligned midway between spindle poles.
prophase Stage of mitosis during which chromosomes condense and become attached to a newly forming spindle.
spindle Temporary structure that moves chromosomes during nuclear division; consists of microtubules.
telophase Stage of mitosis during which chromosomes arrive at opposite spindle poles and decondense, and two new nuclei form.

11.3 HOW DOES A EUKARYOTIC CELL DIVIDE?

In most eukaryotes, the cell cytoplasm divides between late anaphase and the end of telophase, so two cells form, each with their own nucleus. The mechanism of cytoplasmic division, which is called **cytokinesis**, differs between plants and animals.

Typical animal cells pinch themselves in two after nuclear division ends (**FIGURE 11.6**). How? The spindle begins to disassemble during telophase ❶. The cell cortex, which is the mesh of cytoskeletal elements just under the plasma membrane (Section 4.9), includes a band of actin and myosin filaments that wraps around the cell's midsection. The band is called a contractile ring because it contracts when its component proteins are energized by phosphate-group transfers from ATP. When the ring contracts, it drags the attached plasma membrane inward ❷. The sinking plasma membrane becomes visible on the outside of the cell as an indentation between the former spindle poles ❸. The indentation, which is called a **cleavage furrow**, advances around the cell and deepens until the cytoplasm (and the cell) is pinched in two ❹. Each of the two cells formed by this division has its own nucleus and some of the parent cell's cytoplasm, and each is enclosed by a plasma membrane.

Dividing plant cells face a particular challenge because a stiff cell wall surrounds their plasma membrane (Section 4.10). Accordingly, plant cells have their own mechanism of cytokinesis. By the end of anaphase, a set of short microtubules has formed on either side of the future plane of division. These microtubules guide vesicles from Golgi bodies and the cell surface to the division plane ❺. After mitosis, the vesicles start to fuse into a disk-shaped **cell plate** ❻. The plate expands at its edges until it reaches the plasma membrane and attaches to it, thus partitioning the cytoplasm ❼. In time, the cell plate will develop into two new cell walls, so each of the descendant cells will be enclosed by its own plasma membrane and wall ❽.

Animal cell

❶ In a dividing animal cell, the spindle begins to disassemble after mitosis is completed.

❷ At the midpoint of the former spindle, a ring of actin and myosin filaments attached to the plasma membrane contracts.

❸ This contractile ring pulls the cell surface inward as it shrinks.

❹ The ring contracts until it pinches the cell in two.

Plant cell

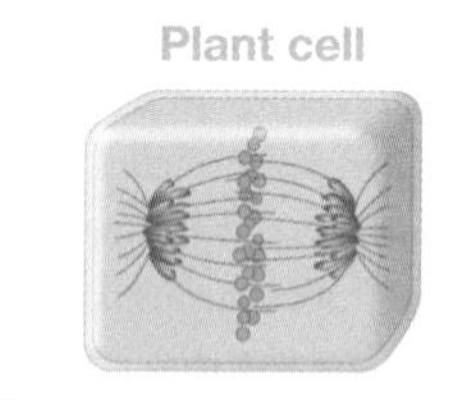

❺ In a dividing plant cell, vesicles cluster at the future plane of division before mitosis ends.

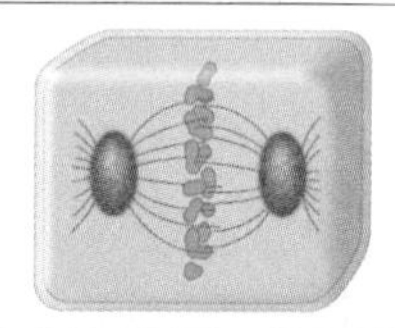

❻ The vesicles fuse with each other, forming a cell plate along the plane of division.

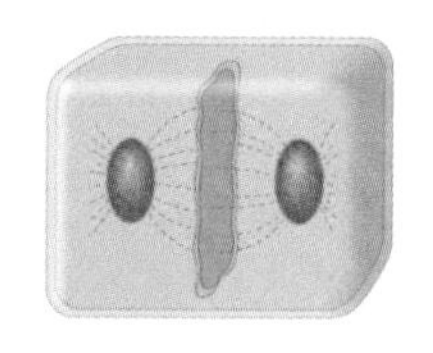

❼ The cell plate expands outward along the plane of division. When it reaches the plasma membrane, it attaches to the membrane and partitions the cytoplasm.

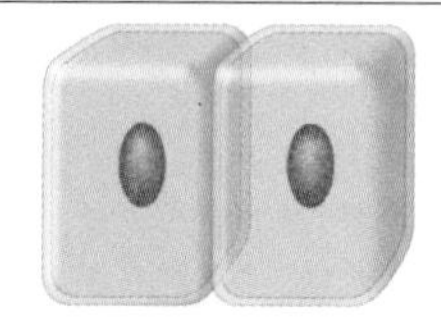

❽ The cell plate matures as two new cell walls. These walls join with the parent cell wall, so each descendant cell becomes enclosed by its own wall.

FIGURE 11.6 {Animated} Cytoplasmic division of animal cells (*top*) and plant cells (*bottom*).

cell plate A disk-shaped structure that forms during cytokinesis in a plant cell; matures as a cross-wall between the two new nuclei.
cleavage furrow In a dividing animal cell, the indentation where cytoplasmic division will occur.
cytokinesis Cytoplasmic division.

TAKE-HOME MESSAGE 11.3

In most eukaryotes, the cell cytoplasm divides between late anaphase and the end of telophase. Two descendant cells form, each with its own nucleus.

The mechanism of cell division differs between plants and animals.

In animal cells, a contractile ring pinches the cytoplasm in two. In plant cells, a cell plate that forms in the middle of the cell partitions the cytoplasm when it reaches and connects to the parent cell wall.

11.4 WHAT IS THE FUNCTION OF TELOMERES?

FIGURE 11.7 Telomeres. The bright dots at the end of each DNA strand in these duplicated chromosomes show telomere sequences.

In 1997, Scottish geneticist Ian Wilmut made worldwide headlines after his team cloned the first mammal from an adult somatic cell (SCNT, Section 8.6). The animal, a lamb named Dolly, was genetically identical to the sheep that had donated an udder cell. At first, Dolly looked and acted like a normal sheep, but she died early. By the time Dolly was five, she was as fat and arthritic as a twelve-year-old sheep. The following year, she contracted a lung disease that is typical of geriatric sheep, and had to be euthanized.

Dolly's early demise may have been the result of abnormally short telomeres. **Telomeres** are noncoding DNA sequences that occur at the ends of eukaryotic chromosomes (**FIGURE 11.7**). Vertebrate telomeres consist of a short DNA sequence, 5′-TTAGGG-3′, repeated perhaps thousands of times. These "junk" repeats provide a buffer against the loss of more valuable DNA internal to the chromosomes.

A telomere buffer is particularly important because, under normal circumstances, a eukaryotic chromosome shortens by about 100 nucleotides with each DNA replication. When a cell's offspring receive chromosomes with too-short telomeres, checkpoint gene products halt the cell cycle, and the descendant cells die shortly thereafter. Most body cells can divide only a certain number of times before this happens. This cell division limit may be a fail-safe mechanism in case a cell loses control over the cell cycle and begins to divide again and again. A limit on the number of divisions keeps such cells from overrunning the body (an outcome that, as you will see in the next section, has dangerous consequences to health). The cell division limit varies by species, and it may be part of the mechanism that sets an organism's life span. When Dolly was only two years old, her telomeres were as short as those of a six-year-old sheep—the exact age of the adult animal that had been her genetic donor. Scientists are careful to point out that shortening telomeres could be an effect of aging rather than a cause.

A few normal cells in an adult retain the ability to divide indefinitely. Their descendants replace cell lineages that eventually die out when they reach their division limit. These cells, which are called stem cells, are immortal because they continue to make an enzyme called telomerase. Telomerase reverses the telomere shortening that normally occurs after DNA replication.

Mice that have had their telomerase enzyme knocked out age prematurely. Their tissues degenerate much more quickly than those of normal mice, and their life expectancy declines to about half that of a normal mouse. When one of these knockout mice is close to the end of its shortened life span, rescuing the function of its telomerase enzyme results in lengthened telomeres. The rescued mouse also regains vitality: Decrepit tissue in its brain and other organs repairs itself and begins to function normally, and the once-geriatric individual even begins to reproduce again. While telomerase holds therapeutic promise for rejuvenation of aged tissues, it can also be dangerous: Cancer cells characteristically express high levels of the molecule.

TAKE-HOME MESSAGE 11.4

Telomeres protect eukaryotic chromosomes from losing genetic information at their ends.

Telomeres shorten with every cell division in normal body cells. When they are too short, the cell stops dividing and dies. Thus, telomeres are associated with aging.

telomere Noncoding, repetitive DNA sequence at the end of chromosomes; protects the coding sequences from degradation.

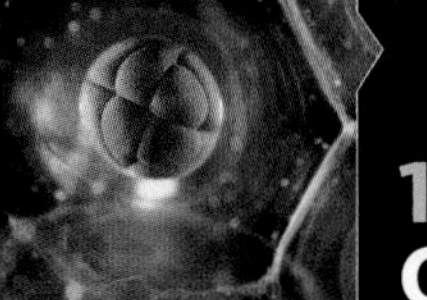

11.5 WHAT HAPPENS WHEN CONTROL OVER THE CELL CYCLE IS LOST?

THE ROLE OF MUTATIONS

Sometimes a checkpoint gene mutates so that its protein product no longer works properly. In other cases, the controls that regulate its expression fail, and a cell makes too much or too little of its product. When enough checkpoint mechanisms fail, a cell loses control over its cell cycle. Interphase may be skipped, so division occurs over and over with no resting period. Signaling mechanisms that cause abnormal cells to die may stop working. The problem is compounded because checkpoint malfunctions are passed along to the cell's descendants, which form a **neoplasm**, an accumulation of cells that lost control over how they grow and divide.

A neoplasm that forms a lump in the body is called a **tumor**, but the two terms are sometimes used interchangeably. Once a tumor-causing mutation has occurred, the gene it affects is called an oncogene. An **oncogene** is any gene that can transform a normal cell into a tumor cell (Greek *onkos*, or bulging mass). Oncogene mutations in reproductive cells can be passed to offspring, which is a reason that some types of tumors run in families.

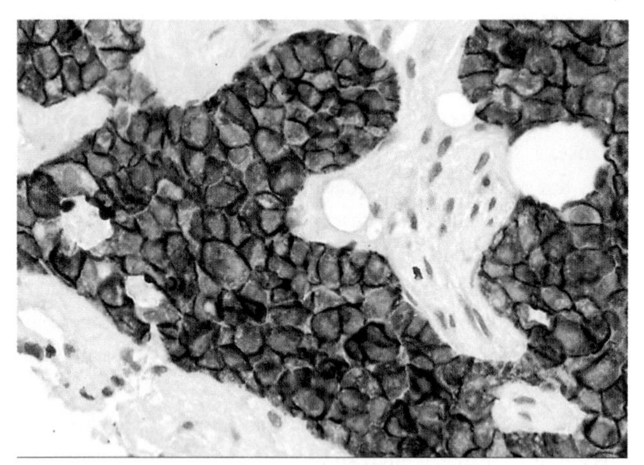

FIGURE 11.8 Effects of an oncogene. In this section of human breast tissue, a brown-colored tracer shows the active form of the EGF receptor. Normal cells are lighter in color. The dark cells have an overactive EGF receptor that is constantly stimulating mitosis; these cells have formed a neoplasm.

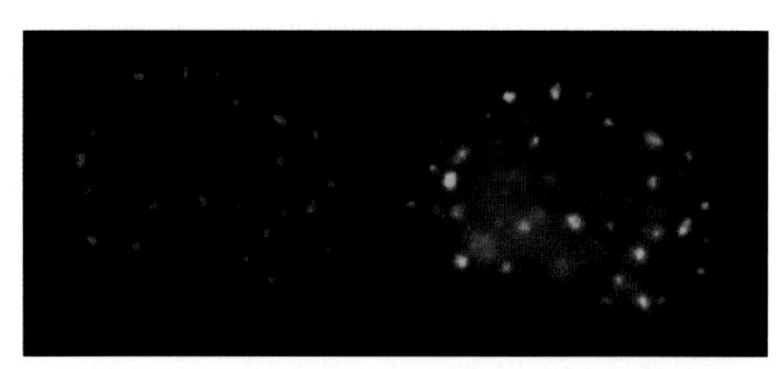

A Red dots show the location of the *BRCA1* gene product.

B Green dots pinpoint the location of another checkpoint gene product.

FIGURE 11.9 Checkpoint genes in action. Radiation damaged the DNA inside this nucleus. Two proteins have clustered around the same chromosome breaks in the same nucleus; both function to recruit DNA repair enzymes. The integrated action of these and other checkpoint gene products blocks mitosis until the DNA breaks are fixed.

Genes encoding proteins that promote mitosis are called **proto-oncogenes** because mutations can turn them into oncogenes. A gene that encodes the epidermal growth factor (EGF) receptor is an example of a proto-oncogene. **Growth factors** are molecules that stimulate a cell to divide and differentiate. The EGF receptor is a plasma membrane protein; when it binds to EGF, it becomes activated and triggers the cell to begin mitosis. Mutations can result in an EGF receptor that stimulates mitosis even when EGF is not present. Most neoplasms carry mutations resulting in an overactivity or overabundance of this particular receptor (**FIGURE 11.8**).

Checkpoint gene products that inhibit mitosis are called tumor suppressors because tumors form when they are missing. The products of the *BRCA1* and *BRCA2* genes (Section 10.6) are examples of tumor suppressors. These proteins regulate, among other things, the expression of DNA repair enzymes (**FIGURE 11.9**). Tumor cells often have mutations in their *BRCA* genes.

Viruses such as HPV (human papillomavirus) cause a cell to make proteins that interfere with its own tumor suppressors. Infection with HPV causes skin growths called warts, and some kinds are associated with neoplasms that form on the cervix.

CANCER

Benign neoplasms such as warts are not usually dangerous (**FIGURE 11.10**). They grow very slowly, and their cells retain the plasma membrane adhesion proteins that keep them properly anchored to the other cells in their home tissue ❶.

A malignant neoplasm is one that gets progressively worse, and is dangerous to health. Malignant cells typically display the following three characteristics:

First, like cells of all neoplasms, malignant cells grow and divide abnormally. Controls that usually keep cells from

cancer Disease that occurs when a malignant neoplasm physically and metabolically disrupts body tissues.
growth factor Molecule that stimulates mitosis and differentiation.
metastasis The process in which malignant cells spread from one part of the body to another.
neoplasm An accumulation of abnormally dividing cells.
oncogene Gene that helps transform a normal cell into a tumor cell.
proto-oncogene Gene that, by mutation, can become an oncogene.
tumor A neoplasm that forms a lump.

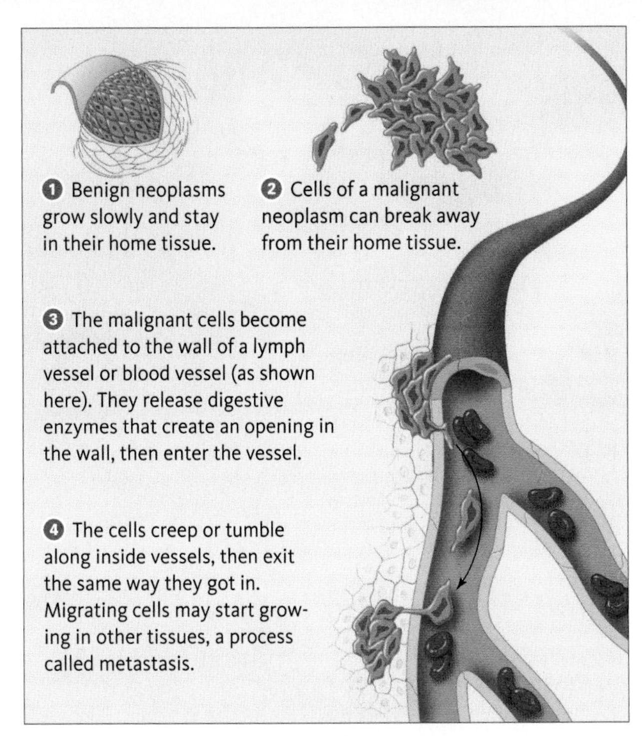

FIGURE 11.10 {Animated} Neoplasms and malignancy.

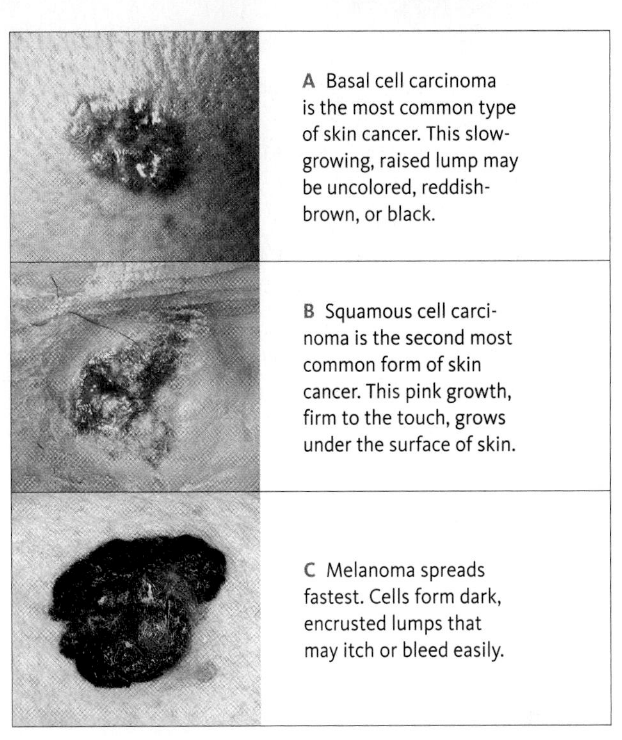

FIGURE 11.11 Skin cancer can be detected with early screening.

getting overcrowded in tissues are lost in malignant cells, so their populations may reach extremely high densities with cell division occurring very rapidly. The number of small blood vessels that transport blood to the growing cell mass also increases abnormally.

Second, the cytoplasm and plasma membrane of malignant cells are altered. The cytoskeleton may be shrunken, disorganized, or both. Malignant cells typically have an abnormal chromosome number, with some chromosomes present in multiple copies, and others missing or damaged. The balance of metabolism is often shifted, as in an amplified reliance on ATP formation by fermentation rather than aerobic respiration.

Altered or missing proteins impair the function of the plasma membrane of malignant cells. For example, these cells do not stay anchored properly in tissues because their plasma membrane adhesion proteins are defective or missing ❷. Malignant cells can slip easily into and out of vessels of the circulatory and lymphatic systems ❸. By migrating through these vessels, the cells can establish neoplasms elsewhere in the body ❹. The process in which malignant cells break loose from their home tissue and invade other parts of the body is called **metastasis**. Metastasis is the third hallmark of malignant cells.

The disease called **cancer** occurs when the abnormally dividing cells of a malignant neoplasm disrupt body tissues, both physically and metabolically. Unless chemotherapy, surgery, or another procedure eliminates malignant cells from the body, they can put an individual on a painful road to death. Each year, cancer causes 15 to 20 percent of all human deaths in developed countries. The good news is that mutations in multiple checkpoint genes are required to transform a normal cell into a malignant one, and such mutations may take a lifetime to accumulate. Lifestyle choices such as not smoking and avoiding exposure of unprotected skin to sunlight can reduce one's risk of acquiring mutations in the first place. Some neoplasms can be detected with periodic screening such as gynecology or dermatology exams (**FIGURE 11.11**). If detected early enough, many types of malignant neoplasms can be removed before metastasis occurs.

TAKE-HOME MESSAGE 11.5

- Neoplasms form when cells lose control over their cell cycle and begin dividing abnormally.
- Mutations in multiple checkpoint genes can give rise to a malignant neoplasm that gets progressively worse.
- Cancer is a disease that occurs when the abnormally dividing cells of a malignant neoplasm physically and metabolically disrupt body tissues.
- Although some mutations are inherited, lifestyle choices and early intervention can reduce one's risk of cancer.

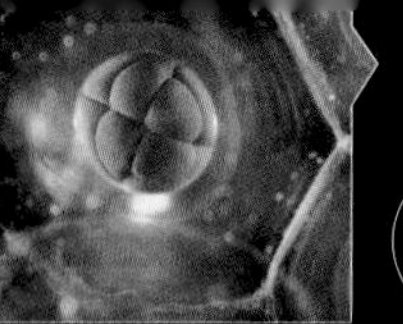

11.6 Application: HENRIETTA'S IMMORTAL CELLS

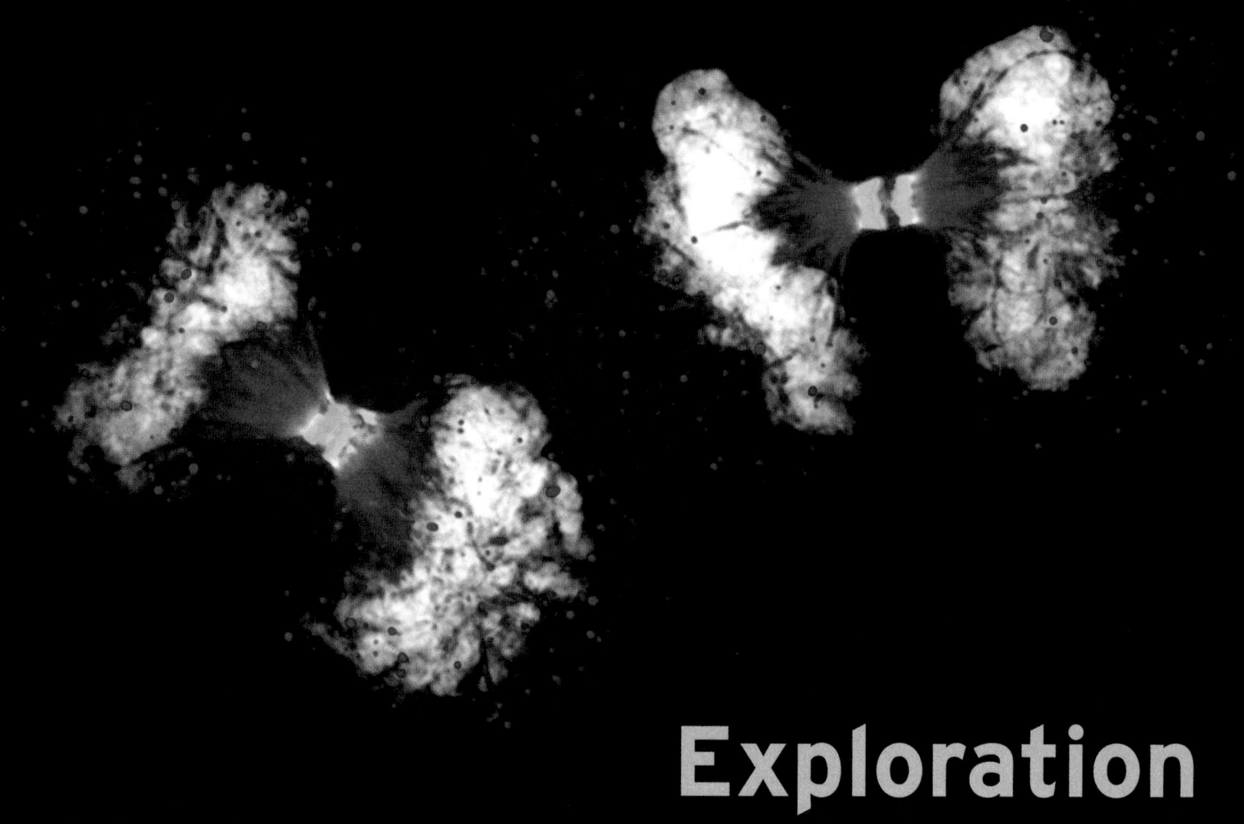

Exploration

FIGURE 11.12 HeLa cells, a legacy of cancer victim Henrietta Lacks. The cells in this fluorescent micrograph are undergoing mitosis.

FINDING HUMAN CELLS THAT GROW IN A LABORATORY TOOK GEORGE AND MARGARET GEY NEARLY 30 YEARS. In 1951, their assistant Mary Kubicek prepared yet another sample of human cancer cells. Mary named the cells HeLa, after the first and last names of the patient from whom the cells had been taken. The HeLa cells began to divide, again and again. The cells were astonishingly vigorous, quickly coating the inside of their test tube and consuming their nutrient broth. Four days later, there were so many cells that the researchers had to transfer them to more tubes. The cell populations increased at a phenomenal rate. The cells were dividing every twenty-four hours and coating the inside of the tubes within days.

Sadly, cancer cells in the patient were dividing just as fast. Only six months after she had been diagnosed with cervical cancer, malignant cells had invaded tissues throughout her body. Two months after that, Henrietta Lacks, a young African American woman from Baltimore, was dead.

Although Henrietta passed away, her cells lived on in the Geys' laboratory. The Geys were able to grow poliovirus in HeLa cells, a practice that enabled them to determine which strains of the virus cause polio. That work was a critical step in the development of polio vaccines, which have since saved millions of lives.

Henrietta Lacks was just thirty-one, a wife and mother of five, when runaway cell divisions of cancer killed her. Her cells, however, are still dividing, again and again, more than fifty years after she died. Frozen away in tiny tubes and packed in Styrofoam boxes, HeLa cells continue to be shipped among laboratories all over the world. They are still widely used to investigate cancer (FIGURE 11.12), viral growth, protein synthesis, the effects of radiation, and countless other processes important in medicine and research. HeLa cells helped several researchers win Nobel Prizes, and some even traveled into space for experiments on satellites.

Henrietta Lacks

In these mitotic HeLa cells, chromosomes appear white and the spindle is red. Green identifies an enzyme that helps attach spindle microtubules to centromeres. Blue pinpoints a protein that helps sister chromatids stay attached to one another at the centromere. At this stage of telophase, the blue and green proteins should be closely associated midway between the two clusters of chromosomes. The abnormal distribution means that the spindle microtubules are not properly attached to the centromeres.

These days, physicians and researchers are required to obtain a signed consent form before they take tissue samples from a patient. No such requirement existed in the 1950s. It was common at that time for doctors to experiment on patients without their knowledge or consent. Thus, the young resident who was treating Henrietta Lacks's cancerous cervix probably never even thought about asking permission before he took a sample of it. That sample was the one that the Geys used to establish the HeLa cell line. No one in Henrietta's family knew about the cells until 25 years after her death. HeLa cells are still being sold worldwide, but her family has not received any share of profits.

Ongoing research with HeLa cells may one day allow researchers to identify drugs that target and destroy malignant cells or stop them from dividing. The research is far too late to have saved Henrietta Lacks, but it may one day yield drugs that put the brakes on cancer.

PEOPLE MATTER

National Geographic Explorer
DR. IAIN COUZIN

Can a million migrating wildebeests help explain why cancer spreads? Ask Iain Couzin. He is exploring how collective behavior in animals can be quantified and analyzed to give us new insights into the patterns of nature—and ourselves.

"We're realizing that animals have highly coordinated social systems and make decisions together," Couzin says. "They can do things collectively that no individual could do alone. It's still a very unexplored area of animal behavior." Couzin blends fieldwork, lab experiments, computer simulations, and complex mathematical models to test theories about how and why cells, animals, and humans organize and work together.

Couzin recently coauthored a paper hypothesizing that cancer cells migrating during metastasis—when the cells leave the primary tumor and journey elsewhere in the body—may have some parallels to animal swarms. As animals do in swarms, metastatic cells collectively can sense the environment and make decisions in response to it. It is very early theoretical work, but it does point to a new angle for cancer research: trying to knock out the mechanisms behind such collective migration.

CREDITS: (12 inset) Courtesy of the family of Henrietta Lacks; (right inset) Courtesy © Iain Couzin.

Summary

SECTION 11.1 A **cell cycle** includes all the stages through which a eukaryotic cell passes during its lifetime; it starts when a new cell forms, and ends when the cell reproduces. Most of a cell's activities, including replication of the cell's **homologous chromosomes**, occur during **interphase**.

A eukaryotic cell reproduces by dividing: nucleus first, then cytoplasm. **Mitosis** is a mechanism of nuclear division that maintains the chromosome number. It is the basis of growth, cell replacements, and tissue repair in multicelled species, and **asexual reproduction** in many species.

SECTION 11.2 Mitosis proceeds in four stages.

In **prophase**, the duplicated chromosomes start to condense. Microtubules assemble and form a **spindle**, and the nuclear envelope breaks up. Some microtubules that extend from one spindle pole attach to one chromatid of each chromosome; some that extend from the opposite spindle pole attach to its sister chromatid. These microtubules drag each chromosome toward the center of the cell.

At **metaphase**, all chromosomes are aligned at the spindle's midpoint.

During **anaphase**, the sister chromatids of each chromosome detach from each other, and the spindle microtubules move them toward opposite spindle poles.

During **telophase**, a complete set of chromosomes reaches each spindle pole. A nuclear envelope forms around each cluster. Two new nuclei, each with the parental chromosome number, are the result.

SECTION 11.3 **Cytokinesis** typically follows nuclear division. In animal cells, a contractile ring of microfilaments pulls the plasma membrane inward, forming a **cleavage furrow** that pinches the cytoplasm in two. In plant cells, vesicles guided by microtubules to the future plane of division merge as a **cell plate**. The plate expands until it fuses with the parent cell wall, thus becoming a cross-wall that partitions the cytoplasm.

SECTION 11.4 **Telomeres** that protect the ends of eukaryotic chromosomes shorten with every DNA replication. Cells that inherit too-short telomeres die, and in most cells this limits the number of divisions that can occur.

SECTION 11.5 The products of checkpoint genes, including receptors for **growth factors**, work together to control the cell cycle. These molecules monitor the integrity of the cell's DNA, and can pause the cycle until breaks or other problems are fixed. When checkpoint mechanisms fail, a cell loses control over its cell cycle, and the cell's descendants form a **neoplasm**. Neoplasms may form lumps called **tumors**.

Checkpoint genes are examples of **proto-oncogenes**, which means mutations can turn them into tumor-causing **oncogenes**. Mutations in multiple checkpoint genes can transform benign neoplasms into malignant ones. Cells of malignant neoplasms can break loose from their home tissues and colonize other parts of the body, a process called **metastasis**. **Cancer** occurs when malignant neoplasms physically and metabolically disrupt normal body tissues.

SECTION 11.6 An immortal line of human cells (HeLa) is a legacy of cancer victim Henrietta Lacks. Researchers all over the world continue to work with these cells as they try to unravel the mechanisms of cancer.

Self-Quiz

Answers in Appendix VII

1. Mitosis and cytoplasmic division function in _______ .
   a. asexual reproduction of single-celled eukaryotes
   b. growth and tissue repair in multicelled species
   c. gamete formation in bacteria and archaea
   d. sexual reproduction in plants and animals
   e. both a and b

2. A duplicated chromosome has _______ chromatid(s).
   a. one c. three
   b. two d. four

3. Homologous chromosomes _______ .
   a. carry the same genes c. are the same length
   b. are the same shape d. all of the above

4. Most cells spend the majority of their lives in _______ .
   a. prophase d. telophase
   b. metaphase e. interphase
   c. anaphase f. d and e

5. The spindle attaches to chromosomes at the _______ .
   a. centriole c. centromere
   b. contractile ring d. centrosome

6. Only _______ is not a stage of mitosis.
   a. prophase c. interphase
   b. metaphase d. anaphase

7. In intervals of interphase, G stands for _______ .
   a. gap b. growth c. Gey d. gene

8. Interphase is the part of the cell cycle when _______ .
   a. a cell ceases to function
   b. a cell forms its spindle apparatus
   c. a cell grows and duplicates its DNA
   d. mitosis proceeds

9. After mitosis, the chromosome number of a descendant cell is _______ the parent cell's.
   a. the same as c. rearranged compared to
   b. one-half of d. doubled compared to

10. A plant cell divides by the process of _______ .
   a. telekinesis c. fission
   b. nuclear division d. cytokinesis

Data Analysis Activities

HeLa Cells Are a Genetic Mess HeLa cells continue to be an extremely useful tool in cancer research. One early finding was that HeLa cells can vary in chromosome number. Defects in proteins that orchestrate cell division result in descendant cells with too many or too few chromosomes, an outcome that is one of the hallmarks of cancer cells.

The panel of chromosomes in **FIGURE 11.13**, originally published in 1989 by Nicholas Popescu and Joseph DiPaolo, shows all of the chromosomes in a single metaphase HeLa cell.

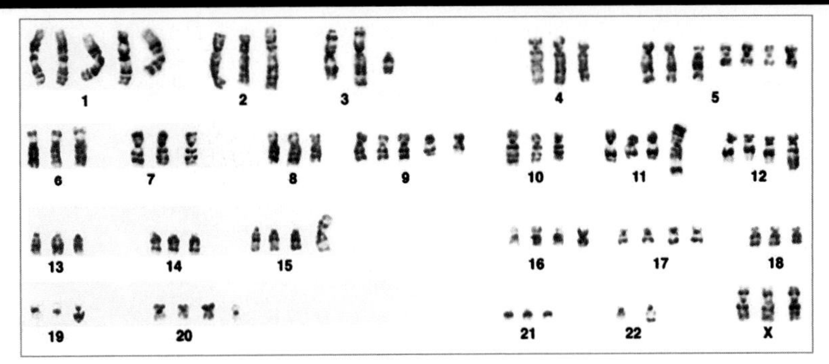

FIGURE 11.13 Chromosomes in a HeLa cell.

1. What is the chromosome number of this HeLa cell?
2. How many extra chromosomes does this cell have, compared to a normal human body cell?
3. Can you tell that this cell came from a female? How?

11. In the diagram of the nucleus *below*, fill in the blanks with the name of each interval.

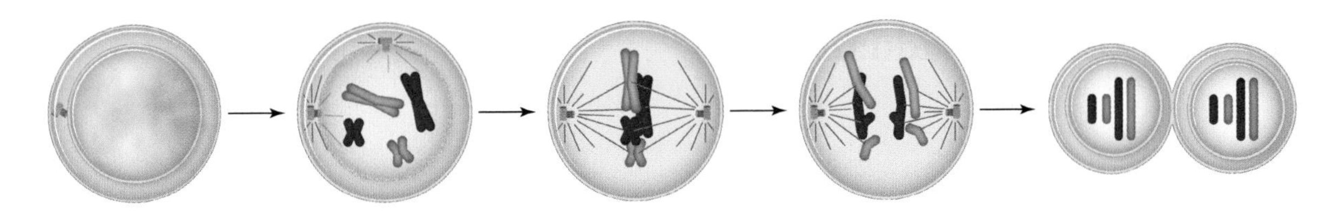

__________ __________ __________ __________ __________

12. *BRCA1* and *BRCA2* _______ .
    a. are checkpoint genes
    b. are proto-oncogenes
    c. encode tumor suppressors
    d. all of the above

13. _______ are characteristic of cancer.
    a. Malignant cells
    b. Neoplasms
    c. Tumors

14. Match each term with its best description.

___ cell plate	a. lump of cells
___ spindle	b. made of microfilaments
___ tumor	c. divides plant cells
___ cleavage furrow	d. organize(s) the spindle
___ contractile ring	e. caused by metastatic cells
___ cancer	f. made of microtubules
___ centrosomes	g. indentation
___ telomere	h. shortens with age

15. Match each stage with the events listed.

___ metaphase	a. sister chromatids move apart
___ prophase	b. chromosomes start to condense
___ telophase	c. new nuclei form
___ interphase	d. all duplicated chromosomes are aligned at the spindle equator
___ anaphase	e. DNA replication
___ cytokinesis	f. cytoplasmic division

Critical Thinking

1. When a cell reproduces by mitosis and cytoplasmic division, does its life end?

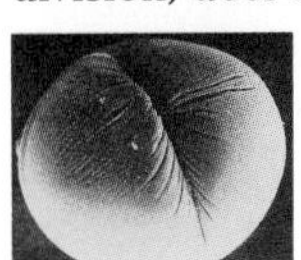

2. The eukaryotic cell in the photo on the *left* is in the process of cytoplasmic division. Is this cell from a plant or an animal? How do you know?

3. Exposure to radioisotopes or other sources of radiation can damage DNA. Humans exposed to high levels of radiation face a condition called radiation poisoning. Why do you think that hair loss and damage to the lining of the gut are early symptoms of radiation poisoning? Speculate about why exposure to radiation is used as a therapy to treat some kinds of cancers.

4. Suppose you have a way to measure the amount of DNA in one cell during the cell cycle. You first measure the amount at the G1 phase. At what points in the rest of the cycle will you see a change in the amount of DNA per cell?

CREDITS: (13) © Dr. Thomas Ried, NIH and the American Association for Cancer Research. (S-Q11) © Cengage Learning; (CT2) D. M. Phillips/Visuals Unlimited.

New Zealand mud snails can reproduce on their own (asexually) or with a partner (sexually). Natural populations of this species vary in the frequency of sexual and asexual individuals.

MEIOSIS AND SEXUAL REPRODUCTION

Links to Earlier Concepts

Before you begin this chapter, be sure you understand how eukaryotic chromosomes are organized (Section 8.3) and how genes work (9.1). You will draw on your knowledge of DNA replication (8.4), cytoplasmic division (11.3), and cell cycle controls (11.5) as we compare meiosis with mitosis (11.2). This chapter also revisits clones (8.6), the effects of mutation (9.5), and the function of telomeres (11.4).

KEY CONCEPTS

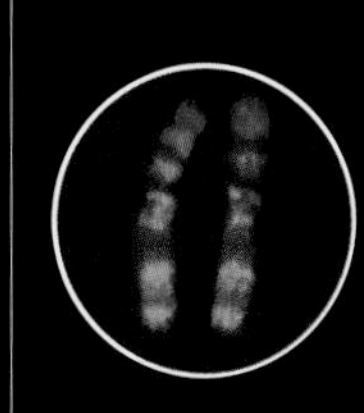

SEX AND ALLELES

In asexual reproduction, one parent transmits its genes to offspring. In sexual reproduction, offspring inherit genes from two parents who usually differ in some number of alleles.

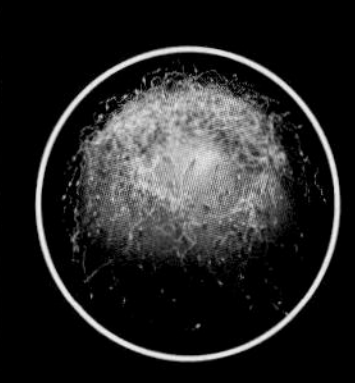

MEIOSIS IN REPRODUCTION

Meiosis is a nuclear division process that reduces the chromosome number. It occurs only in cells that play a role in sexual reproduction in eukaryotes.

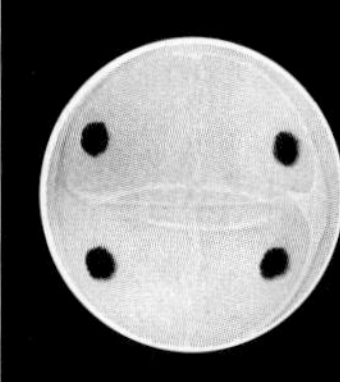

STAGES OF MEIOSIS

The chromosome number becomes reduced by the two nuclear divisions of meiosis. During this process, the chromosomes are sorted into four new nuclei.

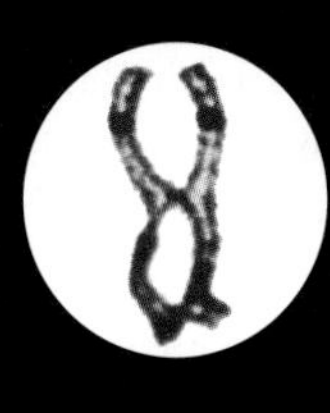

SHUFFLING PARENTAL DNA

During meiosis, homologous chromosomes swap segments, then are randomly sorted into separate nuclei. Both processes lead to novel combinations of alleles among offspring.

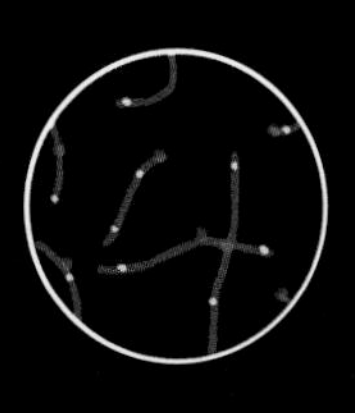

MITOSIS AND MEIOSIS COMPARED

Similarities between mitosis and meiosis suggest meiosis originated by evolutionary remodeling of mechanisms that already existed for mitosis and, before that, for repairing damaged DNA.

12.1 WHY SEX?

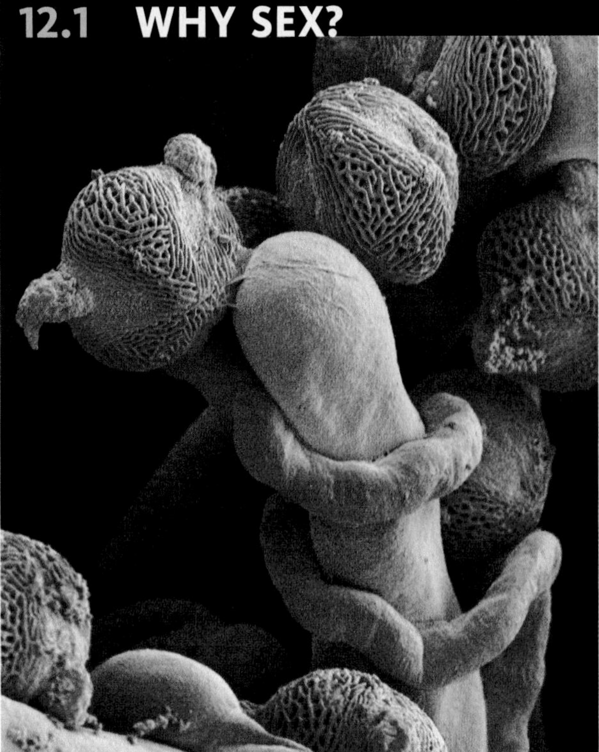

FIGURE 12.1 A moment in sexual reproduction. Sex mixes up the genetic material of two individuals. In flowering plants, pollen grains (orange) germinate on flower carpels (yellow). Pollen tubes with male gametes inside grow from the grains down into tissues of the ovary, which house the flower's female gametes.

INTRODUCING ALLELES

In asexual reproduction, a single individual gives rise to offspring that are identical to itself and to one another. By contrast, **sexual reproduction** involves two individuals and mixes their genetic material (**FIGURE 12.1**).

In your somatic (body) cells and in those of many other sexually reproducing eukaryotes, one chromosome of each homologous pair is maternal, and the other is paternal (Section 8.3). Homologous chromosomes carry the same genes (**FIGURE 12.2A**). However, the corresponding genes on maternal and paternal chromosomes often vary—just a bit—in DNA sequence. Over evolutionary time, unique mutations accumulate in separate lines of descent, and some of those mutations occur in genes. Thus, the DNA sequence of any gene may differ from the corresponding gene on the homologous chromosome (**FIGURE 12.2B**). Different forms of the same gene are called **alleles**.

Alleles may encode slightly different forms of the gene's product. Such differences influence the details of traits shared by a species. Consider that one of the approximately 20,000 genes in human chromosomes encodes beta globin (Section 9.5). Like most human genes, the beta globin gene has multiple alleles—more than 700 in this case. A few beta globin alleles cause sickle-cell anemia, several cause beta thalassemia, and so on. Allele differences among individuals are one reason that the members of a sexually reproducing species are not identical. Offspring of sexual reproducers inherit new combinations of alleles, which is the basis of new combinations of traits.

ON THE ADVANTAGES OF SEX

If the function of reproduction is the perpetuation of one's genes, then an organism that reproduces only by mitosis would seem to win the evolutionary race. When it reproduces, it passes all of its genes to every one of its offspring. Only about half of a sexual reproducer's genes are passed to each offspring. Yet most eukaryotes reproduce sexually, at least some of the time. Why?

Consider that all offspring of an asexual reproducer are clones. They have the same alleles—and the same traits—as their parent. Consistency is a good thing if an organism lives in a favorable, unchanging environment. Alleles that help it survive and reproduce do the same for its descendants. However, most environments are constantly changing. Offspring of asexual reproducers are equally vulnerable to unfavorable environmental change. In changing environments, sexual reproducers have the evolutionary edge. Their offspring inherit different combinations of alleles, so they vary in the details of their shared traits. Some may have a particular combination of traits that suits them perfectly to a change. As a group, their diversity offers them a better chance of surviving an environmental challenge than clones.

Perhaps the most important advantage of sexual reproduction involves the inevitable occurrence of harmful mutations. A population of sexual reproducers has a better chance of weathering the effects of such mutations. With asexual reproduction, individuals bearing a harmful mutation necessarily pass it to all of their offspring. This outcome would be rare in sexual reproduction, because each offspring of a sexual union has a 50 percent chance of inheriting a parent's mutation. Thus, all else being equal, harmful mutations accumulate in an asexually reproducing population more quickly than in a sexually reproducing one.

Environmental change is the norm. Consider, for example, the interaction between a predatory species and its prey. In each generation, prey individuals with traits that allow them to hide from, fend off, or escape the predator will leave the most young. However, the predator is constantly changing too: In each generation, predators best able to find, capture, and overcome prey leave the most descendants. Thus predators and prey are locked in a constant race, with each genetic improvement in one species

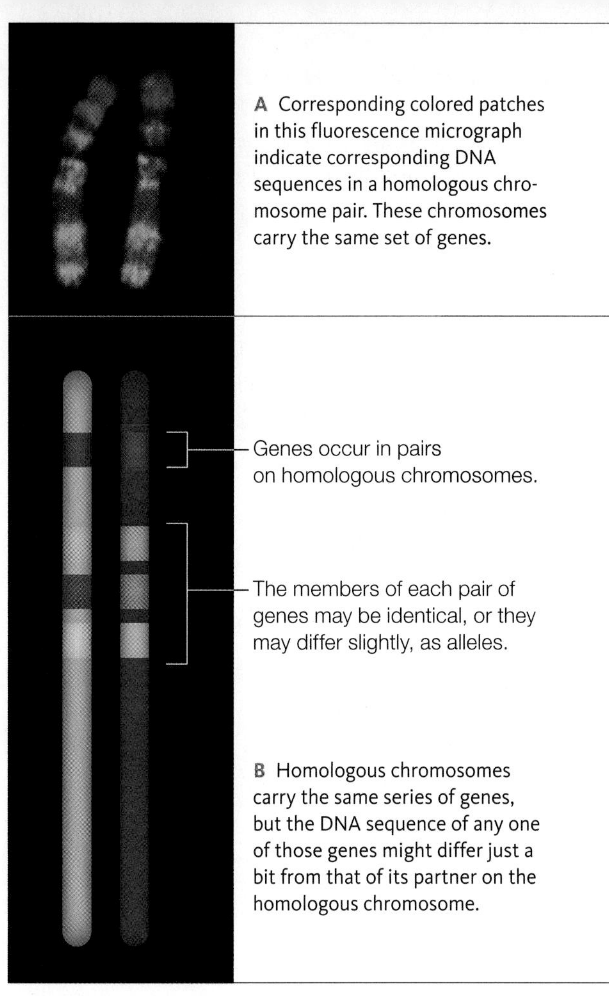

FIGURE 12.2 Genes on chromosomes. Different forms of a gene are called alleles.

countered by an improvement in the other. The Red Queen hypothesis holds that sexual reproduction is widespread because of the pressure for constant change resulting from such species interactions. The name of the hypothesis is a reference to Lewis Carroll's book *Through the Looking Glass*. In the book, the Queen of Hearts tells Alice, "It takes all the running you can do, to keep in the same place."

alleles Forms of a gene with slightly different DNA sequences; may encode slightly different versions of the gene's product.
sexual reproduction Reproductive mode by which offspring arise from two parents and inherit genes from both.

TAKE-HOME MESSAGE 12.1

Paired genes on homologous chromosomes may vary in DNA sequence as alleles. Alleles arise by mutation.

Alleles are the basis of differences in shared traits. Offspring of sexual reproducers inherit new combinations of parental alleles—thus new combinations of such traits.

PEOPLE MATTER

National Geographic Grantee
MAURINE NEIMAN

Animals that reproduce solely by asexual means are very rare. However, some populations of the New Zealand freshwater snail pictured in the chapter opener do just that. Maurine Neiman compares sexually reproducing and asexual populations of these snails to determine the costs and benefits of sex. Neiman and her collaborators Amy Krist and Adam Kay have discovered that sexual and asexual snails might differ in their need for phosphorus. Like most sexual organisms, the sexual snails have two chromosome sets; like most animals that cannot reproduce sexually, the asexual snails have at least three. DNA has a high phosphorus content, so having the extra sets of chromosomes multiplies each organism's requirement for this nutrient, which could add up to a big minus in the cost-of-asexuality equation.

Neiman's research has shown that extra chromosome sets translate into a disadvantage in low-phosphorus environments for the asexual snails, setting the stage for the possibility that the sexual snails—with the fewest chromosome sets of all—are likely to beat out asexuals when phosphorus is scarce. Follow-up experiments will extend beyond explaining the predominance of sex. These tiny snails have spread far beyond their native New Zealand, and huge populations of them are now disrupting ecosystems all over the world. The invasive populations are always asexual (and in fact many other invasive species have three or more sets of chromosomes). Fertilizers and detergents contain a high level of phosphorus, so agricultural runoff and other types of water pollution may be fueling population explosions of these species.

FIGURE 12.3 **Gametes.** This illustration shows a human egg (female gamete) surrounded by sperm (male gametes).

MEIOSIS HALVES THE CHROMOSOME NUMBER

Sexual reproduction involves the fusion of mature reproductive cells—**gametes**—from two parents (FIGURE 12.3). Gametes have a single set of chromosomes, so they are **haploid** (n): Their chromosome number is half of the diploid ($2n$) number (Section 8.3). **Meiosis**, the nuclear division mechanism that halves the chromosome number, is necessary for gamete formation. Meiosis also gives rise to new combinations of parental alleles.

Gametes arise by division of **germ cells**, which are immature reproductive cells that form in special reproductive organs (FIGURE 12.4). Animals and plants make gametes somewhat differently. In animals, meiosis in diploid germ cells gives rise to eggs (female gametes) or sperm (male gametes). In plants, haploid germ cells form by meiosis. Gametes form when these cells divide by mitosis.

The first part of meiosis is similar to mitosis. A cell duplicates its DNA before either nuclear division process begins. As in mitosis, a spindle forms, and its microtubules move the duplicated chromosomes to opposite spindle poles. However, meiosis sorts the chromosomes into new nuclei not once, but twice, so it results in the formation of four haploid nuclei. The two consecutive nuclear divisions are called meiosis I and meiosis II:

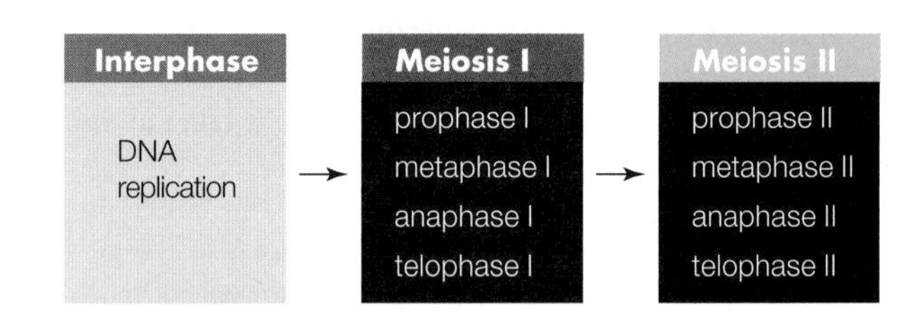

In some cells, meiosis II occurs immediately after meiosis I. In others, a period of protein synthesis—but no DNA replication—intervenes between the divisions.

During meoisis I, every duplicated chromosome aligns with its homologous partner (FIGURE 12.5 ❶). Then the homologous chromosomes are pulled away from one another and packaged into separate nuclei ❷. At this stage of meiosis, the chromosome number has been reduced. Each

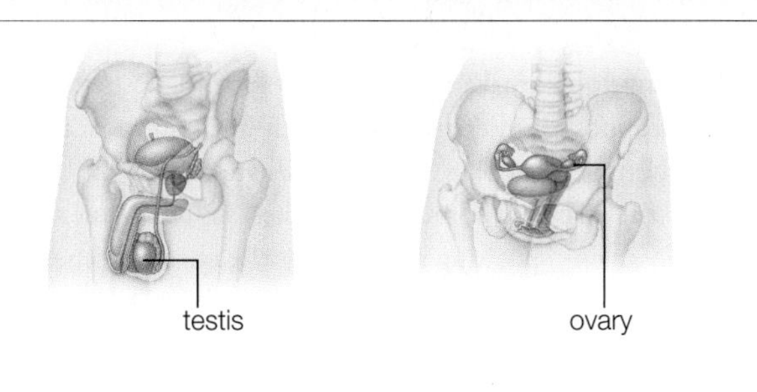

A Reproductive organs of humans. Meiosis in germ cells inside testes and ovaries produces gametes (sperm and eggs).

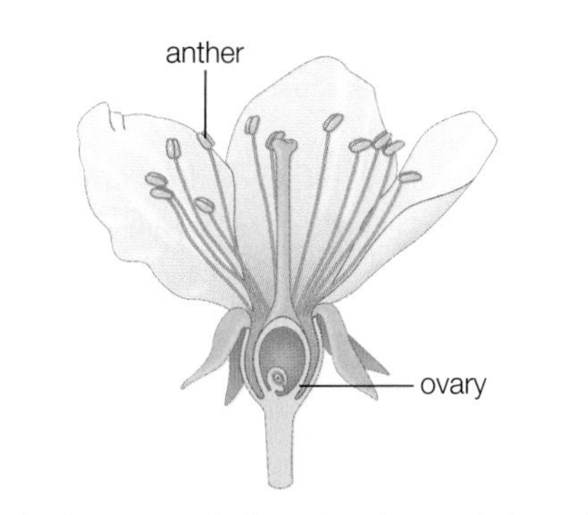

B Reproductive organs of a flowering plant. Meiosis produces haploid germ cells inside anthers and ovaries. These cells divide by mitosis to give rise to gametes (sperm and eggs).

FIGURE 12.4 {Animated} Examples of reproductive organs in A animals and B plants.

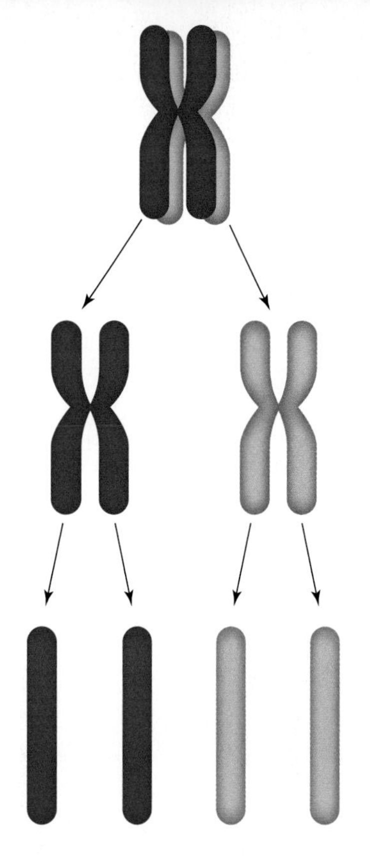

❶ Chromosomes are duplicated before meiosis begins. During meiosis I, each chromosome in the nucleus pairs with its homologous partner. The nucleus contains two of each chromosome, so it is diploid (2*n*).

❷ Homologous partners separate and are packaged into two new nuclei. Each new nucleus contains one of each chromosome, so it is haploid (*n*). The chromosomes are still duplicated.

❸ Sister chromatids separate in meiosis II and are packaged into four new nuclei. Each new nucleus contains one of each chromosome, so it is haploid (*n*). The chromosomes are now unduplicated.

FIGURE 12.5 How meiosis halves the chromosome number.

of the two new nuclei has one copy of each chromosome, so it is haploid (*n*). The chromosomes are still duplicated (the sister chromatids remain attached to one another).

During meiosis II, the sister chromatids are pulled apart, and each becomes an individual, unduplicated chromosome ❸. The chromosomes are sorted into four new nuclei. Each new nucleus still has one copy of each chromosome, so it is haploid (*n*).

Thus, meiosis partitions the chromosomes of one diploid nucleus (2*n*) into four haploid (*n*) nuclei. The next section zooms in on the details of this process.

FERTILIZATION RESTORES THE CHROMOSOME NUMBER

Haploid gametes form by meiosis. The diploid chromosome number is restored at **fertilization**, when two haploid gametes fuse to form a **zygote**, the first cell of a new individual. Thus, meiosis halves the chromosome number, and fertilization restores it.

If meiosis did not precede fertilization, the chromosome number would double with every generation. As you will see in Chapter 14, chromosome number changes can have drastic consequences, particularly in animals. An individual's set of chromosomes is like a fine-tuned blueprint that must be followed exactly, page by page, in order to build a body that functions normally.

fertilization Fusion of two gametes to form a zygote.
gamete Mature, haploid reproductive cell; e.g., an egg or a sperm.
germ cell Immature reproductive cell that gives rise to haploid gametes when it divides.
haploid Having one of each type of chromosome characteristic of the species.
meiosis Nuclear division process that halves the chromosome number. Basis of sexual reproduction.
zygote Cell formed by fusion of two gametes at fertilization; the first cell of a new individual.

TAKE-HOME MESSAGE 12.2

The nuclear division process of meiosis is the basis of sexual reproduction in plants and animals.

Meiosis halves the diploid (2*n*) chromosome number, to the haploid number (*n*), for forthcoming gametes.

When two gametes fuse at fertilization, the diploid chromosome number is restored in the resulting zygote.

12.3 WHAT HAPPENS TO A CELL DURING MEIOSIS?

FIGURE 12.6 shows the stages of meiosis in a diploid (2*n*) cell, which contains two sets of chromosomes. DNA replication occurs before meiosis I, so each chromosome has two sister chromatids.

Meiosis I The first stage of meiosis I is prophase I ❶. During this phase, the chromosomes condense, and homologous chromosomes align tightly and swap segments (more about segment-swapping in the next section). The nuclear envelope breaks up. A spindle forms, and by the end of prophase I, microtubules attach one chromosome of each homologous pair to one spindle pole, and the other to the opposite spindle pole. These microtubules grow and shrink, pushing and pulling the chromosomes as they do. At metaphase I ❷, all of the microtubules are the same length, and the chromosomes are aligned in the middle of the cell. In anaphase I ❸, the spindle pulls the homologous chromosomes of each pair apart and toward opposite spindle poles. The two sets of chromosomes reach the spindle poles during telophase I ❹, and a new nuclear envelope forms around each cluster of chromosomes as the DNA loosens up. The two new nuclei are haploid (*n*); each contains one set of (duplicated) chromosomes. The

FIGURE 12.6 {Animated} Meiosis. Two pairs of chromosomes are illustrated in a diploid (2*n*) cell. Homologous chromosomes are indicated in blue and pink. Micrographs show meiosis in a lily plant cell (*Lilium regale*).

FIGURE IT OUT: During which stage of meiosis does the chromosome number become reduced?

Answer: Anaphase I

MEIOSIS I: ONE DIPLOID NUCLEUS TO TWO HAPLOID NUCLEI

① Prophase I
Homologous chromosomes condense, pair up, and swap segments. Spindle microtubules attach to them as the nuclear envelope breaks up.

② Metaphase I
Homologous chromosome pairs are aligned between spindle poles. Spindle microtubules attach the two chromosomes of each pair to opposite spindle poles.

③ Anaphase I
All of the homologous chromosomes separate and begin heading toward the spindle poles.

④ Telophase I
A complete set of chromosomes clusters at both ends of the cell. A nuclear envelope forms around each set, so two haploid (*n*) nuclei form.

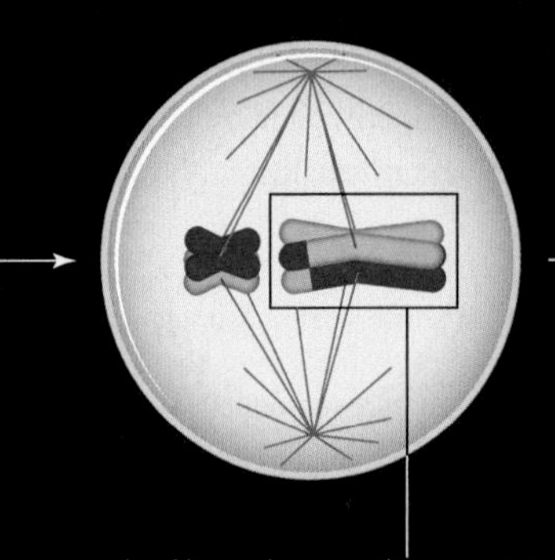

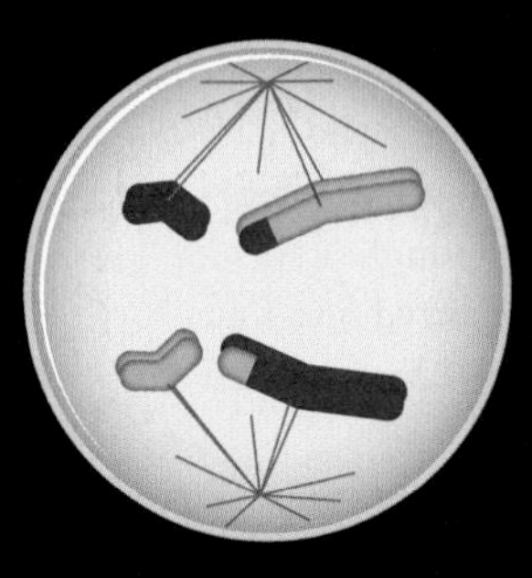

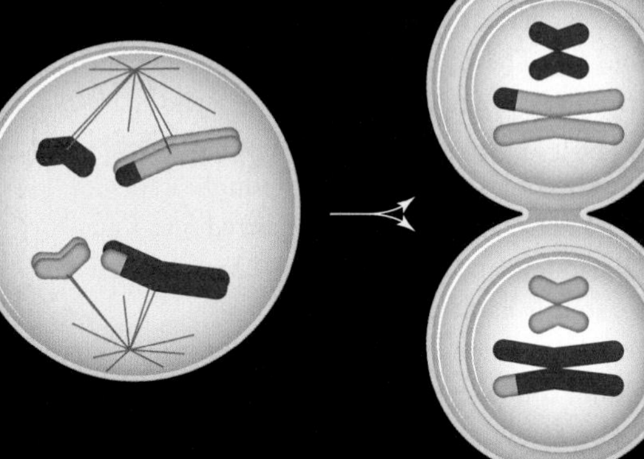

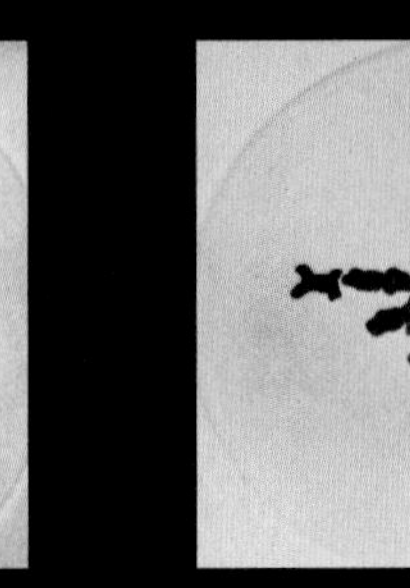

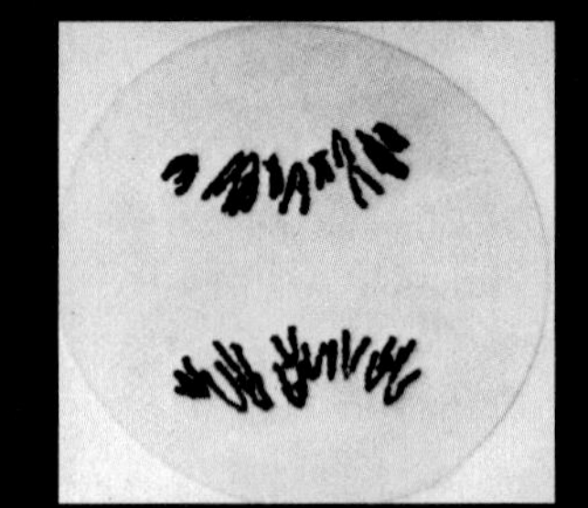

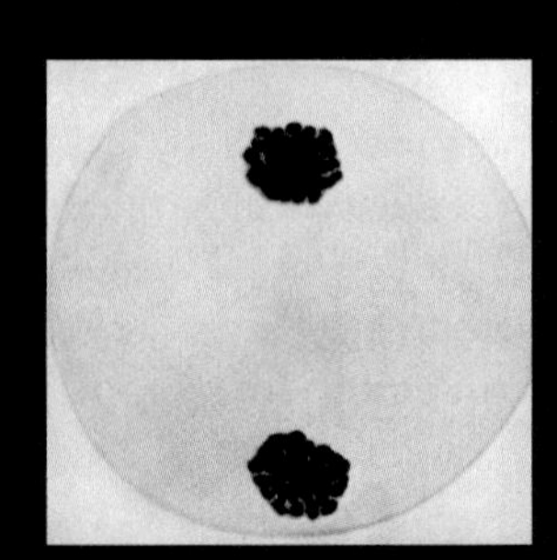

cytoplasm often divides at this point. Each chromosome is still duplicated (it consists of two sister chromatids).

Meiosis II DNA replication does not occur before meiosis II, which proceeds simultaneously in both nuclei that formed in meiosis I. In prophase II ⑤, the chromosomes condense and the nuclear envelope breaks up. A new spindle forms. By the end of prophase II, spindle microtubules attach each chromatid to one spindle pole, and its sister chromatid to the opposite spindle pole. These microtubules push and pull the chromosomes, aligning them in the middle of the cell at metaphase II ⑥. In anaphase II ⑦, the spindle microtubules pull the sister chromatids apart and toward opposite spindle poles. Each chromosome is now unduplicated (it consists of one molecule of DNA). During telophase II ⑧, these chromosomes reach the spindle poles. New nuclear envelopes form around the four clusters of chromosomes as the DNA loosens up. The cytoplasm often divides at this point to form four haploid (n) cells whose nuclei contain one set of (unduplicated) chromosomes.

TAKE-HOME MESSAGE 12.3

During meiosis, the nucleus of a diploid ($2n$) cell divides twice. Four haploid (n) nuclei form, each with a full set of chromosomes—one of each type.

MEIOSIS II: TWO HAPLOID NUCLEI TO FOUR HAPLOID NUCLEI

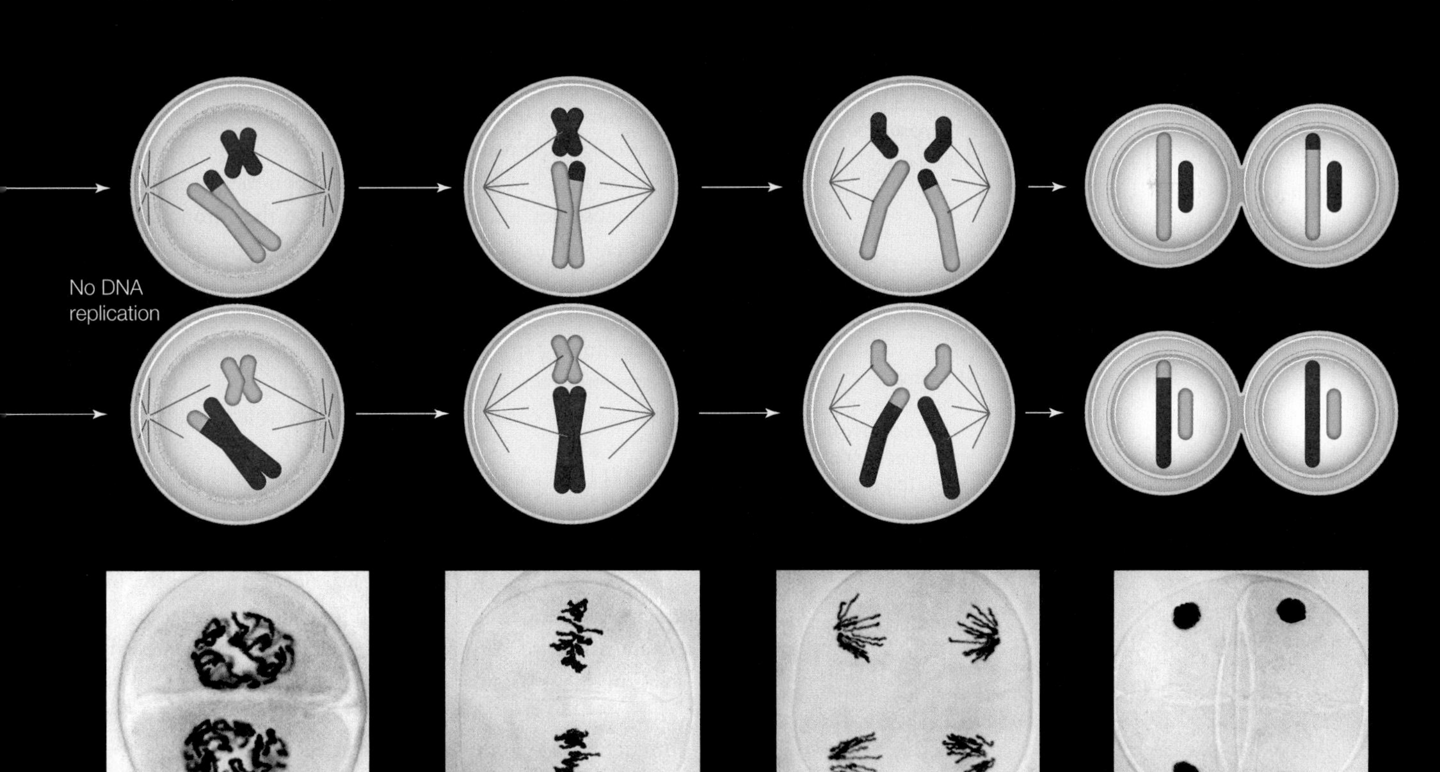

CREDITS: (4) Bottom photos, With thanks to the John Innes Foundation Trustees, computer enhanced by Gary Head; Top art, © Cengage Learning.

12.4 HOW DOES MEIOSIS GIVE RISE TO NEW COMBINATIONS OF PARENTAL ALLELES?

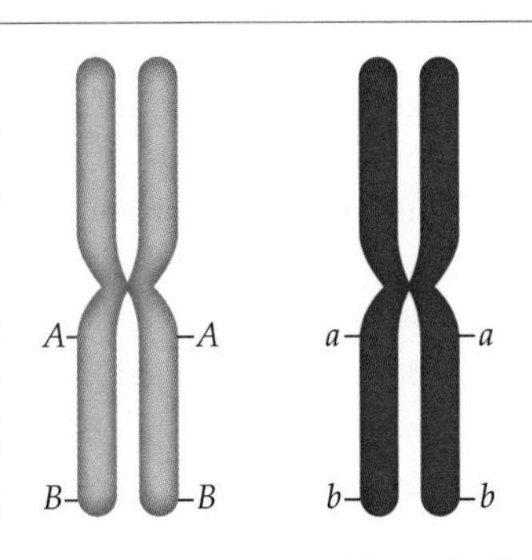

A Here, we focus on only two of the many genes on a chromosome. In this example, one gene has alleles *A* and *a*; the other has alleles *B* and *b*.

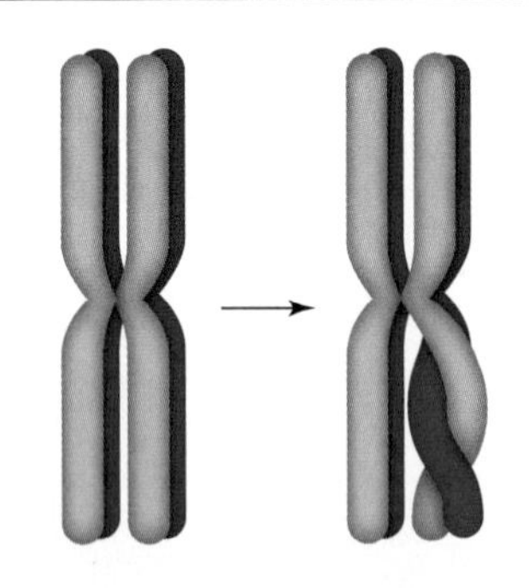

B Close contact between homologous chromosomes promotes crossing over between nonsister chromatids. Paternal and maternal chromatids exchange corresponding pieces.

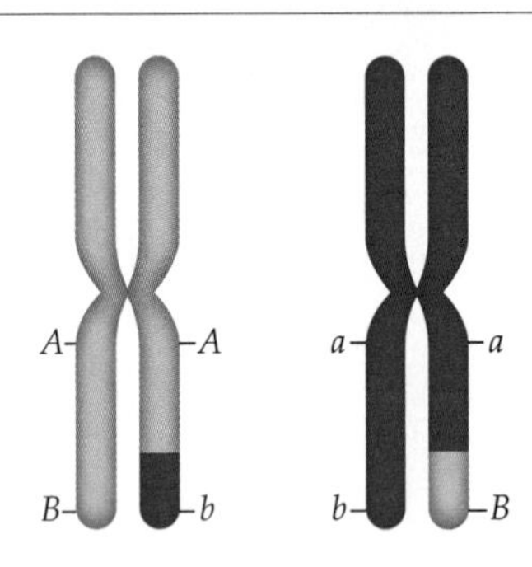

C Crossing over mixes up paternal and maternal alleles on homologous chromosomes.

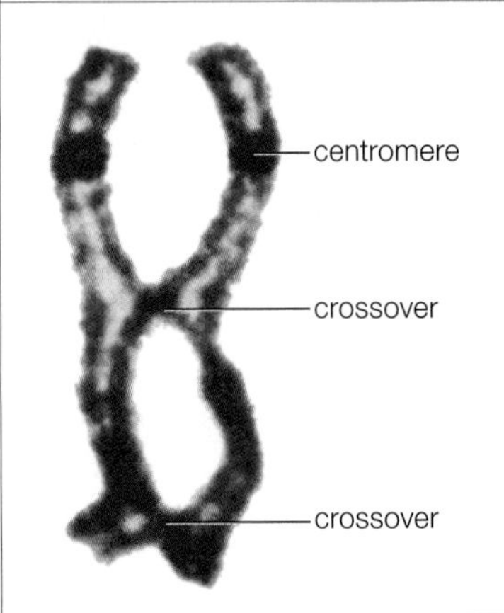

D Each pair of homologous chromosomes can cross over multiple times. This is a normal and common process of meiosis.

FIGURE 12.7 {Animated} Crossing over. Blue signifies a paternal chromosome, and pink, its maternal homologue. For clarity, we show only one pair of homologous chromosomes.

FIGURE IT OUT: In how many places is the chromosome pictured in D crossing over?

Answer: Two

The previous section mentioned briefly that duplicated chromosomes swap segments with their homologous partners during prophase I. It also showed how spindle microtubules align and then separate homologous chromosomes during anaphase I. These events, along with fertilization, contribute to the variation in combinations of traits among the offspring of sexually reproducing species.

CROSSING OVER IN PROPHASE I

Early in prophase I of meiosis, all chromosomes in the cell condense. When they do, each is drawn close to its homologous partner, so that the chromatids align along their length:

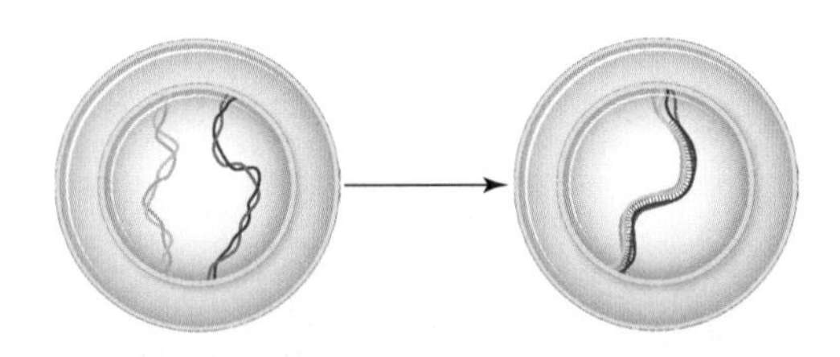

This tight, parallel orientation favors **crossing over**, a process by which a chromosome and its homologous partner exchange corresponding pieces of DNA during meiosis (**FIGURE 12.7**). Homologous chromosomes may swap any segment of DNA along their length, although crossovers tend to occur more frequently in certain regions.

Swapping segments of DNA shuffles alleles between homologous chromosomes. It breaks up the particular combinations of alleles that occurred on the parental chromosomes, and makes new ones on the chromosomes that end up in gametes. Thus, crossing over introduces novel combinations of traits among offspring. It is a normal and frequent process in meiosis, but the rate of crossing over varies among species and among chromosomes. In humans, between 46 and 95 crossovers occur per meiosis, so on average each chromosome crosses over at least once.

CHROMOSOME SEGREGATION

Normally, all of the new nuclei that form in meiosis I receive a complete set of chromosomes. However, whether a new nucleus ends up with the maternal or paternal version of a chromosome is entirely random. The chance that the maternal or the paternal version of any chromosome will end up in a particular nucleus is 50 percent. Why? The answer has to do with the way the spindle segregates the homologous chromosomes during meiosis I.

The process of chromosome segregation begins in prophase I. Imagine one of your own germ cells undergoing meiosis. Crossovers have already made genetic mosaics of

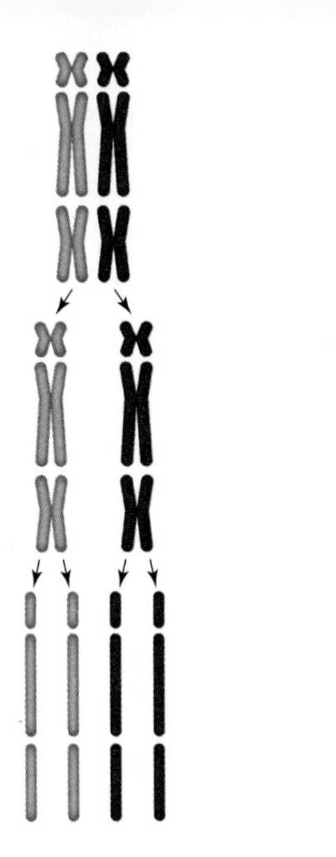
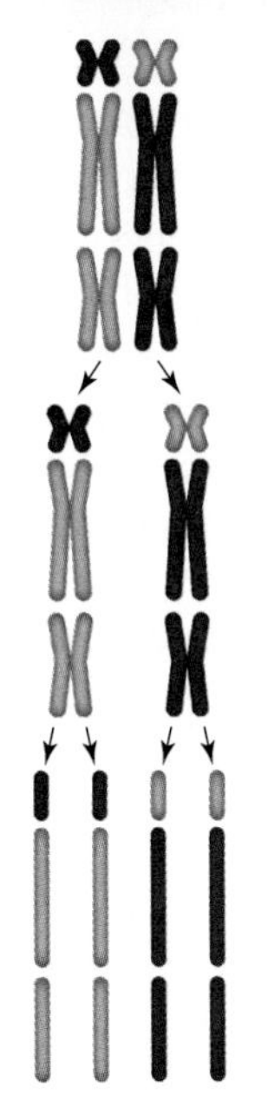
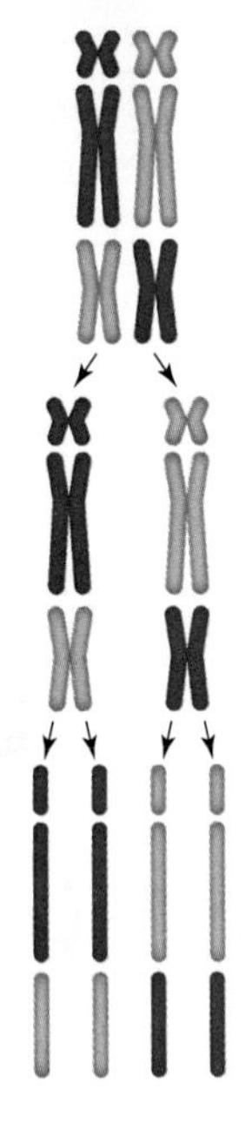
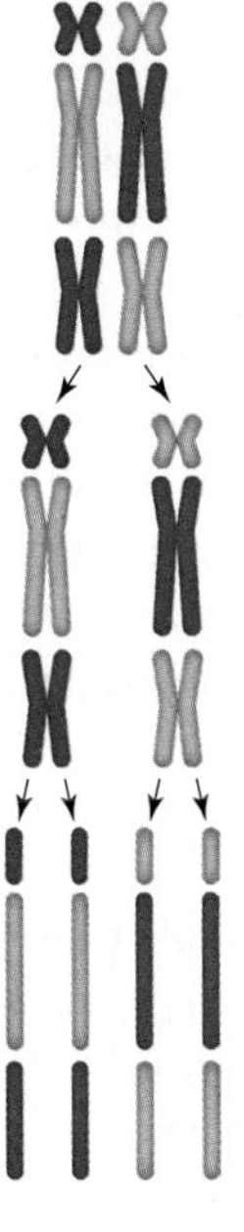

❶ The four possible alignments of three pairs of chromosomes in a nucleus at metaphase I.

❷ Resulting combinations of maternal and paternal chromosomes in the two nuclei that form at telophase I.

❸ Resulting combinations of maternal and paternal chromosomes in the four nuclei that form at telophase II. Eight different combinations are possible.

FIGURE 12.8 {Animated} Hypothetical segregation of three pairs of chromosomes in meiosis I. Maternal chromosomes are pink; paternal, blue. Which chromosome of each pair gets packaged into which of the two new nuclei that form at telophase I is random. For simplicity, no crossing over occurs in this example, so all sister chromatids are identical.

its chromosomes, but for simplicity let's put crossing over aside for a moment. Just call the twenty-three chromosomes you inherited from your mother the maternal ones, and the twenty-three from your father the paternal ones.

During prophase I, microtubules fasten your cell's chromosomes to the spindle poles. Chances are very low that all of the maternal chromosomes get attached to one pole and all of the paternal chromosomes get attached to the other. Microtubules extending from a spindle pole bind to the centromere of the first chromosome they contact, regardless of whether it is maternal or paternal. Though each homologous partner becomes attached to the opposite spindle pole, there is no pattern to the attachment of the maternal or paternal chromosomes to a particular pole.

Now imagine that your germ cell has just three pairs of chromosomes (**FIGURE 12.8**). By metaphase I, those three pairs of maternal and paternal chromosomes have been divided up between the two spindle poles in one of four ways ❶. In anaphase I, homologous chromosomes separate and are pulled toward opposite spindle poles. In telophase I, a new nucleus forms around the chromosomes that cluster at each spindle pole. Each nucleus contains one of eight possible combinations of maternal and paternal chromosomes ❷.

crossing over Process by which homologous chromosomes exchange corresponding segments of DNA during prophase I of meiosis.

In telophase II, each of the two nuclei divides and gives rise to two new haploid nuclei. The two new nuclei are identical because no crossing over occurred in our hypothetical example, so all of the sister chromatids were identical. Thus, at the end of meiosis in this cell, two (2) spindle poles have divvied up three (3) chromosome pairs. The resulting four nuclei have one of eight (2^3) possible combinations of maternal and paternal chromosomes ❸.

Cells that give rise to human gametes have twenty-three pairs of homologous chromosomes, not three. Each time a human germ cell undergoes meiosis, the four gametes that form end up with one of 8,388,608 (or 2^{23}) possible combinations of homologous chromosomes. That number does not even take into account crossing over, which mixes up the alleles on maternal and paternal chromosomes, or fusion with another gamete at fertilization.

TAKE-HOME MESSAGE 12.4

Crossing over—recombination between nonsister chromatids of homologous chromosomes—occurs during prophase I.

Homologous chromosomes can get attached to either spindle pole in prophase I, so each chromosome of a homologous pair can end up in either one of the two new nuclei.

Crossing over and random sorting of chromosomes into gametes give rise to new combinations of alleles—thus new combinations of traits—among offspring of sexual reproducers.

12.5 ARE THE PROCESSES OF MITOSIS AND MEIOSIS RELATED?

This chapter opened with hypotheses about evolutionary advantages of asexual and sexual reproduction. It seems like a giant evolutionary step from producing clones to producing genetically varied offspring, but was it really?

By mitosis and cytoplasmic division, one cell becomes two new cells that have the parental chromosomes. Mitotic (asexual) reproduction results in clones of the parent. Meiosis results in the formation of haploid gametes. Gametes of two parents fuse to form a zygote, which is a cell of mixed parentage. Meiotic (sexual) reproduction results in offspring that differ genetically from the parent, and from one another.

Though the end results differ, there are striking parallels between the four stages of mitosis and meiosis II (**FIGURE 12.9**). As one example, a spindle forms and separates chromosomes during both processes. There are many more similarities at the molecular level.

Long ago, the molecular machinery of mitosis may have been remodeled into meiosis. Evidence for this hypothesis includes a host of shared molecules, including the products

FIGURE 12.9 {Animated} Comparing meiosis II with mitosis.

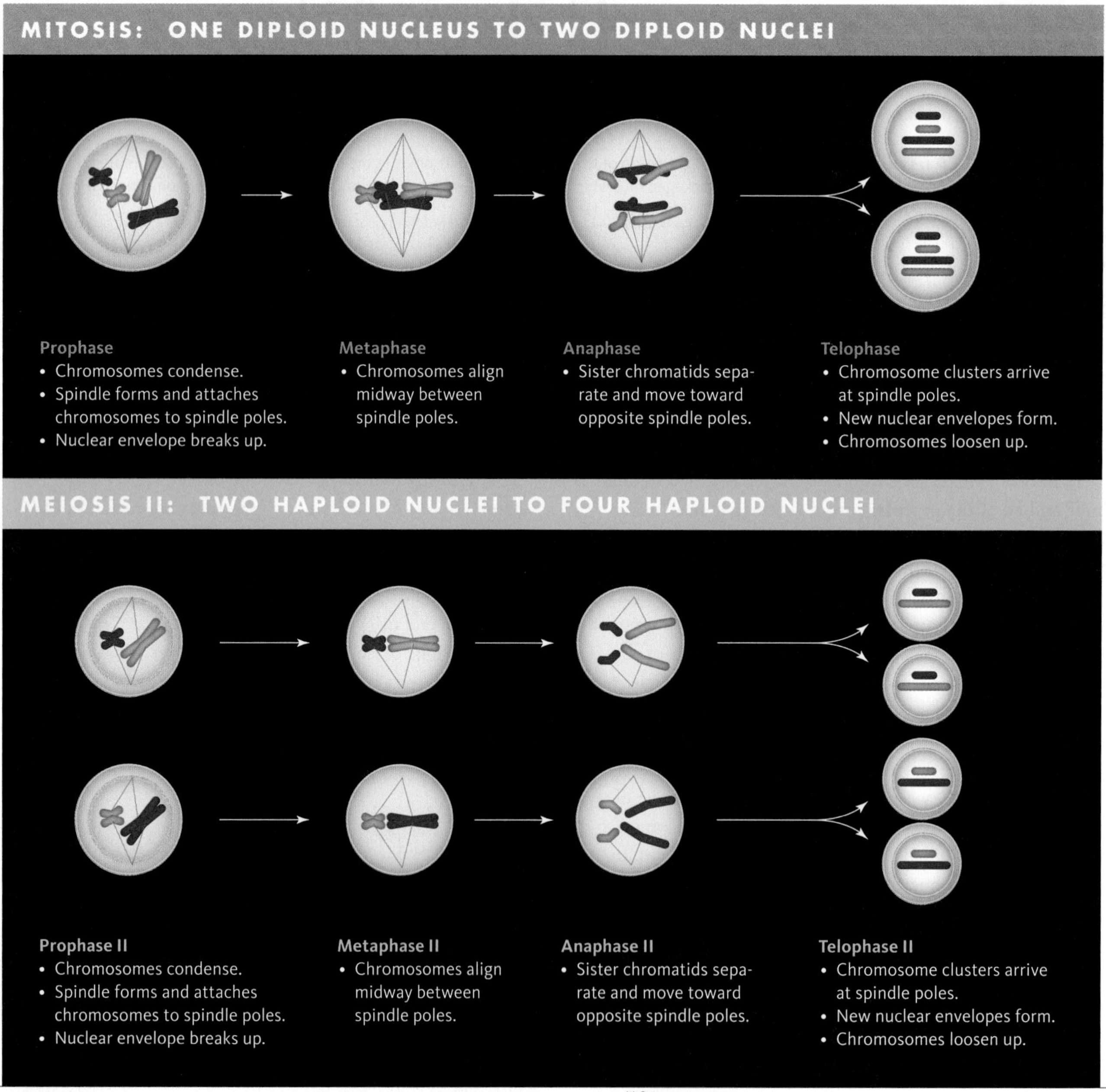

MITOSIS: ONE DIPLOID NUCLEUS TO TWO DIPLOID NUCLEI

Prophase
- Chromosomes condense.
- Spindle forms and attaches chromosomes to spindle poles.
- Nuclear envelope breaks up.

Metaphase
- Chromosomes align midway between spindle poles.

Anaphase
- Sister chromatids separate and move toward opposite spindle poles.

Telophase
- Chromosome clusters arrive at spindle poles.
- New nuclear envelopes form.
- Chromosomes loosen up.

MEIOSIS II: TWO HAPLOID NUCLEI TO FOUR HAPLOID NUCLEI

Prophase II
- Chromosomes condense.
- Spindle forms and attaches chromosomes to spindle poles.
- Nuclear envelope breaks up.

Metaphase II
- Chromosomes align midway between spindle poles.

Anaphase II
- Sister chromatids separate and move toward opposite spindle poles.

Telophase II
- Chromosome clusters arrive at spindle poles.
- New nuclear envelopes form.
- Chromosomes loosen up.

of the *BRCA* genes (Sections 10.6 and 11.5) that are made by all modern eukaryotes. By monitoring and fixing problems with the DNA—such as damaged or mismatched bases (Section 8.5)—these molecules actively maintain the integrity of a cell's chromosomes. It turns out that many of the same molecules help homologous chromosomes cross over in prophase I of meiosis (**FIGURE 12.10**). Some proteins function as part of checkpoints in both mitosis and meiosis, so mutations that affect them or the rate at which they are made can affect the outcomes of both nuclear division processes.

In anaphase of mitosis, sister chromatids are pulled apart. What would happen if the connections between the sisters did not break? Each duplicated chromosome would be pulled to one or the other spindle pole—which is exactly what happens in anaphase I of meiosis.

The shared molecules and mechanisms imply a shared evolutionary history; sexual reproduction probably originated with mutations that affected processes of mitosis. As you will see in later chapters, the remodeling of existing processes into new ones is a common evolutionary theme.

FIGURE 12.10 Example of a molecule that functions in mitosis and meiosis. This fluorescence micrograph shows homologous chromosome pairs (red) in the nucleus of a human cell during prophase I of meiosis. Centromeres are blue. Yellow pinpoints the location of a protein called MLH1 assisting with crossovers. MLH1 also helps repair mismatched bases during mitosis.

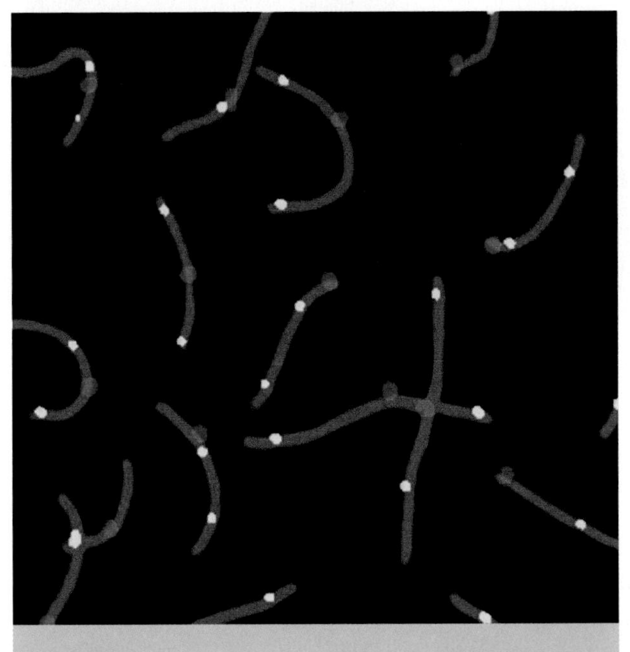

TAKE-HOME MESSAGE 12.5

Meiosis may have evolved by the remodeling of existing mechanisms of mitosis.

12.6 HOW TO SURVIVE FOR 80 MILLION YEARS WITHOUT SEX

FIGURE 12.11 A bdelloid rotifer. All of these tiny animals are female.

WHY DO MALES EXIST? **No male has ever been found among the tiny freshwater creatures called bdelloid rotifers (FIGURE 12.11). Females have been reproducing for 80 million years solely through cloning themselves. Bdelloids are one of the few groups of animals to have completely abandoned sex.**

Compared to sex, asexual reproduction is often seen as a poor long-term strategy because it lacks crossing over—the chromosomal shuffling that brings about genetic diversity thought to give species an adaptive edge in the face of new challenges. Bdelloids have contradicted this theory by being very successful; over 360 species are alive today.

A newly discovered ability may help to explain the success of the bdelloids despite their rejection of sex. These rotifers can apparently import genes from bacteria, fungi, protists, and even plants. If the main advantage of sex is that it promotes genetic diversity, why worry about it when you have the genes of entire kingdoms available to you?

The direct swapping of genetic material is incredibly rare in animals, but bdelloids are bringing in external genes to an extent completely unheard of in complex organisms. Each rotifer is a genetic mosaic whose DNA spans almost all the major kingdoms of life: About 10 percent of its active genes have been pilfered from other organisms.

Summary

SECTION 12.1 **Sexual reproduction** mixes up the genetic information of two parents. The offspring of sexual reproducers typically vary in shared, inherited traits. This variation in traits can offer an evolutionary advantage over genetically identical offspring produced by asexual reproduction.

Sexual reproduction produces offspring with pairs of chromosomes, one of each homologous pair from the mother and the other from the father. The two chromosomes of a pair carry the same genes. The DNA sequence of paired genes often varies slightly, in which case they are called **alleles**. Alleles are the basis of differences in shared, heritable traits. They arise by mutation.

SECTION 12.2 **Meiosis**, the basis of sexual reproduction in eukaryotes, is a nuclear division mechanism that halves the chromosome number for forthcoming **gametes**. **Haploid** (n) gametes are mature reproductive cells that form from **germ cells.** The fusion of two haploid gametes during **fertilization** restores the diploid parental chromosome number in the **zygote**, the first cell of the new individual.

SECTION 12.3 DNA replication occurs before meiosis, so each chromosome consists of two molecules of DNA (sister chromatids). Two nuclear divisions (I and II) occur during meiosis. Meiosis I begins when the chromosomes condense and align tightly with their homologous partners during prophase I. Microtubules then extend from the spindle poles, penetrate the nuclear region, and attach to one or the other chromosome of each homologous pair. At metaphase I, all chromosomes are lined up at the spindle equator. During anaphase I, homologous chromosomes separate and move to opposite spindle poles. Two nuclear envelopes form around the two sets of chromosomes during telophase I. The cytoplasm may divide at this point. There may be a resting period before meiosis resumes, but DNA replication does not occur.

The second nuclear division, meiosis II, occurs in both haploid nuclei that formed in meiosis I. The chromosomes are still duplicated; each still consists of two sister chromatids. The chromosomes condense in prophase II, and align in metaphase II. Sister chromatids of each chromosome are pulled apart from each other in anaphase II, so at the end of meiosis each chromosome consists of one molecule of DNA. By the end of telophase II, four haploid nuclei have typically formed, each with a complete set of (unduplicated) chromosomes.

SECTION 12.4 Meiosis shuffles parental alleles, so offspring inherit non-parental combinations of them. During prophase I, homologous chromosomes exchange corresponding segments. This **crossing over** mixes up the alleles on maternal and paternal chromosomes, thus giving rise to combinations of alleles not present in either parental chromosome. The random segregation of maternal and paternal chromosomes into gametes also contributes to variation in traits among offspring of sexual reproducers. Microtubules can attach the maternal or the paternal chromosome of each pair to one or the other spindle pole. Either chromosome may end up in any new nucleus, and in any gamete.

SECTION 12.5 Like mitosis, meiosis requires a spindle to move and sort duplicated chromosomes, but meiosis occurs only in cells that are involved in sexual reproduction. The process of meiosis resembles that of mitosis, and may have evolved from it. Many of the same molecules function the same way in both processes.

SECTION 12.6 A few groups of animals have survived for millions of years by reproducing only asexually, despite the lack of chromosome shufflings that bring about genetic diversity in offspring. Bdelloid rotifers may have offset this disadvantage by picking up new genes from organisms in other kingdoms.

Self-Quiz

Answers in Appendix VII

1. The main evolutionary advantage of sexual over asexual reproduction is that it produces ________ .
   a. more offspring per individual
   b. more variation among offspring
   c. healthier offspring

2. Meiosis is a necessary part of sexual reproduction because it ________ .
   a. divides two nuclei into four new nuclei
   b. reduces the chromosome number for gametes
   c. produces clones that can cross over

3. Meiosis ________ .
   a. occurs in all eukaryotes
   b. supports growth and tissue repair in multicelled species
   c. gives rise to genetic diversity among offspring
   d. is part of the life cycle of all cells

4. Sexual reproduction in animals requires ________ .
   a. meiosis
   b. fertilization
   c. germ cells
   d. all of the above

5. Meiosis ________ the parental chromosome number.
   a. doubles
   b. halves
   c. maintains
   d. mixes up

6. Dogs have a diploid chromosome number of 78. How many chromosomes do their gametes have?
   a. 39
   b. 78
   c. 156
   d. 234

7. The cell in the diagram to the *right* is in anaphase I, not anaphase II. I know this because ________ .

Data Analysis Activities

BPA and Abnormal Meiosis In 1998, researchers at Case Western University were studying meiosis in mouse oocytes when they saw an unexpected and dramatic increase of abnormal meiosis events (**FIGURE 12.12**). Improper segregation of chromosomes during meiosis is one of the main causes of human genetic disorders, which we will discuss in Chapter 14.

The researchers discovered that the spike in meiotic abnormalities began immediately after the mouse facility started washing the animals' plastic cages and water bottles in a new, alkaline detergent. The detergent had damaged the plastic, which as a result was leaching bisphenol A (BPA). BPA is a synthetic chemical that mimics estrogen, the main female sex hormone in animals. BPA is still widely used to manufacture polycarbonate plastic items (including water bottles) and epoxies (including the coating on the inside of metal cans of food).

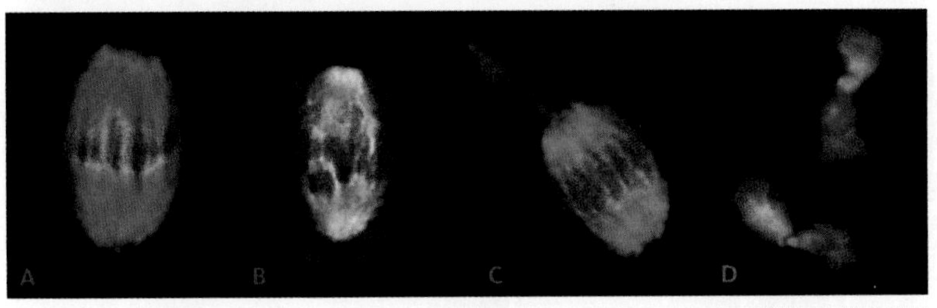

Caging materials	Total number of oocytes	Abnormalities
Control: New cages with glass bottles	271	5 (1.8%)
Damaged cages with glass bottles		
Mild damage	401	35 (8.7%)
Severe damage	149	30 (20.1%)
Damaged bottles	197	53 (26.9%)
Damaged cages with damaged bottles	58	24 (41.4%)

FIGURE 12.12 Meiotic abnormalities associated with exposure to damaged plastic caging. Fluorescent micrographs show nuclei of single mouse oocytes in metaphase I. **A** Normal metaphase; **B–D** examples of abnormal metaphase. Chromosomes appear red; spindle fibers, green.

1. What percentage of mouse oocytes displayed abnormalities of meiosis with no exposure to damaged caging?
2. Which group of mice showed the most meiotic abnormalities in their oocytes?
3. What is abnormal about metaphase I as it is occurring in the oocytes shown in the micrographs in **FIGURE 12.12B, C**, and **D**?

8. The cell pictured to the *right* is in which stage of nuclear division?
   a. anaphase c. anaphase II
   b. anaphase I d. none of the above

9. Crossing over mixes up _______ .
   a. chromosomes c. zygotes
   b. alleles d. gametes

10. Crossing over happens during which phase of meiosis?
   a. prophase I c. anaphase I
   b. prophase II d. anaphase II

11. _______ contributes to variation in traits among the offspring of sexual reproducers.
   a. Crossing over c. Fertilization
   b. Random attachment of chromosomes to spindle poles d. both a and b
   e. all are factors

12. Which of the following is one of the very important differences between mitosis and meiosis?
   a. Chromosomes align midway between spindle poles only in meiosis.
   b. Homologous chromosomes pair up only in meiosis.
   c. DNA is replicated only in mitosis.
   d. Sister chromatids separate only in meiosis.
   e. Interphase occurs only in mitosis.

13. Match each term with its description.

___ interphase	a. different forms of a gene
___ metaphase I	b. useful for varied offspring
___ alleles	c. may be none between meiosis I and meiosis II
___ zygotes	d. chromosomes lined up
___ gametes	e. haploid
___ males	f. form at fertilization
___ prophase I	g. mash-up time

Critical Thinking

1. In your own words, explain why sexual reproduction tends to give rise to greater genetic diversity among offspring in fewer generations than asexual reproduction.

2. Make a simple sketch of meiosis in a cell with a diploid chromosome number of 4. Now try it when the chromosome number is 3.

3. The diploid chromosome number for the body cells of a frog is 26. What would the frog chromosome number be after three generations if meiosis did not occur before gamete formation?

CREDITS: (12) Reprinted from *Current Biology*, Vol 13, (Apr 03), Authors Hunt, Koehler, Susiarjo, Hodges, Ilagan, Voigt, Thomas, Thomas and Hassold, Bisphenol A Exposure Causes Meiotic Aneuploidy in the Female Mouse, pp. 546–553, © 2003 Cell Press. Published by Elsevier Ltd. With permission from Elsevier; (in text S-Q 8) Michael Clayton/University of Wisconsin, Department of Botany.

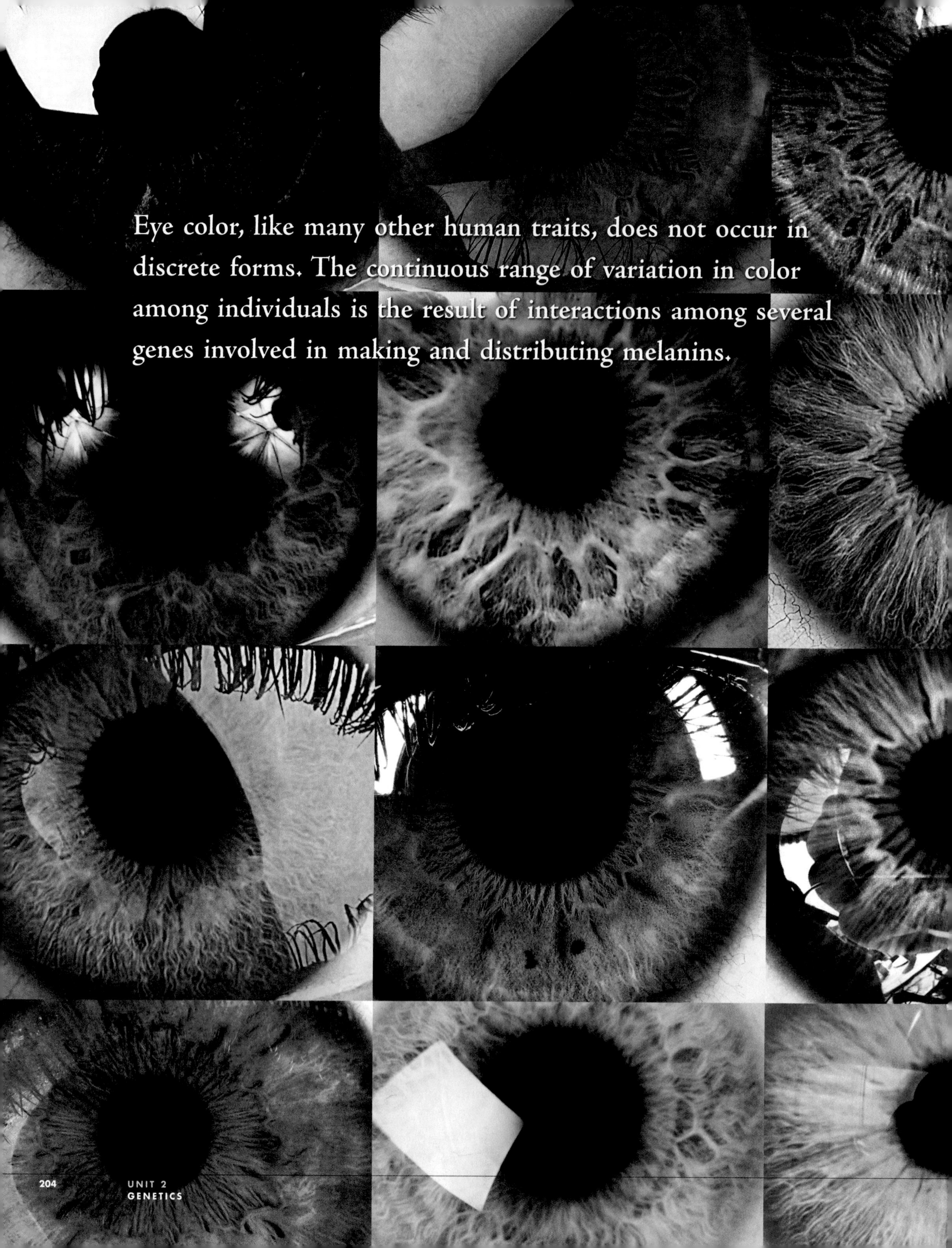

Eye color, like many other human traits, does not occur in discrete forms. The continuous range of variation in color among individuals is the result of interactions among several genes involved in making and distributing melanins.

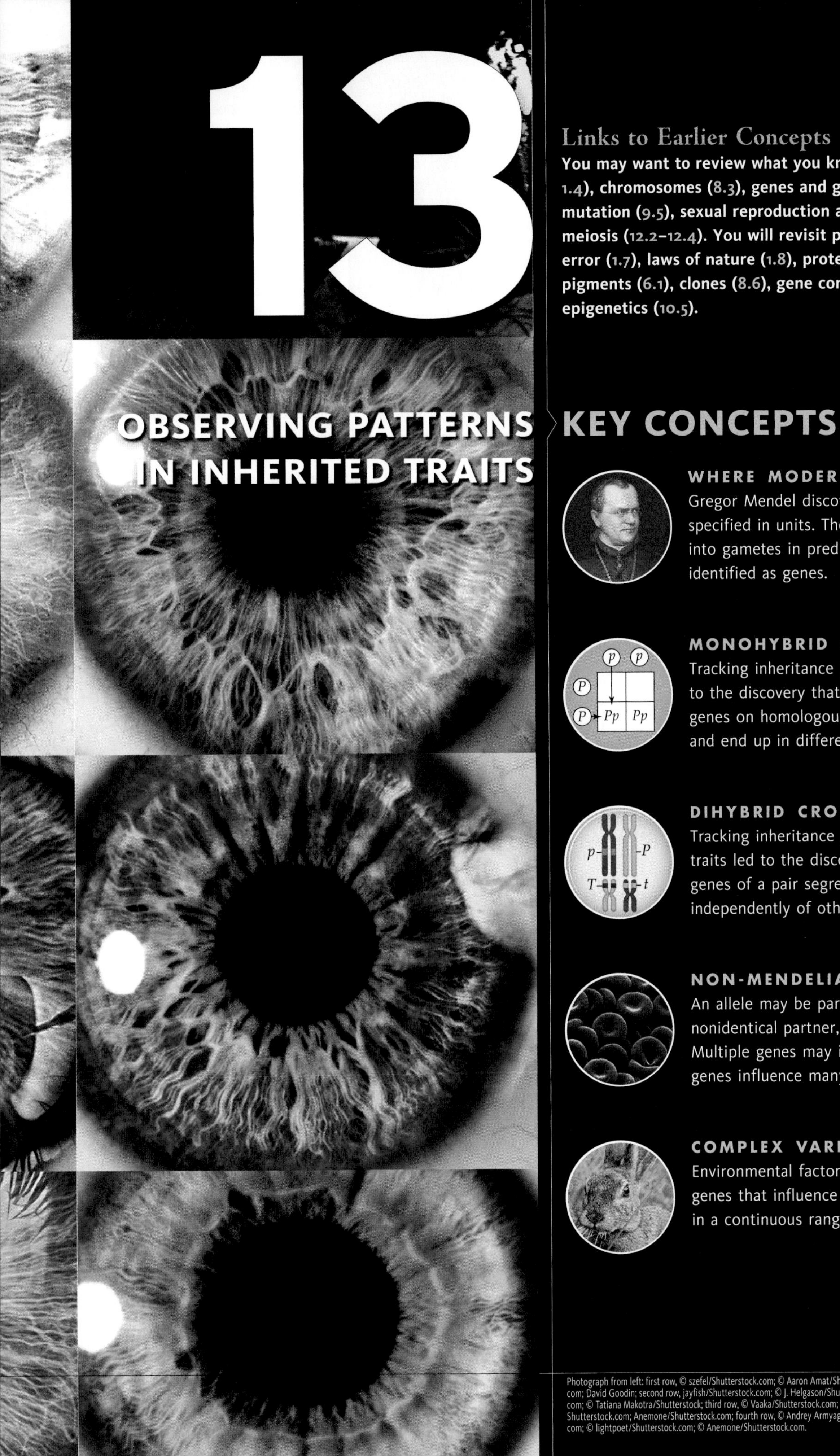

13

OBSERVING PATTERNS IN INHERITED TRAITS

Links to Earlier Concepts

You may want to review what you know about traits (Section 1.4), chromosomes (8.3), genes and gene expression (9.1), mutation (9.5), sexual reproduction and alleles (12.1), and meiosis (12.2–12.4). You will revisit probability and sampling error (1.7), laws of nature (1.8), protein structure (3.4,3.5), pigments (6.1), clones (8.6), gene control (10.1, 10.2, 11.5), and epigenetics (10.5).

KEY CONCEPTS

WHERE MODERN GENETICS STARTE

Gregor Mendel discovered that inherited traits ar specified in units. The units, which are distributec into gametes in predictable patterns, were later identified as genes.

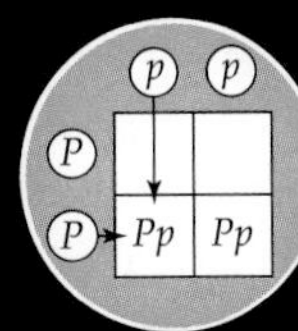

MONOHYBRID CROSSES

Tracking inheritance patterns of single traits led to the discovery that during meiosis, pairs of genes on homologous chromosomes separate and end up in different gametes.

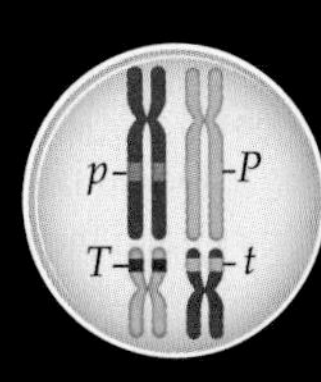

DIHYBRID CROSSES

Tracking inheritance patterns of two unrelated traits led to the discovery that in most cases, genes of a pair segregate into gametes independently of other gene pairs.

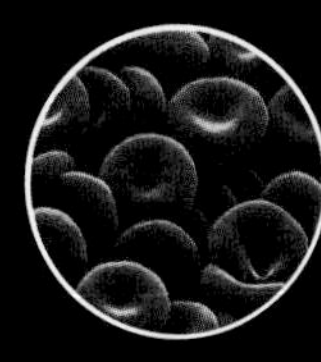

NON-MENDELIAN INHERITANCE

An allele may be partly dominant over a nonidentical partner, or codominant with it. Multiple genes may influence a trait; some genes influence many traits.

COMPLEX VARIATIONS IN TRAITS

Environmental factors can alter the expression of genes that influence a trait. Many traits appear in a continuous range of forms.

In the nineteenth century, people thought that hereditary material must be some type of fluid, with fluids from both parents blending at fertilization like milk into coffee. However, the idea of "blending inheritance" failed to explain what people could see with their own eyes. Children sometimes have traits such as freckles that do not appear in either parent. A cross between a black horse and a white one does not produce gray offspring.

The naturalist Charles Darwin did not accept the idea of blending inheritance, but he could not come up with an alternative hypothesis even though inheritance was central to his theory of natural selection. (We return to Darwin and his

theory of natural selection in Chapter 16.) At the time, no one knew that hereditary information (DNA) is divided into discrete units (genes), an insight that is critical to understanding how traits are inherited. However, even before Darwin presented his theory, someone had been gathering evidence that would support it. Gregor Mendel (*above*), an Austrian monk, had been carefully breeding thousands of pea plants. By keeping detailed records of how traits passed from one generation to the next, Mendel had been collecting evidence of how inheritance works.

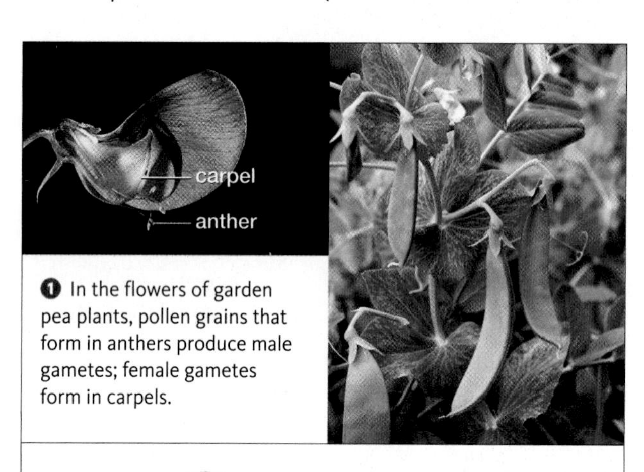

❶ In the flowers of garden pea plants, pollen grains that form in anthers produce male gametes; female gametes form in carpels.

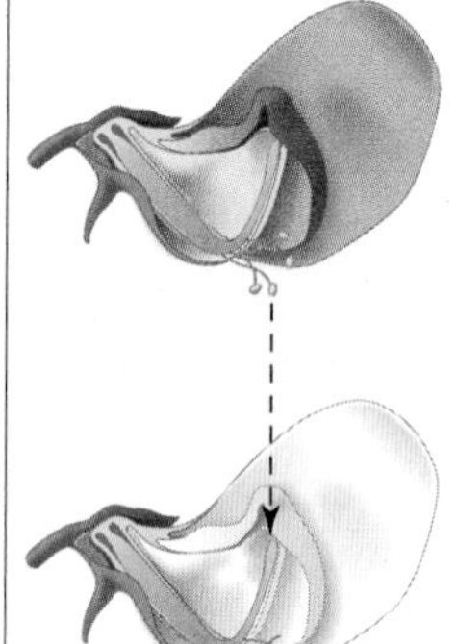

❷ Experimenters can control the transfer of hereditary material from one pea plant to another by snipping off a flower's anthers (to prevent the flower from self-fertilizing), and then brushing pollen from another flower onto its carpel.

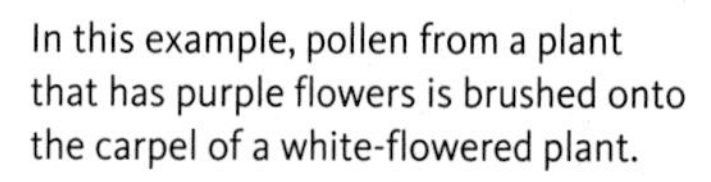

In this example, pollen from a plant that has purple flowers is brushed onto the carpel of a white-flowered plant.

❸ Later, seeds develop inside pods of the cross-fertilized plant. An embryo in each seed develops into a mature pea plant.

❹ Every plant that arises from the cross has purple flowers. Predictable patterns such as this are evidence of how inheritance works.

FIGURE 13.1 {Animated} Breeding garden pea plants.

MENDEL'S EXPERIMENTS

Mendel cultivated the garden pea plant (**FIGURE 13.1**). This species is naturally self-fertilizing, which means its flowers produce male and female gametes ❶ that form viable embryos when they meet up. In order to study inheritance, Mendel had to carry out controlled matings between individuals with specific traits, then observe and document the traits of their offspring. To keep an individual pea plant from self-fertilizing, Mendel removed the pollen-bearing anthers from its flowers. He then cross-fertilized the plant by brushing its egg-bearing carpels with pollen from another plant ❷. He collected the seeds ❸ from the cross-fertilized individual, and recorded the traits of the new pea plants that grew from them ❹.

Many of Mendel's experiments started with plants that "breed true" for particular traits such as white flowers or purple flowers. Breeding true for a trait means that, new mutations aside, all offspring have the same form of the trait as the parent(s), generation after generation. For example, all offspring of pea plants that breed true for white flowers also have white flowers. As you will see in the next section, Mendel cross-fertilized pea plants that breed true for different forms of a trait, and discovered that the traits of the offspring often appear in predictable patterns. Mendel's meticulous work tracking pea plant traits led him to conclude (correctly) that hereditary information passes from one generation to the next in discrete units.

INHERITANCE IN MODERN TERMS

DNA was not proven to be hereditary material until the 1950s (Section 8.1), but Mendel discovered its units, which we now call genes, almost a century before then. Today, we know that individuals of a species share certain traits because their chromosomes carry the same genes.

Each gene occurs at a specific location, or **locus** (plural, loci), on a particular chromosome (**FIGURE 13.2**). The somatic cells of humans and other animals are diploid,

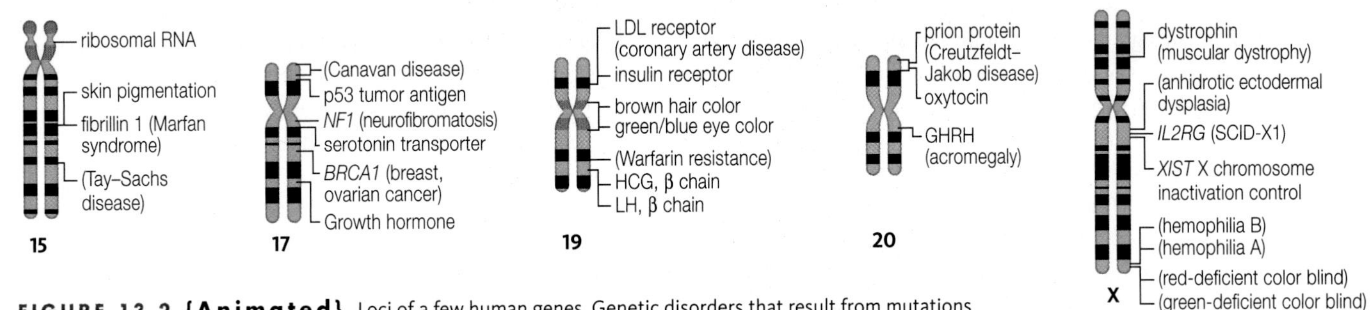

FIGURE 13.2 {Animated} Loci of a few human genes. Genetic disorders that result from mutations in the genes are shown in parentheses. The number or letter below each chromosome is its name; the characteristic banding patterns appear after staining. A similar map of all 23 human chromosomes is in Appendix IV.

so they have pairs of genes, on pairs of homologous chromosomes. In most cases, both genes of a pair are expressed. Genes at the same locus on a pair of homologous chromosomes may be identical, or they may vary as alleles (Section 12.1). Organisms breed true for a trait because they carry identical alleles of genes governing that trait. An individual with two identical alleles of a gene is **homozygous** for the allele. By contrast, an individual with two different alleles of a gene is **heterozygous** (*hetero–* means mixed).

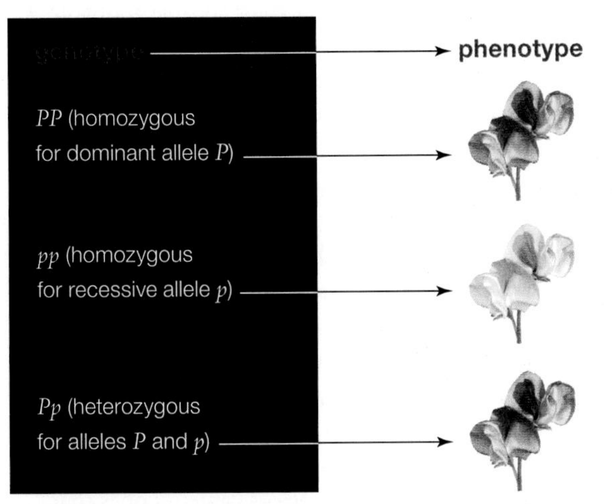

FIGURE 13.3 Genotype gives rise to phenotype. In this example, the dominant allele *P* specifies purple flowers; the recessive allele *p*, white flowers.

dominant Refers to an allele that masks the effect of a recessive allele paired with it in heterozygous individuals.
genotype The particular set of alleles that is carried by an individual's chromosomes.
heterozygous Having two different alleles of a gene.
homozygous Having identical alleles of a gene.
hybrid The heterozygous offspring of a cross or mating between two individuals that breed true for different forms of a trait.
locus Location of a gene on a chromosome.
phenotype An individual's observable traits.
recessive Refers to an allele with an effect that is masked by a dominant allele on the homologous chromosome.

Hybrids are heterozygous offspring of a cross or mating between individuals that breed true for different forms of a trait.

When we say that an individual is homozygous or heterozygous, we are discussing its **genotype**, the particular set of alleles it carries. Genotype is the basis of **phenotype**, which refers to the individual's observable traits. "White-flowered" and "purple-flowered" are examples of pea plant phenotypes that arise from differences in genotype.

The phenotype of a heterozygous individual depends on how the products of its two different alleles interact. In many cases, the effect of one allele influences the effect of the other, and the outcome of this interaction is reflected in the individual's phenotype. An allele is **dominant** when its effect masks that of a **recessive** allele paired with it. Usually, a dominant allele is represented by an italic capital letter such as *A*; a recessive allele, with a lowercase italic letter such as *a*. Consider the purple- and white-flowered pea plants that Mendel studied. In these plants, the allele that specifies purple flowers (let's call it *P*) is dominant over the allele that specifies white flowers (*p*). Thus, a pea plant homozygous for the dominant allele (*PP*) has purple flowers; one homozygous for the recessive allele (*pp*) has white flowers (**FIGURE 13.3**). A heterozygous plant (*Pp*) has purple flowers.

TAKE-HOME MESSAGE 13.1

Gregor Mendel indirectly discovered the role of alleles in inheritance by carefully breeding pea plants and tracking traits of their offspring.

Genotype refers to the particular set of alleles that an individual carries. Genotype is the basis of phenotype, which refers to the individual's observable traits.

A homozygous individual has two identical alleles of a gene. A heterozygous individual has two nonidentical alleles of the gene.

A dominant allele masks the effect of a recessive allele paired with it in a heterozygous individual.

13.2 HOW ARE ALLELES DISTRIBUTED INTO GAMETES?

When homologous chromosomes separate during meiosis (Section 12.3), the gene pairs on those chromosomes separate too. Each gamete that forms carries only one of the two genes of a pair (**FIGURE 13.4**). Thus, plants homozygous for the dominant allele (*PP*) can only make gametes that carry the dominant allele *P* ❶. Plants homozygous for the recessive allele (*pp*) can only make gametes that carry the recessive allele *p* ❷. If these homozygous plants are crossed (*PP* × *pp*), only one outcome is possible: A gamete carrying allele *P* meets up with a gamete carrying allele *p* ❸. All offspring of this cross will have both alleles—they will be heterozygous (*Pp*).

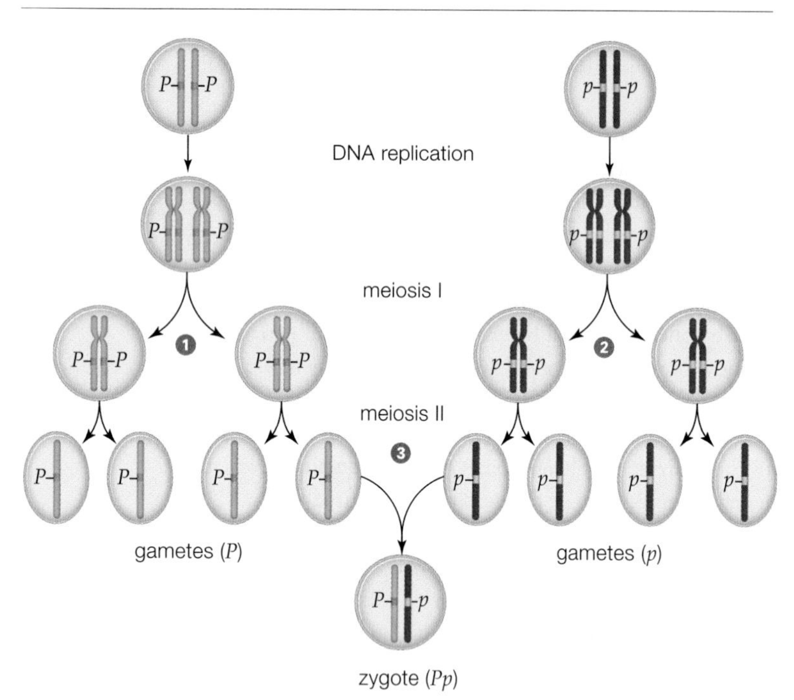

FIGURE 13.4 Segregation of genes on homologous chromosomes into gametes. Homologous chromosomes separate during meiosis, so the pairs of genes they carry separate too. Each of the resulting gametes carries one of the two members of each gene pair. For clarity, only one set of chromosomes is illustrated.

❶ All gametes made by a parent homozygous for a dominant allele carry that allele.

❷ All gametes made by a parent homozygous for a recessive allele carry that allele.

❸ If these two parents are crossed, the union of any of their gametes at fertilization produces a zygote with both alleles. All offspring of this cross will be heterozygous.

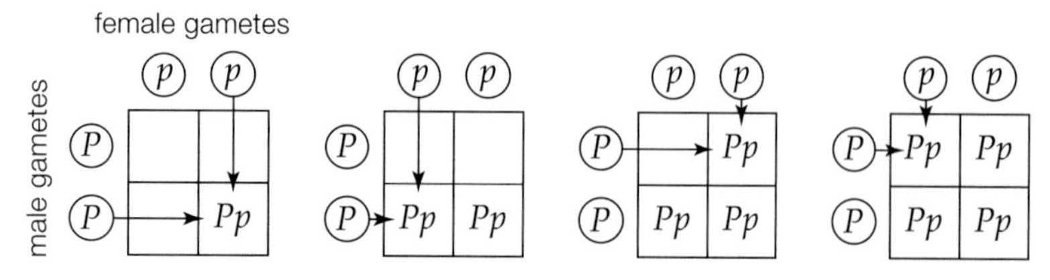

FIGURE 13.5 Making a Punnett square. Parental gametes are listed in circles on the top and left sides of a grid. Each square is filled with the combination of alleles that would result if the gametes in the corresponding row and column met up.

A grid called a **Punnett square** is helpful for predicting the outcomes of such crosses (**FIGURE 13.5**).

Our example illustrated a pattern so predictable that it can be used as evidence of a dominance relationship between alleles. In a **testcross**, an individual that has a dominant trait (but an unknown genotype) is crossed with an individual known to be homozygous for the recessive allele. The pattern of traits among the offspring of the cross can reveal whether the tested individual is heterozygous or homozygous. If all of the offspring of the testcross have the dominant trait, then the parent with the unknown genotype is homozygous for the dominant allele. If any of the offspring have the recessive trait, then it is heterozygous.

Dominance relationships between alleles determine the phenotypic outcome of a **monohybrid cross**, in which individuals that are identically heterozygous for one gene—*Pp* for example—are bred together or self-fertilized. The frequency at which traits associated with the alleles appear among the offspring depends on whether one of the alleles is dominant over the other.

To perform a monohybrid cross, we would start with two individuals that breed true for two different forms of a trait. In garden pea plants, flower color (purple and white) is one example of a trait with two distinct forms, but there are many others. Mendel investigated seven of them: stem length (tall and short), seed color (yellow and green), pod texture (smooth and wrinkled), and so on (**TABLE 13.1**). A cross between individuals that breed true for two forms of a trait yields offspring identically heterozygous for the alleles that govern the trait. When these F_1 (first generation) hybrids are crossed, the frequency at which the two traits appear in the F_2 (second generation) offspring offers information about a dominance relationship between the alleles. F is an abbreviation for filial, which means offspring.

A cross between two purple-flowered heterozygous individuals (*Pp*) is an example of a monohybrid cross. Each of these plants can make two types of gametes: ones that carry a *P* allele, and ones that carry a *p* allele (**FIGURE 13.6A**). So, in a monohybrid cross between two *Pp* plants (*Pp* × *Pp*), the two types of gametes can meet up in four possible ways at fertilization:

Possible Event		Probable Outcome
Sperm *P* meets egg *P*	⟶	zygote genotype is *PP*
Sperm *P* meets egg *p*	⟶	zygote genotype is *Pp*
Sperm *p* meets egg *P*	⟶	zygote genotype is *Pp*
Sperm *p* meets egg *p*	⟶	zygote genotype is *pp*

Three out of four possible outcomes of this cross include at least one copy of the dominant allele *P*. Each time

TABLE 13.1

Mendel's Seven Pea Plant Traits

Trait	Dominant Form	Recessive Form
Seed Shape	Round	Wrinkled
Seed Color	Yellow	Green
Pod Texture	Smooth	Wrinkled
Pod Color	Green	Yellow
Flower Color	Purple	White
Flower Position	Along Stem	At Tip
Stem Length	Tall	Short

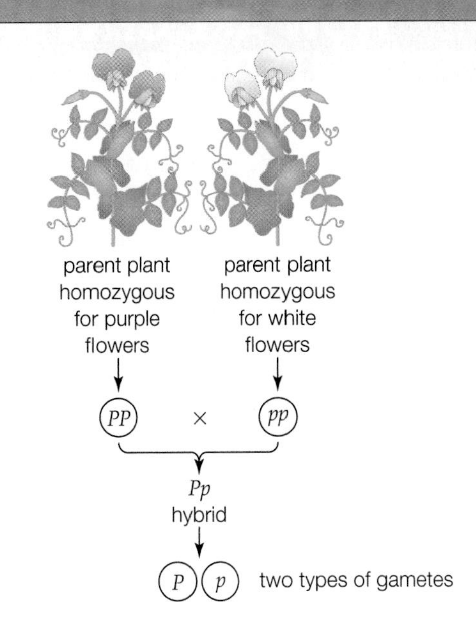

A All of the F_1 offspring of a cross between two plants that breed true for different forms of a trait are identically heterozygous (*Pp*). These offspring make two types of gametes: *P* and *p*.

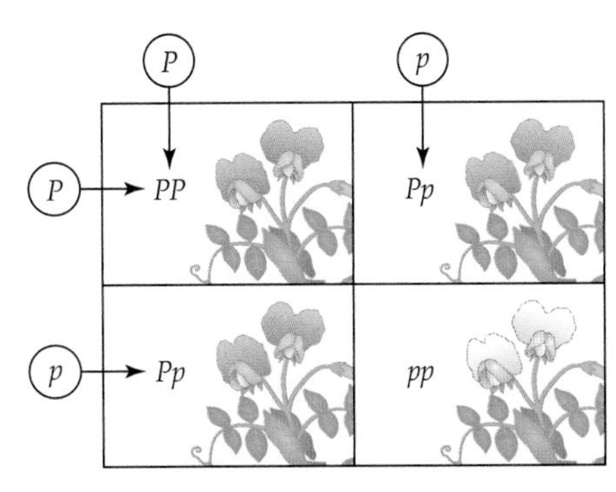

B A monohybrid cross is a cross between these F_1 offspring. In this example, the phenotype ratio in F_2 offspring is 3:1 (3 purple to 1 white).

FIGURE 13.6 {Animated} A monohybrid cross.

FIGURE IT OUT: How many genotypes are possible in the F_2 generation?

Answer: Three: *PP*, *Pp*, and *pp*

fertilization occurs, there are 3 chances in 4 that the resulting offspring will inherit a *P* allele, and have purple flowers. There is 1 chance in 4 that it will inherit two recessive *p* alleles, and have white flowers. Thus, the probability that a particular offspring of this cross will have purple or white flowers is 3 purple to 1 white, which we represent as a ratio of 3:1 (**FIGURE 13.6B**). The 3:1 pattern is an indication that purple and white flower color are specified by alleles with a clear dominance relationship: Purple is dominant; white, recessive. If the probability of one individual inheriting a particular genotype is difficult to imagine, think about probability in terms phenotypes of many offspring. In this example, there will be roughly three purple-flowered plants for every white-flowered one.

The phenotype ratios in the F_2 offspring of Mendel's monohybrid crosses were all close to 3:1. These results became the basis of his **law of segregation**, which we state here in modern terms: Diploid cells carry pairs of genes, on pairs of homologous chromosomes. The two genes of each pair are separated from each other during meiosis, so they end up in different gametes.

TAKE-HOME MESSAGE 13.2

Homologous chromosomes carry pairs of genes. The two genes of each pair are separated from each other during meiosis, so they end up in different gametes.

law of segregation The two members of each pair of genes on homologous chromosomes end up in different gametes during meiosis.
monohybrid cross Cross between two individuals identically heterozygous for one gene; for example *Aa* × *Aa*.
Punnett square Diagram used to predict the genetic and phenotypic outcome of a cross.
testcross Method of determining genotype by tracking a trait in the offspring of a cross between an individual of unknown genotype and an individual known to be homozygous recessive.

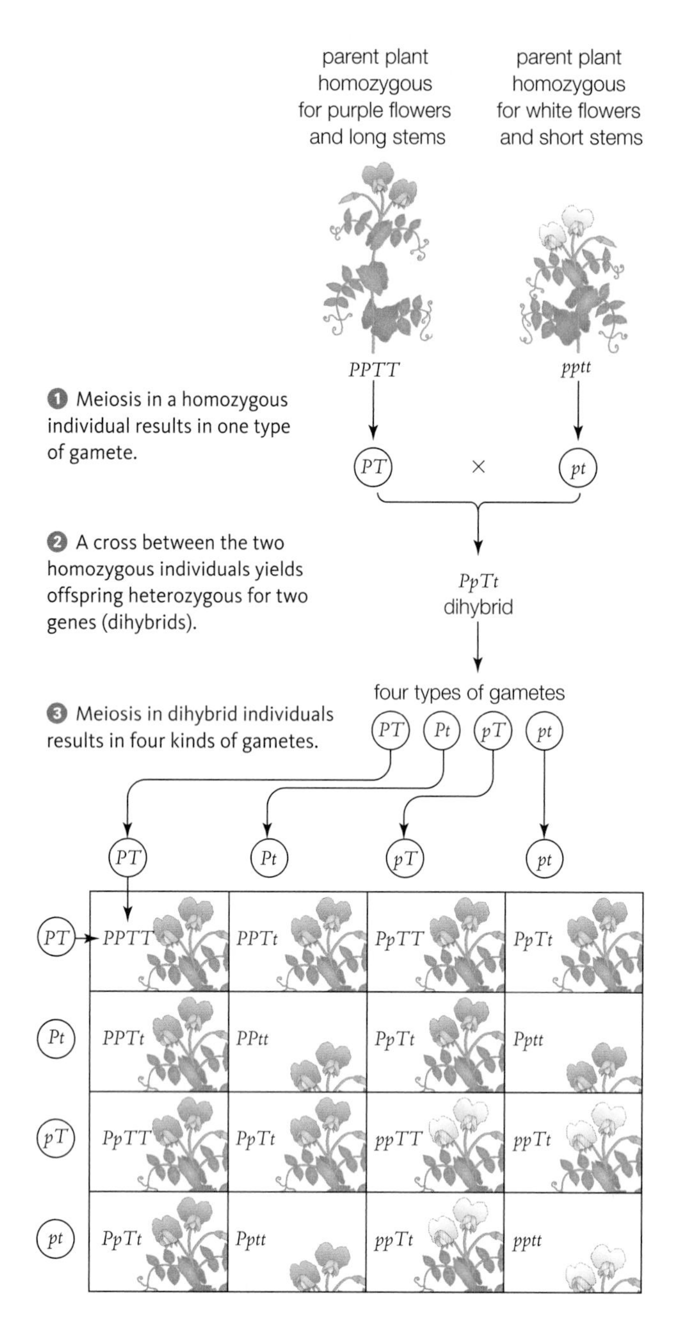

4 If two of the dihybrid individuals are crossed, the four types of gametes can meet up in 16 possible ways. Of 16 possible offspring genotypes, 9 will result in plants that are purple-flowered and tall; 3, purple-flowered and short; 3, white-flowered and tall; and 1, white-flowered and short. Thus, the ratio of phenotypes is 9:3:3:1.

FIGURE 13.7 {Animated} A dihybrid cross between plants that differ in flower color and plant height. In this example, *P* and *p* are dominant and recessive alleles for flower color; *T* and *t* are dominant and recessive alleles for height.

FIGURE IT OUT: What do the flowers inside the boxes represent?

Answer: Phenotypes of the F_2 offspring

A monohybrid cross allows us to track alleles of one gene pair. What about alleles of two gene pairs? In a **dihybrid cross**, individuals identically heterozygous for alleles of two genes (dihybrids) are crossed. As with a monohybrid cross, the pattern of traits seen in the offspring of the cross depends on the dominance relationships between alleles of the genes.

Let's use a gene for flower color (*P*, purple; *p*, white) and one for plant height (*T*, tall; *t*, short) in an example. **FIGURE 13.7** shows a dihybrid cross starting with one parent plant that breeds true for purple flowers and tall stems (*PPTT*), and one that breeds true for white flowers and short stems (*pptt*). The *PPTT* plant only makes gametes with the dominant alleles (*PT*); the *pptt* plant only makes gametes with the recessive alleles (*pt*) 1. So, all offspring from a cross between these parent plants (*PPTT* × *pptt*) will be dihybrids (*PpTt*) with purple flowers and tall stems 2.

Four combinations of alleles are possible in the gametes of *PpTt* dihybrids 3. If two *PpTt* plants are crossed (a dihybrid cross, *PpTt* × *PpTt*), the four types of gametes can combine in sixteen possible ways at fertilization 4. Nine of the sixteen genotypes would give rise to tall plants with purple flowers; three, to short plants with purple flowers; three, to tall plants with white flowers; and one, to short plants with white flowers. Thus, the ratio of phenotypes among the offspring of this dihybrid cross would be 9:3:3:1.

Mendel discovered the 9:3:3:1 ratio of phenotypes among the offspring of his dihybrid crosses, but he had no idea what it meant. He could only say that "units" specifying one trait (such as flower color) are inherited independently of "units" specifying other traits (such as plant height). In time, Mendel's hypothesis became known as the **law of independent assortment**, which we state here in modern terms: During meiosis, the two genes of a pair tend to be sorted into gametes independently of how other gene pairs are sorted into gametes.

Mendel published his results in 1866, but apparently his work was read by few and understood by no one at the time. In 1871 he was promoted, and his pioneering experiments ended. When he died in 1884, he did not know that his work with pea plants would be the starting point for modern genetics.

CONTRIBUTION OF CROSSOVERS

How two genes get sorted into gametes depends partly on whether they are on the same chromosome. When homologous chromosomes separate during meiosis, either member of the pair can end up in either of the two new nuclei that form. This random assortment happens

A This example shows just two pairs of homologous chromosomes in the nucleus of a diploid (2*n*) reproductive cell. Maternal and paternal chromosomes, shown in pink and blue, have already been duplicated.

B Either chromosome of a pair may get attached to either spindle pole during meiosis I. With two pairs of homologous chromosomes, there are two different ways that the maternal and paternal chromosomes can get attached to opposite spindle poles.

C Two nuclei form with each scenario, so there are a total of four possible combinations of parental chromosomes in the nuclei that form after meiosis I.

D Thus, when sister chromatids separate during meiosis II, the gametes that result have one of four possible combinations of maternal and paternal chromosomes.

gamete genotype: *pt* *PT* *pT* *Pt*

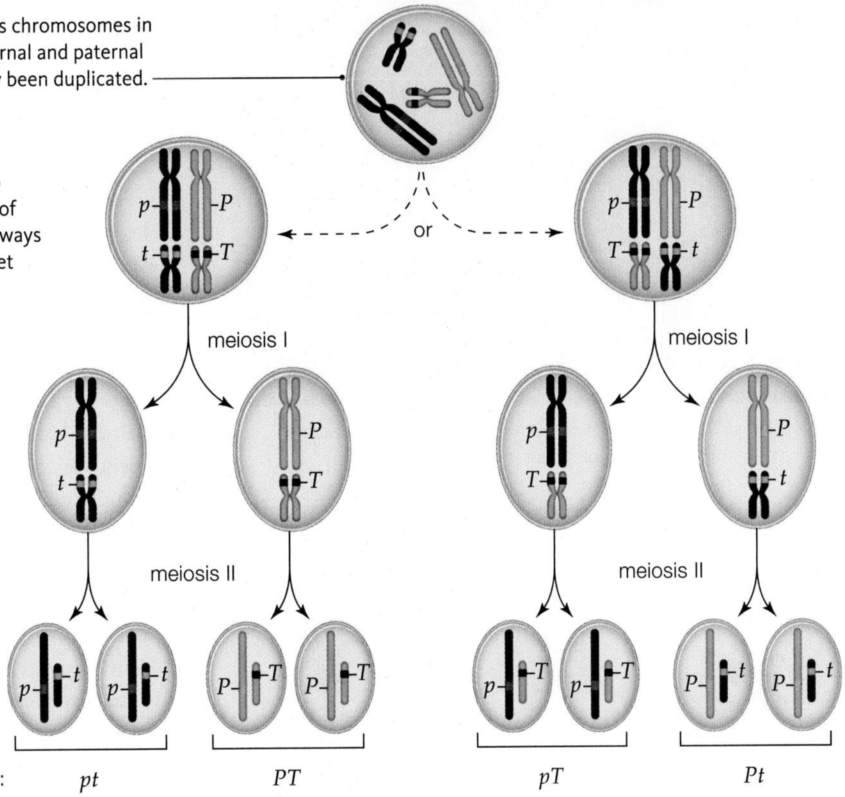

FIGURE 13.8 {Animated} Independent assortment of genes on different chromosomes. Genes that are far apart on the same chromosome usually assort independently too, because crossovers typically separate them.

independently for each pair of homologous chromosomes in the cell. Thus, genes on one chromosome assort into gametes independently of genes on the other chromosomes (**FIGURE 13.8**).

Pea plants have seven chromosomes. Mendel studied seven pea genes, and all of them assorted into gametes independently of one another. Was he lucky enough to choose one gene on each of those chromosomes? As it turns out, some of the genes Mendel studied *are* on the same chromosome. These genes are far enough apart that crossing over occurs between them very frequently—so frequently that they tend to assort into gametes independently, just as if they were on different chromosomes. By contrast, genes that are very close together on a chromosome usually do not assort independently into gametes, because crossing over does not happen between them very often. Thus, gametes usually end up with parental combinations of alleles of these genes.

Genes that do not assort independently into gametes are said to be linked. Linked genes were identified by tracking inheritance in human families over several generations. All of the genes on a chromosome are called a **linkage group**. Peas have 7 different chromosomes, so they have 7 linkage groups. Humans have 23 different chromosomes, so they have 23 linkage groups.

dihybrid cross Cross between two individuals identically heterozygous for two genes; for example *AaBb* × *AaBb*.
law of independent assortment During meiosis, members of a pair of genes on homologous chromosomes tend to be distributed into gametes independently of other gene pairs.
linkage group All genes on a chromosome.

TAKE-HOME MESSAGE 13.3

During meiosis, gene pairs on homologous chromosomes tend to be distributed into gametes independently of how other gene pairs are distributed.

Independent assortment depends on proximity. Genes that are closer together on a chromosome get separated less frequently by crossovers, so gametes often receive parental combinations of alleles of these genes.

13.4 ARE ALL GENES INHERITED IN A MENDELIAN PATTERN?

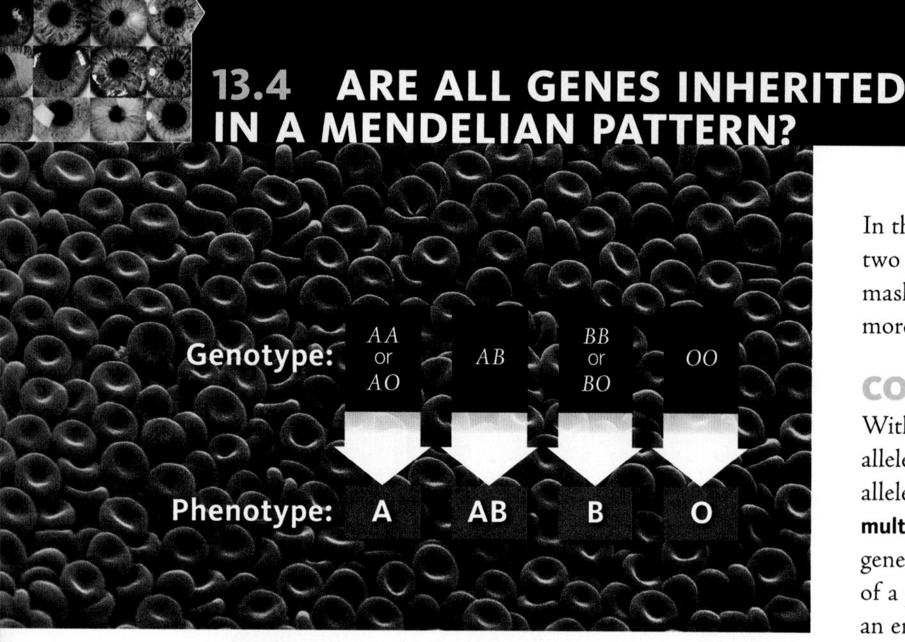

FIGURE 13.9 Combinations of alleles (genotype) that are the basis of human blood type (phenotype).

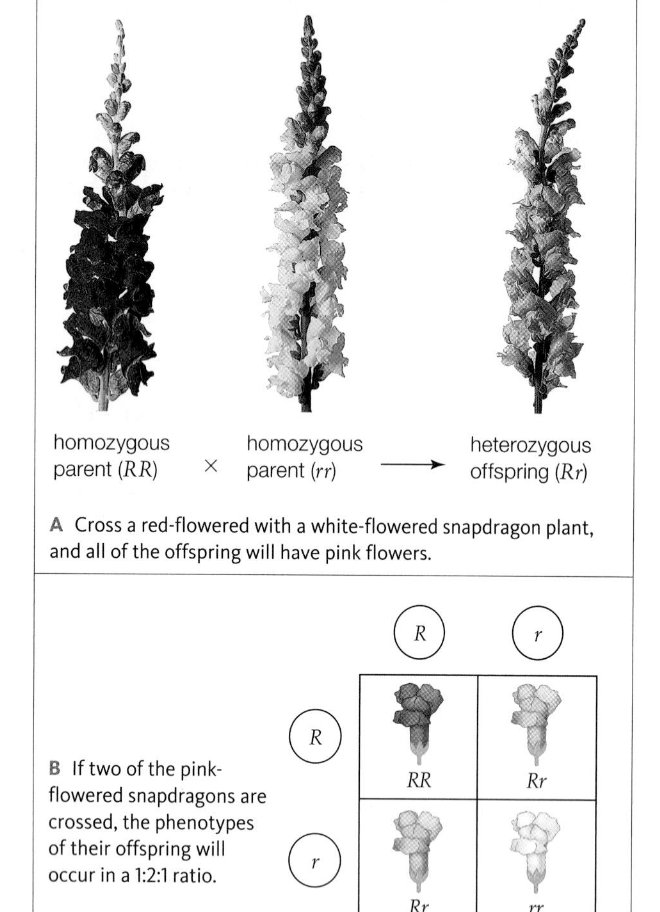

A Cross a red-flowered with a white-flowered snapdragon plant, and all of the offspring will have pink flowers.

B If two of the pink-flowered snapdragons are crossed, the phenotypes of their offspring will occur in a 1:2:1 ratio.

FIGURE 13.10 {Animated} Incomplete dominance in heterozygous (pink) snapdragons. One allele (*R*) results in the production of a red pigment; the other (*r*) results in no pigment.

FIGURE IT OUT: Is the experiment in **B** a monohybrid cross or a dihybrid cross?

Answer: A monohybrid cross

In the Mendelian inheritance patterns discussed in the last two sections, the effect of a dominant allele on a trait fully masks that of a recessive one. Other inheritance patterns are more common, but also more complex.

CODOMINANCE

With **codominance**, traits associated with two nonidentical alleles of a gene are equally apparent in heterozygotes; neither allele is dominant or recessive. Codominance may occur in **multiple allele systems**, in which three or more alleles of a gene persist at relatively high frequency among individuals of a population. Consider the *ABO* gene, which encodes an enzyme that modifies a carbohydrate on the surface of human red blood cells. The *A* and *B* alleles encode slightly different versions of this enzyme, which in turn modify the carbohydrate differently. The *O* allele has a mutation that prevents its enzyme product from becoming active at all.

The two alleles you carry for the *ABO* gene determine the form of the carbohydrate on your blood cells, so they are the basis of your blood type. The *A* and the *B* allele are codominant when paired. If your genotype is *AB*, then you have both versions of the enzyme, and your blood type is AB. The *O* allele is recessive when paired with either the *A* or *B* allele. If your genotype is *AA* or *AO*, your blood type is A. If your genotype is *BB* or *BO*, it is type B. If you are *OO*, it is type O (**FIGURE 13.9**).

Receiving incompatible blood in a transfusion is dangerous because the immune system attacks red blood cells bearing molecules that do not occur in one's own body. The attack can cause the blood cells to clump or burst, with potentially fatal consequences. The blood cells of people with type O blood do not carry the carbohydrate that can trigger this immune response. Thus, people with type O blood can donate blood to anyone; they are called universal donors. However, because their body is unfamiliar with the carbohydrates made by people with type A or B blood, they can receive type O blood only. People with type AB blood can receive a transfusion of any blood type, so they are called universal recipients.

INCOMPLETE DOMINANCE

With **incomplete dominance**, one allele is not fully dominant over the other, so the heterozygous phenotype is an intermediate blend of the two homozygous phenotypes. A gene that affects flower color in snapdragon plants is an example. One allele of the gene (*R*) encodes an enzyme that makes a red pigment. The enzyme encoded by a mutated allele (*r*) cannot make any pigment. Plants homozygous for the *R* allele (*RR*) make a lot of red pigment, so they have red flowers. Plants homozygous for the *r* allele (*rr*) make no pigment, so their flowers are white. Heterozygous plants (*Rr*) make only enough red pigment to color their flowers pink (**FIGURE 13.10**). A cross between two heterozygous plants yields red-, pink-, and white-flowered offspring in a 1:2:1 ratio.

EPISTASIS

Some traits are affected by multiple genes, an effect called polygenic inheritance or **epistasis**. Consider fur color in dogs, which depends on pigments called melanins. A dark brown melanin gives rise to brown or black fur; a reddish melanin is responsible for yellow fur. The production and deposition of melanin pigments in fur depends on several genes. The product of one gene (*TYRP1*) helps make the brown melanin. A dominant allele (*B*) of this gene results in a higher production of this melanin than the recessive allele (*b*). A different gene (*MC1R*) affects which type of melanin is produced. A dominant allele (*E*) of this gene triggers production of the brown melanin; its recessive partner (*e*) carries a mutation that results in production of the reddish form. Dogs homozygous for the *e* allele are yellow because they produce only the reddish melanin (**FIGURE 13.11**).

PLEIOTROPY

In many cases, a single gene influences multiple traits, an effect called **pleiotropy**. Mutations that alter the gene affect all of the traits at once. Many complex genetic disorders, including sickle-cell anemia (Section 9.5) and Marfan syndrome, arise as a result of mutations in pleiotropic genes. Marfan syndrome is caused by mutations that affect fibrillin. Long fibers of this protein impart elasticity to tissues of the heart, skin, blood vessels, tendons, and other body parts. Mutations cause tissues to form with defective fibrillin or none at all. The largest blood vessel leading from the heart, the aorta, is particularly affected. The aorta's thick wall is not as elastic as it should be, and it eventually stretches and becomes leaky. Thinned and weakened, the aorta can rupture during exercise—an abruptly fatal outcome.

About 1 in 5,000 people have Marfan syndrome, and there is no cure. Its effects—and risks—are manageable with early diagnosis, but symptoms are often missed. Many affected people die suddenly and early without ever knowing they had the disorder (**FIGURE 13.12**).

	EB	*Eb*	*eB*	*eb*
EB	*EEBB*	*EEBb*	*EeBB*	*EeBb*
Eb	*EEBb*	*EEbb*	*EeBb*	*Eebb*
eB	*EeBB*	*EeBb*	*eeBB*	*eeBb*
eb	*EeBb*	*Eebb*	*eeBb*	*eebb*

FIGURE 13.11 {Animated} An example of epistasis. Interactions among products of two gene pairs affect coat color in Labrador retrievers. Dogs with alleles *E* and *B* have black fur. Those with an *E* and two recessive *b* alleles have brown fur. Dogs homozygous for the recessive *e* allele have yellow fur.

FIGURE 13.12
A heartbreaker: Marfan syndrome. In 2006, 21-year-old basketball star Haris Charalambous collapsed and died suddenly during warm-up exercises. An autopsy revealed that his aorta had burst, an effect of the Marfan syndrome that Charalambous did not realize he had. Assistant trainer Brian Jones says, "Haris was just the nicest, funniest kid in the world. With his size, he was sort of lovably goofy. He was everybody's best friend."

TAKE-HOME MESSAGE 13.4

Some alleles are not dominant or recessive when paired.

With incomplete dominance, one allele is not fully dominant over another, so the heterozygous phenotype is an intermediate blend of the two homozygous phenotypes.

In codominance, two alleles have full and separate effect, so the phenotype of a heterozygous individual comprises both homozygous phenotypes.

In some cases, one gene influences multiple traits. In other cases, multiple genes influence the same trait.

codominance Effect in which the full and separate phenotypic effects of two alleles are apparent in heterozygous individuals.
epistasis Polygenic inheritance, in which a trait is influenced by multiple genes.
incomplete dominance Effect in which one allele is not fully dominant over another, so the heterozygous phenotype is an intermediate blend between the two homozygous phenotypes.
multiple allele system Gene for which three or more alleles persist in a population at relatively high frequency.
pleiotropy Effect in which a single gene affects multiple traits.

A The color of the snowshoe hare's fur varies by season. In summer, the fur is brown (*left*); in winter, white (*right*). Both forms offer seasonally appropriate camouflage from predators.

B The height of a mature yarrow plant depends on the elevation at which it grows.

C The body form of the water flea on the top develops in environments with few predators. A longer tail spine and a pointy head (*bottom*) develop in response to chemicals emitted by insects that prey on the fleas.

FIGURE 13.13 {Animated} Examples of environmental effects on phenotype.

The phrase "nature versus nurture" refers to a centuries-old debate about whether human behavioral traits arise from one's genetics (nature) or from environmental factors (nurture). It turns out that both play a role. The environment affects the expression of many genes, which in turn affects phenotype—including human behavioral traits. We can summarize this thinking with an equation:

genotype + environment ⟶ **phenotype**

Epigenetics research is revealing that the environment makes an even greater contribution to this equation than most biologists had suspected (Section 10.5).

Environmentally driven changes in gene expression patterns involve gene control (Section 10.1). For example, environmental cues trigger some cell-signaling pathways that end with methyl groups being removed from or added to particular regions of DNA. The change in methylation enhances or suppresses gene expression in those regions.

SOME ENVIRONMENTAL EFFECTS

Mechanisms that adjust phenotype in response to external cues are part of an individual's normal ability to adapt to its environment, as the following examples illustrate.

Seasonal Changes in Coat Color Seasonal changes in temperature and the length of day affect the production of melanin and other pigments that color the skin and fur of many animals. These species have different color phases in different seasons (**FIGURE 13.13A**). Hormonal signals triggered by the seasonal changes cause fur to be shed, and new fur grows back with different types and amounts of pigments deposited in it. The resulting change in phenotype provides these animals with seasonally appropriate camouflage from predators.

Effect of Altitude on Yarrow In plants, a flexible phenotype gives immobile individuals an ability to thrive in diverse habitats. For example, genetically identical yarrow plants grow to different heights at different altitudes (**FIGURE 13.13B**). More challenging temperature, soil, and water conditions are typically encountered at higher altitudes. Differences in altitude are also correlated with changes in the reproductive mode of yarrow: Plants at higher altitude tend to reproduce asexually, and those at lower altitude tend to reproduce sexually.

Alternative Phenotypes in Water Fleas Water fleas have different phenotypes depending on whether the aquatic insects that prey on them are present (**FIGURE 13.13C**). Individuals also switch between asexual and sexual modes of reproduction depending on environmental conditions. During the early spring, competition is scarce in their freshwater pond habitats. At that time, the fleas reproduce rapidly by asexual means, giving birth to large numbers of female offspring that quickly fill the ponds. Later in the season, competition for resources intensifies as the pond water becomes warmer, saltier, and more crowded. Under these conditions, some of the water fleas start giving birth to males, and then reproducing sexually. The increased genetic diversity of sexually produced offspring may offer the population an advantage in a more challenging environment.

Psychiatric Disorders Does the environment affect human genes? Researchers recently discovered that mutations in four human gene regions are associated with five psychiatric disorders: autism, depression, schizophrenia, bipolar disorder, and attention deficit hyperactivity disorder (ADHD). However, there must be an environmental component too, because one person with the mutations might get one type of disorder, while a relative with the same mutations might get another—two different results from the same genetic underpinnings. Moreover, the majority of people who carry these mutations never end up with a psychiatric disorder.

Recent discoveries in animal models are beginning to unravel some of the mechanisms by which environment can influence mental state in humans. For example, we now know that learning and memory are associated with dynamic and rapid DNA modifications in brain cells. Mood is, too. Stress-induced depression causes methylation-based silencing of a particular nerve growth factor gene; some antidepressants work by reversing this methylation. As another example, rats whose mothers are not very nurturing end up anxious and having a reduced resilience for stress as adults. The difference between these rats and ones who had nurturing maternal care is traceable to epigenetic DNA modifications that result in a lower than normal level of another nerve growth factor. Drugs can reverse these modifications—and their effects. We do not yet know all of the genes that influence human mental state, but the implication of such research is that future treatments for many disorders will involve deliberate modification of methylation patterns in an individual's DNA.

TAKE-HOME MESSAGE 13.5

The environment influences gene expression, and therefore can alter phenotype.

Cell-signaling pathways link environmental cues with changes in gene expression.

PEOPLE MATTER

DR. GAY BRADSHAW

The air explodes with the sound of high-powered rifles and the startled infant watches his family fall to the ground, the image seared into his memory. He and other orphans are then transported to distant locales to start new lives. Ten years later, the teenaged orphans begin a killing rampage, leaving more than a hundred victims.

A scene describing post-traumatic stress disorder (PTSD) in Kosovo or Rwanda? The similarities are striking—but the teenagers are young elephants, and the victims, rhinoceroses.

Gay Bradshaw, a psychologist and the director of the Kerulos Center in Oregon, has brought the latest insights from human neuroscience and psychology to bear on startling field observations of elephant behavior. She suspects that some threatened elephant populations might be suffering from chronic stress and trauma brought on by human encroachment and killing. "The loss of older elephants," says Bradshaw, "and the extreme psychological and physical trauma of witnessing the massacres of their family members interferes with a young elephant's normal development."

Under normal conditions, an early and healthy emotional relationship between an infant and its mother fosters the development of self-regulatory structures in the brain's right hemisphere. All mammals share this developmental attachment mechanism. With trauma, a malfunction can develop that makes the individual vulnerable to PTSD and predisposed to violence as an adult. Individuals who survive trauma often face a lifelong struggle with depression, suicide, or behavioral dysfunctions. In addition, children of trauma survivors can exhibit similar symptoms, an effect that is likely to be epigenetic at least in part.

As with humans, an intact, functioning social order helps buffer the effects of trauma in elephants. When park rangers introduced older males into the herd of marauding adolescent orphans, the orphans' violent behavior abruptly stopped.

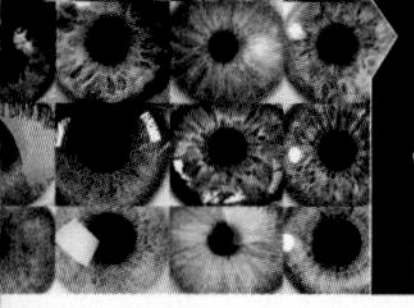

13.6 DO ALL TRAITS OCCUR IN DISTINCT FORMS?

FIGURE 13.14 Face length varies continuously in dogs. A gene with 12 alleles influences this trait.

A To see if human height varies continuously, male biology students at the University of Florida were divided into categories of one-inch increments in height and counted.

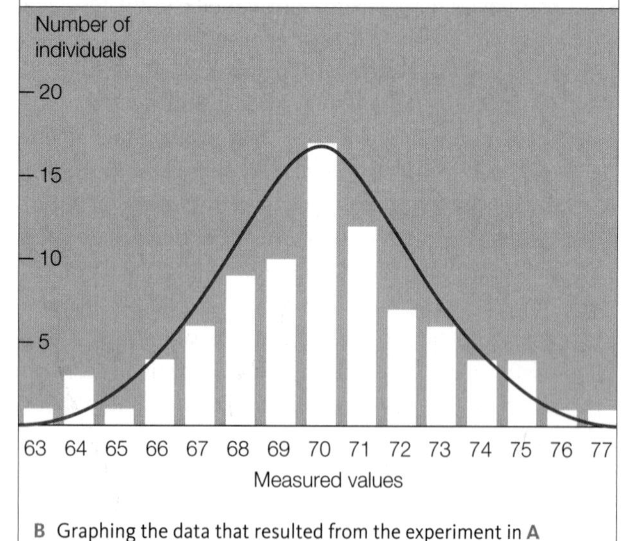

B Graphing the data that resulted from the experiment in **A** produces a bell-shaped curve, an indication that height does vary continuously in humans.

FIGURE 13.15 {Animated} Continuous variation.

The pea plant phenotypes that Mendel studied appeared in two or three forms, which made them easy to track through generations. However, many other traits do not appear in distinct forms. Such traits are often the result of complex genetic interactions—multiple genes, multiple alleles, or both—with added environmental influences (we return to this topic in Chapter 17, as we consider some evolutionary consequences of variation in phenotype). Tracking traits with complex variation presents a special challenge, which is why the genetic basis of many of them has not yet been completely unraveled.

Some traits occur in a range of small differences that is called **continuous variation**. Continuous variation can be an outcome of epistasis, in which multiple genes affect a single trait. The more genes that influence a trait, the more continuous is its variation. Traits that arise from genes with a lot of alleles may also vary continuously. Some genes have regions of DNA in which a series of 2 to 6 nucleotides is repeated hundreds or thousands of times in a row. These **short tandem repeats** can spontaneously expand or contract very quickly compared with the rate of mutation, and the resulting changes in the gene's DNA sequence may be preserved as alleles. For example, short tandem repeats have given rise to 12 alleles of a homeotic gene that influences the length of the face in dogs, with longer repeats associated with longer faces (**FIGURE 13.14**).

Human skin color varies continuously (a topic that we return to in Chapter 14), as does human eye color (shown in the chapter opener). How do we determine whether a particular trait varies continuously? Let's use another human trait, height, as an example. First, the total range of phenotypes is divided into measurable categories—inches, in this case (**FIGURE 13.15A**). Next, the individuals in each category are counted; these counts reveal the relative frequencies of phenotypes across the range of values. Finally, the data is plotted as a bar chart (**FIGURE 13.15B**). A graph line around the top of the bars shows the distribution of values for the trait. If the line is a bell-shaped curve, or **bell curve**, the trait varies continuously.

bell curve Bell-shaped curve; typically results from graphing frequency versus distribution for a trait that varies continuously.
continuous variation Range of small differences in a shared trait.
short tandem repeat In chromosomal DNA, sequences of a few nucleotides repeated multiple times in a row.

TAKE-HOME MESSAGE 13.6

The more genes and other factors that influence a trait, the more continuous is its range of variation.

13.7 Application: MENACING MUCUS

Lindsay, 22
Savannah, 19
Ben, 23
Jeff, 21
Brandon, 18
Cody, 23

FIGURE 13.16 Cystic fibrosis. These are a few of the many young victims of cystic fibrosis, which occurs most often in people of northern European ancestry. At least one young person dies every day in the United States from complications of this disease.

CYSTIC FIBROSIS (CF) IS THE MOST COMMON FATAL GENETIC DISORDER IN THE UNITED STATES. It occurs in people homozygous for a mutated allele of the *CFTR* gene. This gene encodes an active transport protein that moves chloride ions out of epithelial cells. Sheets of these cells line the passageways and ducts of the lungs, liver, pancreas, intestines, reproductive system, and skin. When chloride ions leave these cells, water follows by osmosis. The process maintains a thin film of water on the surface of the epithelial sheets.

The allele most commonly associated with CF has a 3 base pair deletion. It is called *ΔF508* because the protein it encodes is missing the normal 508th amino acid, a phenylalanine (F). The deletion prevents proper membrane trafficking of newly assembled polypeptides, which are left stranded in endoplasmic reticulum. The altered protein can function properly, but it never reaches the cell surface to do its job.

Epithelial cell membranes that lack the CFTR protein cannot transport chloride ions. Too few chloride ions leave these cells. Not enough water leaves them either, so the surfaces of epithelial cell sheets are too dry. Mucus that normally slips through the body's tubes sticks to the walls of the tubes instead. This outcome has pleiotropic effects because thick globs of mucus accumulate and clog passageways and ducts throughout the body. Breathing becomes difficult as the mucus obstructs the smaller airways of the lungs. Digestive problems arise as ducts that lead to the gut get clogged with mucus. Males are typically infertile because their sperm flow is hampered.

CFTR also helps alert the immune system to the presence of disease-causing bacteria in the lungs. The CFTR protein functions as a receptor: It binds directly to bacteria and triggers endocytosis. Endocytosis of bacteria into epithelial cells lining the respiratory tract initiates an immune response. When the cells lack CFTR, this early alert system fails, so bacteria have time to multiply before being detected by the immune system. Thus, chronic bacterial infections of the lungs are a hallmark of cystic fibrosis. Antibiotics help control infections, but there is no cure. Most affected people die before age thirty, when their tormented lungs fail (FIGURE 13.16).

The *ΔF508* allele is at least 50,000 years old and very common: in some populations, 1 in 25 people are heterozygous for it. Why does the allele persist if it is so harmful? *ΔF508* is codominant with the normal allele. Heterozygous individuals typically have no symptoms of cystic fibrosis; their cells have plasma membranes with enough CFTR to transport chloride ions normally. The *ΔF508* allele may offer these individuals an advantage in surviving certain deadly infectious diseases. CFTR's receptor function is an essential part of the immune response to bacteria in the respiratory tract. However, the same function allows bacteria to enter cells of the gastrointestinal tract, where they can be deadly. For example, endocytosis of *Salmonella typhi* bacteria into epithelial cells lining the gut results in a dangerous infection called typhoid fever. Cells lacking CFTR do not take up these bacteria. Thus, people who carry *ΔF508* are probably less susceptible to typhoid fever and other bacterial diseases that begin in the intestinal tract.

Summary

SECTION 13.1 Gregor Mendel indirectly discovered the role of alleles in inheritance by breeding pea plants and tracking traits of the offspring. Each gene occurs at a **locus**, or location, on a chromosome. Individuals with identical alleles are **homozygous** for the allele. **Heterozygous** individuals, or **hybrids**, have two nonidentical alleles. A **dominant** allele masks the effect of a **recessive** allele on the homologous chromosome. **Genotype** (an individual's particular set of alleles) gives rise to **phenotype**, which refers to an individual's observable traits.

SECTION 13.2 Crossing individuals that breed true for two forms of a trait yields identically heterozygous offspring. A cross between such offspring is a **monohybrid cross**. The frequency at which the traits appear in offspring of such **testcrosses** can reveal dominance relationships among the alleles associated with those traits.

Punnett squares are useful for determining the probability of offspring genotype and phenotype. Mendel's monohybrid cross results led to his **law of segregation** (stated here in modern terms): Diploid cells have pairs of genes on homologous chromosomes. The two genes of a pair separate from each other during meiosis, so they end up in different gametes.

SECTION 13.3 Crossing individuals that breed true for two forms of two traits yields F_1 offspring identically heterozygous for alleles governing those traits. A cross between such offspring is a **dihybrid cross**. The frequency at which the two traits appear in F_2 offspring can reveal dominance relationships between alleles associated with those traits. Mendel's dihybrid cross results led to his **law of independent assortment** (stated here in modern terms): Paired genes on homologous chromosomes tend to sort into gametes independently of other gene pairs during meiosis. Crossovers can break up **linkage groups**.

SECTION 13.4 With **incomplete dominance**, the phenotype of heterozygous individuals is an intermediate blend of the two homozygous phenotypes. With **codominant** alleles, heterozygous individuals have both homozygous phenotypes. Codominance may occur in **multiple allele systems** such as the one underlying ABO blood typing. With **epistasis**, two or more genes affect the same trait. A **pleiotropic** gene affects two or more traits.

SECTION 13.5 An individual's phenotype is influenced by environmental factors. Environmental cues alter gene expression by way of cell signaling pathways that ultimately affect gene controls.

SECTION 13.6 A trait that is influenced by multiple genes often occurs in a range of small increments of phenotype called **continuous variation**. Continuous variation typically occurs as a **bell curve** in the range of values. Multiple alleles such as those that arise in regions of **short tandem repeats** can give rise to continuous variation.

SECTION 13.7 Cystic fibrosis occurs in people homozygous for a mutated allele of the *CFTR* gene. The allele persists at high frequency despite its devastating effects. Carrying the allele may offer heterozygous individuals protection from dangerous gastrointestinal tract infections.

Self-Quiz

Answers in Appendix VII

1. A heterozygous individual has a _______ for a trait being studied.
   a. pair of identical alleles
   b. pair of nonidentical alleles
   c. haploid condition, in genetic terms

2. An organism's observable traits constitute its _______.
   a. phenotype
   b. variation
   c. genotype
   d. pedigree

3. In genetics, F stands for filial, which means _______.
   a. friendly
   b. offspring
   c. final
   d. hairlike

4. The second-generation offspring of a cross between individuals who are homozygous for different alleles of a gene are called the _______.
   a. F_1 generation
   b. F_2 generation
   c. hybrid generation
   d. none of the above

5. F_1 offspring of the cross $AA \times aa$ are _______.
   a. all AA
   b. all aa
   c. all Aa
   d. 1/2 AA and 1/2 aa

6. Refer to question 5. Assuming complete dominance, the F_2 generation will show a phenotypic ratio of _______.
   a. 3:1
   b. 9:1
   c. 1:2:1
   d. 9:3:3:1

7. A testcross is a way to determine _______.
   a. phenotype
   b. genotype
   c. both a and b

8. Assuming complete dominance, crosses between two dihybrid F_1 pea plants, which are offspring from a cross $AABB \times aabb$, result in F_2 phenotype ratios of _______.
   a. 1:2:1
   b. 3:1
   c. 1:1:1:1
   d. 9:3:3:1

9. The probability of a crossover occurring between two genes on the same chromosome _______.
   a. is unrelated to the distance between them
   b. decreases with the distance between them
   c. increases with the distance between them

10. A gene that affects three traits is _______.
    a. epistatic
    b. a multiple allele system
    c. pleiotropic
    d. dominant

11. The phenotype of individuals heterozygous for _______ alleles comprises both homozygous phenotypes.
    a. epistatic
    b. codominant
    c. pleiotropic
    d. hybrid

Data Analysis Activities

Carrying the Cystic Fibrosis Allele Offers Protection from Typhoid Fever Epithelial cells that lack the CFTR protein cannot take up bacteria by endocytosis. Endocytosis is an important part of the respiratory tract's immune defenses against common *Pseudomonas* bacteria, which is why *Pseudomonas* infections of the lungs are a chronic problem in cystic fibrosis patients. Endocytosis is also the way that *Salmonella typhi* enter cells of the gastrointestinal tract, where internalization of this bacteria can result in typhoid fever.

Typhoid fever is a common worldwide disease. Its symptoms include extreme fever and diarrhea, and the resulting dehydration causes delirium that may last several weeks. If untreated, it kills up to 30 percent of those infected. Around 600,000 people die annually from typhoid fever. Most of them are children.

In 1998, Gerald Pier and his colleagues compared the uptake of *S. typhi* by different types of epithelial cells: those homozygous for the normal allele, and those heterozygous for the *ΔF508* allele associated with CF. (Cells that are homozygous for the mutation do not take up any *S. typhi* bacteria.) Some of the results are shown in **FIGURE 13.17**.

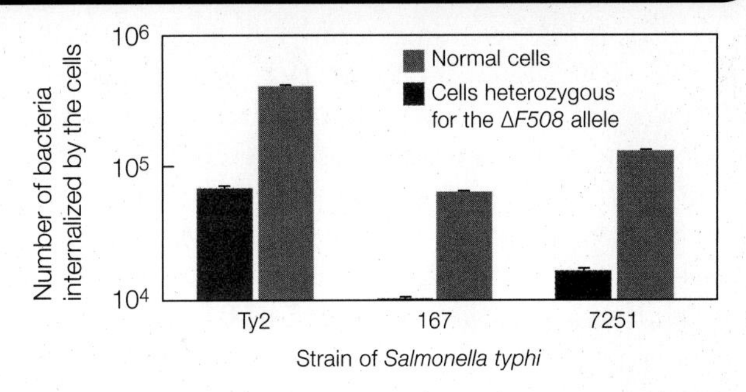

FIGURE 13.17 Effect of the *ΔF508* mutation on the uptake of three different strains of *Salmonella typhi* bacteria by epithelial cells.

1. Regarding the Ty2 strain of *S. typhi*, about how many more bacteria were able to enter normal cells (those heterozygous for the normal allele) than cells heterozygous for the *ΔF508* allele?
2. Which strain of bacteria entered normal epithelial cells most easily?
3. Entry of all three *S. typhi* strains into the heterozygous epithelial cells was inhibited. Is it possible to tell which strain was most inhibited?

12. ________ in a trait is indicated by a bell curve.

13. Match the terms with the best description.

___ dihybrid cross	a. *bb*
___ monohybrid cross	b. *AaBb* × *AaBb*
___ homozygous condition	c. *Aa*
___ heterozygous condition	d. *Aa* × *Aa*

Genetics Problems

Answers in Appendix VII

1. Mendel crossed a true-breeding pea plant with green pods and a true-breeding pea plant with yellow pods. All offspring had green pods. Which color is recessive?

2. Assuming that independent assortment occurs during meiosis, what type(s) of gametes will form in individuals with the following genotypes?

a. *AABB* b. *AaBB* c. *Aabb* d. *AaBb*

3. Determine the predicted genotype frequencies among the offspring of an *AABB* × *aaBB* mating.

4. Heterozygous individuals perpetuate some alleles that have lethal effects in homozygous individuals. A mutated allele (M^L) associated with taillessness in Manx cats (*left*) is an example. Cats homozygous for this allele (M^LM^L) typically die before birth due to severe spinal cord defects. In a case of incomplete dominance, cats heterozygous for the M^L allele and the normal, unmutated allele (M) have a short, stumpy tail or none at all. Two M^LM heterozygous cats mate. What is the probability that any of their kittens will be heterozygous (M^LM)?

5. People homozygous for a base-pair substitution in the beta-globin gene have sickle-cell anemia (Section 9.5). A couple who are planning to have children discover that both of the individuals are heterozygous for the mutated allele (Hb^S) and the normal allele (Hb^A). Calculate the probability that any one of their children will be born with sickle-cell anemia.

6. In sweet pea plants, an allele for purple flowers (P) is dominant to an allele for red flowers (p). An allele for long pollen grains (L) is dominant to an allele for round pollen grains (l). Bateson and Punnett crossed a plant having purple flowers/long pollen grains with one having white flowers/round pollen grains. All F_1 offspring had purple flowers and long pollen grains. Among the F_2 generation, the researchers observed the following phenotypes:

296 purple flowers/long pollen grains
19 purple flowers/round pollen grains
27 red flowers/long pollen grains
85 red flowers/round pollen grains

What is the best explanation for these results?

This family lives in Tanzania, where exposure to intense sunlight is responsible for the skin cancer that kills almost everyone with the albino phenotype. An abnormally low amount of melanin leaves people with this trait defenseless against UV radiation in the sun's rays. Recessive alleles on an autosome give rise to albinism.

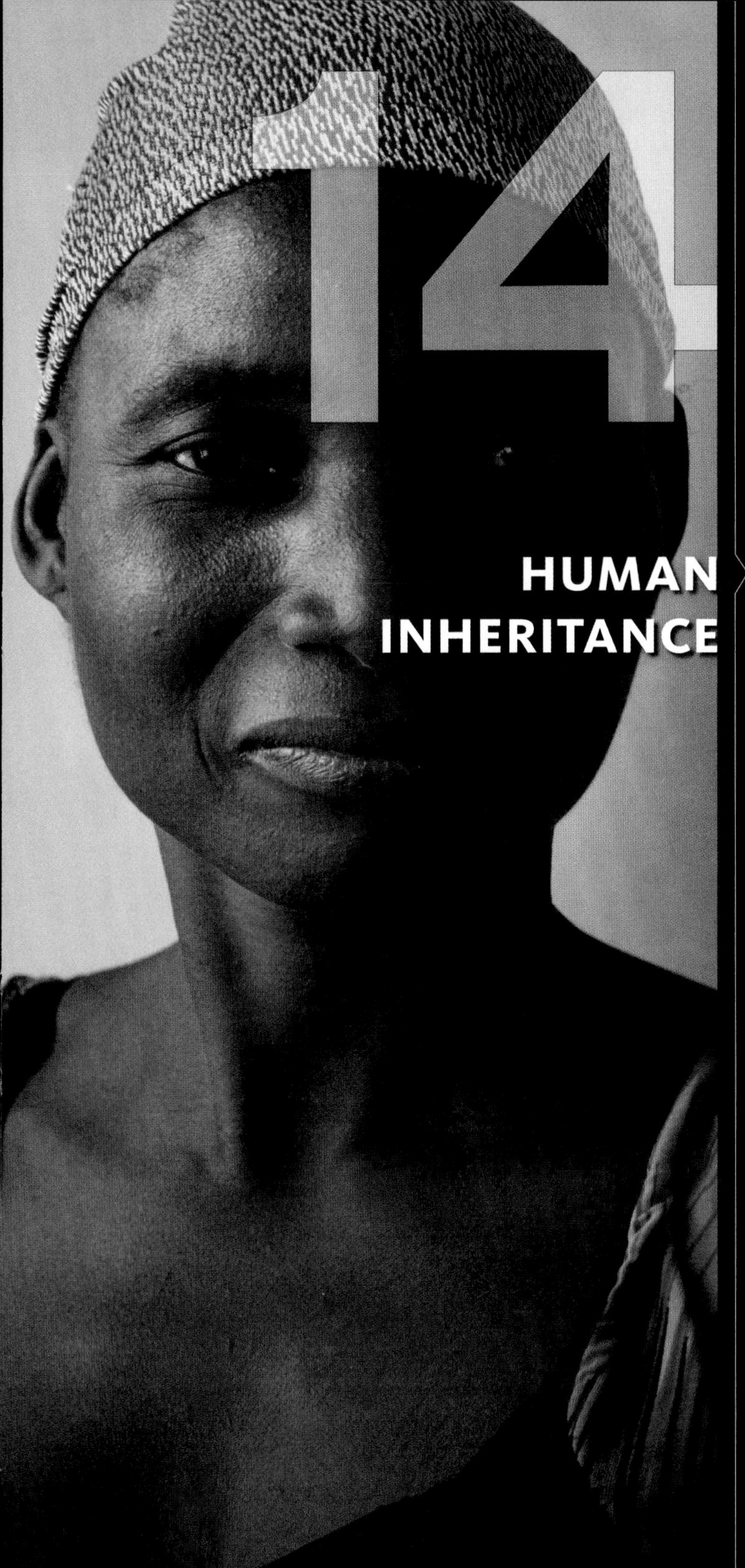

HUMAN INHERITANCE

Links to Earlier Concepts

Be sure you understand dominance relationships (Sections 13.1, 13.4, and 13.5), gene expression (9.1, 9.2), and mutations (9.5). You will use your knowledge of chromosomes (8.3), DNA replication and repair (8.4, 8.5), meiosis (12.2, 12.3), and sex determination (10.3). Sampling error (1.7), proteins (3.4), cell components (4.5, 4.9, 4.10), metabolism (5.4), pigments (6.1), telomeres (11.4), and oncogenes (11.5) will turn up in the context of genetic disorders.

KEY CONCEPTS

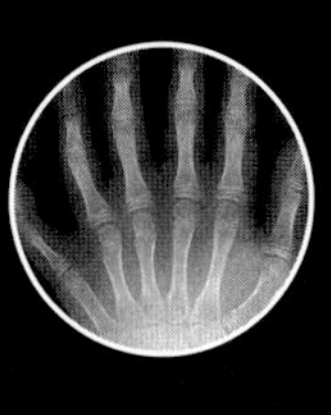

TRACKING TRAITS IN HUMANS

Inheritance patterns in humans are revealed by following traits through generations of a family. Tracked traits are often genetic abnormalities or syndromes associated with a genetic disorder.

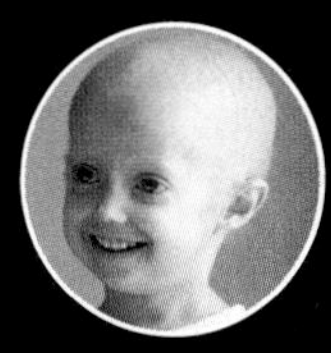

AUTOSOMAL INHERITANCE

Traits associated with dominant alleles on autosomes appear in every generation. Traits associated with recessive alleles on autosomes can skip generations.

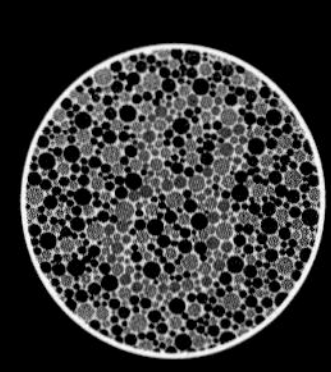

SEX-LINKED INHERITANCE

Traits associated with alleles on the X chromosome tend to affect more men than women. Men cannot pass such alleles to a son; carrier mothers bridge affected generations.

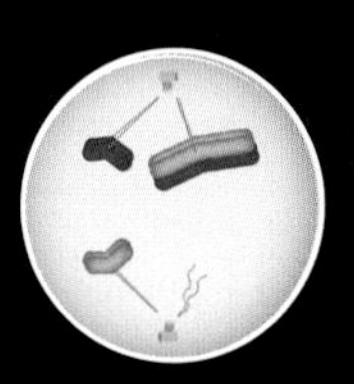

CHROMOSOME CHANGES

Some genetic disorders arise after large-scale change in chromosome structure. With few exceptions, a change in the number of autosomes is fatal in humans.

GENETIC TESTING

TABLE 14.1

Examples of Genetic Abnormalities and Disorders in Humans

Disorder or Abnormality	Main Symptoms
Autosomal dominant inheritance pattern	
Achondroplasia	One form of dwarfism
Aniridia	Defects of the eyes
Camptodactyly	Rigid, bent fingers
Familial hypercholesterolemia	High cholesterol level; clogged arteries
Huntington's disease	Degeneration of the nervous system
Marfan syndrome	Abnormal or missing connective tissue
Polydactyly	Extra fingers, toes, or both
Progeria	Drastic premature aging
Neurofibromatosis	Tumors of nervous system, skin
Autosomal recessive inheritance pattern	
Albinism	Absence of pigmentation
Hereditary methemoglobinemia	Blue skin coloration
Cystic fibrosis	Difficulty breathing; chronic lung infections
Ellis–van Creveld syndrome	Dwarfism, heart defects, polydactyly
Fanconi anemia	Physical abnormalities, marrow failure
Galactosemia	Brain, liver, eye damage
Hereditary hemochromatosis	Joints, organs damaged by iron overload
Phenylketonuria (PKU)	Mental impairment
Sickle-cell anemia	Anemia, pain, swelling, frequent infections
Tay–Sachs disease	Deterioration of mental and physical abilities; early death
X-linked recessive inheritance pattern	
Androgen insensitivity syndrome	XY individual but having some female traits; sterility
Red–green color blindness	Inability to distinguish red from green
Hemophilia	Impaired blood clotting ability
Muscular dystrophies	Progressive loss of muscle function
X-linked anhidrotic dysplasia	Mosaic skin (patches with or without sweat glands); other ill effects
X-linked dominant inheritance pattern	
Fragile X syndrome	Intellectual, emotional disability
Incontinentia pigmenti	Abnormalities of skin, hair, teeth, nails, eyes; neurological problems
Changes in chromosome number	
Down syndrome	Mental impairment; heart defects
Turner syndrome (XO)	Sterility; abnormal ovaries, sexual traits
Klinefelter syndrome	Sterility; mild mental impairment
XXX syndrome	Minimal abnormalities
XYY condition	Mild mental impairment or no effect
Changes in chromosome structure	
Chronic myelogenous leukemia (CML)	Overproduction of white blood cells; organ malfunctions
Cri-du-chat syndrome	Mental impairment; abnormal larynx

Some organisms, including pea plants and fruit flies, are ideal for genetic analysis. They have relatively few chromosomes, they reproduce quickly under controlled conditions, and breeding them poses few ethical problems. It does not take long to follow a trait through many generations. Humans, however, are a different story. Unlike flies grown in laboratories, we humans live under variable conditions, in different places, and we live as long as the geneticists who study us. Most of us select our own mates and reproduce if and when we want to. Our families tend to be on the small side, so sampling error (Section 1.7) is a major factor in studying them.

Because of these and other challenges, geneticists often use historical records to track traits through many generations of a family. These researchers make standardized charts of genetic connections called **pedigrees** (**FIGURE 14.1**). Analysis of a pedigree can reveal whether a trait is associated with a dominant or recessive allele, and whether the allele is on an autosome or a sex chromosome. Pedigree analysis also allows geneticists to determine the probability that a trait will recur in future generations of a family or a population.

TYPES OF GENETIC VARIATION

Some easily observed human traits follow Mendelian inheritance patterns. Like the flower color of pea plants, these traits are controlled by a single gene with alleles that have a clear dominance relationship. (Appendix IV shows a map of human chromosomes with the locations of some of these alleles.) Consider how someone who is homozygous for two recessive alleles of the *MC1R* gene (Section 13.4) makes the reddish melanin but not the brown melanin, so this person has red hair.

Single genes on autosomes or sex chromosomes also govern more than 6,000 genetic abnormalities and disorders. **TABLE 14.1** lists a few examples. A genetic abnormality is a rare or uncommon version of a trait, such as having six fingers on a hand. By contrast, a genetic disorder sooner or later causes medical problems that may be severe. A genetic disorder is often characterized by a specific set of symptoms (a syndrome). Most research in the field of human genetics focuses on disorders, because what we learn may help us develop treatments for affected people.

The next two sections of this chapter focus on inheritance patterns of human single-gene disorders, which affect about 1 in 200 people. Keep in mind that these patterns are the least common. Most human traits are

pedigree Chart showing the pattern of inheritance of a trait through generations in a family.

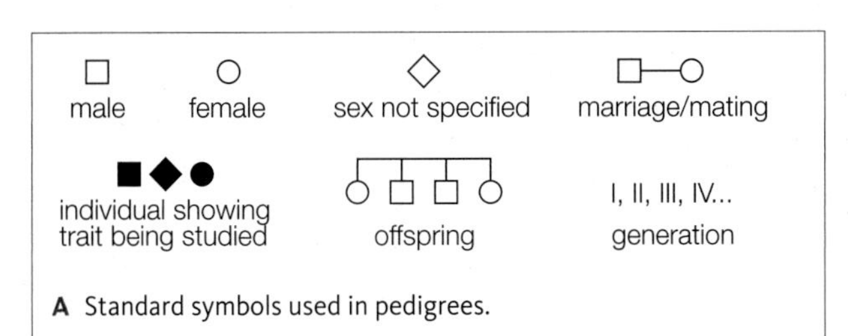

A Standard symbols used in pedigrees.

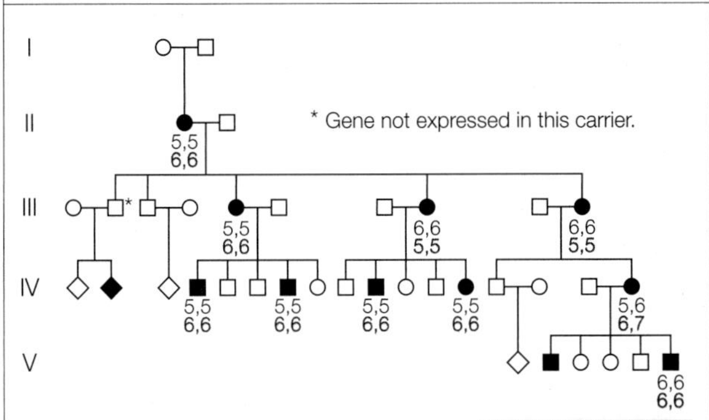

B A pedigree for polydactyly, which is characterized by extra fingers, toes, or both. The black numbers signify the number of fingers on each hand; the red numbers signify the number of toes on each foot. Though it occurs on its own, polydactyly is also one of several symptoms of Ellis–van Creveld syndrome.

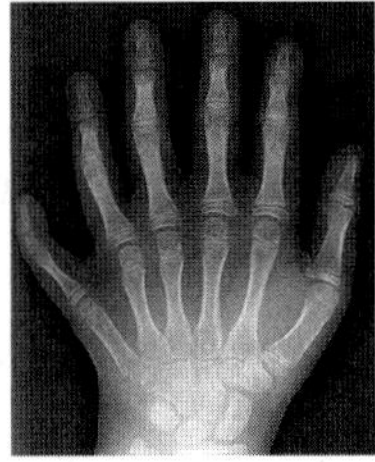

FIGURE 14.1 {Animated} Pedigrees.

polygenic (Section 13.4) and have epigenetic contributions (Section 13.5). Many genetic disorders are like this, including diabetes, asthma, obesity, cancers, heart disease, and multiple sclerosis. The inheritance patterns of these disorders are complex, and despite intense research our understanding of the genetics behind them remains incomplete.

Sections 14.4 and 14.5 explore some causes and effects of major changes in chromosomes—alterations in chromosome number or structure. Such changes occur in about 1 of every 100 births worldwide, and in many cases they have drastic consequences on health.

TAKE-HOME MESSAGE 14.1

Human inheritance patterns are often studied by tracking genetic abnormalities or disorders through family trees.

A genetic disorder is an inherited condition that causes medical problems. A genetic abnormality is a rare version of an inherited trait.

Some human genetic traits are governed by single genes and are inherited in a Mendelian fashion. Many others are influenced by multiple genes and epigenetics.

PEOPLE MATTER

DR. NANCY WEXLER

The village of Barranquitas, Venezuela, has the highest incidence of Huntington's disease in the world. A person with this incurable, fatal hereditary disorder gradually loses muscle control. Eventually, serious problems with swallowing cause many patients to die from choking or malnutrition. Beyond the physical symptoms, deep depression can often take hold.

Huntington's affects 1 in 10,000 people worldwide, but in Barranquitas the rate is more like 1 in 10. Some 1,000 villagers already have full-blown Huntington's; many more carry the gene. Such a high concentration of Huntington's patients made this region the backbone of Nancy Wexler's research. Wexler has been coming here for more than 30 years to study the genetics behind the disorder.

This is more than an academic pursuit or a career goal for her. "My mother died of Huntington's and she was a scientist. My father was a scientist too, and so we said 'Let's find a cure.' And we still say that. You can't get up in the morning without having hope and confidence that the cure is just around the corner."

Wexler and her colleagues collected DNA from and compiled an extended pedigree for nearly 10,000 Venezuelans. Her research was critical to the discovery that a dominant allele on human chromosome 4 causes Huntington's. As the daughter of a Huntington's sufferer, she has a one-in-two chance of carrying the fatal genetic flaw herself.

In 1999, Wexler cofounded a care home for Huntington's sufferers near Lake Maracaibo in Venezuela. The home, Casa Hogar, is a haven for more than 50 people whose families can no longer cope. Casa Hogar is facing a chronic lack of funding, possibly even closure. Still, Wexler remains confident that one day it won't be needed. "We never know when some miraculous discovery is going to be made," she said. "There are science breakthroughs on the horizon and happening now so I am very hopeful about the cure in the near future."

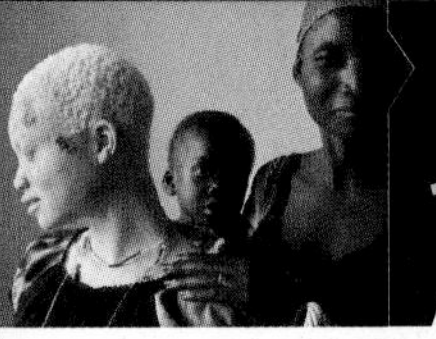

14.2 HOW DO WE KNOW WHEN A TRAIT IS AFFECTED BY AN ALLELE ON AN AUTOSOME?

THE AUTOSOMAL DOMINANT PATTERN

A trait associated with a dominant allele on an autosome appears in people who are heterozygous for it as well as those who are homozygous. Such traits appear in every generation of a family, and they occur with equal frequency in both sexes. When one parent is heterozygous, and the other is homozygous for the recessive allele, each of their children has a 50 percent chance of inheriting the dominant allele and having the associated trait (**FIGURE 14.2A**).

Achondroplasia A form of hereditary dwarfism called achondroplasia offers an example of an autosomal dominant disorder (one caused by a dominant allele on an autosome). Mutations associated with achondroplasia occur in a gene for a growth hormone receptor. The mutations cause the receptor, a regulatory molecule that slows bone development, to be overly active. About 1 in 10,000 people is heterozygous for one of these mutations. As adults, affected people are, on average, about four feet, four inches (1.3 meters) tall, with arms and legs that are short relative to torso size (**FIGURE 14.2B**). An allele that causes achondroplasia can be passed to children because its expression does not interfere with reproduction, at least in heterozygous people. The homozygous condition results in severe skeletal malformations that cause early death.

Huntington's Disease Alleles that cause Huntington's disease are also inherited in an autosomal dominant pattern. Mutations associated with this disorder alter a gene for a cytoplasmic protein whose function is still unknown. The mutations are insertions caused by expansion of a short tandem repeat (Section 13.6), in which the same three nucleotides become repeated many, many times in the gene's sequence. The altered gene encodes an oversized protein product that gets chopped into pieces inside nerve cells of the brain. The pieces accumulate in cytoplasm as large clumps that eventually prevent the cells from functioning properly. Brain cells involved in movement, thinking, and emotion are particularly affected. In the most common form of Huntington's, symptoms do not start until after the age of thirty, and affected people die during their forties or fifties. With this and other late-onset disorders, people tend to reproduce before symptoms appear, so the allele is often passed unknowingly to children.

Hutchinson–Gilford Progeria Hutchinson–Gilford progeria is an autosomal dominant disorder characterized by drastically accelerated aging. It is usually caused by a mutation in the gene for lamin A, a protein subunit of intermediate filaments that support the nuclear envelope (Section 4.9). Lamins also have roles in mitosis, DNA synthesis and repair, and transcriptional regulation. The

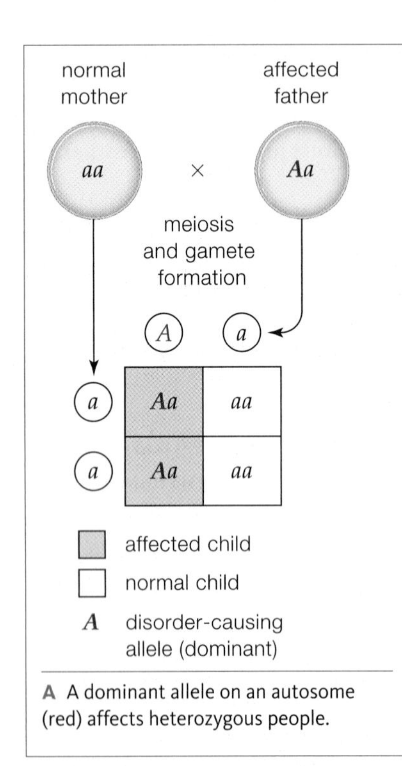

A A dominant allele on an autosome (red) affects heterozygous people.

B Achondroplasia affects Ivy Broadhead (*left*), as well as her brother, father, and grandfather.

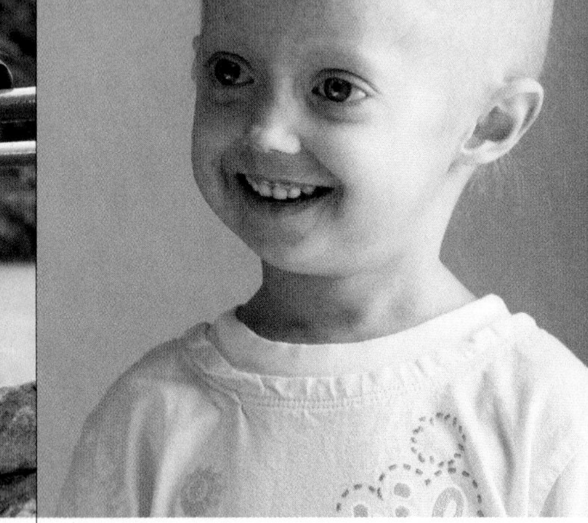

C Five-year-old Megan is already showing symptoms of Hutchinson–Gilford progeria.

FIGURE 14.2 {Animated} Autosomal dominant inheritance.

mutation, a base-pair substitution, adds a signal for a splice site (Section 9.2). The resulting lamin A protein is too short and cannot be processed correctly after translation. Cells that carry this mutation have a nucleus that is grossly abnormal, with nuclear pore complexes that do not assemble properly and membrane proteins localized to the wrong side of the nuclear envelope. The function of the nucleus as protector of chromosomes and gateway of transcription is severely impaired, so DNA damage accumulates quickly. The effects are pleiotropic. Outward symptoms begin to appear before age two, as skin that should be plump and resilient starts to thin, muscles weaken, and bones soften. Premature baldness is inevitable (**FIGURE 14.2C**). Most people with the disorder die in their early teens as a result of a stroke or heart attack brought on by hardened arteries, a condition typical of advanced age. Progeria does not run in families because affected people do not live long enough to reproduce.

THE AUTOSOMAL RECESSIVE PATTERN

A recessive allele on an autosome is expressed only in homozygous people, so traits associated with the allele tend to skip generations. Both sexes are equally affected. People heterozygous for the allele are carriers, which means that they have the allele but not the trait. Any child of two carriers has a 25 percent chance of inheriting the allele from both parents (**FIGURE 14.3A**). Being homozygous for the allele, such children would have the trait.

Tay–Sachs Disease Alleles associated with Tay–Sachs disease are inherited in an autosomal recessive pattern. In the general population, about 1 in 300 people is a carrier for one of these allele, but the incidence is ten times higher in some groups, such as Jews of eastern European descent. The gene altered in Tay–Sachs encodes a lysosomal enzyme responsible for breaking down a particular type of lipid. Mutations cause this enzyme to misfold and become destroyed, so cells make the lipid but cannot break it down. Typically, newborns homozygous for a Tay–Sachs allele seem normal, but within three to six months they become irritable, listless, and may have seizures as the lipid accumulates in their nerve cells. Blindness, deafness, and paralysis follow. Affected children usually die by age five (**FIGURE 14.3B**).

Albinism Albinism, a genetic abnormality characterized by an abnormally low level of the pigment melanin, is also inherited in an autosomal recessive pattern. Mutations associated with the albino phenotype occur in genes involved in melanin synthesis. Skin, hair, or eye pigmentation may be reduced or missing, as shown in the chapter opening photo. In the most dramatic form of the phenotype, the skin is very white and does not tan, and the hair is white. The lack of pigment in the irises of the eyes allows underlying blood vessels to show through, so the irises appear red. Melanin plays a role in the retina, so people with the albino phenotype tend to have vision problems.

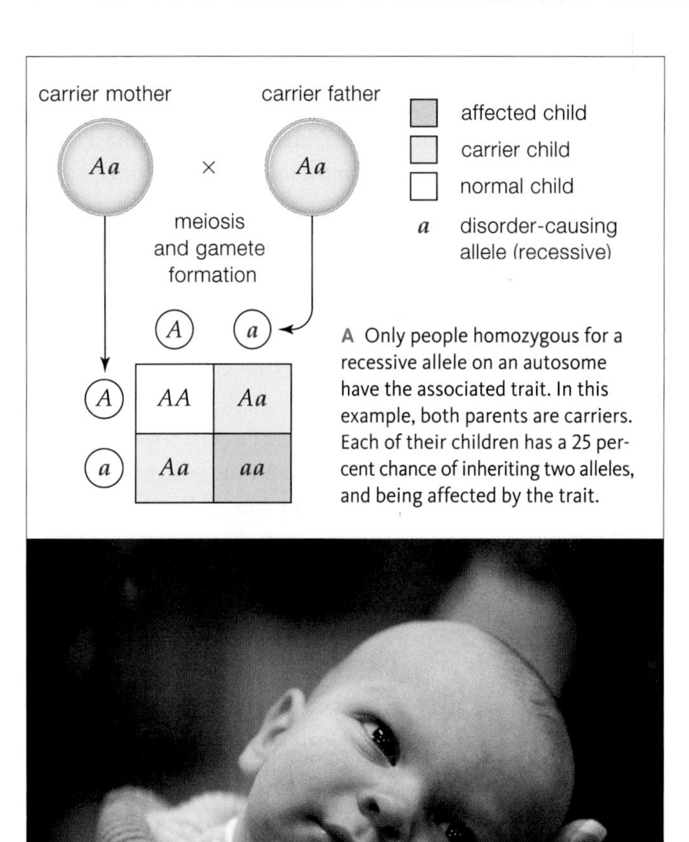

A Only people homozygous for a recessive allele on an autosome have the associated trait. In this example, both parents are carriers. Each of their children has a 25 percent chance of inheriting two alleles, and being affected by the trait.

B Conner Hopf was diagnosed with Tay–Sachs disease at age 7½ months. He died before his second birthday.

FIGURE 14.3 {Animated} Autosomal recessive inheritance.

TAKE-HOME MESSAGE 14.2

With an autosomal dominant inheritance pattern, anyone with the allele, homozygous or heterozygous, has the associated trait. The trait tends to appear in every generation.

With an autosomal recessive inheritance pattern, only persons who are homozygous for an allele have the associated trait. The trait tends to skip generations.

14.3 HOW DO WE KNOW WHEN A TRAIT IS AFFECTED BY AN ALLELE ON AN X CHROMOSOME?

Many genetic disorders are associated with alleles on the X chromosome (**FIGURE 14.4**). Almost all of them are inherited in a recessive pattern, probably because those caused by dominant X chromosome alleles tend to be lethal in male embryos.

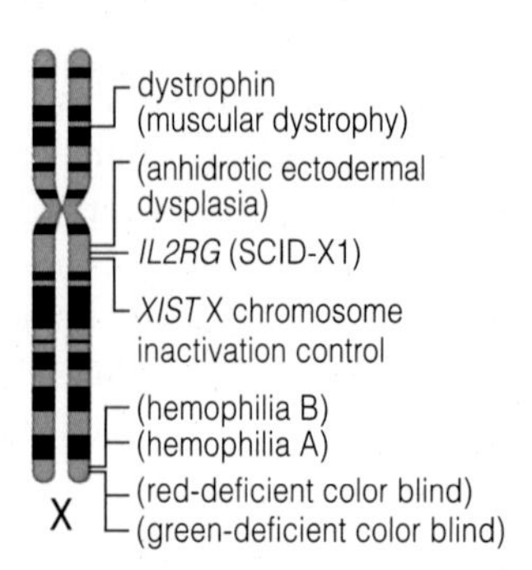

FIGURE 14.4 The human X chromosome. This chromosome carries about 2,000 genes—almost 10 percent of the total. Most X chromosome alleles that cause genetic disorders are inherited in a recessive pattern. A few disorders are listed (in parentheses).

THE X-LINKED RECESSIVE PATTERN

A recessive allele on the X chromosome (an X-linked recessive allele) leaves two clues when it causes a genetic disorder. First, an affected father never passes an X-linked recessive allele to a son, because all children who inherit their father's X chromosome are female (**FIGURE 14.5A**). Thus, a heterozygous female is always the bridge between an affected male and his affected grandson. Second, the disorder appears in males more often than in females. This is because all males who carry the allele have the disorder, but not all heterozygous females do. Remember that one of the two X chromosomes in each cell of a female is inactivated as a Barr body (Section 10.3). As a result, only about half of a heterozygous female's cells express the recessive allele. The other half of her cells express the dominant, normal allele that she carries on her other X chromosome, and this expression can mask the phenotypic effects of the recessive allele.

Red–Green Color Blindness Color blindness refers to a range of conditions in which an individual cannot distinguish among some or all colors in the spectrum of visible light. These conditions are typically inherited in an X-linked recessive pattern, because most of the genes involved in color vision are on the X chromosome.

Humans can sense the differences among 150 colors, and this perception depends on pigment-containing receptors

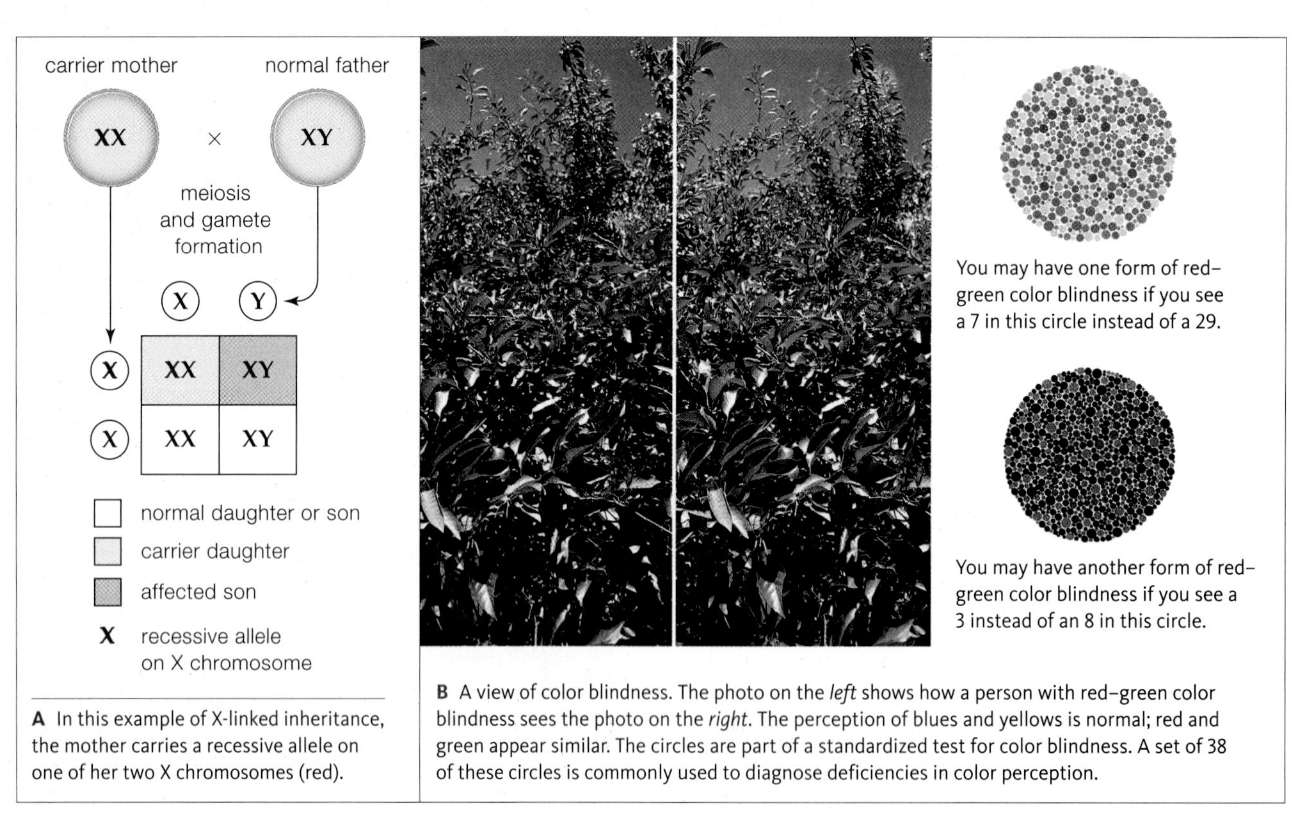

A In this example of X-linked inheritance, the mother carries a recessive allele on one of her two X chromosomes (red).

B A view of color blindness. The photo on the *left* shows how a person with red–green color blindness sees the photo on the *right*. The perception of blues and yellows is normal; red and green appear similar. The circles are part of a standardized test for color blindness. A set of 38 of these circles is commonly used to diagnose deficiencies in color perception.

FIGURE 14.5 {Animated} X-linked recessive inheritance.

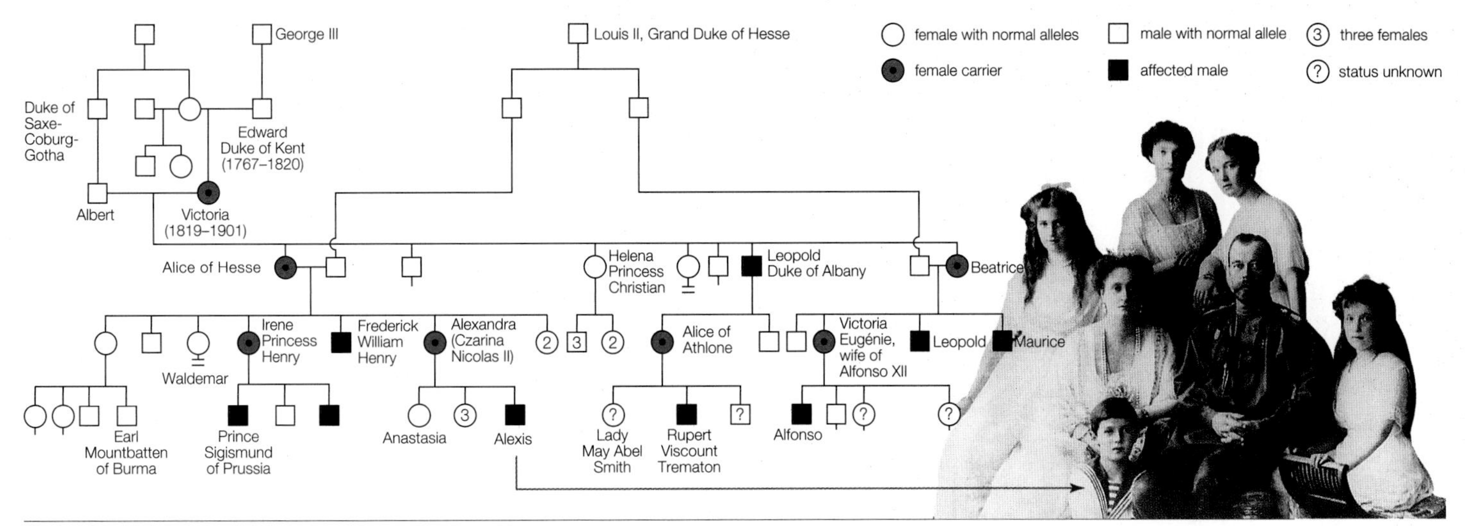

FIGURE 14.6 {Animated} A classic case of X-linked recessive inheritance: a partial pedigree of the descendants of Queen Victoria of England. At one time, the recessive X-linked allele that resulted in hemophilia was present in eighteen of Victoria's sixty-nine descendants, who sometimes intermarried. Of the Russian royal family members shown, the mother (Alexandra Czarina Nicolas II) was a carrier.

FIGURE IT OUT: How many of Alexis's siblings were affected by hemophilia A?

Answer: None

in the eyes. Mutations that result in altered or missing receptors affect color vision. For example, people who have red–green color blindness see fewer than 25 colors because receptors that respond to red and green wavelengths are weakened or absent (**FIGURE 14.5B**). Some confuse red and green; others see green as gray.

Duchenne Muscular Dystrophy An X-linked recessive disorder, Duchenne muscular dystrophy (DMD), is characterized by muscle degeneration. It is caused by mutations in the X chromosome gene for dystrophin, a cytoskeletal protein that links actin microfilaments in cytoplasm to a complex of proteins in the plasma membrane. This complex structurally and functionally links the cell to extracellular matrix. When dystrophin is absent, the entire protein complex is unstable. Muscle cells, which are subject to stretching, are particularly affected. Their plasma membrane is easily damaged, and they become flooded with calcium ions. Eventually, the muscle cells die and become replaced by fat cells and connective tissue.

DMD affects about 1 in 3,500 people, almost all of them boys. Symptoms begin between ages three and seven. Anti-inflammatory drugs can slow the progression of DMD, but there is no cure. When an affected boy is about twelve years old, he will begin to use a wheelchair and his heart muscle will start to fail. Even with the best care, he will probably die before the age of thirty, from a heart disorder or respiratory failure (suffocation).

Hemophilia A Hemophilia A is an X-linked recessive disorder that interferes with blood clotting. Most of us have a blood clotting mechanism that quickly stops bleeding from minor injuries. That mechanism involves factor VIII, a protein product of a gene on the X chromosome. Bleeding can be prolonged in males who carry a mutation in this gene. Females who have two mutated alleles are also affected (heterozygous females make enough factor VIII to have a clotting time that is close to normal). Affected people tend to bruise very easily, but internal bleeding is their most serious problem. Repeated bleeding inside the joints disfigures them and causes chronic arthritis.

In the nineteenth century, the incidence of hemophilia A was relatively high in royal families of Europe and Russia, probably because the common practice of inbreeding kept the allele in their family trees (**FIGURE 14.6**). Today, about 1 in 7,500 people in the general population is affected. That number may be rising because the disorder is now a treatable one. More affected people are living long enough to transmit the mutated allele to children.

TAKE-HOME MESSAGE 14.3

Men who have an X-linked allele have the associated trait, but not all heterozygous women do. Thus, the trait appears more often in men.

Men transmit an X-linked allele to their daughters, but not to their sons.

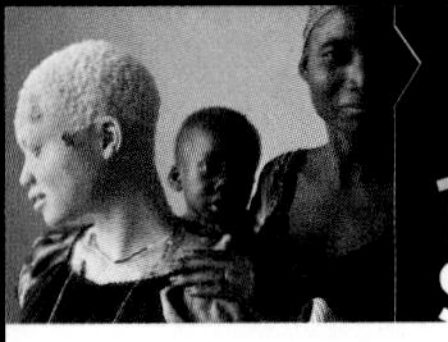

14.4 HOW DOES CHROMOSOME STRUCTURE CHANGE?

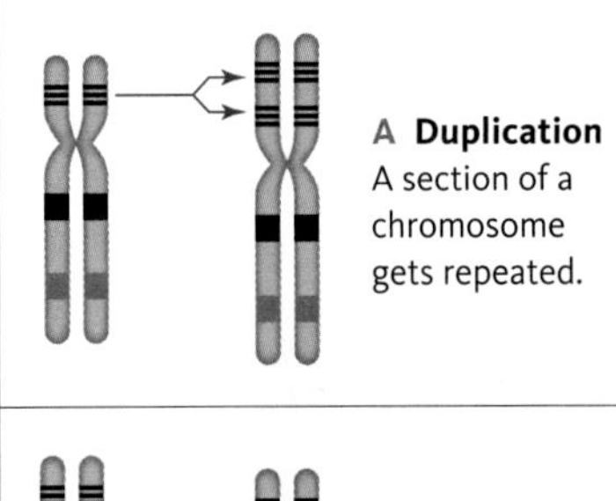

A Duplication A section of a chromosome gets repeated.

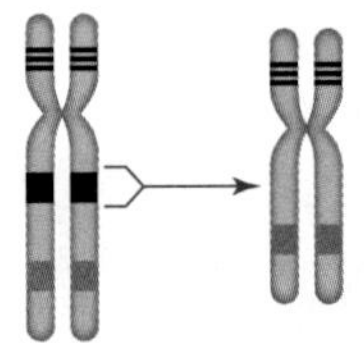

B Deletion A section of chromosome gets lost.

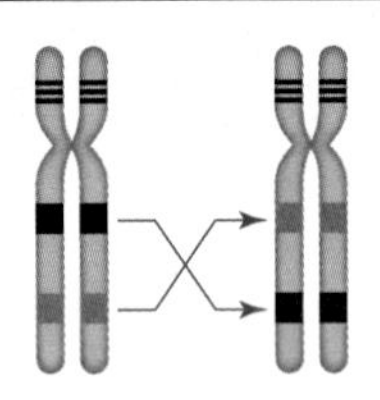

C Inversion A section of a chromosome gets flipped so it runs in the opposite orientation.

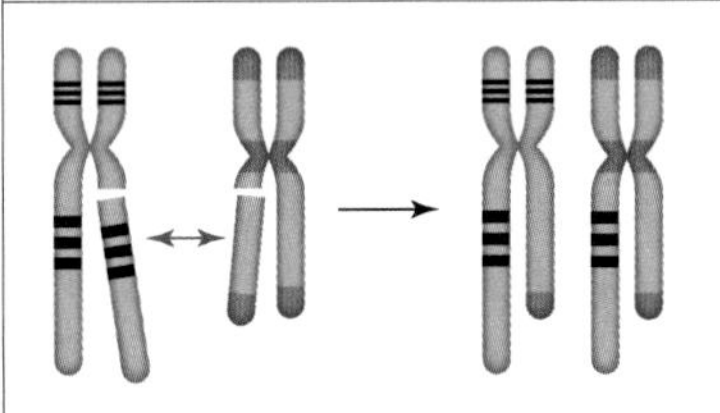

D Translocation A broken piece of a chromosome gets reattached in the wrong place. This example shows a reciprocal transloca-tion, in which two nonhomologous chromosomes exchange chunks.

FIGURE 14.7 {Animated} Major changes in chromosome structure.

Mutation is a term that generally refers to small-scale changes in DNA sequence—one or a few nucleotides. Chromosome changes on a larger scale also occur. Like mutations, these changes may be induced by exposure to chemicals or radiation. Others are an outcome of faulty crossing over during prophase I of meiosis. For example, nonhomologous chromosomes may align and swap segments at spots where the DNA sequence is similar. Homologous chromosomes sometimes misalign along their length. In both cases, crossing over results in the exchange of segments that are not equivalent.

Chromosome structural changes also result from the activity of **transposable elements**, which are segments of DNA, hundreds to thousands of nucleotides long, that can move spontaneously within or between chromosomes. Repeated DNA sequences at their ends allow the segments to move during mitosis or meiosis. Transposable elements are common in the DNA of all species; about 45 percent of human DNA consists of them or their remnants.

TYPES OF CHROMOSOMAL CHANGE

Regardless of the cause, large-scale changes in chromosome structure can be categorized into several groups (**FIGURE 14.7**). In most cases, these changes have drastic effects on health; about half of all miscarriages are due to chromosome abnormalities of the developing embryo.

Duplication Even normal chromosomes have DNA sequences that are repeated two or more times. These repetitions are called **duplications** (**FIGURE 14.7A**). Some newly occurring duplications, such as the expansion mutations that cause Huntington's disease, cause genetic abnormalities or disorders. Others, as you will see shortly, have been evolutionarily important.

Deletion Large-scale deletions in a chromosome (**FIGURE 14.7B**) often have severe consequences. Duchenne muscular dystrophy most often arises from deletions in the X chromosome. A different deletion in chromosome 5 shortens life span, impairs mental functioning, and results in an abnormally shaped larynx. This disorder, cri-du-chat (French for "cat's cry"), is named for the sound that affected infants make when they cry.

Inversion With an **inversion**, a segment of chromosomal DNA becomes oriented in the reverse direction, with no molecular loss (**FIGURE 14.7C**). An inversion may not affect a carrier's health if it does not interrupt a gene or gene control region, because the individual's cells still contain their full complement of genetic material. However, fertility may be compromised because a chromosome with an inversion does not pair properly with its homologous partner during meiosis. Crossovers between these mispaired chromosomes can produce other chromosome abnormalities that reduce the viability of forthcoming embryos. People who carry an inversion may not know about it until they are diagnosed with infertility and their karyotype is checked.

Translocation If a chromosome breaks, the broken part may get attached to a different chromosome, or to a different part of the same one. This type of structural change is called a **translocation**. Most translocations are reciprocal, in which two nonhomologous chromosomes exchange broken parts (**FIGURE 14.7D**). A reciprocal translocation between chromosomes 8 and 14 is the usual cause of Burkitt's lymphoma, an aggressive cancer of the immune system. This translocation moves a proto-oncogene to a region that is vigorously transcribed in immune cells, with the result being uncontrolled cell divisions that are characteristic of cancer (Section 11.5). Many other reciprocal translocations have no adverse effects on health, but, like inversions, they can compromise fertility. During meiosis, translocated chromosomes pair abnormally and segregate improperly; about half of the resulting gametes carry major duplications or deletions. If one of these gametes unites with a normal gamete at fertilization, the resulting embryo almost always dies. As with inversions, people who carry a translocation may not know about it until they have difficulty with fertility.

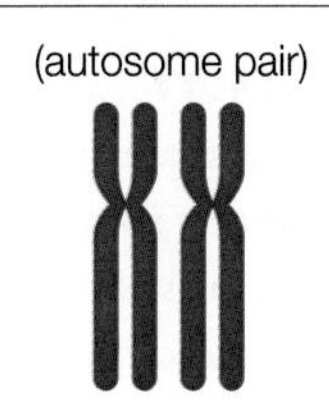

Ancestral reptiles
>350 mya

A Before 350 mya, sex was determined by temperature, not by chromosome differences.

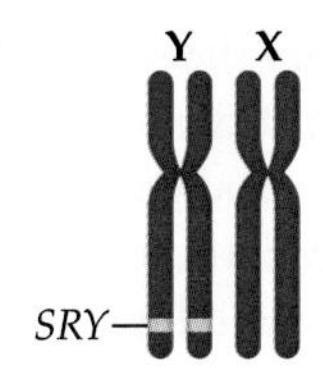

Ancestral reptiles
350 mya

B The *SRY* gene begins to evolve 350 mya. The DNA sequences of the chromosomes diverge as other mutations accumulate.

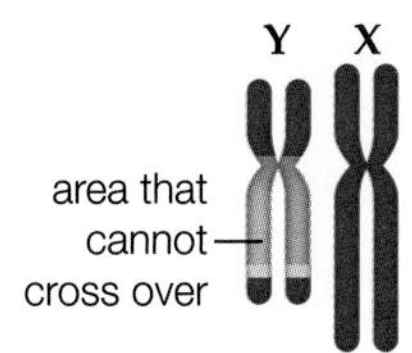

Monotremes
320–240 mya

C By 320–240 mya, the DNA sequences of the chromosomes are so different that the pair can no longer cross over in one region. The Y chromosome begins to shorten.

Marsupials
170–130 mya

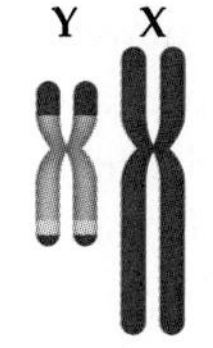

Monkeys
130–80 mya

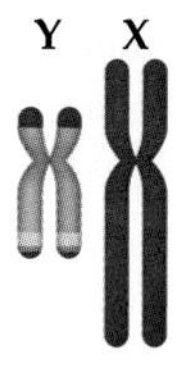

Humans
50–30 mya

D Three more times, the pair stops crossing over in yet another region. Each time, the DNA sequences of the chromosomes diverge, and the Y chromosome shortens. Today, the pair crosses over only at a small region near the ends.

FIGURE 14.8 Evolution of the Y chromosome. Today, the *SRY* gene determines male sex. Homologous regions of the chromosomes are shown in pink; mya, million years ago. Monotremes are egg-laying mammals; marsupials are pouched mammals.

CHROMOSOME CHANGES IN EVOLUTION

There is evidence of major structural alterations in the chromosomes of all known species. For example, duplications have often allowed a copy of a gene to mutate while the original carried out its unaltered function. The multiple and strikingly similar globin chain genes of mammals apparently evolved by this process. Globin chains, remember, associate to form molecules of hemoglobin (Section 9.5). Two identical genes for the alpha chain—and five other slightly different versions of it—form a cluster on chromosome 16. The gene for the beta chain clusters with four other slightly different versions on chromosome 11.

As another example, X and Y chromosomes were once homologous autosomes in ancient, reptilelike ancestors of mammals (**FIGURE 14.8**). Ambient temperature probably determined the gender of those organisms, as it still does in turtles and some other modern reptiles. About 350 million years ago, a gene on one of the two homologous chromosomes mutated. The change, which was the beginning of the male sex determination gene *SRY*, interfered with crossing over during meiosis. A reduced frequency of crossing over allowed the chromosomes to diverge around the changed region as mutations began to accumulate separately in the two chromosomes. Over evolutionary time, the chromosomes became so different that they no longer crossed over at all in the changed region, so they diverged even more. Today, the Y chromosome is much smaller than the X, and is homologous with it only in a tiny part. The Y crosses over mainly with itself—by translocating duplicated regions of its own DNA.

Some chromosome structure changes contributed to differences among closely related organisms, such as apes and humans. Human somatic cells have twenty-three pairs of chromosomes, but cells of chimpanzees, gorillas, and orangutans have twenty-four. Thirteen human chromosomes are almost identical with chimpanzee chromosomes. Nine more are similar, except for some inversions. One human chromosome matches up with two in chimpanzees and the other great apes (**FIGURE 14.9**). During human evolution, two chromosomes evidently fused end to end and formed our chromosome 2. How do we know? The region where the fusion occurred contains remnants of a telomere (Section 11.4).

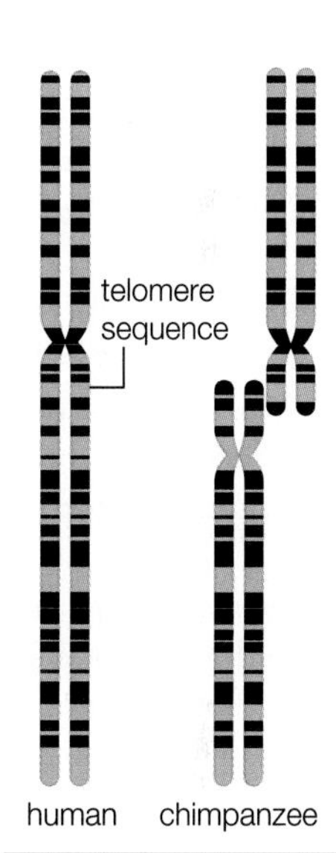

FIGURE 14.9 Human chromosome 2 compared with chimpanzee chromosomes 2A and 2B.

duplication Repeated section of a chromosome.
inversion Structural rearrangement of a chromosome in which part of the DNA becomes oriented in the reverse direction.
translocation Structural change of a chromosome in which a broken piece gets reattached in the wrong location.
transposable element Segment of DNA that can move spontaneously within or between chromosomes.

TAKE-HOME MESSAGE 14.4

A segment of a chromosome may be duplicated, deleted, inverted, or translocated. Any of these changes are usually harmful or lethal, but may be conserved in the rare circumstance that it has a neutral or beneficial effect.

14.5 WHAT ARE THE EFFECTS OF CHROMOSOME NUMBER CHANGES IN HUMANS?

About 70 percent of flowering plant species, and some insects, fishes, and other animals, are **polyploid**, which means that they have three or more complete sets of chromosomes. Cells in some adult human tissues are normally polyploid, but inheriting more than two full sets of chromosomes is invariably fatal in humans.

An **aneuploid** individual inherited too many or too few copies of a particular chromosome. Less than 1 percent of children are born with a diploid chromosome number that differs from the normal 46.

Changes in chromosome number are usually the outcome of **nondisjunction**, the failure of chromosomes to separate properly during nuclear division—mitosis or meiosis. Nondisjunction during meiosis (**FIGURE 14.10**) can affect chromosome number at fertilization. For example, if a normal gamete (n) fuses with another gamete that has an extra chromosome ($n+1$), the resulting zygote will have three copies of one type of chromosome and two of every other type ($2n+1$), a type of aneuploidy called trisomy. If an $n-1$ gamete fuses with a normal n gamete, the new individual will be $2n-1$, a condition called monosomy.

AUTOSOMAL ANEUPLOIDY AND DOWN SYNDROME

In most cases, autosomal aneuploidy in humans is fatal before birth or shortly thereafter. An important exception is trisomy 21. A person born with three chromosomes 21 has a high likelihood of surviving infancy, and will develop Down syndrome. Mild to moderate mental impairment and health problems such as heart disease are hallmarks of this disorder. Other effects may include a somewhat flattened facial profile, a fold of skin that starts at the inner corner of each eyelid, white spots on the iris (**FIGURE 14.11**), and one deep crease (instead of two shallow creases) across each palm. The skeleton grows and develops abnormally, so older children have short body parts, loose joints, and misaligned bones of the fingers, toes, and hips. The muscles and reflexes are weak, and motor skills such as speech develop slowly. With medical care, affected individuals live about fifty-five years. Early training can help these individuals learn to care for themselves and to take part in normal activities. Down syndrome occurs once in about 700 births, and the risk increases with maternal age.

SEX CHROMOSOME ANEUPLOIDY

Nondisjunction also causes alterations in the number of X and Y chromosomes, with a frequency of about 1 in 400 live births. Most often, such alterations lead to mild difficulties in learning and impaired motor skills such as a speech delay. These problems may be very subtle.

Turner Syndrome Individuals with Turner syndrome have an X chromosome and no corresponding X or Y chromosome (XO). The syndrome is thought to arise most frequently as an outcome of inheriting an unstable Y chromosome from the father. The zygote starts out being genetically male, with an X and a Y chromosome. Sometime during early development, the Y chromosome breaks up and is lost, so the embryo continues to develop as a female.

There are fewer people affected by Turner syndrome than other chromosome abnormalities: Only about 1 in 2,500 newborn girls has it. XO individuals grow up well proportioned but short, with an average height of four feet, eight inches (1.4 meters). Their ovaries do not develop

FIGURE 14.10 {Animated} An example of nondisjunction during meiosis. Of the two pairs of homologous chromosomes shown here, one fails to separate during anaphase I. The chromosome number is altered in the resulting gametes.

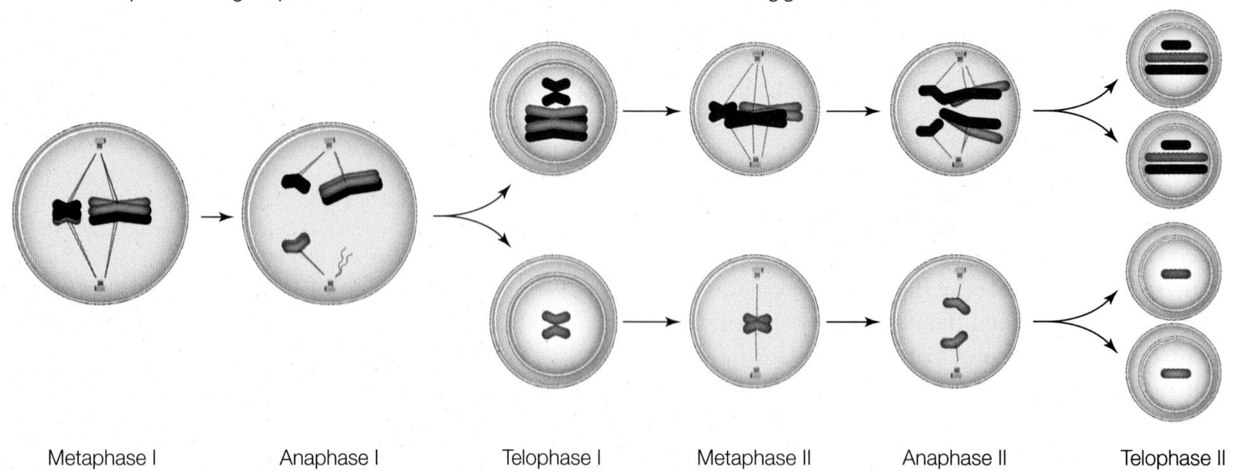

FIGURE 14.11 A Down syndrome phenotype. Excess tissue deposits on the colored part of the eye give rise to a ring of starlike white speckles, a lovely phenotypic effect of the chromosome number change that causes Down syndrome.

properly, so they do not make enough sex hormones to become sexually mature and do not develop secondary sexual traits such as enlarged breasts.

XXX Syndrome A female may inherit multiple X chromosomes, a condition called XXX syndrome. XXX syndrome occurs in about 1 of 1,000 births. As with Down syndrome, risk increases with maternal age. Only one X chromosome is typically active in female cells, so having extra X chromosomes usually does not cause physical or medical problems, but mild mental impairment may occur.

Klinefelter Syndrome About 1 out of every 500 males has an extra X chromosome (XXY). The resulting disorder, Klinefelter syndrome, develops at puberty. XXY males tend to be overweight, tall, and have mild mental impairment. They make more estrogen and less testosterone than normal males. This hormone imbalance causes affected men to have small testes and a small prostate gland, a low sperm count, sparse facial and body hair, a high-pitched voice, and enlarged breasts. Testosterone injections during puberty can minimize some of these traits.

XYY Syndrome About 1 in 1,000 males is born with an extra Y chromosome (XYY), a result of nondisjunction of the Y chromosome during sperm formation. Adults tend to be taller than average and have mild mental impairment, but most are otherwise normal. XYY men were once thought to be predisposed to a life of crime. This misguided view was based on sampling error (too few cases in narrowly chosen groups such as prison inmates) and bias (the researchers who gathered the karyotypes also took the personal histories of the participants). That view has since been disproven: Men with XYY syndrome are only slightly more likely to be convicted for crimes than unaffected men. Researchers now believe this slight increase can be explained by poor socioeconomic conditions related to the effects of the syndrome.

aneuploid Having too many or too few copies of a particular chromosome.
nondisjunction Failure of sister chromatids or homologous chromosomes to separate during nuclear division.
polyploid Having three or more of each type of chromosome characteristic of the species.

TAKE-HOME MESSAGE 14.5

Polyploidy is fatal in humans, but not in flowering plants and some other organisms.

Aneuploidy can arise from nondisjunction during meiosis. In humans, most cases of aneuploidy are associated with some degree of mental impairment.

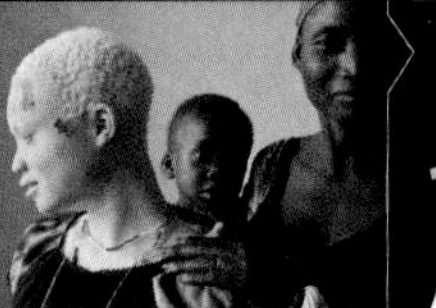

14.6 HOW DO WE USE WHAT WE KNOW ABOUT HUMAN INHERITANCE?

Studying human inheritance patterns has given us many insights into how genetic disorders arise and progress, and how to treat them. Some disorders can be detected early enough to start countermeasures before symptoms develop. For this reason, most hospitals in the United States now screen newborns for mutations that cause phenylketonuria, or PKU. The mutations affect an enzyme that converts the amino acid phenylalanine to tyrosine. Without this enzyme, the body becomes deficient in tyrosine, and phenylalanine accumulates to high levels. The imbalance inhibits protein synthesis in the brain, which in turn results in severe neurological symptoms. Restricting all intake of phenylalanine can slow the progression of PKU, so routine early screening has resulted in fewer individuals suffering from the symptoms of the disorder.

The probability that a child will inherit a genetic disorder can often be estimated by testing prospective parents for alleles known to be associated with genetic disorders. Karyotypes and pedigrees are also useful in this type of screening, which can help the parents make decisions about family planning. Genetic screening can also be done post-conception, in which case it is called prenatal diagnosis (prenatal means before birth). Prenatal diagnosis checks for physical and genetic abnormalities in an embryo or fetus. More than 30 conditions are detectable prenatally, including aneuploidy, hemophilia, Tay–Sachs disease, sickle-cell anemia, muscular dystrophy, and cystic fibrosis. If the disorder is treatable, early detection allows the newborn to receive prompt and appropriate treatment. A few defects are even surgically correctable before birth. Prenatal diagnosis also gives parents time to prepare for the birth of an affected child, and an opportunity to decide whether to continue with the pregnancy or terminate it.

As an example of how prenatal diagnosis works, consider a woman who becomes pregnant at age thirty-five. Her doctor will probably perform a procedure called obstetric sonography, in which ultrasound waves directed across the woman's abdomen form images of the fetus's limbs and internal organs. If the images reveal a physical defect that may be the result of a genetic disorder, a more invasive technique would be recommended for further diagnosis. With fetoscopy, sound waves pulsed from inside the mother's uterus yield images much higher in resolution than ultrasound. Samples of tissue or blood are often taken at the same time, and some corrective surgeries can be performed.

Human genetics studies show that our thirty-five-year-old woman has about a 1 in 80 chance that her baby will be born with a chromosomal abnormality, a risk more than six times greater than when she was twenty years old. Thus, even if no abnormalities are detected by ultrasound, she probably will be offered a more thorough diagnostic procedure, amniocentesis, in which a small sample of fluid is drawn from the amniotic sac enclosing the fetus (**FIGURE 14.12**). The fluid contains cells shed by the fetus, and those cells can be tested for genetic disorders. Chorionic villus sampling (CVS) can be performed earlier than amniocentesis. With this technique, a few cells from the chorion are removed and tested for genetic disorders. (The chorion is a membrane that surrounds the amniotic sac and helps form the placenta, an organ that allows substances to be exchanged between mother and embryo.)

An invasive procedure often carries a risk to the fetus. The risks vary by the procedure. Amniocentesis has improved so much that, in the hands of a skilled physician, the procedure no longer increases the risk of miscarriage. CVS occasionally disrupts the placenta's development and thus causes underdeveloped or missing fingers and toes in 0.3 percent of newborns. Fetoscopy raises the miscarriage risk by a whopping 2 to 10 percent.

FIGURE 14.12 **{Animated}** An 8-week-old fetus. With amniocentesis, fetal cells shed into the fluid inside the amniotic sac are tested for genetic disorders. Chorionic villus sampling tests cells of the chorion, which is part of the placenta.

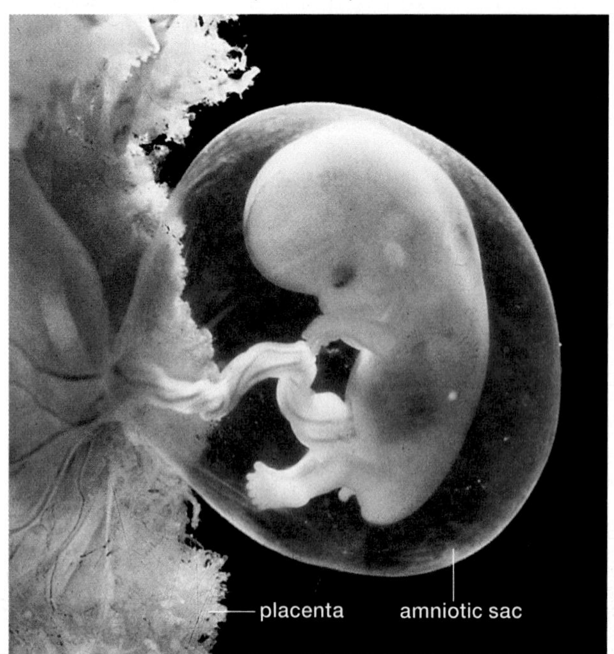

TAKE-HOME MESSAGE 14.6

Studying inheritance patterns for genetic disorders has helped researchers develop treatments for some of them.

Genetic testing can provide prospective parents with information about the health of their future children.

14.7 Application: SHADES OF SKIN

FIGURE 14.13 Fraternal twins Kian and Remee. Both of the children's grandmothers are of European descent, and have pale skin. Both of their grandfathers are of African descent, and have dark skin. The twins inherited different alleles of some genes that affect skin color from their parents, who, given the appearance of their children, must be heterozygous for those alleles.

SKIN COLOR, LIKE MOST OTHER HUMAN TRAITS, HAS A GENETIC BASIS. The color of human skin begins with melanosomes, which are organelles that make melanin pigments. Most people have about the same number of melanosomes in their skin cells. Variations in skin color arise from differences in formation and deposition of melanosomes in the skin, as well as in the kinds and amounts of melanins they make. More than 100 genes are involved in these processes.

Human skin color variation may have evolved as a balance between vitamin production and protection against harmful UV radiation in the sun's rays. Dark skin would have been beneficial under the intense sunlight of African savannas where humans first evolved. Melanin is a natural sunscreen: It prevents UV radiation from breaking down folate, a vitamin essential for normal sperm formation and embryonic development.

Early human groups that migrated to regions with cold climates were exposed to less sunlight. In these regions, lighter skin color is beneficial. Why? UV radiation stimulates skin cells to make a molecule the body converts to vitamin D. Where sunlight exposure is minimal, UV radiation is less of a risk than vitamin D deficiency, which has serious health consequences for developing fetuses and children.

The evolution of regional variations in human skin color began with mutations. Consider a gene on chromosome 15 that encodes a transport protein in melanosome membranes. Nearly all people of African, Native American, or east Asian descent carry the same allele of this gene. Between 6,000 and 10,000 years ago, a mutation gave rise to a different allele. The mutation, a single base-pair substitution, changed the 111th amino acid of the transport protein from alanine to threonine. The change results in less melanin—and lighter skin color—than the original African allele does. Today, nearly all people of European descent carry this mutated allele.

A person of mixed ethnicity may make gametes that contain different combinations of alleles for dark and light skin. It is fairly rare that one of those gametes contains mainly alleles for dark skin, or mainly alleles for light skin, but it happens (FIGURE 14.13). Skin color is only one of many human traits that vary as a result of single nucleotide mutations. The small scale of such changes offers a reminder that all of us share the genetic legacy of common ancestry.

Summary

SECTION 14.1 Geneticists study inheritance patterns in humans by tracking genetic disorders and abnormalities through families. A genetic abnormality is an uncommon version of a heritable trait that does not result in medical problems. A genetic disorder is a heritable condition that sooner or later results in mild or severe medical problems. Geneticists make **pedigrees** to reveal inheritance patterns for alleles that can be predictably associated with specific phenotypes.

SECTION 14.2 An allele is inherited in an autosomal dominant pattern if the trait it specifies appears in everyone who carries it, and both sexes are affected with equal frequency. Such traits appear in every generation of families that carry the allele. An allele is inherited in an autosomal recessive pattern if the trait it specifies appears only in homozygous people. Such traits also appear in both sexes equally, but they can skip generations.

SECTION 14.3 An allele is inherited in an X-linked pattern when it occurs on the X chromosome. Most X-linked disorders are inherited in a recessive pattern, and these tend to appear in men more often than in women. Heterozygous women have a dominant, normal allele that can mask the effects of the recessive one; men do not. Men can transmit an X-linked allele to their daughters, but not to their sons. Only a woman can pass an X-linked allele to a son.

SECTION 14.4 Faulty crossovers and the activity of **transposable elements** can give rise to major changes in chromosome structure, including **duplications**, **inversions**, and **translocations**. Some of these changes are harmful or lethal in humans; others affect fertility. Even so, major structural changes have accumulated in the chromosomes of all species over evolutionary time.

SECTION 14.5 Occasionally, abnormal events occur before or during meiosis, and new individuals end up with the wrong chromosome number. Consequences range from minor to lethal changes in form and function.

Chromosome number change is usually an outcome of **nondisjunction**, in which chromosomes fail to separate properly during meiosis. **Polyploid** individuals have three or more of each type of chromosome. Polyploidy is lethal in humans, but not in flowering plants, and some insects, fishes, and other animals.

Aneuploid individuals have too many or too few copies of a chromosome. In humans, most cases of autosomal aneuploidy are lethal. Trisomy 21, which causes Down syndrome, is an exception. A change in the number of sex chromosomes usually results in some degree of impairment in learning and motor skills.

SECTION 14.6 Prospective parents can estimate their risk of transmitting a harmful allele to offspring with genetic screening, in which their pedigrees and genotype are analyzed by a genetic counselor. Prenatal genetic testing can reveal a genetic disorder before birth.

SECTION 14.7 Like most other human traits, skin color has a genetic basis. Minor differences in the alleles that govern melanin production and the deposition of melanosomes affect skin color. The differences probably evolved as a balance between vitamin production and protection against harmful UV radiation.

Self-Quiz

Answers in Appendix VII

1. Constructing a family pedigree is particularly useful when studying inheritance patterns in organisms that ______ .
   a. produce many offspring per generation
   b. produce few offspring per generation
   c. have a very large chromosome number
   d. reproduce asexually
   e. have a fast life cycle

2. Pedigree analysis is necessary when studying human inheritance patterns because ______ .
   a. humans have more than 20,000 genes
   b. of ethical problems with human experimentation
   c. inheritance in humans is more complicated than in other organisms
   d. genetic disorders occur in humans
   e. all of the above

3. A recognized set of symptoms that characterize a genetic disorder is a(n) ______ .
   a. syndrome b. disease c. abnormality

4. If one parent is heterozygous for a dominant allele on an autosome and the other parent does not carry the allele, any child of theirs has a ______ chance of having the associated trait.
   a. 25 percent c. 75 percent
   b. 50 percent d. no chance; it will die

5. Is this statement true or false? A son can inherit an X-linked recessive allele from his father.

6. A trait that is present in a male child but not in either of his parents is characteristic of ______ inheritance.
   a. autosomal dominant
   b. autosomal recessive
   c. X-linked recessive
   d. It is not possible to answer this question without more information.

7. Color blindness is inherited in an ______ pattern.
   a. autosomal dominant c. X-linked dominant
   b. autosomal recessive d. X-linked recessive

8. A female child inherits one X chromosome from her mother and one from her father. What sex chromosome does a male child inherit from each of his parents?

Data Analysis Activities

Skin Color Survey of Native Peoples In 2000, researchers measured the average amount of UV radiation received in more than fifty regions of the world, and correlated it with the average skin reflectance of people native to those regions (reflectance is a way to measure the amount of melanin pigment in skin). Some of the results of this study are shown in **FIGURE 14.14**.

1. Which country receives the most UV radiation?
2. Which country receives the least UV radiation?
3. People native to which country have the darkest skin?
4. People native to which country have the lightest skin?
5. According to these data, how does the skin color of indigenous peoples correlate with the amount of UV radiation incident in their native regions?

Country	Skin Reflectance	UVMED
Australia	19.30	335.55
Kenya	32.40	354.21
India	44.60	219.65
Cambodia	54.00	310.28
Japan	55.42	130.87
Afghanistan	55.70	249.98
China	59.17	204.57
Ireland	65.00	52.92
Germany	66.90	69.29
Netherlands	67.37	62.58

FIGURE 14.14 Skin color of indigenous peoples and regional incident UV radiation. Skin reflectance measures how much light of 685-nanometer wavelength is reflected from skin; UVMED is the annual average UV radiation received at Earth's surface.

9. Alleles for Tay–Sachs disease are inherited in an autosomal recessive pattern. Why would two parents with a normal phenotype have a child with Tay–Sachs?
   a. Both parents are homozygous for a Tay–Sachs allele.
   b. Both parents are heterozygous for a Tay–Sachs allele.
   c. A new mutation gave rise to Tay–Sachs in the child.
   d. b or c

10. The *SRY* gene gives rise to the male phenotype in humans (Sections 10.3 and 14.4). What do you think the inheritance pattern of *SRY* alleles is called?

11. Nondisjunction may occur during _______ .
   a. mitosis c. fertilization
   b. meiosis d. both a and b

12. Nondisjunction can result in _______ .
   a. duplications c. crossing over
   b. aneuploidy d. pleiotropy

13. Is this statement true or false? Inheriting three or more of each type of chromosome characteristic of the species results in a condition called polyploidy.

14. Klinefelter syndrome (XXY) can be easily diagnosed by _______ .
   a. pedigree analysis c. karyotyping
   b. aneuploidy d. phenotypic treatment

15. Match the chromosome terms appropriately.

___ polyploidy	a. symptoms of a genetic disorder
___ deletion	b. segment of a chromosome moves to a nonhomologous chromosome
___ aneuploidy	c. extra sets of chromosomes
___ translocation	d. gets around
___ syndrome	e. a chromosome segment lost
___ transposable element	f. one extra chromosome

Genetics Problems

Answers in Appendix VII

1. Does the phenotype indicated by the red circles and squares in this pedigree show an inheritance pattern that is autosomal dominant, autosomal recessive, or X-linked?

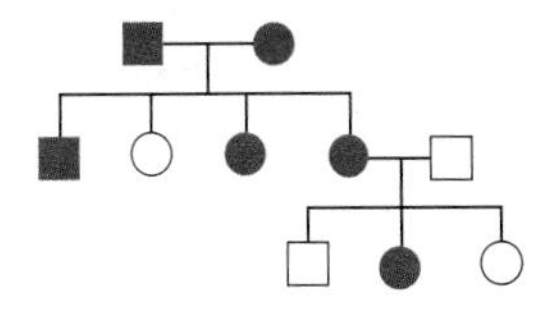

2. Human females have two X chromosomes (XX); males have one X and one Y chromosome (XY).
   a. Does a male inherit an X chromosome from his mother or his father?
   b. With respect to X-linked alleles, how many different types of gametes can a male produce?
   c. A female homozygous for an X-linked allele can produce how many types of gametes with respect to that allele?
   d. A female heterozygous for an X-linked allele can produce how many types of gametes with respect to that allele?

3. Somatic cells of individuals with Down syndrome usually have an extra chromosome 21; they contain forty-seven chromosomes. A few individuals with Down syndrome have forty-six chromosomes: two normal-appearing chromosomes 21, and a longer-than-normal chromosome 14. Speculate on how this chromosome abnormality arises.

4. An allele responsible for Marfan syndrome (Section 13.4) is inherited in an autosomal dominant pattern. What is the chance that a child will inherit the allele if one parent does not carry it and the other is heterozygous?

CREDITS: (14) © Cengage Learning, based on *Journal of Human Evolution* (2000)39, 57–106 doi: 10.1006/jhev.2000.0403 © 2000 Academic Press; (in text) © Cengage Learning.

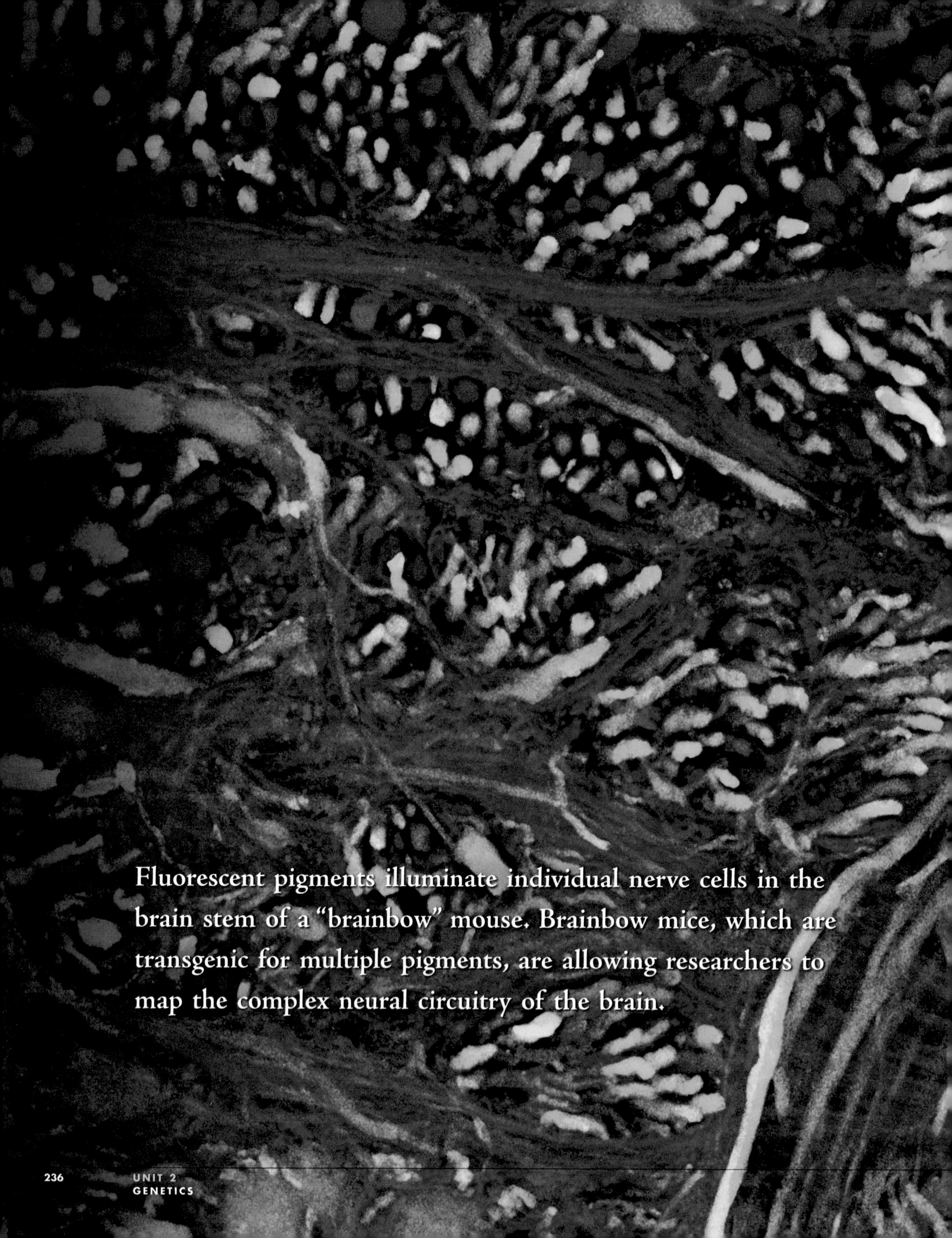

Fluorescent pigments illuminate individual nerve cells in the brain stem of a "brainbow" mouse. Brainbow mice, which are transgenic for multiple pigments, are allowing researchers to map the complex neural circuitry of the brain.

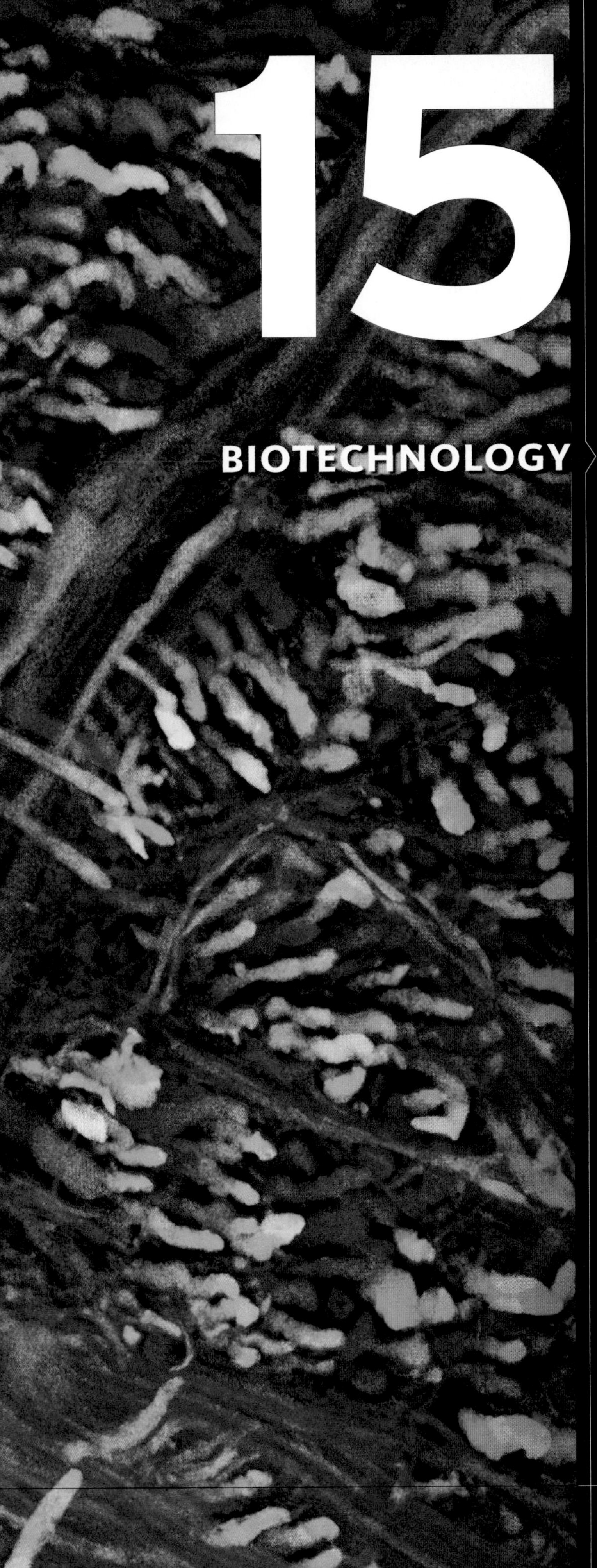

15 BIOTECHNOLOGY

Links to Earlier Concepts

This chapter builds on your understanding of DNA (Sections 8.2, 8.3, 13.1, 14.6) and DNA replication (8.4). Clones (8.6), gene expression (9.1, 9.2), and knockouts (10.2) are important in genetic engineering, particularly in research on human traits (13.6) and genetic disorders (Chapter 14). You will revisit tracers (2.1), triglycerides (3.3), denaturation (3.5), bacteria (4.4), β-carotene (6.1), mutations (9.5), the *lac* operon (10.4), cancer (11.5), and alleles (12.1).

KEY CONCEPTS

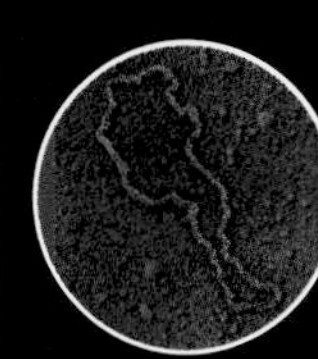

DNA CLONING

Researchers make recombinant DNA by cutting and pasting together DNA from different species. Plasmids and other vectors can carry foreign DNA into host cells.

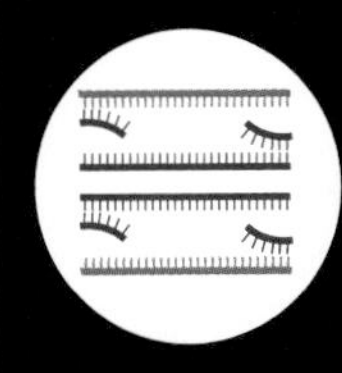

FINDING NEEDLES IN HAYSTACKS

Genetic engineering, the directed modification of an organism's genes, relies on laboratory techniques for isolating and identifying particular fragments of DNA.

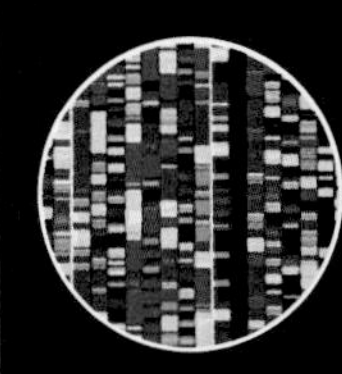

DNA SEQUENCING

Sequencing reveals the linear order of nucleotides in DNA. Comparing genomes offers insights into human genes and evolution. DNA sequence can be used to identify individuals.

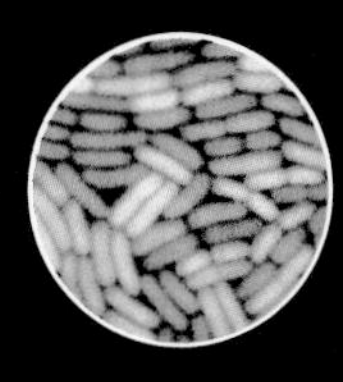

GENETIC ENGINEERING

Genetic engineering is now a routine part of research and industrial applications. Genetically modified organisms are used to produce food, medicines, and other products.

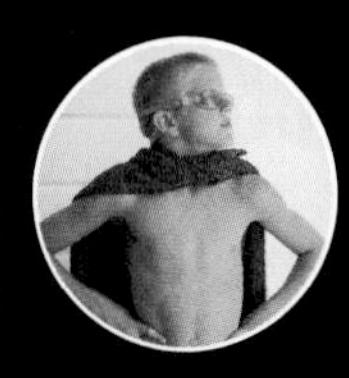

GENE THERAPY

The directed modification of human DNA continues to be tested in medical applications. It also continues to raise ethical questions about modifying the human genome.

Photograph courtesy of © Dr. Jean Levit. The Brainbow technique was developed in the laboratories of Jeff W. Lichtman and Joshua R. Sanes at Harvard University. This image has received the Bioscape imaging competition 2007 prize.

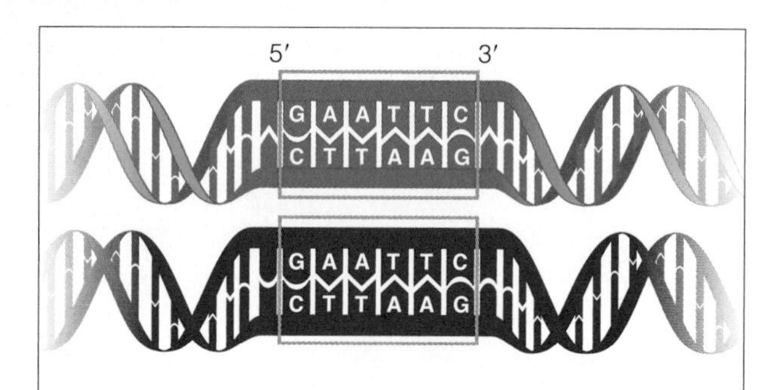

❶ The restriction enzyme *Eco* RI (named after *E. coli*, the bacteria from which it was isolated) recognizes a specific base sequence (GAATTC) in DNA from two different sources.

❷ The enzyme cuts the DNA into fragments. *Eco* RI leaves single-stranded tails ("sticky ends") where it cuts DNA.

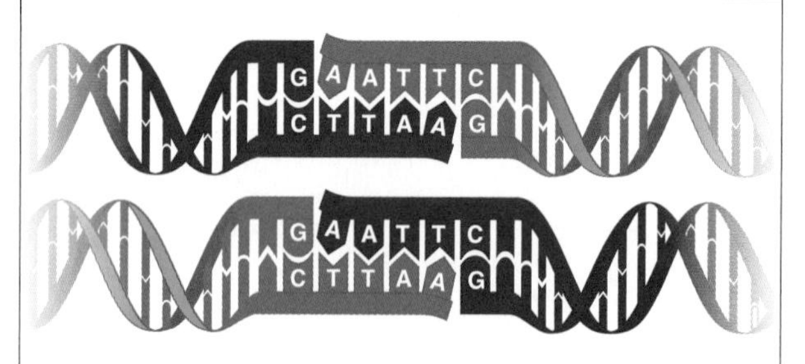

❸ When the DNA fragments from the two sources are mixed together, matching sticky ends base-pair with each other.

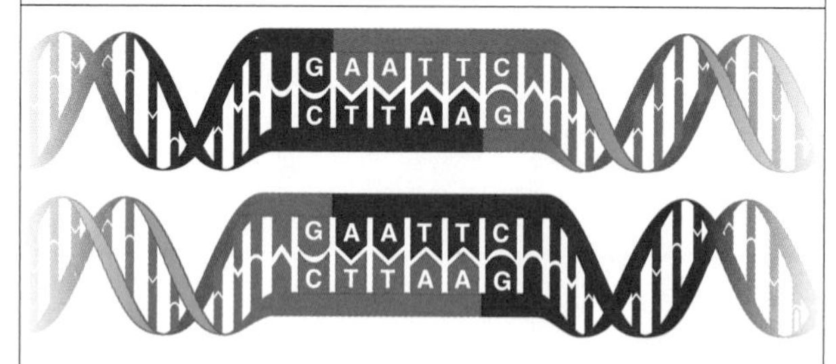

❹ DNA ligase joins the base-paired DNA fragments. Molecules of recombinant DNA are the result.

FIGURE 15.1 {Animated} Making recombinant DNA.

FIGURE IT OUT: Why did the enzyme cut both strands of DNA?

Answer: Because the recognition sequence occurs on both strands.

CUT AND PASTE

In the 1950s, excitement over the discovery of DNA's structure (Section 8.2) gave way to frustration: No one could determine the order of nucleotides in a molecule of DNA. Identifying a single base among thousands or millions of others turned out to be a huge technical challenge.

Research in a seemingly unrelated field yielded a solution when Werner Arber, Hamilton Smith, and their coworkers discovered how some bacteria resist infection by bacteriophage (Section 8.1). These bacteria have enzymes that chop up any injected viral DNA before it has a chance to integrate into the bacterial chromosome. The enzymes restrict viral growth; hence their name, restriction enzymes. A **restriction enzyme** cuts DNA wherever a specific nucleotide sequence occurs (**FIGURE 15.1** ❶). The discovery of restriction enzymes allowed researchers to cut chromosomal DNA into manageable chunks. It also allowed them to combine DNA fragments from different organisms. How? Many restriction enzymes leave single-stranded tails on DNA fragments ❷. Researchers realized that complementary tails will base-pair, regardless of the source of DNA ❸. The tails are called "sticky ends," because two DNA fragments stick together when their matching tails base-pair. The enzyme DNA ligase (Section 8.4) can be used to seal the gaps between base-paired sticky ends, so continuous DNA strands form ❹. Thus, using appropriate restriction enzymes and DNA ligase, researchers can cut and paste DNA from different sources. The result, a hybrid molecule that consists of genetic material from two or more organisms, is called **recombinant DNA**.

Making recombinant DNA is the first step in **DNA cloning**, a set of laboratory methods that uses living cells to

FIGURE 15.2 Plasmid cloning vectors. **A** Micrograph of a plasmid. **B** A commercial plasmid cloning vector. Restriction enzyme recognition sequences are indicated on the right by the name of the enzyme that cuts them. Researchers insert foreign DNA into the vector at these sequences. Bacterial genes (gold) help researchers identify host cells that take up a vector with inserted DNA. This vector carries two antibiotic resistance genes and the *lac* operon (Section 10.4).

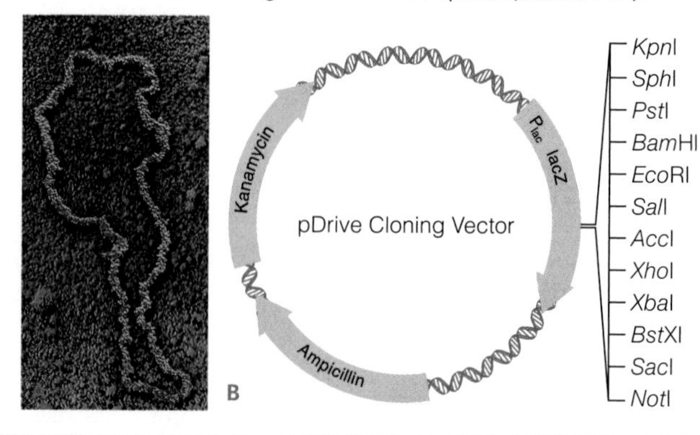

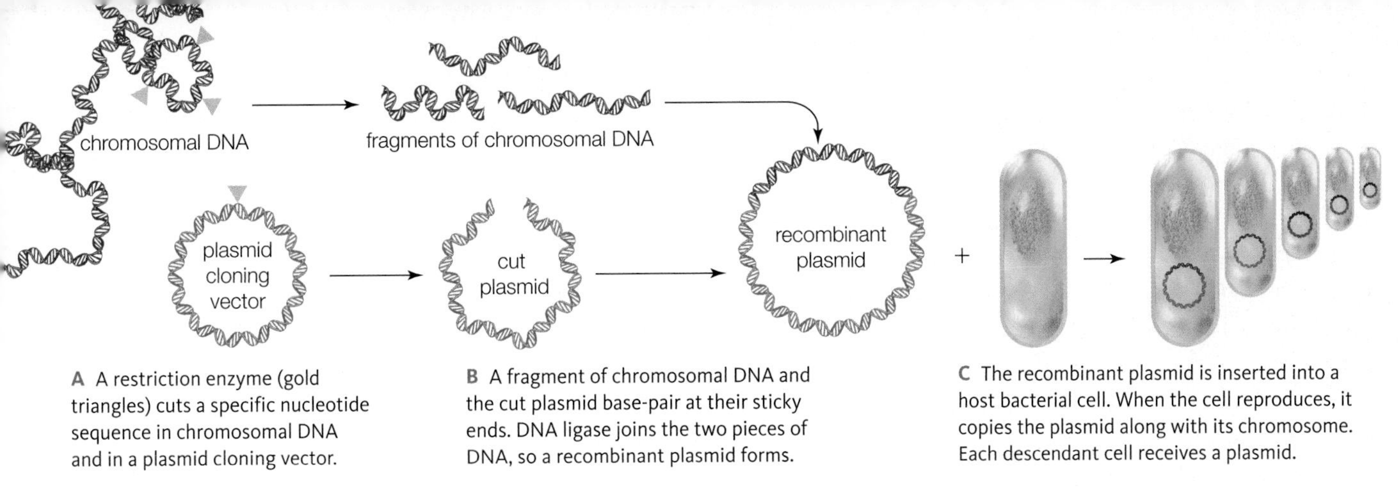

A A restriction enzyme (gold triangles) cuts a specific nucleotide sequence in chromosomal DNA and in a plasmid cloning vector.

B A fragment of chromosomal DNA and the cut plasmid base-pair at their sticky ends. DNA ligase joins the two pieces of DNA, so a recombinant plasmid forms.

C The recombinant plasmid is inserted into a host bacterial cell. When the cell reproduces, it copies the plasmid along with its chromosome. Each descendant cell receives a plasmid.

FIGURE 15.3 {Animated} An example of cloning. Here, a fragment of chromosomal DNA is inserted into a bacterial plasmid.

mass-produce specific DNA fragments. Researchers clone a fragment of DNA by inserting it into a **cloning vector**, which is a molecule that can carry foreign DNA into host cells. Bacterial plasmids (Section 4.4) may be used as cloning vectors (**FIGURE 15.2**). A bacterium copies all of its DNA before it divides, so its offspring inherit plasmids along with chromosomes. If a plasmid carries a fragment of foreign DNA, that fragment gets copied and distributed to descendant cells along with the plasmid (**FIGURE 15.3**).

A host cell into which a cloning vector has been inserted can be grown in the laboratory (cultured) to yield a huge population of genetically identical cells, or clones (Section 8.6). Each clone contains a copy of the vector and the inserted DNA fragment. The hosted DNA fragment can be harvested in large quantities from the clones.

cDNA CLONING

Remember from Section 9.2 that eukaryotic DNA contains introns. Unless you are a eukaryotic cell, it is not very easy to determine which parts of eukaryotic DNA encode gene products. Thus, researchers who study gene expression in eukaryotes often start with mature mRNA, because introns are removed during post-transcriptional processing.

An mRNA cannot be cut with restriction enzymes or pasted with DNA ligase, because these enzymes work only on double-stranded DNA. Thus, cloning with mRNA requires **reverse transcriptase**, a replication enzyme that uses an RNA template to assemble a strand of complementary DNA, or **cDNA**:

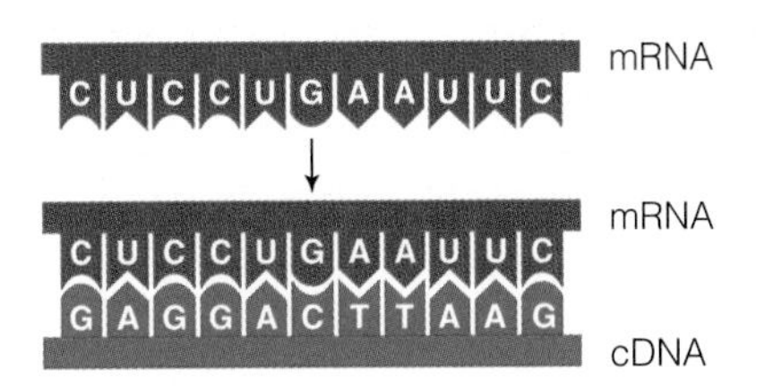

DNA polymerase is used to copy the cDNA into a second strand of DNA. The outcome is a double-stranded DNA version of the original mRNA:

Like any other double-stranded DNA, this fragment may be cut with restriction enzymes and pasted into a cloning vector using DNA ligase.

cDNA Complementary strand of DNA synthesized from an RNA template by the enzyme reverse transcriptase.
cloning vector A DNA molecule that can accept foreign DNA and be replicated inside a host cell.
DNA cloning Set of methods that uses living cells to make many identical copies of a DNA fragment.
recombinant DNA A DNA molecule that contains genetic material from more than one organism.
restriction enzyme Type of enzyme that cuts DNA at a specific nucleotide sequence.
reverse transcriptase An enzyme that uses mRNA as a template to make a strand of cDNA.

TAKE-HOME MESSAGE 15.1

DNA cloning uses living cells to mass-produce particular DNA fragments. Restriction enzymes cut DNA into fragments, then DNA ligase seals the fragments into cloning vectors. Recombinant DNA molecules result.

A cloning vector that holds foreign DNA can be introduced into a living cell. When the host cell divides, it gives rise to huge populations of genetically identical cells (clones), each with a copy of the foreign DNA.

A Individual bacterial cells from a DNA library are spread over the surface of a solid growth medium. The cells divide repeatedly and form colonies—clusters of millions of genetically identical descendant cells.

B Special paper is pressed onto the surface of the growth medium. Some cells from each colony stick to the paper.

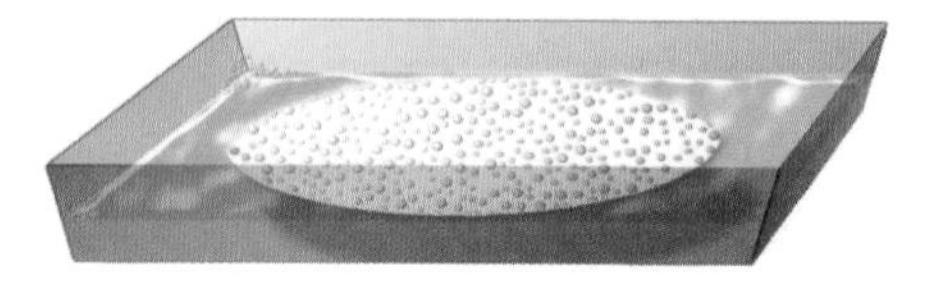

C The paper is soaked in a solution that ruptures the cells and makes the released DNA single-stranded. The DNA clings to the paper in spots mirroring the distribution of colonies.

D A radioactive probe is added to the liquid bathing the paper. The probe hybridizes with any spot of DNA that contains a complementary sequence.

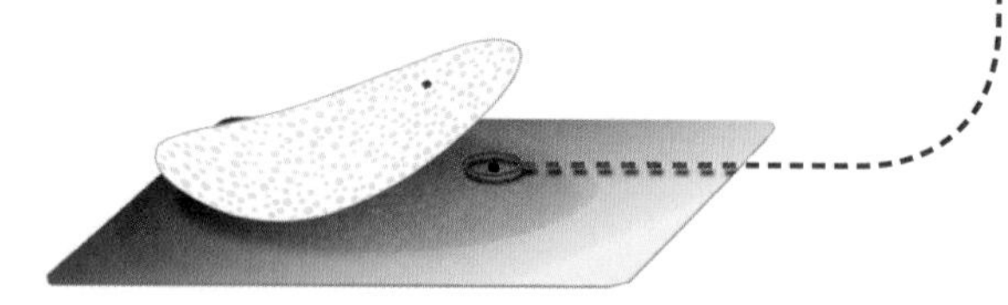

E The paper is pressed against x-ray film. The radioactive probe darkens the film in a spot where it has hybridized. The spot's position is compared to the positions of the original bacterial colonies. Cells from the colony that corresponds to the spot are cultured, and their DNA is harvested.

FIGURE 15.4 {Animated} Nucleic acid hybridization. In this example, a radioactive probe helps identify a bacterial colony that contains a targeted sequence of DNA.

DNA LIBRARIES

The entire set of genetic material—the **genome**—of most organisms consists of thousands of genes. To study or manipulate a single gene, researchers must first separate it from all of the other genes in a genome. They often begin by cutting an organism's DNA into pieces, and then cloning all the pieces. The result is a genomic library, a set of clones that collectively contain all of the DNA in a genome. Researchers may also harvest mRNA, make cDNA copies of it, and then clone the cDNA. The resulting cDNA library represents only those genes being expressed at the time the mRNA was harvested.

Genomic and cDNA libraries are **DNA libraries**, sets of cells that host various cloned DNA fragments. In such libraries, a cell that contains a particular DNA fragment of interest is mixed up with thousands or millions of others that do not—a needle in a genetic haystack. One way to find that clone among the others involves the use of a **probe**, which is a fragment of DNA or RNA labeled with a tracer (Section 2.1). For example, to find a particular gene, researchers may use radioactive nucleotides to synthesize a short strand of DNA complementary in sequence to a similar gene. Because the nucleotide sequences of the probe and the gene are complementary, the two can hybridize. (Remember from Section 8.4 that nucleic acid hybridization is the establishment of base pairing between nucleic acid strands.) When the probe is mixed with DNA from a library, it will hybridize with the gene, but not with other DNA (**FIGURE 15.4**). Researchers can pinpoint a clone that hosts the gene by detecting the label on the probe. That clone is isolated and cultured, and DNA can be extracted in bulk from the cultured cells for research or other purposes.

PCR

The **polymerase chain reaction** (**PCR**) is a technique used to mass-produce copies of a particular section of DNA without having to clone it in living cells (**FIGURE 15.5**). The reaction can transform a needle in a haystack—that one-in-a-million fragment of DNA—into a huge stack of needles with a little hay in it.

The starting material for PCR is any sample of DNA with at least one molecule of a targeted sequence. It might be DNA from a mixture of 10 million different clones, a sperm, a hair left at a crime scene, or a mummy—essentially any sample that has DNA in it.

The PCR reaction is similar to DNA replication (Section 8.4). It requires two primers. Each base-pairs with one end of the section of DNA to be amplified,

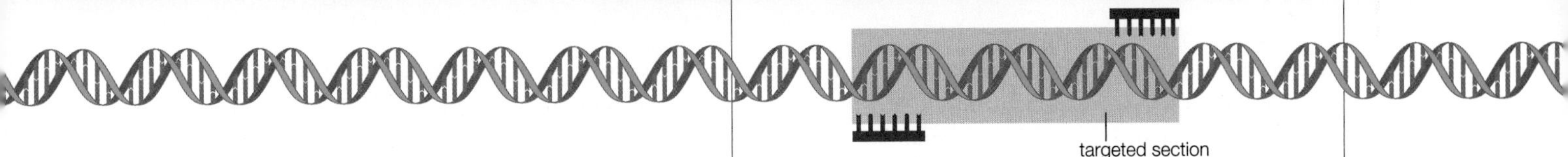

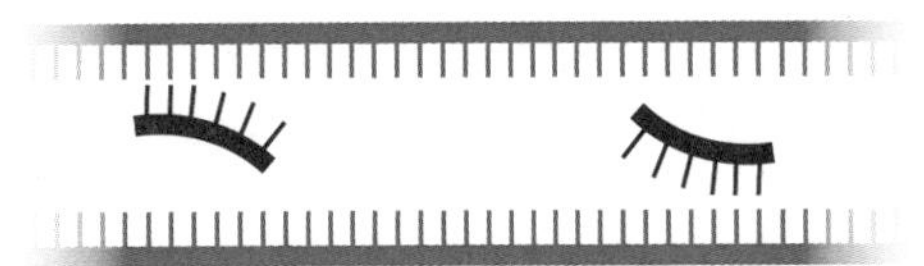

❶ DNA template (blue) is mixed with primers (pink), nucleotides, and heat-tolerant *Taq* DNA polymerase.

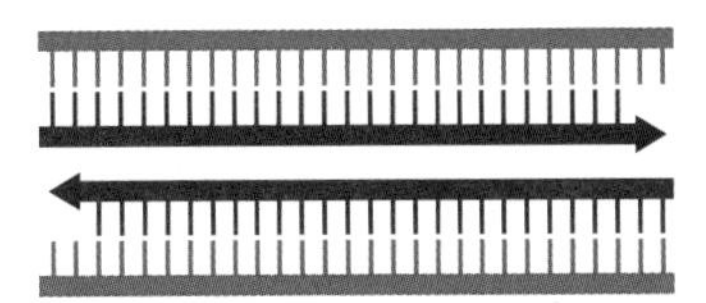

❷ When the mixture is heated, the double-stranded DNA separates into single strands. When the mixture is cooled, some of the primers base-pair with the DNA at opposite ends of the targeted sequence.

❸ *Taq* polymerase begins DNA synthesis at the primers, so it produces complementary strands of the targeted DNA sequence.

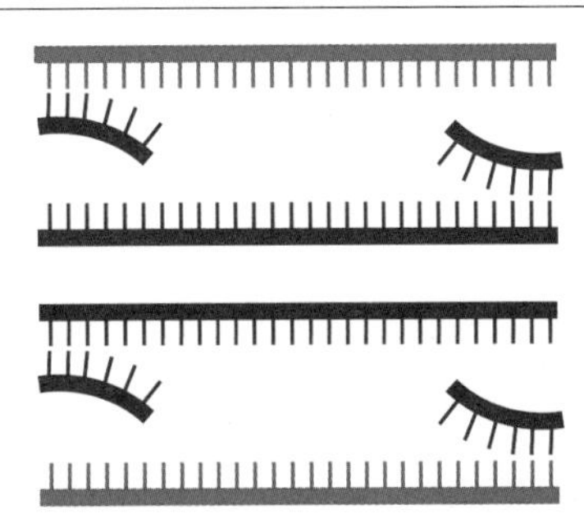

❹ The mixture is heated again, so all double-stranded DNA separates into single strands. When it is cooled, primers base-pair with the targeted sequence in the original template DNA and in the new DNA strands.

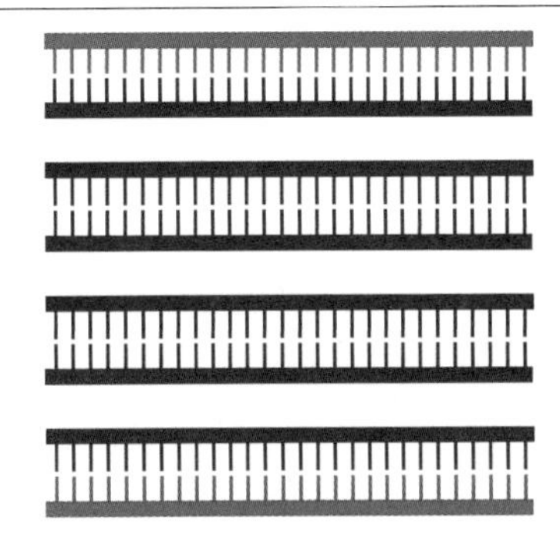

❺ Each cycle of heating and cooling can double the number of copies of the targeted DNA section.

FIGURE 15.5 {Animated} Two rounds of PCR. Each cycle of this reaction can double the number of copies of a targeted section of DNA. Thirty cycles can make a billion copies.

or mass-produced ❶. Researchers mix these primers with the starting (template) DNA, nucleotides, and DNA polymerase, then expose the reaction mixture to repeated cycles of high and low temperatures. A few seconds at high temperature disrupts the hydrogen bonds that hold the two strands of a DNA double helix together (Section 8.2), so every molecule of DNA unwinds and becomes single-stranded. As the temperature of the reaction mixture is lowered, the single DNA strands hybridize with the primers ❷.

The DNA polymerases of most organisms denature at the high temperature required to separate DNA strands. The kind that is used in PCR reactions, *Taq* polymerase, is from *Thermus aquaticus*. This bacterial species lives in hot springs and hydrothermal vents, so its DNA polymerase necessarily tolerates heat. *Taq* polymerase, like other DNA polymerases, recognizes hybridized primers as places to start DNA synthesis ❸. Synthesis proceeds along the template strand until the temperature rises and the DNA separates into single strands ❹. The newly synthesized DNA is a copy of the targeted section. When the mixture is cooled, the primers rehybridize, and DNA synthesis begins again. Each cycle of heating and cooling takes only a few minutes, but it can double the number of copies of the targeted section of DNA ❺. Thirty PCR cycles may amplify that number a billionfold.

DNA library Collection of cells that host different fragments of foreign DNA, often representing an organism's entire genome.
genome An organism's complete set of genetic material.
polymerase chain reaction (PCR) Method that rapidly generates many copies of a specific section of DNA.
probe Short fragment of DNA labeled with a tracer; designed to hybridize with a nucleotide sequence of interest.

TAKE-HOME MESSAGE 15.2

Researchers isolate one gene from the many other genes in a genome by making a DNA library or with PCR.

Probes may be used to identify one clone that hosts a particular DNA fragment of interest among many other clones in a DNA library.

PCR quickly mass-produces copies of a particular section of DNA.

15.3 HOW IS DNA SEQUENCE DETERMINED?

Researchers use a technique called **sequencing** to determine the order of nucleotides in a fragment of DNA that has been isolated by cloning or PCR. The most common method is similar to DNA replication (Section 8.4). The DNA to be sequenced (the template) is mixed with nucleotides, a primer, and DNA polymerase. Starting at the primer, the polymerase joins the nucleotides into a new strand of DNA, in the order dictated by the sequence of the template (**FIGURE 15.6**).

Remember that DNA polymerase can add a nucleotide only to the hydroxyl group on the 3′ carbon of a DNA strand. The sequencing reaction mixture includes four kinds of modified nucleotides that lack the hydroxyl group on their 3′ carbon ❶. Each kind (A, C, G, or T) is labeled with a different colored pigment. During the reaction, the polymerase randomly adds either a regular nucleotide or a modified nucleotide to the end of a growing DNA strand. If it adds a modified nucleotide, the 3′ carbon of the strand will not have a hydroxyl group, so synthesis of the strand ends there ❷. The reaction produces millions of DNA fragments of different lengths—incomplete, complementary copies of the starting DNA ❸. Each fragment of a given length ends with the same modified nucleotide. For example, if the tenth base in the template DNA was thymine, then any newly synthesized fragment that is 10 bases long ends with a modified adenine.

The DNA fragments are then separated by length. Using a technique called **electrophoresis**, an electric field pulls the fragments through a semisolid gel. Fragments of different sizes move through the gel at different rates. The shorter the fragment, the faster it moves, because shorter fragments slip through the tangled molecules of the gel faster than longer fragments do. All fragments of the same length move through the gel at the same speed, so they gather into bands. All fragments in a given band have the same modified nucleotide at their ends, and the pigment labels now impart distinct colors to the bands ❹. Each color designates one of the four modified nucleotides, so the order of colored bands in the gel represents the DNA sequence ❺.

FIGURE 15.6 {Animated} DNA sequencing, in which DNA polymerase is used to incompletely replicate a section of DNA.

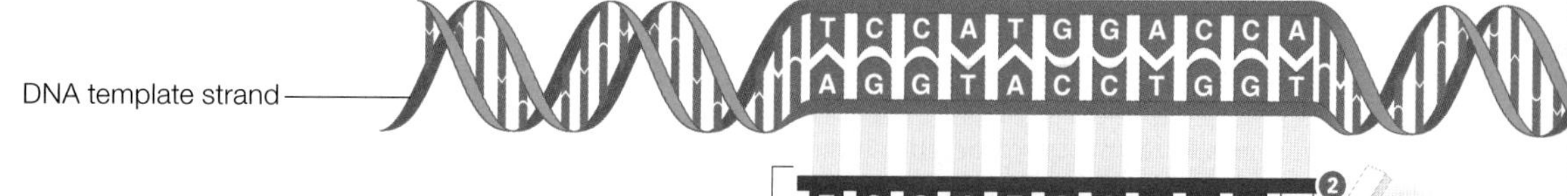

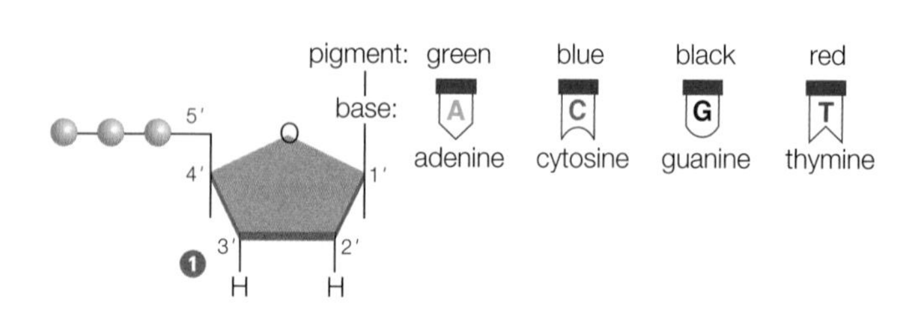

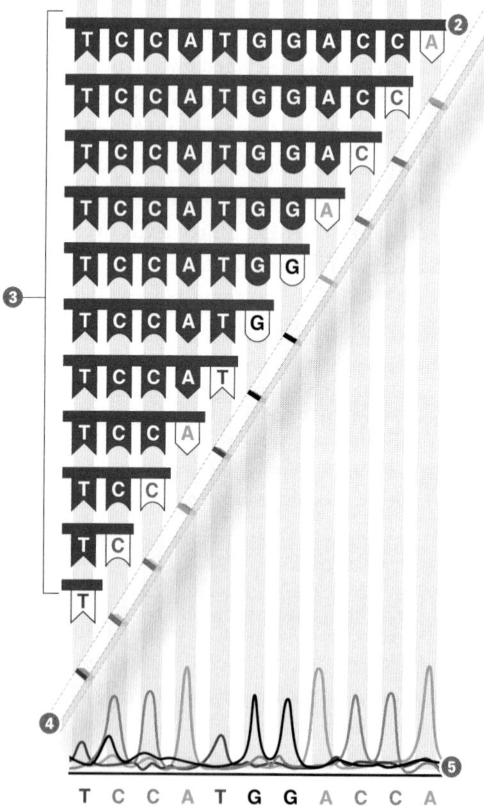

❶ Sequencing depends on modified nucleotides that have a hydrogen atom instead of a hydroxyl group on the 3′ carbon (compare the structure with those in **FIGURE 8.4**). Each is labeled with a colored pigment.

❷ DNA polymerase uses a section of DNA as a template to synthesize new strands of DNA. Synthesis of each new strand stops when a modified nucleotide is added.

❸ At the end of the reaction, there are many incomplete copies of the template DNA in the mixture.

❹ Electrophoresis separates the copied DNA fragments into bands according to their length. All of the DNA strands in each band end with the same nucleotide; thus, each band is the color of that nucleotide's tracer pigment.

❺ A computer detects and records the color of successive bands on the gel (see **FIGURE 15.7** for an example). The order of colors of the bands represents the sequence of the template DNA.

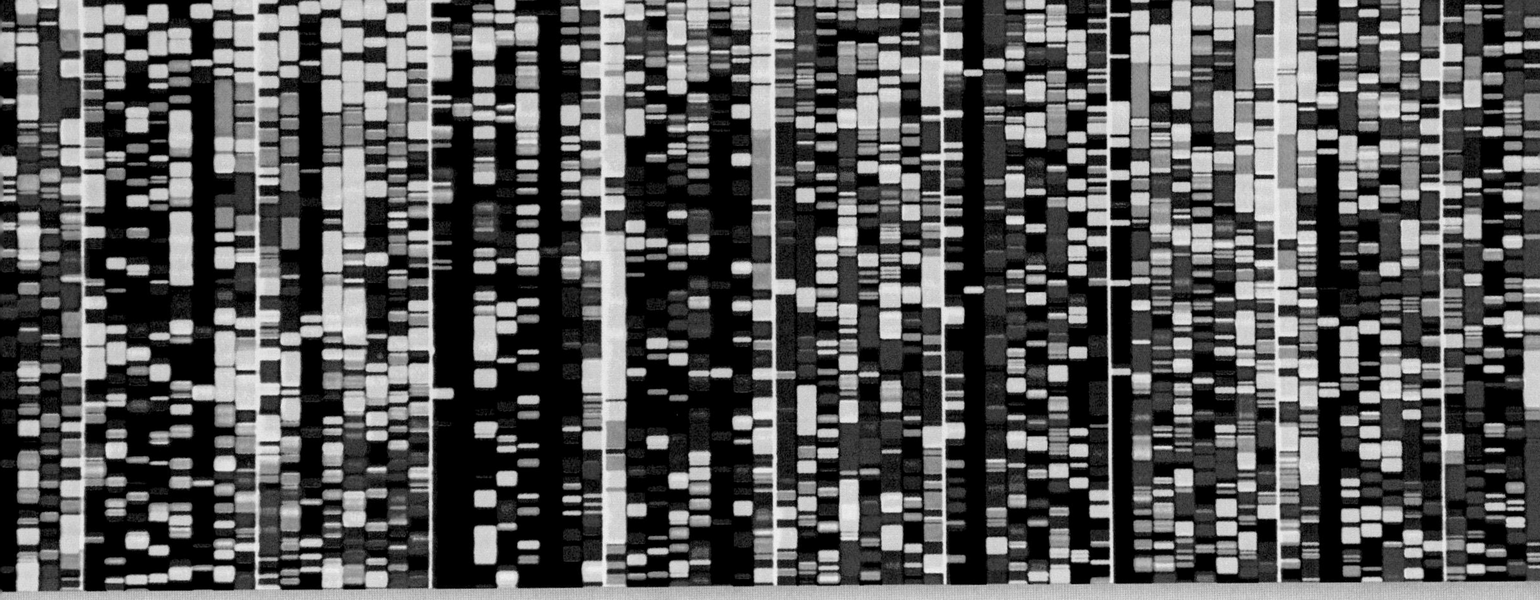

FIGURE 15.7 Human DNA sequence data. The order of colors in each vertical lane reveals one part of the DNA sequence.

THE HUMAN GENOME PROJECT

The sequencing method we have just described was invented in 1975. Ten years later, it had become so routine that scientists began to consider sequencing the entire human genome—all 3 billion nucleotides. Proponents of the idea said it could provide huge payoffs for medicine and research. Opponents said this daunting task would divert attention and funding from more urgent research. It would require 50 years to sequence the human genome given the techniques of the time. However, the techniques continued to improve rapidly, and with each improvement more nucleotides could be sequenced in less time. Automated (robotic) DNA sequencing and PCR had just been invented. Both were still too cumbersome and expensive to be useful in routine applications, but they would not be so for long. Waiting for faster technologies seemed the most efficient way to sequence the genome, but just how fast did they need to be before the project should begin?

A few privately owned companies decided not to wait, and started sequencing. One of them intended to determine the genome sequence in order to patent it. The idea of patenting the human genome provoked widespread outrage, but it also spurred commitments in the public sector. In 1988, the National Institutes of Health (NIH) essentially took over the project by hiring James Watson (of DNA structure fame) to head an official Human Genome Project, and providing $200 million per year to fund it.

A partnership formed between the NIH and international institutions that were sequencing different parts of the genome. Watson set aside 3 percent of the funding for studies of ethical and social issues arising from the work. He later resigned over a patent disagreement, and geneticist Francis Collins took his place.

Amid ongoing squabbles over patent issues, Celera Genomics formed in 1998. With biologist Craig Venter at its helm, the company intended to commercialize human genetic information. Celera invented faster techniques for sequencing genomic DNA, because the first to have the complete sequence had a legal basis for patenting it. The competition motivated the international partnership to accelerate its efforts. Then, in 2000, U.S. President Bill Clinton and British Prime Minister Tony Blair jointly declared that the sequence of the human genome could not be patented. Celera kept sequencing anyway, and, in 2001, the competing governmental and corporate teams published about 90 percent of the sequence. In 2003, fifty years after the discovery of the structure of DNA, the sequence of the human genome was officially completed (**FIGURE 15.7**).

TAKE-HOME MESSAGE 15.3

With DNA sequencing, a strand of DNA is partially replicated. Electrophoresis is used to separate the resulting fragments by length.

Improved sequencing techniques and worldwide efforts allowed the human genome sequence to be determined.

electrophoresis Technique that separates DNA fragments by size.
sequencing Method of determining the order of nucleotides in DNA.

CREDIT: (7) Patrick Landmann/Science Source.

15.4 HOW DO WE USE WHAT RESEARCHERS DISCOVER ABOUT GENOMES?

It took 15 years to sequence the human genome for the first time, but the techniques have improved so much that sequencing an entire genome now takes about a day. Anyone can now pay to have their genome sequenced (a cost of $5,000 to $8,000, at this writing).

Despite our ability to determine the sequence of an individual's genome, however, it will be a long time before we understand all the information coded within that sequence. The human genome contains a massive amount of seemingly cryptic data. One way to decipher it is by comparing it to the genomes of other species, the premise being that all organisms are descended from shared ancestors, so all genomes are related to some extent. We see evidence of such genetic relationships simply by comparing the raw sequence data, which, in some regions, is extremely similar across many species (**FIGURE 15.8**).

The study of genomes is called **genomics**, a broad field that encompasses whole-genome comparisons, structural analysis of gene products, and surveys of small-scale variations in sequence. Genomics is providing powerful insights into evolution, and it has many medical benefits. We have learned the function of many human genes by studying their counterpart genes in other species. For instance, researchers comparing human and mouse genomes discovered a human version of a mouse gene, *APOA5*, that encodes a lipoprotein (Section 3.4). Mice with an *APOA5* knockout have four times the normal level of triglycerides in their blood. The researchers then looked for—and found—a correlation between *APOA5* mutations and high triglyceride levels in humans. High triglycerides are a risk factor for coronary artery disease.

DNA PROFILING

About 99 percent of your DNA is exactly the same as everyone else's. The shared part is what makes you human; the differences make you a unique member of the species. If you compared your DNA with your neighbor's, about 2.97 billion nucleotides of the two sequences would be identical; the remaining 30 million or so nonidentical nucleotides are sprinkled throughout your chromosomes. The sprinkling is not entirely random because some regions of DNA vary less than others. Such conserved regions are of particular interest to researchers because they are the ones most likely to have an essential function. When a conserved sequence does vary among people, the variation tends to be in nucleotides at a particular location. A base-pair substitution carried by a measurable percentage of a population, usually above 1 percent, is called a **single-nucleotide polymorphism**, or **SNP** (pronounced "snip").

Alleles of most genes differ by single nucleotides, and differences in alleles are the basis of the variation in human traits that makes each individual unique (Section 12.1). In fact, those differences are so unique that they can be used to identify you. Identifying an individual by his or her DNA is called **DNA profiling**.

One type of DNA profiling involves SNP-chips (**FIGURE 15.9**). A SNP-chip is a tiny glass plate with

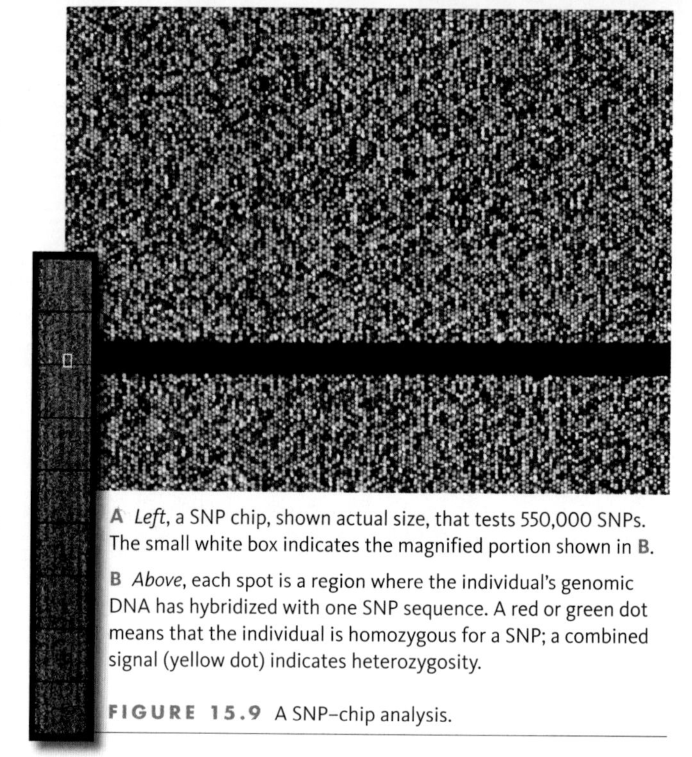

A *Left*, a SNP chip, shown actual size, that tests 550,000 SNPs. The small white box indicates the magnified portion shown in B.

B *Above*, each spot is a region where the individual's genomic DNA has hybridized with one SNP sequence. A red or green dot means that the individual is homozygous for a SNP; a combined signal (yellow dot) indicates heterozygosity.

FIGURE 15.9 A SNP-chip analysis.

FIGURE 15.8 Genomic DNA alignment. This is a region of the gene for a DNA polymerase. Nucleotides that differ from those in the human sequence are highlighted. The chance that any two of these sequences would randomly match is about 1 in 10^{46}.

```
758 GATAATCCTGTTTTGAACAAAAGGTCAAATTGCTGAATAGAAA-GTCTTGATTAACTAAAAGATGTACAAAGTGGAATTA 836  Human
752 GATAATCCTGTTTTGAACAAAAGGTCAAATTGCTGAATAGAAA-GTCTTGATTAACTAAAAGATGTACAAAGTGGAATTA 830  Mouse
751 GATAATCCTGTTTTGAACAAAAGGTCAAATTGCTGAATAGAAA-GTCTTGATTAACTAAAAGATGTACAAAGTGGAATTA 829  Rat
754 GATAATCCTGTTTTGAACAAAAGGTCAAATTGCTGAATAGAAA-GTCTTGATTAACTAAAAGATGTACAAAGTGGAATTA 832  Dog
782 GATAATCCTGTTTTGAACAAAAGGTCAAATTGCTGAATAGAAA-GTCTTGATTAACTAAAAGATGTACAAAGTGGAATTA 860  Chicken
758 GATAATCCTGTTTTGAACAAAAGGTCAAATTGCTGAATAGAAA-GTCTTGATTAAGTAAAAGATGTACAAAGTGGAATTA 836  Frog
823 GATAATCCTGTTTTGAACAAAAGGTCAGATTGCTGAATAGAAAAGGCTTGATTAAAGCAGAGATGTACAAAGTGGACGCA 902  Zebrafish
763 GATAATCCTGTTTTGAACAAAAGGTCAAATTGTTGAATAGAGACGCTTTGATAAAGCGGAGGAGGTACAAAGTGGGACC- 841  Pufferfish
```


CREDITS: (8) © Cengage Learning; (9A) The Sanger Institute. Wellcome Images; (9B) Wellcome Trust Sanger Institute.

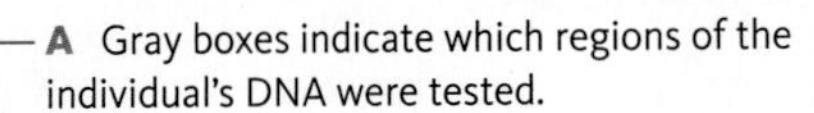

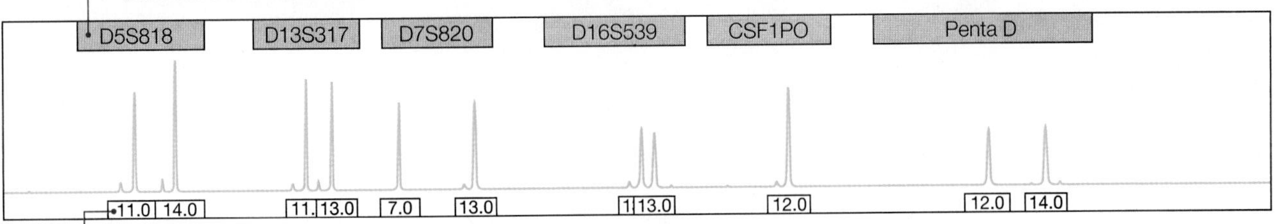

B The number of repeats is shown in a box below each peak. A peak's location on the x-axis corresponds to the length of the DNA fragment amplified (a measure of the number of repeats). Peak size reflects the amount of DNA.

FIGURE 15.10 {Animated} An individual's (partial) short tandem repeat profile. Remember, human body cells are diploid. Double peaks appear on a profile when the two members of a chromosome pair carry a different number of repeats.

FIGURE IT OUT: How many repeats does this individual have at the Penta D region?

Answer: 12 on one chromosome; 14 on the other

microscopic spots of DNA stamped on it. The DNA sample in each spot is a short, synthetic single strand with a unique SNP sequence. When an individual's genomic DNA is washed over a SNP-chip, it hybridizes only with DNA spots that have a matching SNP sequence. Probes reveal where the genomic DNA has hybridized—and which SNPs are carried by the individual.

Another method of DNA profiling involves analysis of short tandem repeats in an individual's chromosomes (Section 13.6). Short tandem repeats tend to occur in predictable spots, but the number of times a sequence is repeated in each spot differs among individuals. For example, one person's DNA may have fifteen repeats of the nucleotides TTTTC at a certain spot on one chromosome. Another person's DNA may have this sequence repeated only twice in the same location. Such repeats slip spontaneously into DNA during replication, and their numbers grow or shrink over generations. Unless two people are identical twins, the chance that they have identical short tandem repeats in even three regions of DNA is 1 in a quintillion (10^{18}), which is far more than the number of people who have ever lived. Thus, an individual's array of short tandem repeats is, for all practical purposes, unique.

Analyzing a person's short tandem repeats begins with PCR, which is used to copy ten to thirteen particular regions of chromosomal DNA known to have repeats. The lengths of the copied DNA fragments differ among most individuals, because the number of tandem repeats in those regions also differs. Thus, electrophoresis can be used to reveal an individual's unique array of short tandem repeats (**FIGURE 15.10**).

Short tandem repeat analysis will soon be replaced by full genome sequencing, but for now it continues to be a common DNA profiling method. Geneticists compare short tandem repeats on Y chromosomes to determine relationships among male relatives, and to trace an individual's ethnic heritage. They also track mutations that accumulate in populations over time by comparing DNA profiles of living humans with those of ancient ones. Such studies are allowing us to reconstruct population dispersals that happened in the ancient past.

Short tandem repeat profiles are routinely used to resolve kinship disputes, and as evidence in criminal cases. Within the context of a criminal or forensic investigation, DNA profiling is called DNA fingerprinting. As of January 2013, the database of DNA fingerprints maintained by the Federal Bureau of Investigation (the FBI) contained the short tandem repeat profiles of 10.1 million convicted offenders, and had been used in over 190,000 criminal investigations. DNA fingerprints have also been used to identify the remains of over 470,000 people, including the individuals who died in the World Trade Center on September 11, 2001.

DNA profiling Identifying an individual by analyzing the unique parts of his or her DNA.
genomics The study of genomes.
single-nucleotide polymorphism (SNP) One-nucleotide DNA sequence variation carried by a measurable percentage of a population.

TAKE-HOME MESSAGE 15.4

Analysis of the human genome sequence is yielding new information about our genes and how they work.

DNA profiling identifies individuals by the unique parts of their DNA.

15.5 WHAT IS GENETIC ENGINEERING?

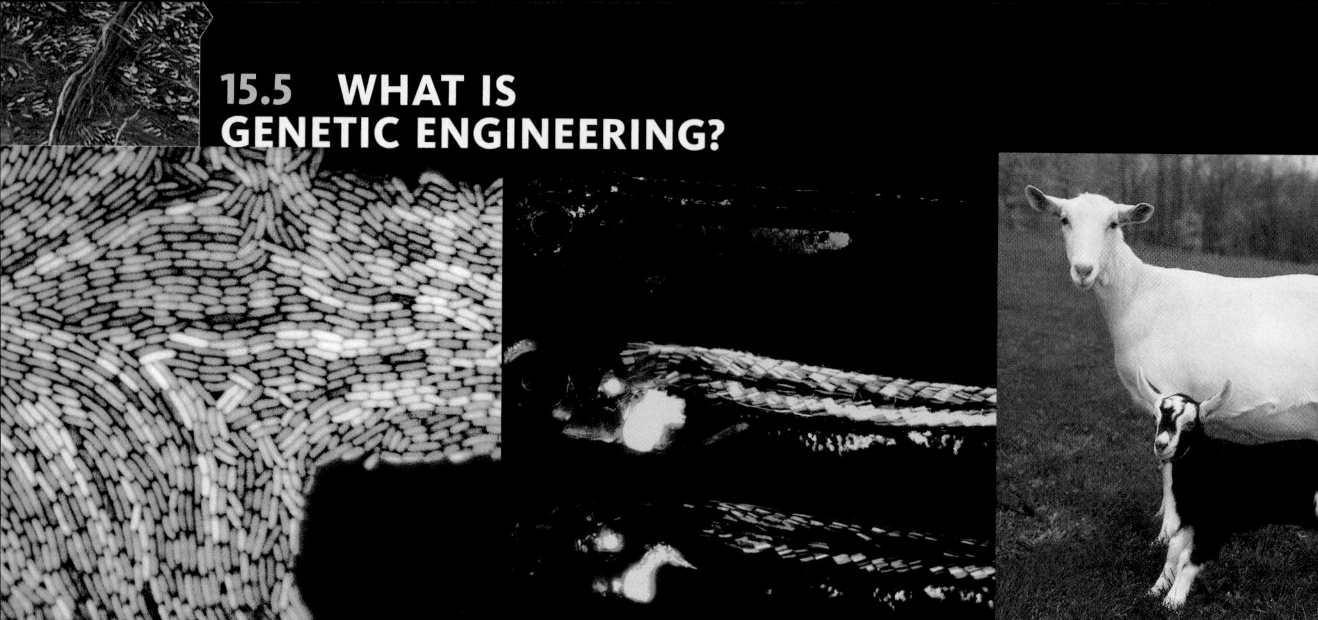

A *E. coli* bacteria transgenic for a fluorescent jellyfish protein. Variation in fluorescence among the genetically identical cells reveals differences in gene expression that may help us understand why some bacteria become dangerously resistant to antibiotics, and others do not.

B Zebrafish engineered to glow in places where BPA, an endocrine-disrupting chemical, is present. The fish are literally illuminating where this pollutant acts in the body—and helping researchers discover what it does when it gets there.

C Transgenic goats produce human antithrombin, an anti-clotting protein. Antithrombin harvested from their milk is used as a drug during surgery or childbirth to prevent blood clotting in people with hereditary antithrombin deficiency. This genetic disorder carries a high risk of life-threatening clots.

FIGURE 15.11 Examples of GMOs.

Genetic engineering is a process by which an individual's genome is deliberately modified. A gene from one species may be transferred to another to produce an organism that is **transgenic**, or a gene may be altered and reinserted into an individual of the same species. Both methods result in a **genetically modified organism**, or **GMO**.

The most common GMOs are bacteria (**FIGURE 15.11A**) and yeast. These cells have the metabolic machinery to make complex organic molecules, and they are easily modified to produce, for example, medically important proteins. People with diabetes were among the first beneficiaries of such organisms. Insulin for their injections was once extracted from animals, but it provoked an allergic reaction in some people. Human insulin, which does not provoke allergic reactions, has been produced by transgenic *E. coli* since 1982. Slight modifications of the gene have yielded fast-acting and slow-release forms of human insulin.

Genetically engineered microorganisms also produce proteins used in foods. For example, cheese is traditionally made with an extract of calf stomachs, which contain the enzyme chymotrypsin. Most cheese manufacturers now use chymotrypsin produced by genetically engineered bacteria. Other enzymes produced by GMOs improve the taste and clarity of beer and fruit juice, slow bread staling, or modify certain fats.

The first genetically modified animals were mice. Today, engineered mice are commonplace, and they are invaluable in research (an example is shown in the chapter opener). We have discovered the function of human genes (including the *APOA5* gene discussed in Section 15.4) by inactivating their counterparts in mice. Genetically modified mice are also used as models of human diseases. For example, researchers inactivated the molecules involved in the control of glucose metabolism, one by one. Studying the effects of the knockouts in mice has resulted in much of our current understanding of how diabetes works in humans.

Other genetically modified animals are useful in research (**FIGURE 15.11B**), and some make molecules that have medical and industrial applications. Various transgenic goats produce proteins used to treat cystic fibrosis, heart attacks, blood clotting disorders (**FIGURE 15.11C**), and even nerve gas exposure. Goats transgenic for a spider silk gene produce the silk protein in their milk; researchers can spin this protein into nanofibers that are useful in medical and electronics applications. Rabbits make human interleukin-2, a protein that triggers divisions of immune cells. Genetic engineering has also given us pigs with heart-healthy fat and environmentally friendly low-phosphate feces, muscle-bound trout, chickens that do not transmit bird flu, and cows that do not get mad cow disease.

As crop production expands to keep pace with human population growth, many farmers have begun to rely on genetically modified crop plants. Genes are often introduced into plant cells by way *Agrobacterium tumefaciens*. These bacteria carry a plasmid with genes that cause tumors to form on infected plants; hence the name Ti plasmid (for

Tumor-inducing). Researchers replace the tumor-inducing genes with foreign or modified genes, then use the plasmid as a vector to deliver the genes into plant cells. Whole plants can be grown from cells that integrate a recombinant plasmid into their chromosomes (**FIGURE 15.12**).

Many genetically modified crops carry genes that impart resistance to devastating plant diseases. Others offer improved yields. GMO crops such as Bt corn and soy help farmers use smaller amounts of toxic pesticides (**FIGURE 15.13**). Organic farmers often spray their crops with spores of Bt (*Bacillus thuringiensis*), a bacterial species that makes a protein toxic only to some insect larvae. Researchers transferred the gene encoding the Bt protein into plants. The engineered plants produce the Bt protein, and larvae die shortly after eating their first and only GMO meal.

Transgenic crop plants are also being developed for impoverished regions of the world. Genes that confer drought tolerance, insect resistance, and enhanced nutritional value are being introduced into plants such as corn, rice, beans, sugarcane, cassava, cowpeas, banana, and wheat. The resulting GMO crops may help people who rely on agriculture for food and income.

At this writing, ninety-two crops have been approved for unrestricted use in the United States. Worldwide, more than 330 million acres are currently planted in GMO crops, the majority of which are corn, sorghum, cotton, soy, canola, and alfalfa engineered for resistance to the herbicide glyphosate. Rather than tilling the soil to control weeds, farmers spray their fields with glyphosate, which kills the weeds but not the engineered crops.

Many people worry that our ability to tinker with genetics has surpassed our ability to understand the impact of the tinkering. Controversy raised by GMO use invites you to read the research and form your own opinions. The alternative is to be swayed by media hype (the term "Frankenfood," for instance), or by reports from potentially biased sources (such as herbicide manufacturers).

genetic engineering Process by which deliberate changes are introduced into an individual's genome.
genetically modified organism (GMO) Organism whose genome has been modified by genetic engineering.
transgenic Refers to a genetically modified organism that carries a gene from a different species.

TAKE-HOME MESSAGE 15.5

Genetic engineering is the deliberate alteration of an individual's genome, and it results in a genetically modified organism (GMO).

A transgenic organism carries a gene from a different species.

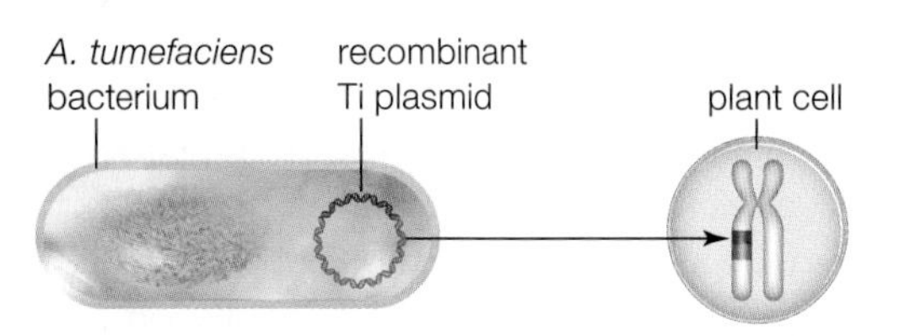

A A Ti plasmid carrying a foreign gene is inserted into an *Agrobacterium tumefaciens* bacterium.

B The bacterium infects a plant cell and transfers the Ti plasmid into it. The plasmid DNA, along with the foreign gene, becomes integrated into one of the cell's chromosomes.

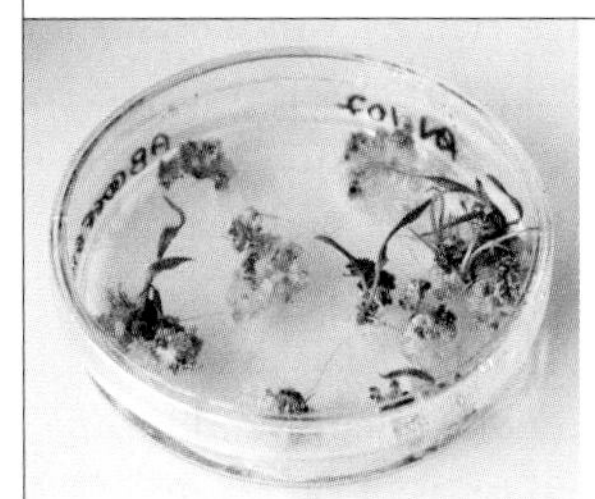

C The infected plant cell divides, and its descendants form an embryo, then a plant (left). Cells of the transgenic plant carry and express the foreign gene.

FIGURE 15.12 {Animated} Using the Ti plasmid to make transgenic plants.

FIGURE 15.13 Farmers can use much less pesticide on crops that make their own. The genetically modified plants that produced the row of corn on the *top* carry a gene from the bacteria *Bacillus thuringensis* (Bt) that conferred insect resistance. Compare the corn from unmodified plants, *bottom*. No pesticides were used on either crop.

15.6 CAN WE GENETICALLY MODIFY PEOPLE?

GENE THERAPY

We know of more than 15,000 serious genetic disorders. Collectively, they cause 20 to 30 percent of infant deaths each year, and account for half of all mentally impaired patients and a fourth of all hospital admissions. They also contribute to many age-related disorders, including cancer, Parkinson's disease, and diabetes. Drugs and other treatments can minimize the symptoms of some genetic disorders, but gene therapy is the only cure. **Gene therapy** is the transfer of recombinant DNA into an individual's body cells, with the intent to correct a genetic defect or treat a disease. The transfer, which occurs by way of lipid clusters or genetically engineered viruses, inserts an unmutated gene into an individual's chromosomes.

Human gene therapy is a compelling reason to embrace genetic engineering research. It is now being tested as a treatment for AIDS, muscular dystrophy, heart attack, sickle-cell anemia, cystic fibrosis, hemophilia A, Parkinson's disease, Alzheimer's disease, several types of cancer, and inherited diseases of the eye, the ear, and the immune system. Results have been encouraging. For example, gene therapy has recently been used to treat acute lymphoblastic leukemia, a typically fatal cancer of bone marrow cells. A viral vector was used to insert a gene into immune cells extracted from patients. When the engineered cells were reintroduced into the patients' bodies, the inserted gene directed the destruction of the cancer cells. The therapy worked astonishingly well: In one patient, all traces of the leukemia vanished in eight days. However, the outcome of manipulating a gene in a living individual can be unpredictable. In an early trial, for example, twenty boys were treated with gene therapy for a severe X-linked genetic disorder called SCID-X1. Five of them developed leukemia, and one died. The researchers had wrongly predicted that cancer related to the therapy would be rare. Research now implicates the gene targeted for repair, especially when combined with the virus that delivered it. Integration of the modified viral DNA activated nearby proto-oncogenes (Section 11.5) in the children's chromosomes.

EUGENICS

The idea of selecting the most desirable human traits, **eugenics**, is an old one. It has been used as a justification for some of the most horrific episodes in human history, including the genocide of 6 million Jews during World War II. Thus, it continues to be a hotly debated social issue. For example, using gene therapy to cure human genetic disorders seems like a socially acceptable goal to most people, but imagine taking this idea a bit further. Would it also be acceptable to engineer the genome of an individual who is within a normal range of phenotype in order to modify a particular trait? Researchers have already produced mice that have improved memory, enhanced learning ability, bigger muscles, and longer lives. Why not people?

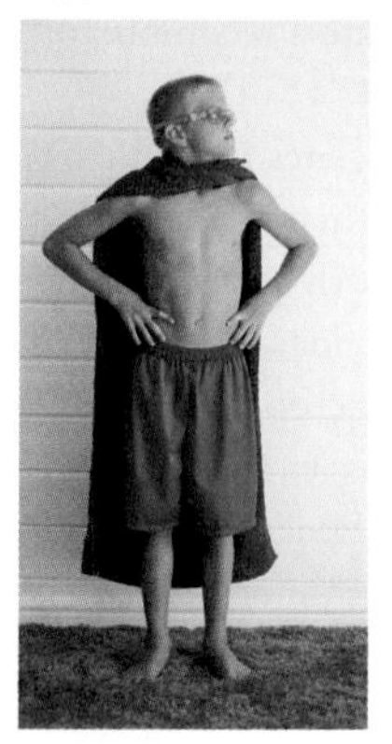

Given the pace of genetics research, the debate is no longer about how we would engineer desirable traits, but how we would choose the traits that are desirable. Realistically, cures for many severe but rare genetic disorders will not be found, because the financial return would not cover the cost of the research. Eugenics, however, may be profitable. How much would potential parents pay to be sure that their child will be tall or blue-eyed? Would it be okay to engineer "superhumans" with breathtaking strength or intelligence? How about a treatment that can help you lose that extra weight, and keep it off permanently? The gray area between interesting and abhorrent can be very different depending on who is asked. In a survey conducted in the United States, more than 40 percent of those interviewed said it would be fine to use gene therapy to make smarter and cuter babies. In one poll of British parents, 18 percent would be willing to use it to keep a child from being aggressive, and 10 percent would use it to keep a child from growing up to be homosexual.

Some people are concerned that gene therapy puts us on a slippery slope that may result in irreversible damage to ourselves and to the biosphere. We as a society may not have the wisdom to know how to stop once we set foot on that slope; one is reminded of our peculiar human tendency to leap before we look. And yet, something about the human experience allows us to dream of such things as wings of our own making, a capacity that carried us into space. In this brave new world, the questions before you are these: What do we stand to lose if serious risks are not taken? And, do we have the right to impose the potential consequences on people who would choose not to take those risks?

---

eugenics Idea of deliberately improving the genetic qualities of the human race.

gene therapy Treating a genetic defect or disorder by transferring a normal or modified gene into the affected individual.

TAKE-HOME MESSAGE 15.6

Genes can be transferred into a person's cells to correct a genetic defect or treat a disease. However, the outcome of altering a person's genome has been unpredictable.

We as a society must continue to work our way through the ethical implications of applying DNA technologies.

15.7 Application: PERSONAL GENETIC TESTING

Actress Angelina Jolie discovered via genetic testing that she carries a *BRCA1* mutation associated with an 87% lifetime risk of developing breast cancer. Even though she did not yet have cancer, Jolie underwent a double mastectomy. By doing so, she reduced her risk of breast cancer to 5%.

Education

FIGURE 15.14 Celebrity Angelina Jolie chose preventive treatment after genetic testing showed she had a very high risk of breast cancer.

DO YOU WANT TO KNOW YOUR SNP PROFILE? Finding out which of about 1 million SNPs you carry has never been easier. Genetic testing companies can extract your DNA from a few drops of spit, then analyze it using a SNP-chip. Results typically include estimated risks of developing conditions associated with your particular set of SNPs. For example, the test will probably determine whether you are homozygous for one allele of the *MC1R* gene (Section 13.4). If you are, the company's report will tell you that you have red hair. Few SNPs have such a clear effect, however. Consider the lipoprotein particles that carry fats and cholesterol through our bloodstreams. These particles consist of variable amounts and types of lipids and proteins, one of which is specified by the gene *APOE*. About one in four people carries an allele of this gene, *ε4*, that increases one's risk of developing Alzheimer's disease later in life. If you are heterozygous for this allele, a DNA testing company will report that your lifetime risk of developing the disease is about 29 percent, as compared with about 9 percent for someone who has no *ε4* allele.

What, exactly, does a 29 percent lifetime risk of Alzheimer's disease mean? The number is a probability statistic; it means, on average, 29 of every 100 people who have the *ε4* allele eventually get the disease. However, a risk is just that. Not everyone who has the *ε4* allele develops Alzheimer's, and not everyone who develops the disease has the allele. Other unknown factors, including epigenetic modifications of DNA, contribute to the disease. We still have a limited understanding of how genes contribute to many health conditions, particularly age-related ones such as Alzheimer's disease

Geneticists believe that it will be at least five to ten more years before genotyping can be used to accurately predict an individual's future health problems. Nonetheless, we are at a tipping point; personalized genetic testing is already beginning to revolutionize medicine. Cancer treatments are now being tailored to fit the genetic makeup of individual patients. People who discover they carry alleles associated with a heightened risk of a medical condition are being encouraged to make lifestyle changes that could delay the condition's onset or prevent it entirely. Preventive treatments based on personal genetics are becoming more common—and more mainstream (FIGURE 15.14).

Summary

SECTION 15.1 In **DNA cloning**, researchers use **restriction enzymes** to cut a sample of DNA into pieces, and then use DNA ligase to splice the fragments into plasmids or other **cloning vectors**. The resulting molecules of **recombinant DNA** are inserted into host cells such as bacteria. Division of host cells produces huge populations of genetically identical descendant cells (clones), each with a copy of the cloned DNA fragment.

The enzyme **reverse transcriptase** is used to transcribe RNA into **cDNA** for cloning.

SECTION 15.2 A **DNA library** is a collection of cells that host different fragments of DNA, often representing an organism's entire **genome**. Researchers can use **probes** to identify cells in a library that carry a specific fragment of DNA. The **polymerase chain reaction** (**PCR**) uses primers and a heat-resistant DNA polymerase to rapidly increase the number of copies of a targeted section of DNA.

SECTION 15.3 Advances in **sequencing**, which reveals the order of nucleotides in DNA, allowed the DNA sequence of the entire human genome to be determined. DNA polymerase is used to partially replicate a DNA template. The reaction produces a mixture of DNA fragments of all different lengths; **electrophoresis** separates the fragments by length into bands.

SECTION 15.4 **Genomics** provides insights into the function of the human genome. Similarities between genomes of different organisms are evidence of evolutionary relationships, and can be used as a predictive tool in research.

DNA profiling identifies a person by the unique parts of his or her DNA. An example is the determination of an individual's array of short tandem repeats or **single-nucleotide polymorphisms** (**SNPs**). Within the context of a criminal investigation, a DNA profile is called a DNA fingerprint.

SECTION 15.5 Recombinant DNA technology is the basis of **genetic engineering**, the directed modification of an organism's genetic makeup with the intent to modify its phenotype. A gene from one species is inserted into an individual of a different species to make a **transgenic** organism, or a gene is modified and reinserted into an individual of the same species. The result of either process is a **genetically modified organism** (**GMO**).

Bacteria and yeast, the most common genetically engineered organisms, produce proteins that have medical value. The majority of the animals that are being created by genetic engineering are used for medical applications or research. Most transgenic crop plants, which are now in widespread use worldwide, were created to help farmers produce food more efficiently. Some have enhanced nutritional value.

SECTION 15.6 With **gene therapy**, a gene is transferred into body cells to correct a genetic defect or treat a disease. Potential benefits of genetically modifying humans must be weighed against potential risks. The practice raises ethical issues such as whether **eugenics** is desirable in some circumstances.

SECTION 15.7 Personal genetic testing, which reveals a person's unique array of SNPs, is beginning to revolutionize the way medicine is practiced.

Self-Quiz Answers in Appendix VII

1. _______ cut(s) DNA molecules at specific sites.
   a. DNA polymerase c. Restriction enzymes
   b. DNA probes d. DNA ligase

2. A _______ is a molecule that can be used to carry a fragment of DNA into a host organism.
   a. cloning vector c. GMO
   b. chromosome d. cDNA

3. Reverse transcriptase assembles a(n) _______ on a(n) _______ template.
   a. mRNA; DNA c. DNA; ribosome
   b. cDNA; mRNA d. protein; mRNA

4. For each species, all _______ in the complete set of chromosomes is/are the _______.
   a. genomes; phenotype c. mRNA; start of cDNA
   b. DNA; genome d. cDNA; start of mRNA

5. A set of cells that host various DNA fragments collectively representing an organism's entire set of genetic information is a _______.
   a. genome c. genomic library
   b. clone d. GMO

6. _______ is a technique to determine the order of nucleotide bases in a fragment of DNA.
   a. PCR c. Electrophoresis
   b. Sequencing d. Nucleic acid hybridization

7. Fragments of DNA can be separated by electrophoresis according to _______.
   a. sequence b. length c. species

8. PCR can be used _______.
   a. to increase the number of specific DNA fragments
   b. in DNA fingerprinting
   c. to modify a human genome
   d. a and b are correct

9. An individual's set of unique _______ can be used as a DNA profile.
   a. DNA sequences c. SNPs
   b. short tandem repeats d. all of the above

10. True or false? Some humans are genetically modified.

Data Analysis Activities

Enhanced Spatial Learning Ability in Mice with an Autism Mutation Autism is a neurobiological disorder with symptoms that include impaired social interactions and stereotyped patterns of behavior. Around 10 percent of autistic people have an extraordinary skill or talent such as greatly enhanced memory.

Mutations in neuroligin 3, an adhesion protein that connects brain cells to one another, have been associated with autism. One mutation changes amino acid 451 from arginine to cysteine. In 2007, Katsuhiko Tabuchi and his colleagues genetically modified mice to carry the same arginine-to-cysteine substitution in their neuroligin 3. Mice with the mutation had impaired social behavior. Spatial learning ability was tested in a water maze, in which a platform is submerged a few millimeters below the surface of a deep pool of warm water. The platform is not visible to swimming mice. Mice do not particularly enjoy swimming, so they locate a hidden platform as fast as they can. When tested again, they can remember its location by checking visual cues around the edge of the pool. How quickly they remember the platform's location is a measure of spatial learning ability (**FIGURE 15.15**).

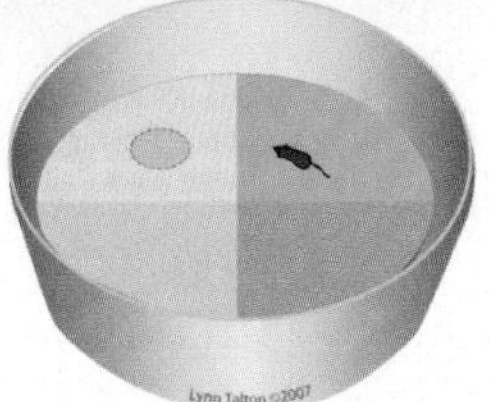

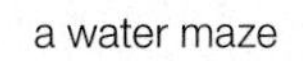

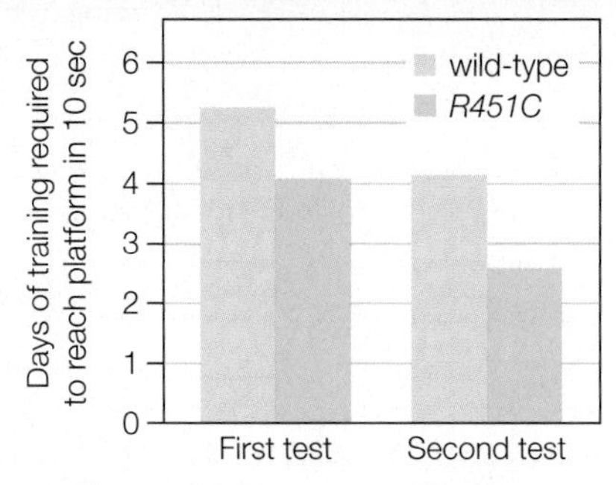

FIGURE 15.15 Spatial learning ability in mice with a mutation in neuroligin 3 (*R451C*), compared with unmodified (wild-type) mice.

1. In the first test, how many days did unmodified mice need to learn to find the location of a hidden platform within 10 seconds?
2. Did the modified or the unmodified mice learn the location of the platform faster in the first test?
3. Which mice learned faster the second time around?
4. Which mice showed the greatest improvement in memory between the first and the second test?

11. Which of the following can be used to carry foreign DNA into host cells? Choose all correct answers.
    a. RNA
    b. viruses
    c. PCR
    d. lipid clusters
    e. bacteria
    f. plasmids

12. Transgenic _______ can pass a foreign gene to offspring.
    a. plants
    b. animals
    c. bacteria
    d. all of the above

13. _______ can correct a genetic defect in an individual.
    a. Cloning vectors
    b. Gene therapy
    c. Eugenics
    d. a and b

14. Match the recombinant DNA method with the appropriate enzyme.
    ___ PCR
    ___ cutting DNA
    ___ cDNA synthesis
    ___ DNA sequencing
    ___ pasting DNA

    a. *Taq* polymerase
    b. DNA ligase
    c. reverse transcriptase
    d. restriction enzyme
    e. DNA polymerase (not *Taq*)

15. Match each term with the most suitable description.
    ___DNA profile
    ___Ti plasmid
    ___eugenics
    ___SNP
    ___transgenic
    ___GMO

    a. GMO with a foreign gene
    b. alleles commonly contain them
    c. a person's unique collection of short tandem repeats
    d. selecting "desirable" traits
    e. genetically modified
    f. used in plant gene transfers

Critical Thinking

1. The results of a paternity test using short tandem repeats are shown in the table below. Who's the daddy? How sure are you?

	Mother	Baby	Alleged Father #1	Alleged Father #2
D3S1358	15, 17	17, 23	23, 27	17, 15
TH01	9, 9	9, 9	9, 12	12, 12
D21S11	29, 29	29, 27	27, 28	29, 28
D18S51	14, 18	18, 20	15, 20	17, 22
Penta E	14, 14	14, 14	14, 14	15, 16
D5D818	11, 14	14, 16	12, 16	14, 20
D13S317	11, 13	10, 13	8, 10	18, 18
D7S820	7, 13	13, 13	13, 19	13, 13
D16S539	13, 13	13, 15	12, 15	10, 12
CSF1PO	12, 12	10, 12	8, 10	12, 17
Penta D	12, 14	5, 12	14, 14	18, 25
amelogenin	X, X	X,Y	X,Y	X,Y
vWa	15, 17	17, 22	15, 22	22, 22
D8S1179	13, 13	8, 13	8, 13	15, 15
TPOX	11, 11	11, 11	10, 11	17, 22
FGA	23, 23	23, 25	18, 25	23, 23

High in the Andes, a scientist infers the stride of an extinct dinosaur by measuring the distance between its fossilized footprints. We often reconstruct history by studying physical evidence of events that took place long ago. This practice relies on a foundational premise of science: Natural phenomena that occurred in the past can be explained by the same physical, chemical, and biological processes operating today.

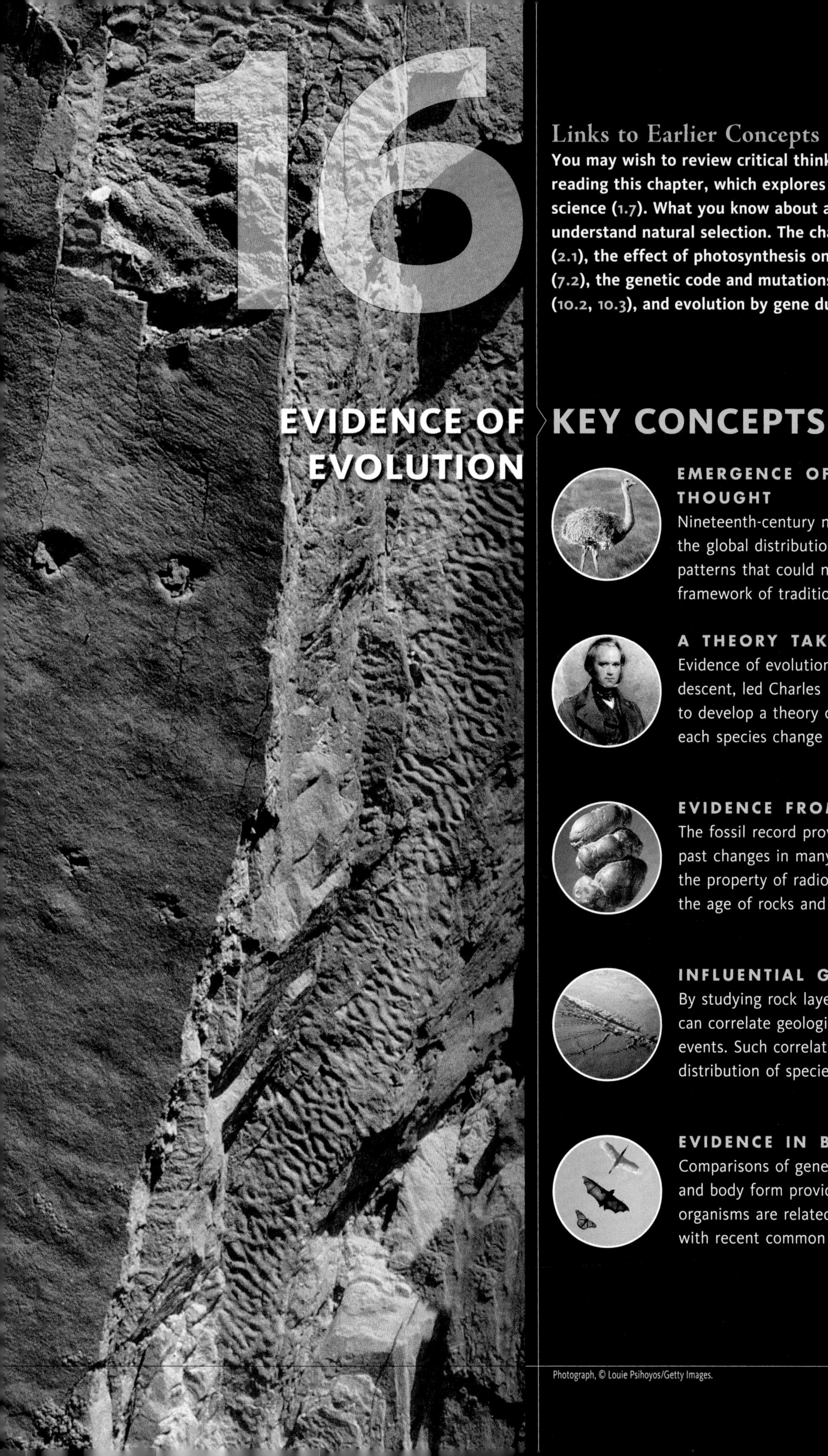

16

EVIDENCE OF EVOLUTION

Links to Earlier Concepts

You may wish to review critical thinking (Section 1.5) before reading this chapter, which explores a clash between belief and science (1.7). What you know about alleles (12.1) will help you understand natural selection. The chapter revisits radioisotopes (2.1), the effect of photosynthesis on Earth's early atmosphere (7.2), the genetic code and mutations (9.3, 9.5), master genes (10.2, 10.3), and evolution by gene duplication (14.4).

KEY CONCEPTS

EMERGENCE OF EVOLUTIONARY THOUGHT

Nineteenth-century naturalists investigating the global distribution of species discovered patterns that could not be explained within the framework of traditional belief systems.

A THEORY TAKES FORM

Evidence of evolution, or change in lines of descent, led Charles Darwin and Alfred Wallace to develop a theory of how traits that define each species change over time.

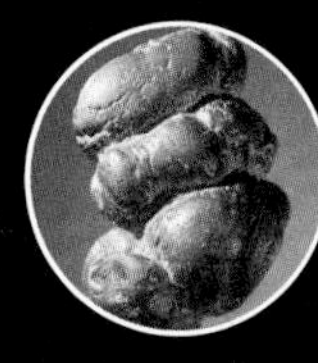

EVIDENCE FROM FOSSILS

The fossil record provides physical evidence of past changes in many lines of descent. We use the property of radioisotope decay to determine the age of rocks and fossils.

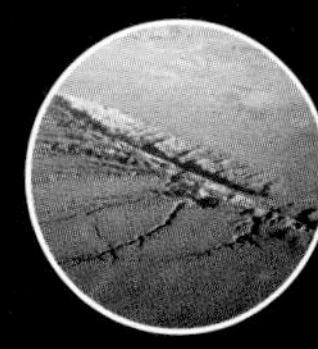

INFLUENTIAL GEOLOGIC FORCES

By studying rock layers and fossils in them, we can correlate geologic events with evolutionary events. Such correlations help explain the distribution of species, past and present.

EVIDENCE IN BODY FORM

Comparisons of genes, developmental patterns, and body form provide information about how organisms are related to one another. Lineages with recent common ancestry are most similar.

16.1 HOW DID OBSERVATIONS OF NATURE CHANGE OUR THINKING IN THE 19TH CENTURY?

A Emu, native to Australia. B Rhea, native to South America. C Ostrich, native to Africa.

FIGURE 16.1 Similar-looking, related species native to distant geographic realms. These birds are unlike most others in several unusual features, including long, muscular legs and an inability to fly. All are native to open grassland regions about the same distance from the equator.

People have long been curious about the natural world and our place in it. About 2,300 years ago, the Greek philosopher Aristotle described nature as a continuum of organization, from lifeless matter through complex plants and animals. Aristotle's work greatly influenced later European thinkers, who adopted his view of nature and modified it in light of their own beliefs. By the fourteenth century, Europeans generally believed that a "great chain of being" extended from the lowest form (snakes), up through humans, to spiritual beings. Each link in the chain was a species, and each was said to have been forged at the same time, in one place and in a perfect state. The chain itself was complete and continuous. Because everything that needed to exist already did, there was no room for change.

European naturalists who embarked on globe-spanning survey expeditions brought back tens of thousands of plants and animals from Asia, Africa, North and South America, and the Pacific Islands. Each newly discovered species was carefully catalogued as another link in the chain of being. By the late 1800s, naturalists were seeing patterns in where species live and similarities in body plans, and had started to think about the natural forces that shape life. These naturalists were pioneers in **biogeography**, the study of patterns in the geographic distribution of species and communities. Some of the patterns raised questions that could not be answered within the framework of prevailing belief systems. For example, globe-trotting explorers had discovered plants and animals living in extremely isolated places. The isolated species looked suspiciously similar to species living across vast expanses of open ocean, or on the other side of impassable mountain ranges. The three birds in **FIGURE 16.1** live on different continents, but they share a set of unusual features. These flightless birds sprint about on long, muscular legs in flat, open grasslands about the same distance from the equator. All raise their long necks to watch for predators. Alfred Wallace, an explorer who was particularly interested in the geographical distribution of animals, thought that the shared set of unusual traits might mean that these three birds descended from a common ancestor (and he was right), but he had no idea how they could have ended up on different continents.

Naturalists of the time also had trouble classifying organisms that are very similar in some features, but different in others. For example, the plants in **FIGURE 16.2** are native to different continents. Both live in hot

FIGURE 16.2 Similar-looking, unrelated species. On the *left*, an African milk barrel cactus (*Euphorbia horrida*), native to the Great Karoo desert of South Africa. On the *right*, saguaro cactus (*Carnegiea gigantea*), native to the Sonoran Desert of Arizona.

deserts where water is seasonally scarce. Both have rows of sharp spines that deter herbivores, and both store water in their thick, fleshy stems. However, their reproductive parts are very different, so these plants cannot be as closely related as their outward appearance might suggest.

Observations such as these are examples of **comparative morphology**, the study of anatomical patterns—similarities and differences among the body plans of organisms. Today, comparative morphology is part of taxonomy (Section 1.4), but in the nineteenth century it was the only way to distinguish species. In some cases, comparative morphology revealed anatomical details—body parts with no apparent function, for example—that added to the mounting confusion. If every species had been created in a perfect state, then why were there useless parts such as wings in birds that do not fly, eyes in moles that are blind, or remnants of a tail in humans (**FIGURE 16.3**)?

Fossils were puzzling too. A **fossil** is physical evidence—remains or traces—of an organism that lived in the ancient past. Geologists mapping rock formations exposed by erosion or quarrying had discovered identical sequences of rock layers in different parts of the world. Deeper layers held fossils of simple marine life. Layers above those held similar but more complex fossils (**FIGURE 16.4**). In higher layers, fossils that were similar but even more complex resembled modern species. What did these sequences mean?

Fossils of many animals unlike any living ones were also being unearthed. If these animals had been perfect at the time of creation, then why had they become extinct?

Taken as a whole, the accumulating findings from biogeography, comparative morphology, and geology did not fit with prevailing beliefs of the nineteenth century. If species had not been created in a perfect state (and extinct species, fossil sequences, and "useless" body parts implied that they had not), then perhaps species had indeed changed over time.

biogeography Study of patterns in the geographic distribution of species and communities.
comparative morphology The scientific study of similarities and differences in body plans.
fossil Physical evidence of an organism that lived in the ancient past.

TAKE-HOME MESSAGE 16.1

Increasingly extensive observations of nature in the nineteenth century did not fit with prevailing belief systems.

Cumulative findings from biogeography, comparative morphology, and geology led naturalists to question traditional ways of interpreting the natural world.

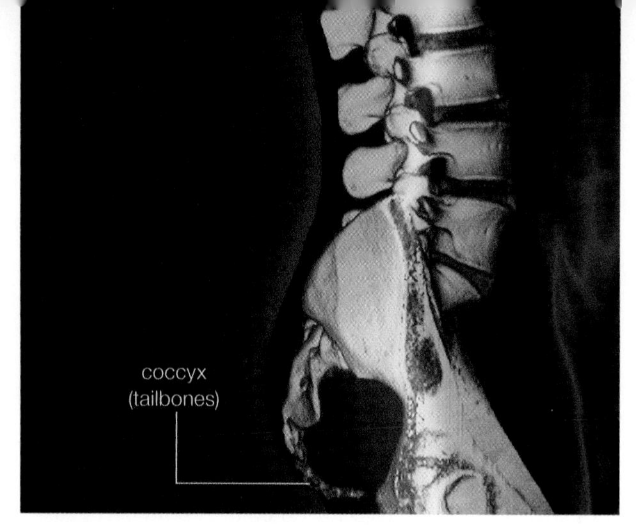

FIGURE 16.3 A vestigial structure: human tailbones. Nineteenth-century naturalists were well aware of—but had trouble explaining—body structures such as human tailbones that had apparently lost most or all function.

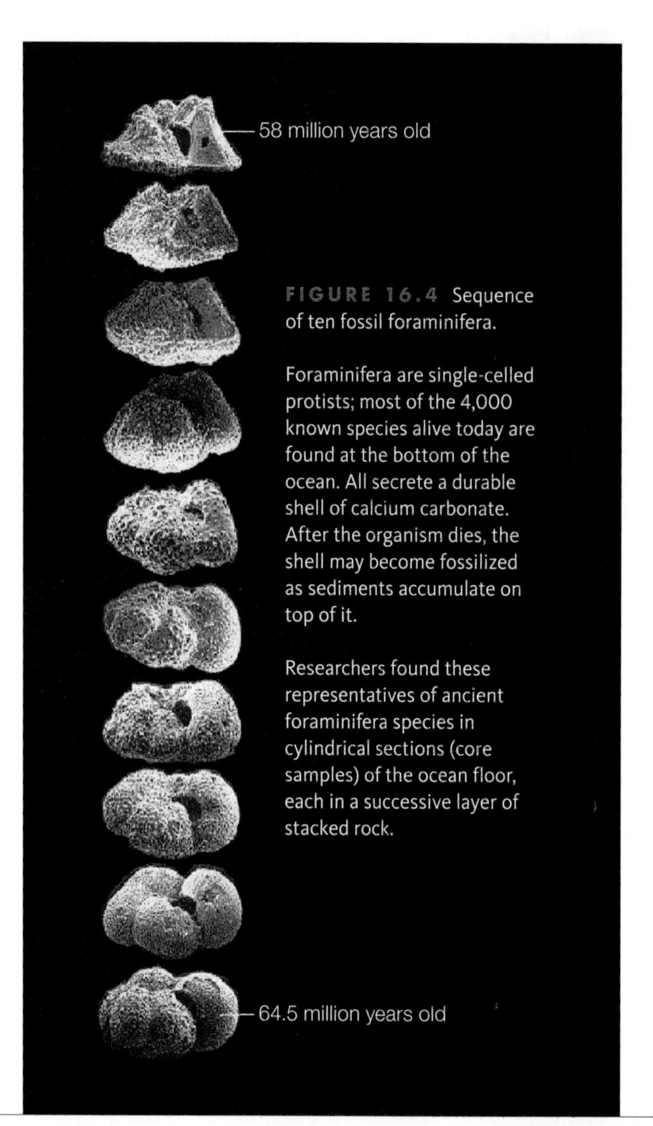

FIGURE 16.4 Sequence of ten fossil foraminifera.

Foraminifera are single-celled protists; most of the 4,000 known species alive today are found at the bottom of the ocean. All secrete a durable shell of calcium carbonate. After the organism dies, the shell may become fossilized as sediments accumulate on top of it.

Researchers found these representatives of ancient foraminifera species in cylindrical sections (core samples) of the ocean floor, each in a successive layer of stacked rock.

CREDITS: (3) © Zephyr/Science Photo Library/Science Source; (4) Courtesy of Daniel C. Kelley, Anthony J. Arnold, and William C. Parker, Florida State University Department of Geological Science.

16.2 WHAT IS NATURAL SELECTION?

Around 1800, naturalists were trying to explain the mounting evidence that life on Earth, and even Earth

itself, had changed over time. Georges Cuvier (*left*), an expert in zoology and paleontology, proposed an idea startling for the time: Many species that had once existed were now extinct. Cuvier knew about evidence that Earth's surface had changed. For example, he had seen fossilized seashells on mountainsides far from modern seas. Like most others of his time, he assumed Earth's age to be in the thousands, not billions, of years. He reasoned that geologic forces unlike any known at the time would have been necessary to raise seafloors to mountaintops in this short time span. Catastrophic geological events would have caused extinctions, after which surviving species repopulated Earth.

Jean-Baptiste Lamarck (*left*) was thinking about processes that drive **evolution**, or change in a line of descent. A line of descent is also called a **lineage**. Lamarck thought that a species gradually improved over generations because of an inherent drive toward perfection, up the chain of being. By Lamarck's hypothesis, environmental pressures cause an internal need for change in an individual's body, and the resulting change is inherited by offspring. (Lamarck was correct in thinking that environmental factors affect traits, but his understanding of how traits are passed to offspring was incomplete.)

Charles Darwin (*left*) had earned a theology degree from Cambridge after an attempt to study medicine. All through school, however, he had spent most of his time with faculty members and other students who embraced natural history. In 1831, when he was 22, Darwin joined a 5-year survey expedition to South America on the ship *Beagle*, and he quickly became an enthusiastic naturalist. During the *Beagle*'s voyage, Darwin found many unusual fossils, and saw diverse species living in environments that ranged from the sandy shores of remote islands to plains high in the Andes. Along the way, he read the first

volume of a new and popular book, Charles Lyell's *Principles of Geology*. Lyell (*left*) was a proponent of what became known as the theory of uniformity, the idea that gradual, everyday geological processes such as erosion could have sculpted Earth's current landscape over great spans of time. The theory challenged the prevailing belief that Earth was 6,000 years old. By Lyell's calculations, it must have taken millions of years to sculpt Earth's surface. Darwin's exposure to Lyell's ideas gave him insights into the history of the regions he would encounter on his journey.

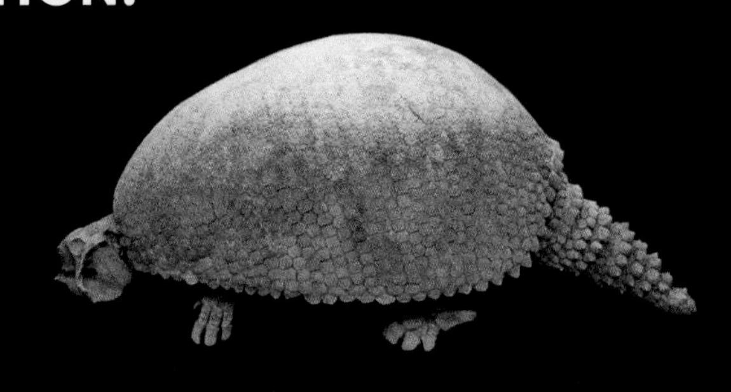

A Fossil of a glyptodon, an automobile-sized mammal that existed from 2 million to 15,000 years ago.

B A modern armadillo, about a foot long.

FIGURE 16.5 Ancient relatives: glyptodon and armadillo. Though widely separated in time, these animals share a restricted distribution and unusual traits, including a shell and helmet of keratin-covered bony plates—a material similar to crocodile and lizard skin. (The fossil in **A** is missing its helmet.)

Among the thousands of specimens Darwin collected during the *Beagle*'s voyage were fossil glyptodons. These armored mammals are extinct, but they have many unusual traits in common with modern armadillos (**FIGURE 16.5**). Armadillos also live only in places where glyptodons once lived. Could the odd shared traits and restricted distribution mean that glyptodons were ancient relatives of armadillos? If so, perhaps traits of their common ancestor had changed in the line of descent that led to armadillos. But why would such changes occur?

Economist Thomas Malthus (*left*) had correlated increases in the size of human populations with episodes of famine, disease, and war. He proposed the idea that humans run out of food, living space, and other resources because they tend to reproduce beyond the capacity of their environment to sustain them. When that happens, the individuals of a population must either compete with one another for the limited resources, or develop technology to increase productivity. Darwin realized that Malthus's ideas had wider application: All populations, not just human ones, must have the capacity to produce more individuals than their environment can support.

Darwin started thinking about how individuals of a species often vary a bit in the details of shared traits such as size, coloration, and so on. He saw such variation among finch species on isolated islands of the Galápagos archipelago. This island chain is separated from South America by 900 kilometers (550 miles) of open ocean, so most species living on the islands did not have the opportunity for interbreeding with mainland populations. The Galápagos island finches resembled finch species in South America, but many of them had unique traits that suited them to their particular island habitat.

Darwin was familiar with dramatic variations in traits of pigeons, dogs, and horses produced through selective breeding. He recognized that a natural environment could similarly select traits that make individuals of a population suited to it. It dawned on Darwin that having a particular form of a shared trait might give an individual an advantage over competing members of its species. In any population, some individuals have forms of shared traits that make them better suited to their environment than others. In other words, individuals of a natural population vary in fitness. Today, we define **fitness** as the degree of adaptation to a specific environment, and measure it by relative genetic contribution to future generations. A trait that enhances an individual's fitness is called an evolutionary **adaptation**, or **adaptive trait**.

Over many generations, individuals with the most adaptive traits tend to survive longer and reproduce more than their less fit rivals. Darwin understood that this process, which he called **natural selection**, could be a mechanism by which evolution occurs. If an individual has a form of a trait that makes it better suited to an environment, then it is better able to survive. If an individual is better able to survive, then it has a better chance of living long enough to produce offspring. If individuals with an adaptive, heritable trait produce more offspring than those that do not, then the frequency of that trait will tend to increase in the population over successive generations. **TABLE 16.1** summarizes this reasoning.

TABLE 16.1

Principles of Natural Selection, in Modern Terms

Observations About Populations

- Natural populations have an inherent capacity to increase in size over time.
- As population size increases, resources that are used by its individuals (such as food and living space) eventually become limited.
- When resources are limited, individuals of a population compete for them.

Observations About Genetics

- Individuals of a species share certain traits.
- Individuals of a natural population vary in the details of those shared traits.
- Shared traits have a heritable basis, in genes. Slightly different forms of those genes (alleles) give rise to variation in shared traits.

Inferences

- A certain form of a shared trait may make its bearer better able to survive.
- Individuals of a population that are better able to survive tend to leave more offspring.
- Thus, an allele associated with an adaptive trait tends to become more common in a population over time.

Darwin wrote out his ideas about natural selection, but let ten years pass without publishing them. In the meantime, Alfred Wallace (*left*), who had been studying wildlife in the Amazon basin and the Malay Archipelago, sent an essay to Darwin for advice. Wallace's essay outlined evolution by natural selection—the very same hypothesis as Darwin's. Wallace had written earlier letters to Darwin and Lyell about patterns in the geographic distribution of species, and had come to the same conclusion. In 1858, the idea of evolution by natural selection was presented at a scientific meeting, with Darwin and Wallace credited as authors. Wallace was in the field and knew nothing about the meeting, which Darwin did not attend. The next year, Darwin published *On the Origin of Species*, which laid out detailed evidence in support of natural selection. Many people had already accepted the idea of descent with modification (evolution). However, there was a fierce debate over the idea that evolution occurs by natural selection. Decades would pass before experimental evidence from the field of genetics led to its widespread acceptance as a theory by the scientific community.

As you will see in the remainder of this chapter, the theory of evolution by natural selection is supported by and helps explain the fossil record as well as similarities in the form, function, and biochemistry of living things.

adaptation (adaptive trait) A heritable trait that enhances an individual's fitness in a particular environment.
evolution Change in a line of descent.
fitness Degree of adaptation to an environment, as measured by an individual's relative genetic contribution to future generations.
lineage Line of descent.
natural selection Differential survival and reproduction of individuals of a population based on differences in shared, heritable traits. Driven by environmental pressures.

TAKE-HOME MESSAGE 16.2

Evidence that Earth and the species on it had changed over very long spans of time led to the theory of evolution by natural selection.

Natural selection is a process in which individuals of a population survive and reproduce with differing success depending on the details of their shared, heritable traits.

Traits favored in a particular environment are adaptive.

16.3 WHY DO BIOLOGISTS STUDY ROCKS AND FOSSILS?

A Fossil skeleton of an ichthyosaur that lived about 200 million years ago. These marine reptiles were about the same size as modern porpoises, breathed air like them, and probably swam as fast, but the two groups are not closely related.

B Extinct wasp encased in amber, which is ancient tree sap. This 9-mm-long insect lived about 20 million years ago.

C Fossilized imprint of a leaf from a 260-million-year-old *Glossopteris*, a type of plant called a seed fern.

D Fossilized footprint of a theropod, a name that means "beast foot." This group of carnivorous dinosaurs, which includes the familiar *Tyrannosaurus rex*, arose about 250 million years ago.

E Coprolite (fossilized feces). Fossilized food remains and parasitic worms inside coprolites offer clues about the diet and health of extinct species. A foxlike animal excreted this one.

FIGURE 16.6 Examples of fossils.

Even before Darwin's time, fossils were recognized as stone-hard evidence of earlier forms of life (**FIGURE 16.6**). Most fossils consist of mineralized bones, teeth, shells, seeds, spores, or other hard body parts. Trace fossils such as footprints and other impressions, nests, burrows, trails, eggshells, or feces are evidence of an organism's activities.

The process of fossilization typically begins when an organism or its traces become covered by sediments, mud, or ash. Groundwater then seeps into the remains, filling spaces around and inside of them. Minerals dissolved in the water gradually replace minerals in bones and other hard tissues. Mineral particles that crystallize and settle out of the groundwater inside cavities and impressions form detailed imprints of internal and external structures. Sediments that slowly accumulate on top of the site exert increasing pressure, and, after a very long time, extreme pressure transforms the mineralized remains into rock.

Most fossils are found in layers of sedimentary rock that forms as rivers wash silt, sand, volcanic ash, and other materials from land to sea. Mineral particles in the materials settle on the seafloor in horizontal layers that vary in thickness and composition. After many millions of years, the layers of sediments become compacted into layered sedimentary rock. Even though most sedimentary rock forms at the bottom of a sea, geologic processes can tilt the rock and lift it far above sea level, where the layers may become exposed by the erosive forces of water and wind (the chapter opening photo shows an example).

Biologists study sedimentary rock formations in order to understand life's historical context. Features of the formations can provide information about conditions in

the environment in which they formed. Consider banded iron, a unique formation named after its distinctive striped appearance (*left*). Huge deposits of this sedimentary rock are the source of most iron we mine for steel today, but they also hold a record of how the evolution of the noncyclic pathway of photosynthesis changed the chemistry of Earth. Banded iron started forming about 2.4 billion years ago, right after photosynthesis evolved (Section 7.2). At that time, Earth's atmosphere and ocean contained very little oxygen, so almost all of the iron on Earth was in a reduced form (Section 5.4). Reduced iron dissolves in water, and ocean water contained a lot of it. Oxygen released into the ocean by early photosynthetic bacteria quickly combined with the dissolved iron. The resulting oxidized iron compounds are completely insoluble in water, and they began to rain down on the ocean floor in massive quantities. These compounds accumulated in sediments that would eventually become compacted into banded iron formations.

The massive sedimentation of oxidized iron continued for about 600 million years. After that, ocean water no longer contained very much dissolved iron, and oxygen gas bubbling out of it had oxidized the iron in rocks exposed to the atmosphere.

THE FOSSIL RECORD

We have fossils for more than 250,000 known species. Considering the current range of biodiversity, there must have been many millions more, but we will never know all of them. Why not? The odds are against finding evidence of an extinct species, because fossils are relatively rare. When an organism dies, its remains are often obliterated quickly by scavengers. Organic materials decompose in the presence of moisture and oxygen, so remains that escape scavenging can endure only if they dry out, freeze, or become encased in an air-excluding material such as sap, tar, or mud. Remains that do become fossilized are often deformed, crushed, or scattered by erosion and other geologic assaults.

In order for us to know about an extinct species that existed long ago, we have to find a fossil of it. At least one specimen had to be buried before it decomposed or something ate it. The burial site had to escape destructive geologic events, and it had to be accessible for us to find.

Most ancient species had no hard parts to fossilize, so we do not find much evidence of them. For example, there are many fossils of bony fishes and mollusks with hard shells, but few fossils of the jellyfishes and soft worms that were probably much more common. Also think about relative numbers of organisms. Fungal spores and pollen grains are typically released by the billions. By contrast, the earliest humans lived in small bands and few of their offspring survived. The odds of finding even one fossilized human bone are much smaller than the odds of finding a fossilized fungal spore. Finally, imagine two species, one that existed only briefly and the other for billions of years. Which is more likely to be represented in the fossil record? Despite these challenges, the fossil record is substantial enough to help us reconstruct large-scale patterns in the history of life.

TAKE-HOME MESSAGE 16.3

Fossils are evidence of organisms that lived in the remote past, a stone-hard historical record of life.

The fossil record will never be complete. Geologic events have obliterated much of it. The rest of the record is slanted toward species that had hard parts, lived in dense populations with wide distribution, and persisted for a long time.

PEOPLE MATTER

National Geographic Explorer-in-Residence
DR. PAUL SERENO

A real-life Indiana Jones, paleontologist Paul Sereno blends his background as an artist with a love for science and history. Sereno's passion carries him to the remote corners of the world to discover new species under the harshest of conditions.

Sereno's fieldwork began in 1988 in the Andes, where his team discovered the first dinosaurs to roam the Earth, including the most primitive of all: *Eoraptor*. This work culminated in the most complete picture yet of the dawn of the dinosaur era, some 225 million years ago. In the 1990s, Sereno's expeditions shifted to the Sahara. There, his teams have since excavated more than 70 tons of fossils representing organisms such as the huge-clawed fish-eater *Suchomimus*, the gigantic *Carcharodontosaurus* (its jaws pictured above, with Sereno), and a series of crocs including the 40-foot-long "SuperCroc" *Sarcosuchus*, the world's largest crocodile. In 2001, a trip to India yielded the Asian continent's first dinosaur skull—a new species of predator, *Rajasaurus*. Also in 2001, Sereno began an ongoing series of expeditions to China, first exploring remote areas of the Gobi in Inner Mongolia and discovering a herd of more than 20 dinosaurs that had died in their tracks. In 2012, he reported the discovery of *Pegomastax*, a bizarre, cat-sized dinosaur with a parrotlike beak and sharp fangs.

16.4 HOW IS THE AGE OF ROCKS AND FOSSILS MEASURED?

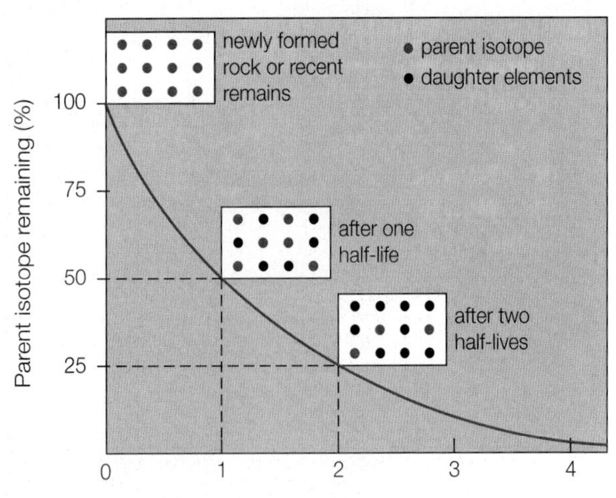

FIGURE 16.7 {Animated} Half-life.

FIGURE IT OUT: How much of any radioisotope remains after two of its half-lives have passed? Answer: 25 percent

A Long ago, ^{14}C and ^{12}C were incorporated into the tissues of a nautilus. The carbon atoms were part of organic molecules in the animal's food. ^{12}C is stable and ^{14}C decays, but the proportion of the two isotopes in the nautilus's tissues remained the same. Why? The nautilus continued to gain both types of carbon atoms in the same proportions from its food.

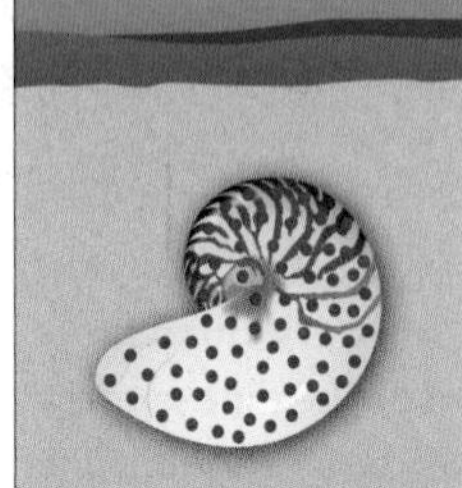

B The nautilus stopped eating when it died, so its body stopped gaining carbon. The ^{12}C atoms in its tissues were stable, but the ^{14}C atoms (represented as red dots) were decaying into nitrogen atoms. Thus, over time, the amount of ^{14}C decreased relative to the amount of ^{12}C. After 5,730 years, half of the ^{14}C had decayed; after another 5,730 years, half of what was left had decayed, and so on.

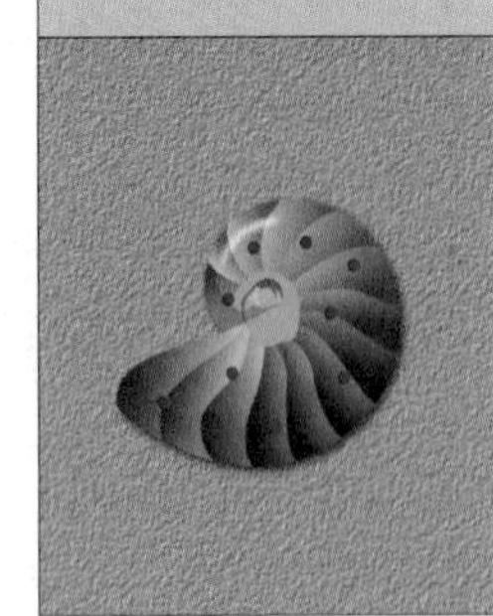

C Fossil hunters discover the fossil and measure its content of ^{14}C and ^{12}C. They use the ratio of these isotopes to calculate how many half-lives passed since the organism died. For example, if its ^{14}C to ^{12}C ratio is one-eighth of the ratio in living organisms, then three half-lives $(½)^3$ must have passed since it died. Three half-lives of ^{14}C is 17,190 years.

FIGURE 16.8 {Animated} Example of how radiometric dating is used to find the age of a carbon-containing fossil. Carbon 14 (^{14}C) is a radioisotope of carbon that decays into nitrogen. It forms in the atmosphere and combines with oxygen to become CO_2, which enters food chains by way of photosynthesis.

Remember from Section 2.1 that a radioisotope is a form of an element with an unstable nucleus. Atoms of a radioisotope become atoms of other elements—daughter elements—as their nucleus disintegrates. This radioactive decay is not influenced by temperature, pressure, chemical bonding state, or moisture; it is influenced only by time. Thus, like the ticking of a perfect clock, each type of radioisotope decays at a constant rate. The time it takes for half of the atoms in a sample of radioisotope to decay is called a **half-life** (**FIGURE 16.7**).

Half-life is a characteristic of each radioisotope. For example, radioactive uranium 238 decays into thorium 234, which decays into something else, and so on until it becomes lead 206. The half-life of the decay of uranium 238 to lead 206 is 4.5 billion years.

The predictability of radioactive decay can be used to find the age of a volcanic rock (the date it solidified). Rock deep inside Earth is hot and molten, so atoms swirl and mix in it. Rock that reaches the surface cools and hardens. As the rock cools, minerals crystallize in it. Each kind of mineral has a characteristic structure and composition. For example, the mineral zircon (*left*) consists mainly of ordered arrays of zirconium silicate molecules ($ZrSiO_4$). Some of the molecules in a newly formed zircon crystal have uranium atoms substituted for zirconium atoms, but never lead atoms. However, uranium decays into lead at a predictable rate. Thus, over time, uranium atoms disappear from a zircon crystal, and lead atoms accumulate in it. The ratio of uranium atoms to lead atoms in a zircon crystal can be measured precisely. That ratio can be used to calculate how long ago the crystal formed (its age).

zircon

We have just described **radiometric dating**, a method that can reveal the age of a material by measuring its content of a radioisotope and daughter elements. The oldest known terrestrial rock, a tiny zircon crystal from the Jack Hills in Western Australia, is 4.404 billion years old.

Recent fossils that still contain carbon can be dated by measuring their carbon 14 content (**FIGURE 16.8**). Most of the ^{14}C in a fossil will have decayed after about 60,000 years. The age of fossils older than that can be estimated by dating volcanic rocks in lava flows above and below the fossil-containing layer of sedimentary rock.

FINDING A MISSING LINK

The discovery of intermediate forms of cetaceans (an order of animals that includes whales, dolphins, and porpoises) provides an example of how scientists use fossil finds and radiometric dating to piece together evolutionary history. For some time, evolutionary biologists predicted that the ancestors of modern cetaceans walked on land, then took up life in the water. Evidence in support of this line of thinking includes a set of distinctive features of the skull and lower jaw that cetaceans share with some kinds of ancient carnivorous land animals. DNA sequence comparisons indicate that the ancient land animals were probably artiodactyls, hooved mammals with an even number of toes (two or four) on each foot (**FIGURE 16.9A**). Modern representatives of the artiodactyl lineage include camels, hippopotamuses, pigs, deer, sheep, and cows.

Until recently, we had no fossils demonstrating gradual changes in skeletal features that accompanied a transition of whale lineages from terrestrial to aquatic life. Researchers knew there were intermediate forms because they had found a representative fossil skull of an ancient whalelike animal, but without a complete skeleton the rest of the story remained speculative.

Then, in 2000, Philip Gingerich and his colleagues unearthed complete skeletons of two ancient whales: a fossil *Rodhocetus kasrani* excavated from a 47-million-year-old rock formation in Pakistan, and a fossil *Dorudon atrox*, from 37-million-year-old rock in Egypt (**FIGURE 16.9B,C**). Both fossil skeletons had whalelike skull bones, as well as intact ankle bones. The ankle bones of both fossils have distinctive features in common with those of extinct and modern artiodactyls. Modern cetaceans do not have even a remnant of an ankle bone (**FIGURE 16.9D**).

Rodhocetus and *Dorudon* were not direct ancestors of modern whales, but their telltale ankle bones mean they are long-lost relatives. Both whales were offshoots of the ancient artiodactyl-to-modern-whale lineage as it transitioned from life on land to life in water. The proportions of limbs, skull, neck, and thorax indicate *Rodhocetus* swam with its feet, not its tail. Like modern whales, the 5-meter (16-foot) *Dorudon* was clearly a fully aquatic tail-swimmer: The entire hindlimb was only about 12 centimeters (5 inches) long, much too small to have supported the animal's tremendous body out of water.

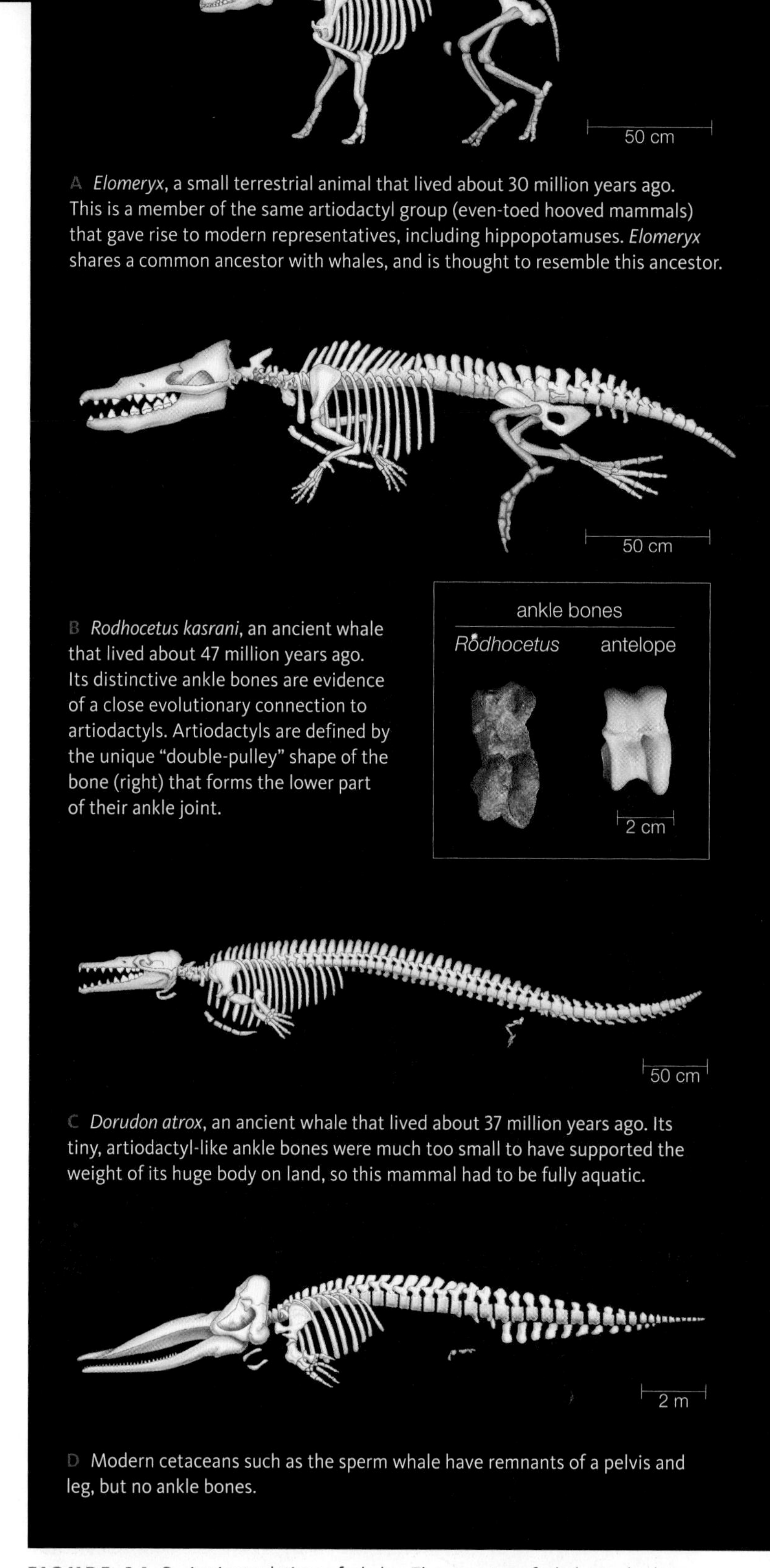

A *Elomeryx*, a small terrestrial animal that lived about 30 million years ago. This is a member of the same artiodactyl group (even-toed hooved mammals) that gave rise to modern representatives, including hippopotamuses. *Elomeryx* shares a common ancestor with whales, and is thought to resemble this ancestor.

B *Rodhocetus kasrani*, an ancient whale that lived about 47 million years ago. Its distinctive ankle bones are evidence of a close evolutionary connection to artiodactyls. Artiodactyls are defined by the unique "double-pulley" shape of the bone (right) that forms the lower part of their ankle joint.

C *Dorudon atrox*, an ancient whale that lived about 37 million years ago. Its tiny, artiodactyl-like ankle bones were much too small to have supported the weight of its huge body on land, so this mammal had to be fully aquatic.

D Modern cetaceans such as the sperm whale have remnants of a pelvis and leg, but no ankle bones.

FIGURE 16.9 Ancient relatives of whales. The ancestor of whales and other cetaceans was an artiodactyl that walked on land. The lineage transitioned from life on land to life in water over millions of years, and as it did, the animals' limb bones became smaller and smaller. Comparable hindlimb bones are highlighted in blue.

---

half-life Characteristic time it takes for half of a quantity of a radioisotope to decay.
radiometric dating Method of estimating the age of a rock or fossil by measuring the content and proportions of a radioisotope and its daughter elements.

TAKE-HOME MESSAGE 16.4

The predictability of radioisotope decay can be used to estimate the age of rock layers and fossils in them.

Radiometric dating helps evolutionary biologists retrace changes in ancient lineages.

16.5 HOW HAS EARTH CHANGED OVER GEOLOGIC TIME?

Wind, water, and other forces continuously sculpt Earth's surface, but they are only part of a much bigger picture of geological change. All continents that exist today were once part of a supercontinent—**Pangea**—that split into fragments and drifted apart.

The idea that continents move around, originally called continental drift, was proposed in the early 1900s to explain why the Atlantic coasts of South America and Africa seem to "fit" like jigsaw puzzle pieces, and why the same types of fossils occur in identical rock formations on both sides of the Atlantic Ocean. It also explained why the magnetic poles of gigantic rock formations point in different directions on different continents. Rock forms when molten lava solidifies on Earth's surface. Some iron-rich minerals become magnetic as they solidify, and their magnetic poles align with Earth's poles when they do. If the continents never moved, then all of these ancient rocky magnets should be aligned north-to-south, like compass needles. Indeed, the magnetic poles of each rock formation are aligned—but they do not always point north-to-south. Either Earth's magnetic poles veer dramatically from their north–south axis, or the continents must wander.

The concept of moving continents was initially greeted with skepticism because there was no known mechanism capable of causing such movements. Then, in the late 1950s, deep-sea explorers found immense ridges and trenches stretching thousands of kilometers across the seafloor (FIGURE 16.10). The discovery led to the **plate tectonics theory**, which explains how continents move: Earth's outer layer of rock is cracked into immense plates, like a huge cracked eggshell. Molten rock streaming from an undersea ridge ❶ or continental rift at one edge of a plate pushes old rock at the opposite edge into a trench ❷. The movement is like that of a colossal conveyor belt that transports continents on top of it to new locations. The plates move no more than 10 centimeters (4 inches) a year—about half as fast as your toenails grow—but it is enough to carry a continent all the way around the world after 40 million years or so (FIGURE 16.11).

The San Andreas Fault, which extends 800 miles through California, marks the boundary between two tectonic plates.

FIGURE 16.10 **Plate tectonics.** Huge pieces of Earth's outer layer of rock slowly drift apart and collide. As these plates move, they convey continents around the globe.

❶ At oceanic ridges, plumes of molten rock welling up from Earth's interior drive the movement of tectonic plates. New crust spreads outward as it forms on the surface, forcing adjacent tectonic plates away from the ridge and into trenches elsewhere.

❷ At trenches, the advancing edge of one plate plows under an adjacent plate and buckles it.

❸ Faults are ruptures in Earth's crust where plates meet. The diagram shows a rift fault, in which plates move apart. The photo above shows a strike-slip fault, in which two abutting plates slip against one another in opposite directions.

❹ Plumes of molten rock rupture a tectonic plate at what are called "hot spots." The Hawaiian Islands have been forming from molten rock that continues to erupt from a hot spot under the Pacific Plate. This and other tectonic plates are shown in Appendix V.

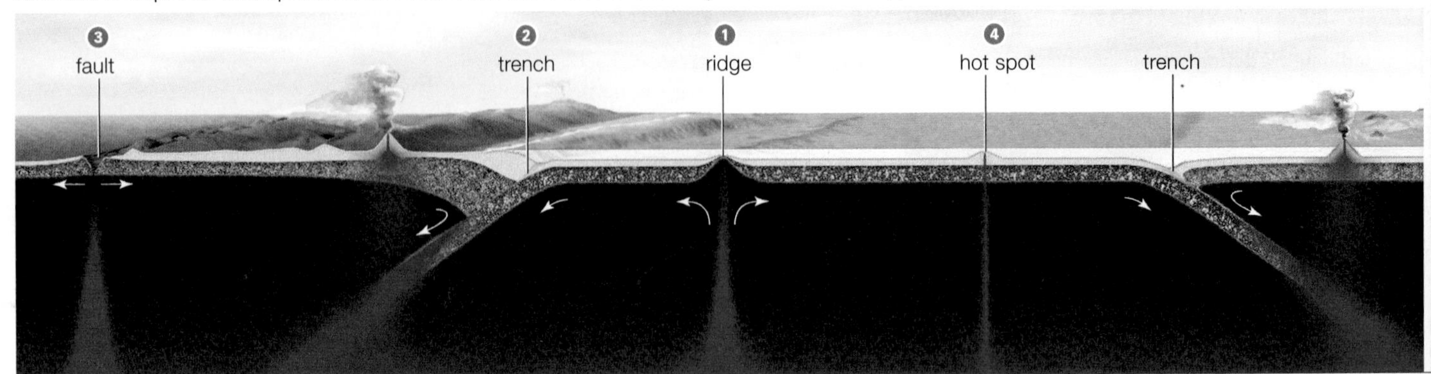

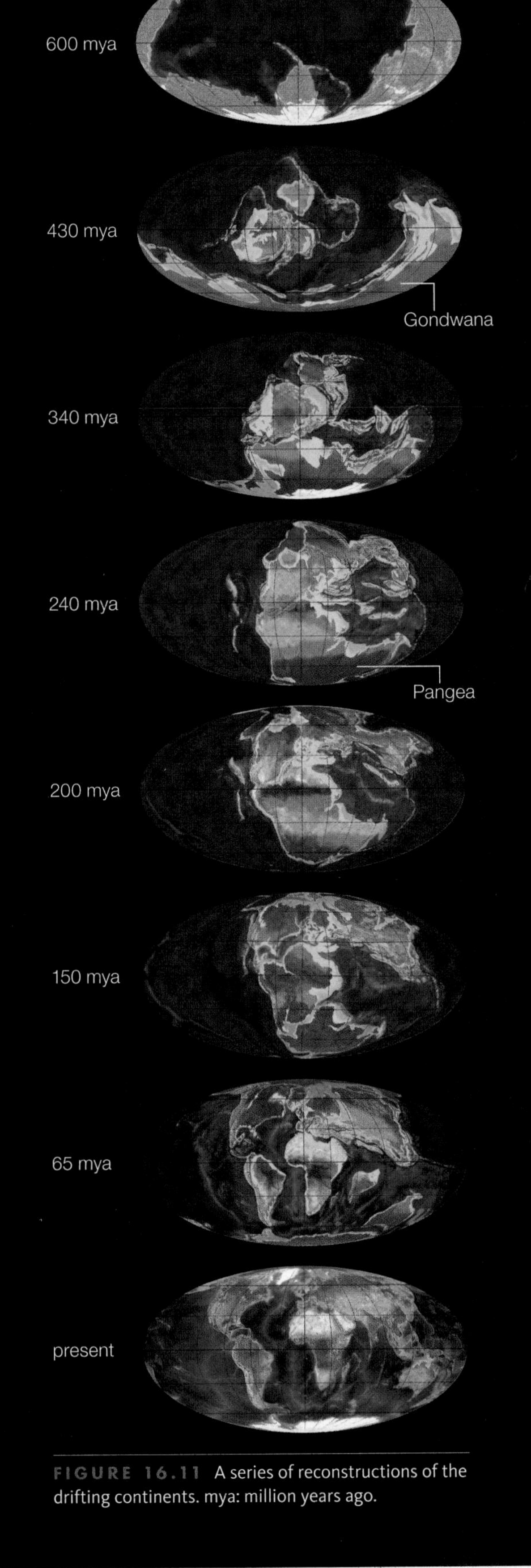

FIGURE 16.11 A series of reconstructions of the drifting continents. mya: million years ago.

Evidence of tectonic movement is all around us, in faults ❸ and other geological features of our landscapes. For example, volcanic island chains (archipelagos) form as a plate moves across an undersea hot spot. These hot spots are places where a plume of molten rock wells up from deep inside Earth and ruptures a tectonic plate ❹.

The fossil record also provides evidence in support of plate tectonics. Consider an unusual rock formation that exists in a huge belt across Africa. The sequence of rock layers in this formation is so complex that it is quite unlikely to have formed more than once, but identical sequences of layers also occur in huge belts that span India, South America, Madagascar, Australia, and Antarctica. Across all of these continents, the layers are the same ages. They also hold fossils found nowhere else, including imprints of the seed fern *Glossopteris* (pictured in **FIGURE 16.6C**). The most probable explanation for these observations is that the layered rock formed in one long belt on a single continent, which later broke up.

We now know that at least five times since Earth's outer layer of rock solidified 4.55 billion years ago, supercontinents formed and then split up again. One called **Gondwana** formed about 500 million years ago. Over the next 230 million years, this supercontinent wandered across the South Pole, then drifted north until it merged with other landmasses to form Pangea (**FIGURE 16.11**). Most of the landmasses currently in the Southern Hemisphere as well as India and Arabia were once part of Gondwana. Many modern species, including the birds pictured in **FIGURE 16.1**, live only in these places.

As you will see in later chapters, the changes brought on by plate tectonics have had a profound impact on life. Colliding continents have physically separated organisms living in oceans, and brought together those that had been living apart on land. As continents broke up, they separated organisms living on land, and brought together ones that had been living in separate oceans. Such changes have been a major driving force of evolution, a topic that we return to in the next chapter.

Gondwana Supercontinent that existed before Pangea, more than 500 million years ago.
Pangea Supercontinent that formed about 270 million years ago.
plate tectonics theory Theory that Earth's outer layer of rock is cracked into plates, the slow movement of which rafts continents to new locations over geologic time.

TAKE-HOME MESSAGE 16.5

Over geologic time, movements of Earth's crust have caused dramatic changes in continents and oceans. These changes profoundly influenced the course of life's evolution.

Eon	Era	Period	Epoch	mya	Major Geologic and Biological Events
Phanerozoic	Cenozoic	Quaternary	Recent		
			Pleistocene	0.01	Modern humans evolve. Major extinction event is now under way.
		Neogene	Pliocene	2.5	Tropics, subtropics extend poleward. Climate cools; dry woodlands and grasslands emerge. Adaptive radiations of mammals, insects, birds.
			Miocene	5.3	
		Paleogene	Oligocene	23.0	
			Eocene	33.9	
			Paleocene	56.0	
				66.0 ◀	Major extinction event
	Mesozoic	Cretaceous	Upper		Flowering plants diversify; sharks evolve. All dinosaurs and many marine organisms disappear at the end of this epoch.
				100.5	
			Lower		Climate very warm. Dinosaurs continue to dominate. Important modern insect groups appear (bees, butterflies, termites, ants, and herbivorous insects including aphids and grasshoppers). Flowering plants originate and become dominant land plants.
				145.0	
		Jurassic			Age of dinosaurs. Lush vegetation; abundant gymnosperms and ferns. Birds appear. Pangea breaks up.
				201.3 ◀	Major extinction event
		Triassic			Recovery from the major extinction at end of Permian. Many new groups appear, including turtles, dinosaurs, pterosaurs, and mammals.
				252 ◀	Major extinction event
	Paleozoic	Permian			Supercontinent Pangea and world ocean form. Adaptive radiation of conifers. Cycads and ginkgos appear. Relatively dry climate leads to drought-adapted gymnosperms and insects such as beetles and flies.
				299	
		Carboniferous			High atmospheric oxygen level fosters giant arthropods. Spore-releasing plants dominate. Age of great lycophyte trees; vast coal forests form. Ears evolve in amphibians; penises evolve in early reptiles (vaginas evolve later, in mammals only).
				359 ◀	Major extinction event
		Devonian			Land tetrapods appear. Explosion of plant diversity leads to tree forms, forests, and many new plant groups including lycophytes, ferns with complex leaves, seed plants.
				419	
		Silurian			Radiations of marine invertebrates. First appearances of land fungi, vascular plants, bony fishes, and perhaps terrestrial animals (millipedes, spiders).
				443 ◀	Major extinction event
		Ordovician			Major period for first appearances. The first land plants, fishes, and reef-forming corals appear. Gondwana moves toward the South Pole and becomes frigid.
				485	
		Cambrian			Earth thaws. Explosion of animal diversity. Most major groups of animals appear (in the oceans). Trilobites and shelled organisms evolve.
				541	
Proterozoic					Oxygen accumulates in atmosphere. Origin of aerobic metabolism. Origin of eukaryotic cells, then protists, fungi, plants, animals. Evidence that Earth mostly freezes over in a series of global ice ages between 750 and 600 mya.
				2,500	
Archaean and earlier					3,800–2,500 mya. Origin of bacteria and archaea. 4,600–3,800 mya. Origin of Earth's crust, first atmosphere, first seas. Chemical, molecular evolution leads to origin of life (from protocells to anaerobic single cells).

FIGURE 16.12 {Animated} The geologic time scale (above) correlated with sedimentary rock exposed by erosion in the Grand Canyon (opposite). Orange triangles mark times of great mass extinctions. "First appearance" refers to appearance in the fossil record, not necessarily the first appearance on Earth. mya: million years ago. Dates are from the International Commission on Stratigraphy, 2013.

Similar sequences of sedimentary rock layers occur around the world. Transitions between the layers mark boundaries between great intervals of time in the **geologic time scale**, which is a chronology of Earth's history (**FIGURE 16.12**). Each layer's composition offers clues about conditions on Earth during the time the layer was deposited. Fossils in the layers are a record of life during that period of time.

geologic time scale Chronology of Earth's history.

TAKE-HOME MESSAGE 16.6

The geologic time scale correlates geological and evolutionary events of the ancient past.

Each rock layer has a composition and set of fossils that reflect events during its deposition. For example, Coconino Sandstone, which stretches from California to Montana, is mainly weathered sand. Ripple marks and reptile tracks are the only fossils in it. Many think it is the remains of a vast sand desert, similar to the modern Sahara.

FIGURE IT OUT: Which formation is marked with the ? ?

Answer: Tapeats sandstone

16.7 WHAT EVIDENCE DOES EVOLUTION LEAVE IN BODY FORM?

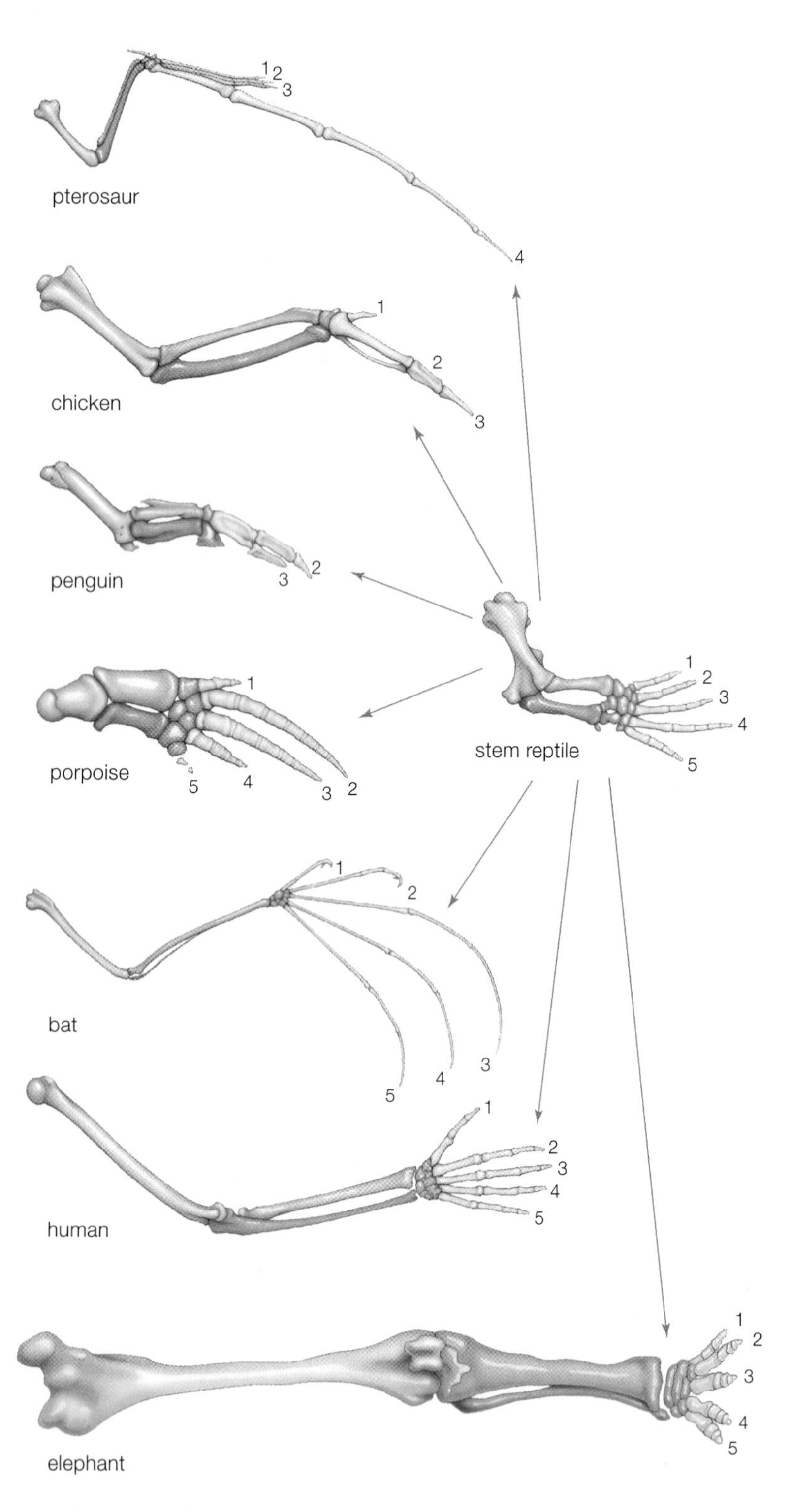

FIGURE 16.13 {Animated} Morphological divergence among vertebrate forelimbs, starting with the bones of an ancient stem reptile. The number and position of many skeletal elements were preserved when these diverse forms evolved; notice the bones of the forearms. Certain bones were lost over time in some of the lineages (compare the digits numbered 1 through 5). Drawings are not to scale.

To biologists, remember, evolution means change in a line of descent. How do they reconstruct evolutionary events that occurred in the ancient past? Evolutionary biologists are a bit like detectives, using clues to piece together a history that they did not witness in person. Fossils provide some clues. The body form and function of organisms that are alive today provide others.

MORPHOLOGICAL DIVERGENCE

Body parts that appear similar in separate lineages because they evolved in a common ancestor are called **homologous structures** (*hom*– means "the same"). Homologous structures may be used for different purposes in different groups, but the very same genes direct their development.

A body part that outwardly appears very different in separate lineages may be homologous in underlying form. Vertebrate forelimbs, for example, vary in size, shape, and function. However, they clearly are alike in the structure and positioning of bony elements, and in their internal patterns of nerves, blood vessels, and muscles.

As you will see in the next chapter, populations that are not interbreeding diverge genetically, and in time these divergences give rise to changes in body form. Change from the body form of a common ancestor is an evolutionary pattern called **morphological divergence**. Consider the limb bones of modern vertebrate animals. Fossil evidence suggests that many vertebrates are descended from a family of ancient "stem reptiles" that crouched low to the ground on five-toed limbs. Descendants of this ancestral group diversified over millions of years, and eventually gave rise to modern reptiles, birds, and mammals. A few lineages that had become adapted to walking on land even returned to life in the seas. During this time, the limbs became adapted for many different purposes (**FIGURE 16.13**). They became modified for flight in extinct reptiles called pterosaurs and in bats and most birds. In penguins and porpoises, the limbs are now flippers useful for swimming. In humans, five-toed forelimbs became arms and hands with four fingers and an opposable thumb. Among elephants, the limbs are now strong and pillarlike, capable of supporting a great deal of weight. Limbs degenerated to nubs in pythons and boa constrictors, and they disappeared entirely in other snakes.

MORPHOLOGICAL CONVERGENCE

Body parts that appear similar in different species are not always homologous; they sometimes evolve independently in lineages subject to the same environmental pressures. The independent evolution of similar body parts in different lineages is **morphological convergence**. Structures that are similar as a result of morphological convergence are called **analogous structures**. Analogous structures look alike but did not evolve in a shared ancestor; they evolved independently after the lineages diverged.

For example, bird, bat, and insect wings all perform the same function, which is flight. However, several clues tell us that the wing surfaces are not homologous. All of the wings are adapted to the same physical constraints that govern flight, but each is adapted in a different way. In the case of birds and bats, the limbs themselves are homologous, but the adaptations that make those limbs useful for flight differ. The surface of a bat wing is a thin, membranous extension of the animal's skin. By contrast, the surface of a bird wing is a sweep of feathers, which are specialized structures derived from skin. Insect wings differ even more. An insect wing forms as a saclike extension of the body wall. Except at forked veins, the sac flattens and fuses into a thin membrane. The sturdy, chitin-reinforced veins structurally support the wing. Unique adaptations for flight are evidence that wing surfaces of birds, bats, and insects are analogous structures that evolved after the ancestors of these modern groups diverged (**FIGURE 16.14**).

As another example of morphological convergence, the similar external structures of American cacti and African euphorbias (see **FIGURE 16.2**) are adaptations to similarly harsh desert environments where rain is scarce. Distinctive accordion-like pleats allow the plant body to swell with water when rain does come. Water stored in the plants' tissues allows them to survive long dry periods. As the stored water is used, the plant body shrinks, and the folded pleats provide it with some shade in an environment that typically has none. Despite these similarities, a closer look reveals many differences that indicate the two types of plants are not closely related. For example, cactus spines have a simple fibrous structure; they are modified leaves that arise from dimples on the plant's surface. Euphorbia spines project smoothly from the plant surface, and they are not modified leaves: In many species the spines are actually dried flower stalks (*below*).

FIGURE 16.14 Morphological convergence in animals. The surfaces of an insect wing, a bat wing, and a bird wing are analogous structures. The diagram shows how the evolution of wings (yellow dots) occurred independently in the three separate lineages that led to bats, birds, and insects. You will read more about diagrams that show evolutionary relationships in Section 17.12.

analogous structures Similar body structures that evolved separately in different lineages.
homologous structures Body structures that are similar in different lineages because they evolved in a common ancestor.
morphological convergence Evolutionary pattern in which similar body parts evolve separately in different lineages.
morphological divergence Evolutionary pattern in which a body part of an ancestor changes in its descendants.

TAKE-HOME MESSAGE 16.7

Body parts are often modified differently in different lines of descent.

Some body parts that appear alike evolved independently in different lineages.

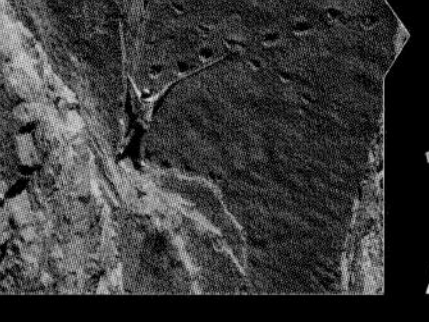

16.8 HOW DO SIMILARITIES IN DNA AND PROTEIN REFLECT EVOLUTION?

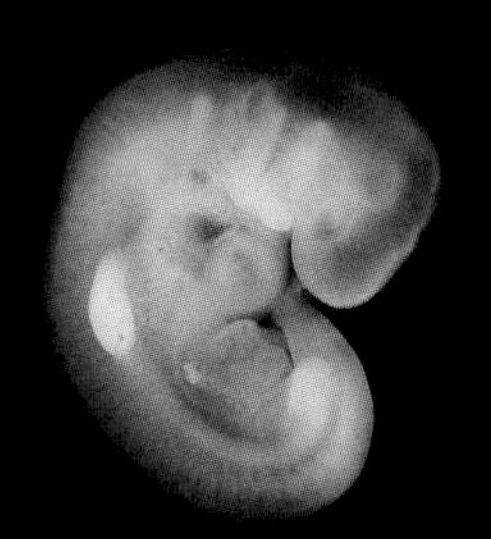
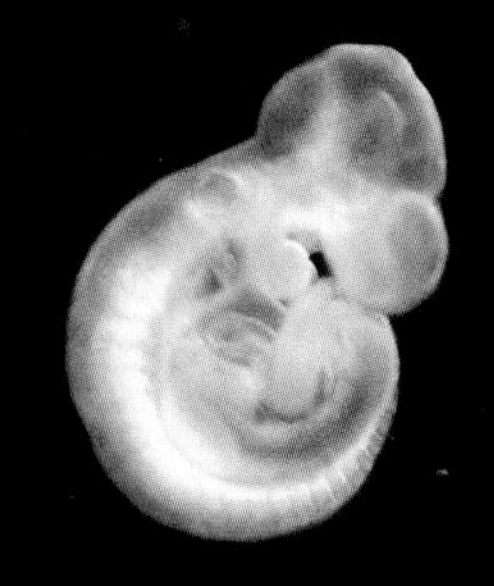
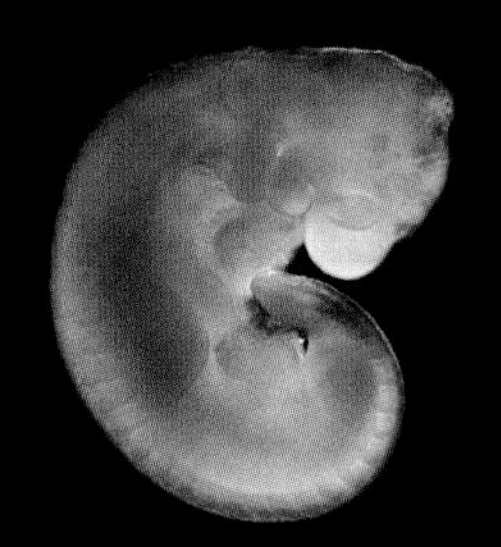
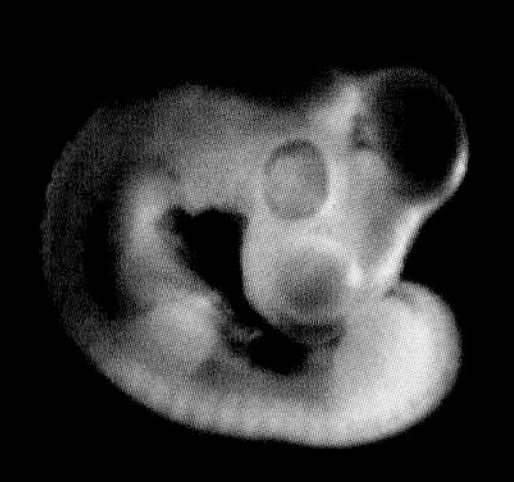
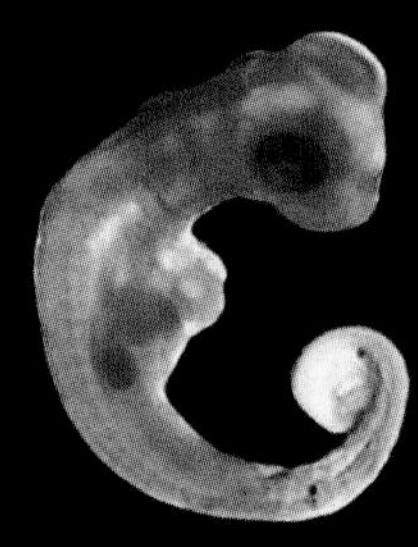

FIGURE 16.15 Visual comparison of vertebrate embryos. All vertebrates go through an embryonic stage in which they have four limb buds, a tail, and divisions called somites along their back. From *left* to *right*: human, mouse, bat, chicken, alligator.

Evolution also leaves clues in patterns of embryonic development: In general, the more closely related animals are, the more similar is their development. For example, all vertebrates go through a stage during which a developing embryo has four limb buds, a tail, and a series of somites—divisions of the body that give rise to the backbone and associated skin and muscle (**FIGURE 16.15**).

Animals have similar patterns of embryonic development because the very same master genes direct the process. Remember from Section 10.2 that the development of an embryo into the body of a plant or animal is orchestrated by layer after layer of master gene expression. The failure of any single master gene to participate in this symphony of expression can result in a drastically altered body plan, typically with devastating consequences. Because a mutation in a master gene typically unravels development completely, these genes tend to be highly conserved. Even among lineages that diverged a very long time ago, such genes often retain similar sequences and functions.

FIGURE 16.16 How differences in body form arise from differences in master gene expression. Expression of the *Hoxc6* gene is indicated by purple stain in two vertebrate embryos, chicken (*left*) and garter snake (*right*). Expression of this gene causes a vertebra to develop ribs as part of the back. Chickens have 7 vertebrae in their back and 14 to 17 vertebrae in their neck; snakes have upwards of 450 back vertebrae and essentially no neck.

Consider homeotic genes called *Hox*. Like other homeotic genes, *Hox* gene expression helps sculpt details of the body's form during embryonic development. Vertebrate animals have multiple sets of the same ten *Hox* genes that occur in insects and other arthropods. You have already read about one of these genes, *antennapedia*, which determines the identity of the thorax (the body part with legs) in fruit flies. One vertebrate version of *antennapedia* is called *Hoxc6*,

FIGURE 16.17 Example of a protein comparison. Here, part of the amino acid sequence of mitochondrial cytochrome *b* from 20 species is aligned. This protein is a crucial component of mitochondrial electron transfer chains. The honeycreeper sequence is identical in ten species of honeycreeper; amino acids that differ in the other species are shown in red. Dashes are gaps in the alignment.

FIGURE IT OUT: Based on this comparison, which species is the most closely related to the honeycreepers?

Answer: The song sparrow

honeycreepers (10)	...CRDVQFGWLIRNLHANGASFFFICIYLHIGRGIYYGSYLNK--ETWNIGVILLLTLMATAFVGYVLPWGQMSFWG...
song sparrow	...CRDVQFGWLIRNLHANGASFFFICIYLHIGRGIYYGSYLNK--ETWNVGIILLLALMATAFVGYVLPWGQMSFWG...
Gough Island finch	...CRDVQFGWLIRNIHANGASFFFICIYLHIGRGLYYGSYLYK--ETWNVGVILLLTLMATAFVGYVLPWGQMSFWG...
deer mouse	...CRDVNYGWLIRYMHANGASMFFICLFLHVGRGMYYGSYTFT--ETWNIGIVLLFAVMATAFMGYVLPWGQMSFWG...
Asiatic black bear	...CRDVHYGWIIRYMHANGASMFFICLFMHVGRGLYYGSYLLS--ETWNIGIILLFTVMATAFMGYVLPWGQMSFWG...
bogue (a fish)	...CRDVNYGWLIRNLHANGASFFFICIYLHIGRGLYYGSYLYK--ETWNIGVVLLLLVMGTAFVGYVLPWGQMSFWG...
human	...TRDVNYGWIIRYLHANGASMFFICLFLHIGRGLYYGSFLYS--ETWNIGIILLLATMATAFMGYVLPWGQMSFWG...
thale cress (a plant)	...MRDVEGGWLLRYMHANGASMFLIVVYLHIFRGLYHASYSSPREFVWCLGVVIFLLMIVTAFIGYVLPWGQMSFWG...
baboon louse	...ETDVMNGWMVRSIHANGASWFFIMLYSHIFRGLWVSSFTQP--LVWLSGVIILFLSMATAFLGYVLPWGQMSFWG...
baker's yeast	...MRDVHNGYILRYLHANGASFFFMVMFMHMAKGLYYGSYRSPRVTLWNVGVIIFTLTIATAFLGYCCVYGQMSHWG...

and it determines the identity of the back (as opposed to the neck or tail). Expression of the *Hoxc6* gene causes ribs to develop on a vertebra. Vertebrae of the neck and tail normally develop with no *Hoxc6* expression, and no ribs (**FIGURE 16.16**).

Given that the very same genes direct development in all of the vertebrate lineages, how do the adult forms end up so different? Part of the answer is that there are differences in the onset, rate, or completion of early steps in development brought about by variations in master gene expression. The variation has arisen at least in part as a result of gene duplications followed by mutation, the same way that multiple globin genes evolved in primates (Section 14.4).

Genes that are not conserved are the basis of major phenotypic differences that define species. Over time, inevitable mutations change the DNA sequence of a lineage's genome. The more recently two lineages diverged, the less time there has been for unique mutations to accumulate in the DNA of each one. That is why the genomes of closely related species tend to be more similar than those of distantly related ones—a general rule that can be used to estimate relative times of divergence. Two species with very few similar genes probably have not shared an ancestor for a long time—long enough for many mutations to have accumulated in the DNA of their separate lineages. Consider that about 88 percent of the mouse genome sequence is identical with the human genome, as is 73 percent of the zebrafish genome, 47 percent of the fruit fly genome, and 25 percent of the rice genome.

Getting useful information from comparing DNA requires a lot more data than comparing proteins. This is because coincidental homologies are statistically more likely to occur with DNA comparisons—there are only four nucleotides in DNA versus twenty amino acids in proteins. Thus, proteins are more commonly compared. By comparing the amino acid sequence of a protein among several species, the number of amino acid differences can be used as a measure of relative relatedness (**FIGURE 16.17**).

TAKE-HOME MESSAGE 16.8

Similarities in patterns of animal development occur because the same genes direct the process. Similar developmental patterns—and shared genes—are evidence of common ancestry, which can be ancient.

Mutations change the nucleotide sequence of each lineage's DNA over time. There are generally fewer differences between the DNA of more closely related lineages.

Similar genes give rise to similar proteins. Fewer differences occur among the proteins of more closely related lineages.

16.9 REFLECTIONS OF A DISTANT PAST

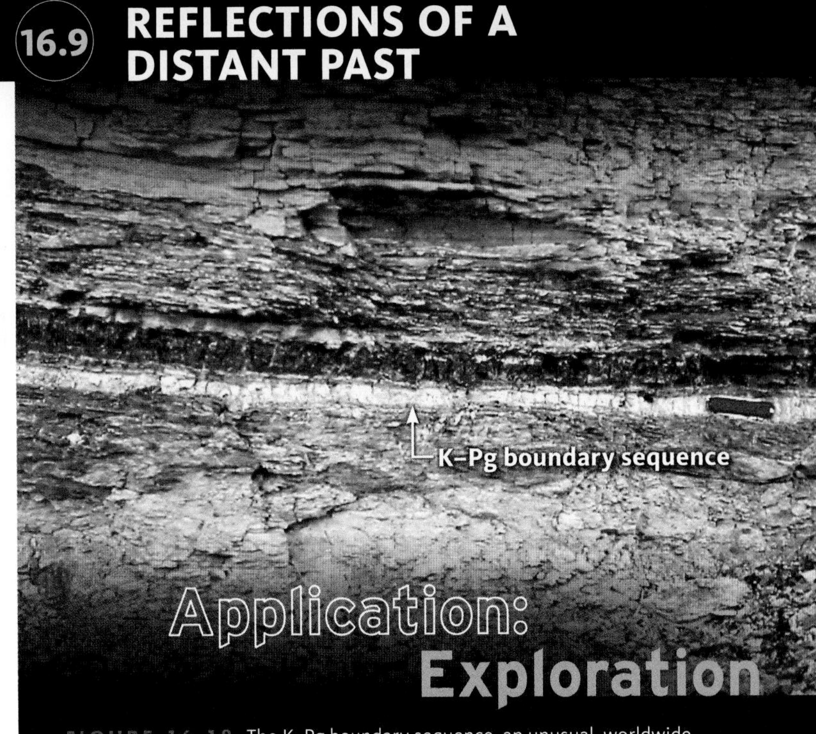

FIGURE 16.18 The K–Pg boundary sequence, an unusual, worldwide sedimentary rock formation that formed 66 million years ago.

WHAT KILLED THE DINOSAURS? Most scientists now think that the dinosaurs perished in the aftermath of a catastrophic meteorite impact. No human witnesses were around at the time, so how do they know what happened? The event is marked by an unusual, worldwide formation of sedimentary rock (FIGURE 16.18). There are plenty of dinosaur fossils below this formation, which is called the K–Pg boundary sequence (formerly known as the K–T boundary). Above it, there are none, anywhere. The rock consists of an unusual clay rich in iridium, an element much more abundant in asteroids than in Earth's crust. It also contains shocked quartz (*left*) and small glass spheres called tektites, minerals that form when quartz or sand undergoes a sudden, violent application of extreme pressure. The only processes on Earth that produce these minerals are atomic bomb explosions and meteorite impacts.

Geologists concluded that the K–Pg boundary layer must have originated with extraterrestrial material, and began looking for evidence of a meteorite that hit Earth 66 million years ago—one big enough to cover the entire planet with its debris. Twenty years later, they found it: an impact crater the size of Ireland off the coast of the Yucatán Peninsula. To make a crater this big, a meteorite 20 km (12 miles) wide would have slammed into Earth with the force of 100 trillion tons of dynamite—enough to cause an ecological disaster of sufficient scale to wipe out almost all life on Earth.

Summary

SECTION 16.1 Expeditions by nineteenth-century explorers yielded increasingly detailed observations of nature. Geology, **biogeography**, and **comparative morphology** of organisms and their **fossils** led to new ways of thinking about the natural world.

SECTION 16.2 Prevailing belief systems may influence interpretation of the underlying cause of a natural event. Nineteenth-century naturalists tried to reconcile traditional belief systems with physical evidence of **evolution,** or change in a **lineage** over time.

Humans select desirable traits in animals by selective breeding. Charles Darwin and Alfred Wallace independently came up with a theory of how environments also select traits, stated here in modern terms: A population tends to grow until it exhausts environmental resources. As that happens, competition for those resources intensifies among the population's members. Individuals with forms of shared, heritable traits that give them an advantage in this competition tend to produce more offspring. Thus, **adaptive traits** (**adaptations**) that impart greater **fitness** to an individual become more common in a population over generations. The process in which environmental pressures result in the differential survival and reproduction of individuals of a population is called **natural selection**. It is one of the processes that drives evolution.

SECTION 16.3 Fossils are typically found in stacked layers of sedimentary rock. Younger fossils usually occur in layers deposited more recently, on top of older fossils in older layers. Fossils are relatively scarce, so the fossil record will always be incomplete.

SECTION 16.4 A radioisotope's characteristic **half-life** can be used to determine the age of rocks and fossils. This technique, **radiometric dating**, helps us understand the ancient history of many lineages.

SECTION 16.5 According to the **plate tectonics theory**, Earth's crust is cracked into giant plates that carry landmasses to new positions as they move. Earth's landmasses have periodically converged as supercontinents such as **Gondwana** and **Pangea**.

SECTION 16.6 Transitions in the fossil record are the boundaries of great intervals of the **geologic time scale**, a chronology of Earth's history that correlates geologic and evolutionary events.

SECTION 16.7 Comparative morphology is one way to study evolutionary connections among lineages. **Homologous structures** are similar body parts that, by **morphological divergence**, became modified differently in different lineages. Such parts are evidence of a common ancestor. **Analogous structures** are body parts that look alike in different lineages but did not evolve in a common ancestor. By the process of **morphological convergence**, they evolved separately after the lineages diverged.

SECTION 16.8 We can discover and clarify evolutionary relationships through comparisons of DNA and protein sequences, because lineages that diverged recently tend to share more sequences than ones that diverged long ago. Master genes that affect development tend to be highly conserved, so similarities in patterns of embryonic development reflect shared ancestry that can be evolutionarily ancient.

SECTION 16.9 A mass extinction 66 million years ago may have been caused by an asteroid impact that left traces in a worldwide sedimentary rock formation.

Self-Quiz

Answers in Appendix VII

1. The number of species on an island usually depends on the size of the island and its distance from a mainland. This statement would most likely be made by _______ .
   a. an explorer
   b. a biogeographer
   c. a geologist
   d. a philosopher

2. The bones of a bird's wing are similar to the bones in a bat's wing. This observation is an example of _______ .
   a. uniformity
   b. evolution
   c. comparative morphology
   d. a lineage

3. Evolution _______ .
   a. is natural selection
   b. is change in a line of descent
   c. can occur by natural selection
   d. b and c are correct

4. A trait is adaptive if it _______ .
   a. arises by mutation
   b. increases fitness
   c. is passed to offspring
   d. occurs in fossils

5. In which type of rock are you more likely to find a fossil?
   a. basalt, a dark, fine-grained volcanic rock
   b. limestone, composed of calcium carbonate sediments
   c. slate, a volcanically melted and cooled shale
   d. granite, which forms by crystallization of molten rock below Earth's surface

6. If the half-life of a radioisotope is 20,000 years, then a sample in which three-quarters of that radioisotope has decayed is _______ years old.
   a. 15,000 b. 26,667 c. 30,000 d. 40,000

7. Did Pangea or Gondwana form first?

8. Forces that cause geologic change include _______ (select all that are correct).
   a. erosion
   b. natural selection
   c. volcanic activity
   d. tectonic plate movement
   e. wind
   f. meteorite impacts

9. Through _______ , a body part of an ancestor is modified differently in different lines of descent.

Data Analysis Activities

Discovery of Iridium in the K–Pg Boundary Layer In the late 1970s, geologist Walter Alvarez was investigating the composition of the K–Pg boundary sequence in different parts of the world. He asked his father, Nobel Prize–winning physicist Luis Alvarez, to help him analyze the elemental composition of the layer (*right*, Luis and Walter Alvarez with a section of the boundary sequence).

The Alvarezes and their colleagues tested the K–Pg boundary sequence in Italy and Denmark. They discovered that it contains a much higher iridium content than the surrounding rock layers. Some of their results are shown in **FIGURE 16.19**.

1. What was the iridium content of the K–Pg boundary sequence?
2. How much higher was the iridium content of the boundary layer than the sample taken 0.7 meter above the sequence?

Sample Depth	Average Abundance of Iridium (ppb)
+ 2.7 m	< 0.3
+ 1.2 m	< 0.3
+ 0.7 m	0.36
boundary layer	41.6
− 0.5 m	0.25
− 5.4 m	0.30

FIGURE 16.19 Abundance of iridium in and near the K-Pg boundary sequence in Stevns Klint, Denmark. Many rock samples taken from above, below, and at the boundary were tested for iridium content. Depths are given as meters above or below the boundary.

The iridium content of an average Earth rock is 0.4 parts per billion (ppb) of iridium. An average meteorite contains about 550 parts per billion of iridium.

10. Homologous structures among major groups of organisms may differ in _______ .
   a. size b. shape c. function d. all of the above

11. By altering steps in the program by which embryos develop, a mutation in a _______ may lead to major differences in body form between related lineages.
   a. derived trait c. homologous structure
   b. homeotic gene d. all of the above

12. The dinosaurs died _______ million years ago.

13. All of the following data types can be used as evidence of shared ancestry except similarities in _______ .
   a. amino acid sequence d. embryonic development
   b. DNA sequence e. form due to convergence
   c. fossil morphology f. all are appropriate

14. Match the terms with the most suitable description.

___ fitness	a.	line of descent
___ fossils	b.	measured by relative genetic contribution to future generations
___ natural selection	c.	human arm and bird wing
___ half-life	d.	evidence of ancient life
___ homologous structures	e.	characteristic of radioisotope
___ analogous structures	f.	insect wing and bird wing
___ lineage	g.	survival of the fittest
___ sedimentary rock	h.	good for finding fossils

Critical Thinking

1. Radiometric dating does not measure the age of an individual atom. It is a measure of the age of a quantity of atoms—a statistic. As with any statistical measure, its values may deviate around an average (see sampling error, Section 1.7). Imagine that one sample of rock is dated ten different ways. Nine of the tests yield an age close to 225,000 years. One test yields an age of 3.2 million years. Do the nine consistent results imply that the one that deviates is incorrect, or does the one odd result invalidate the nine that are consistent?

2. If you think of geologic time spans as minutes, life's history might be plotted on a clock such as the one shown *below*. According to this clock, the most recent epoch started in the last 0.1 second before noon. Where does that put you?

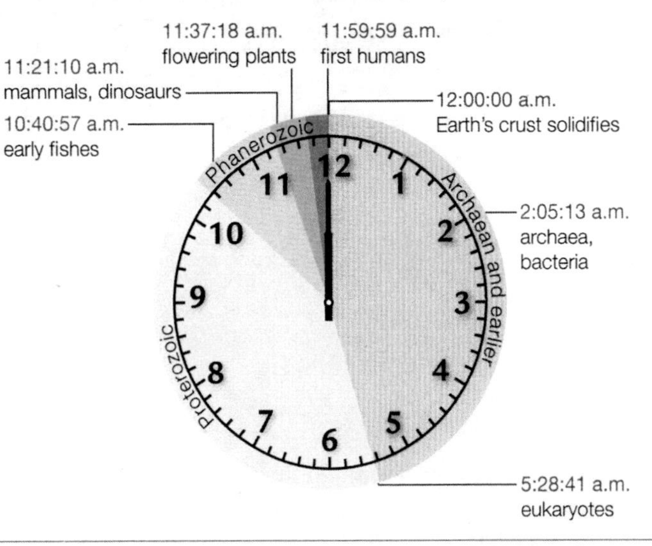

CENGAGE brain To access course materials, please visit www.cengagebrain.com.

CREDITS: (19) left, © Lawrence Berkeley National Laboratory; right, © Cengage Learning; (in text) From Starr/Evers/Starr, Biology Today and Tomorrow with Physiology, 4E. © 2013 Cengage Learning.

Appendix I. Periodic Table of the Elements

The symbol for each element is an abbreviation of its name. Some symbols for elements are abbreviations for their Latin names. For instance, Pb (lead) is short for *plumbum*; the word "plumbing" is related—ancient Romans made their water pipes with lead.

Elements in each vertical column of the table behave in similar ways. For instance, all of the elements in the far right column of the table are inert gases; they do not interact with other atoms. In nature, such elements occur only as solitary atoms.

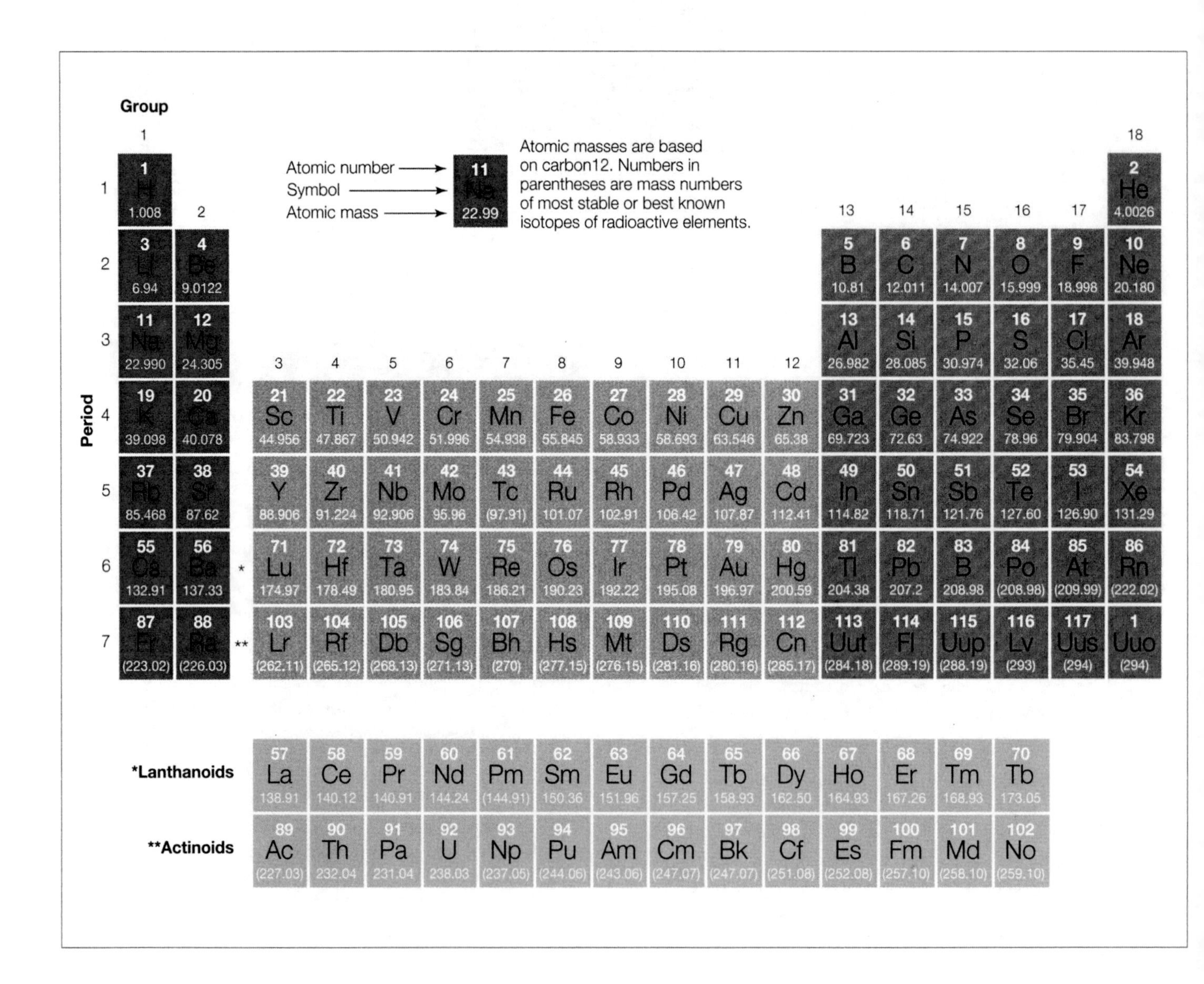

Appendix II. The Amino Acids

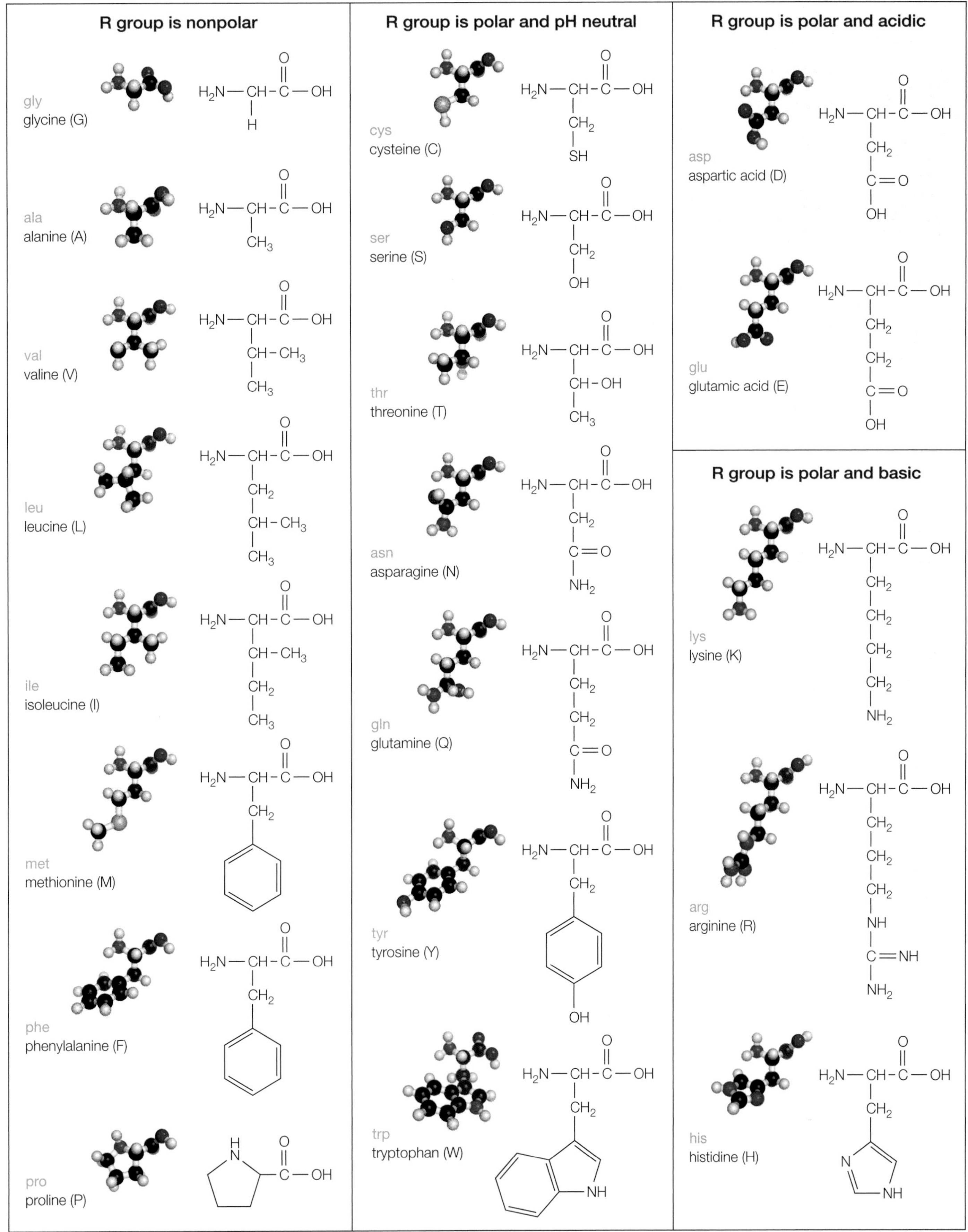

Appendix III. A Closer Look at Some Major Metabolic Pathways

Glycolysis

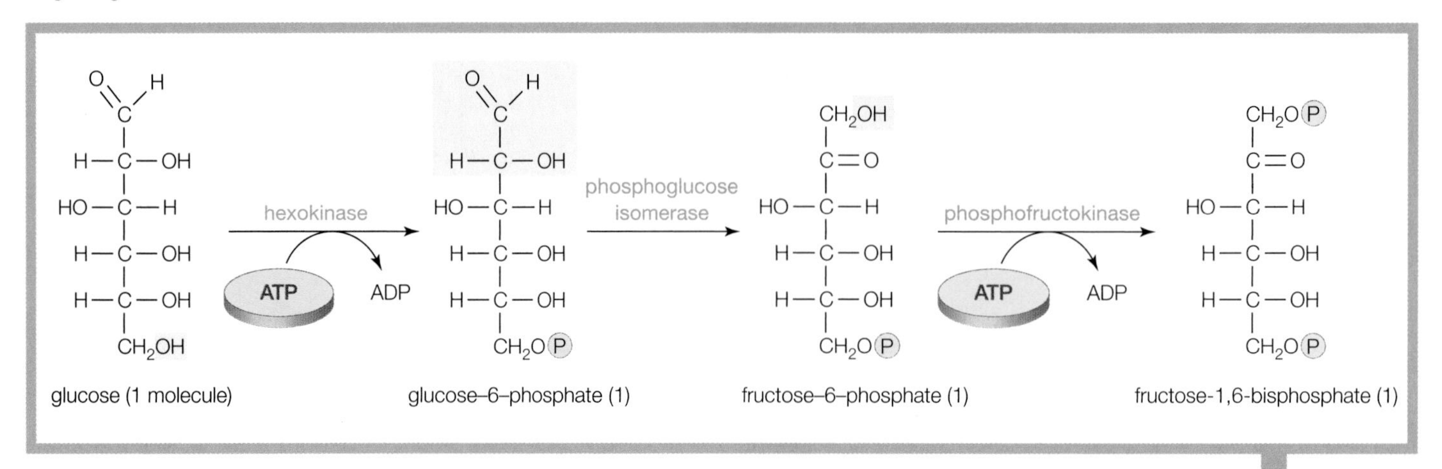

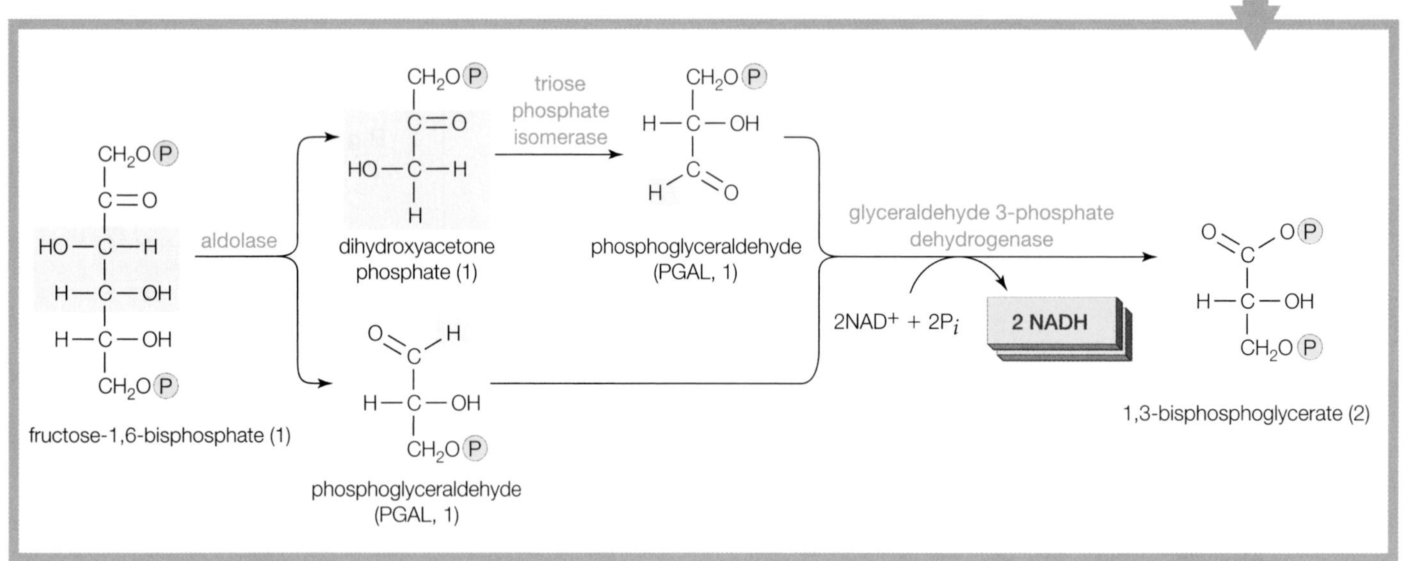

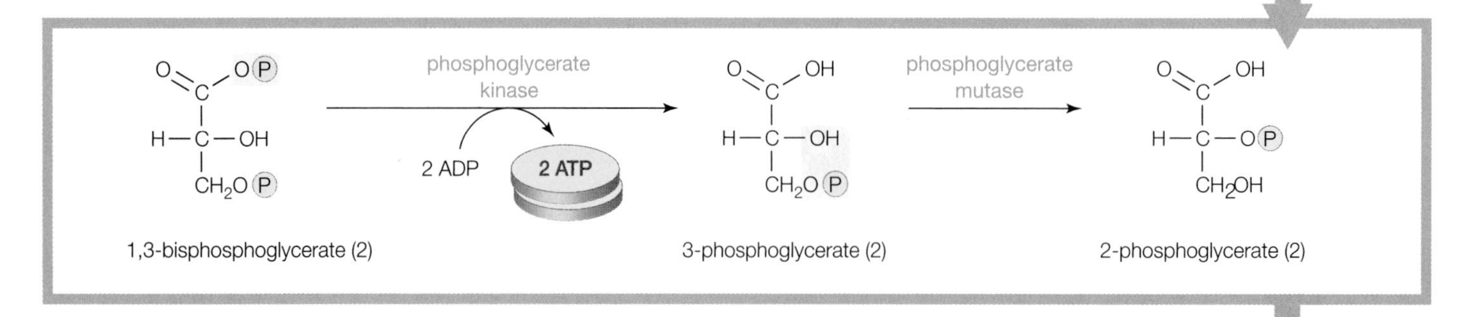

FIGURE A Glycolysis breaks down one glucose molecule into two 3-carbon pyruvate molecules for a net yield of two ATP. Enzyme names are indicated in green; parts of substrate molecules undergoing chemical change are highlighted blue.

enolase
pyruvate kinase
2 ADP
2 ATP
2-phosphoglycerate (2)
phosphoenolpyruvate (2)
pyruvate (2)

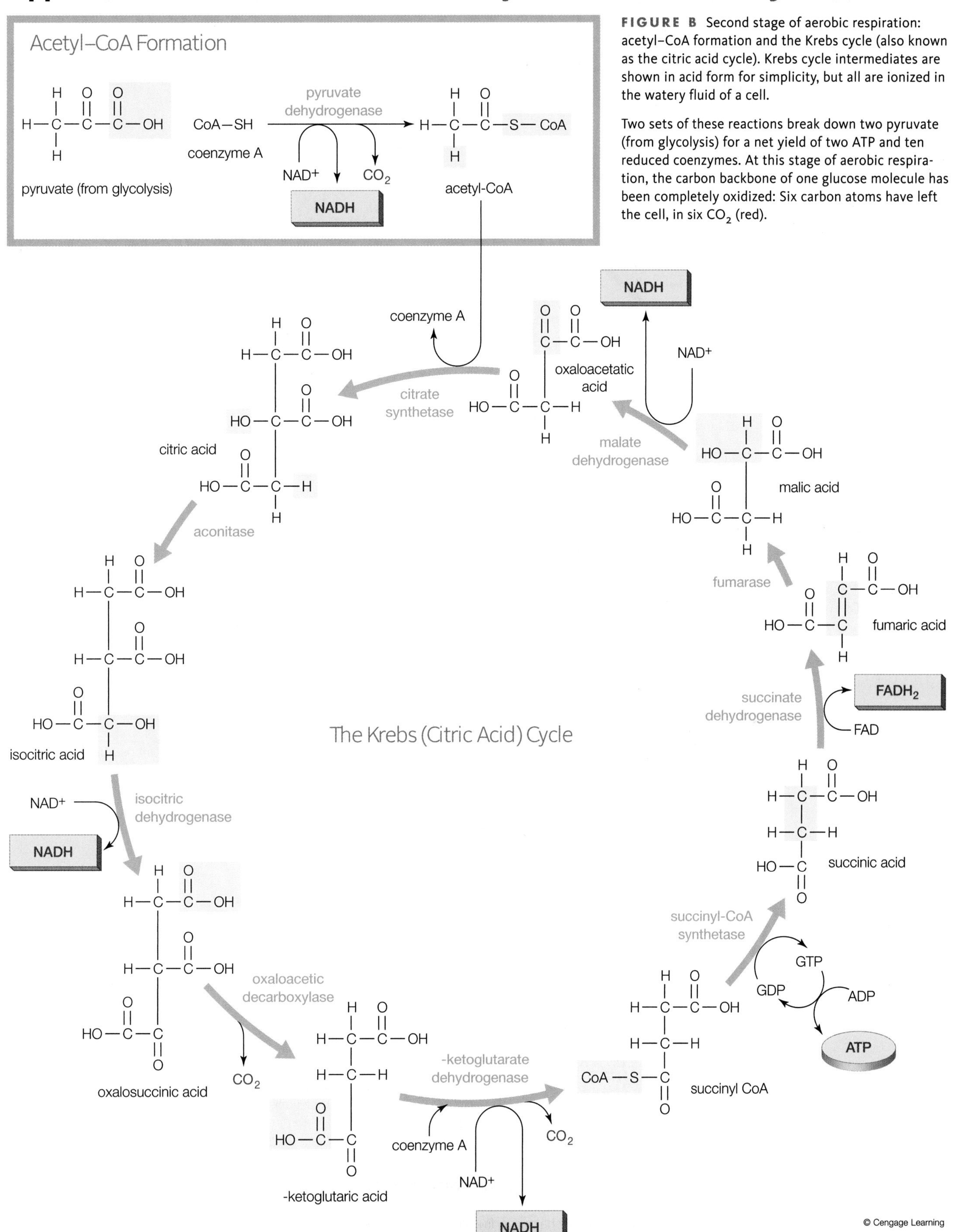

FIGURE B Second stage of aerobic respiration: acetyl–CoA formation and the Krebs cycle (also known as the citric acid cycle). Krebs cycle intermediates are shown in acid form for simplicity, but all are ionized in the watery fluid of a cell.

Two sets of these reactions break down two pyruvate (from glycolysis) for a net yield of two ATP and ten reduced coenzymes. At this stage of aerobic respiration, the carbon backbone of one glucose molecule has been completely oxidized: Six carbon atoms have left the cell, in six CO_2 (red).

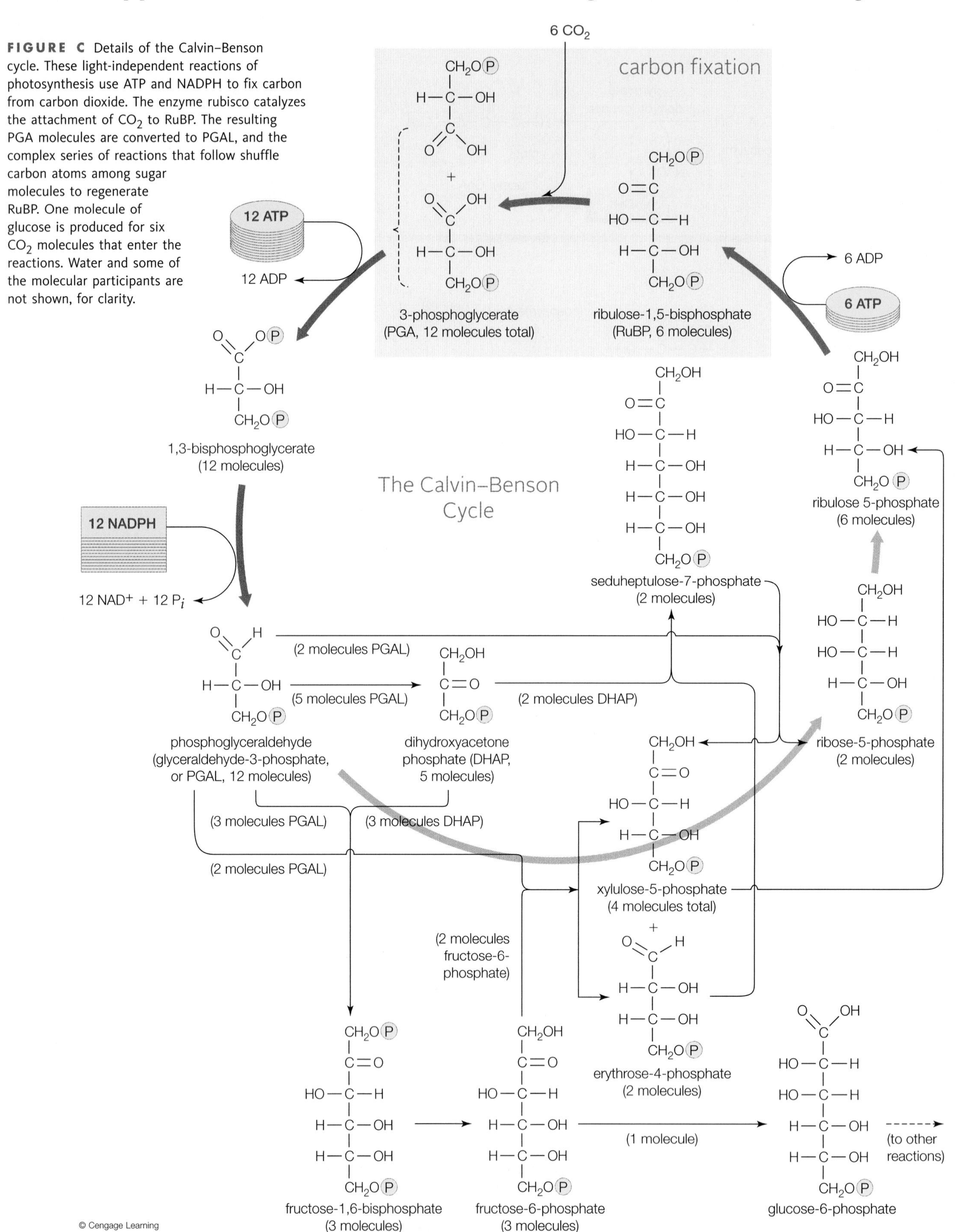

FIGURE C Details of the Calvin–Benson cycle. These light-independent reactions of photosynthesis use ATP and NADPH to fix carbon from carbon dioxide. The enzyme rubisco catalyzes the attachment of CO_2 to RuBP. The resulting PGA molecules are converted to PGAL, and the complex series of reactions that follow shuffle carbon atoms among sugar molecules to regenerate RuBP. One molecule of glucose is produced for six CO_2 molecules that enter the reactions. Water and some of the molecular participants are not shown, for clarity.

Appendix IV. A Plain English Map of the Human Chromosomes

Haploid set of human chromosomes. The banding patterns characteristic of each type of chromosome appear after staining with a reagent called Giemsa. The locations of some of the 20,065 known genes (as of November, 2005) are indicated. Also shown are locations that, when mutated, cause some of the genetic diseases discussed in the text.

Appendix V. Restless Earth—Life's Changing Geologic Stage

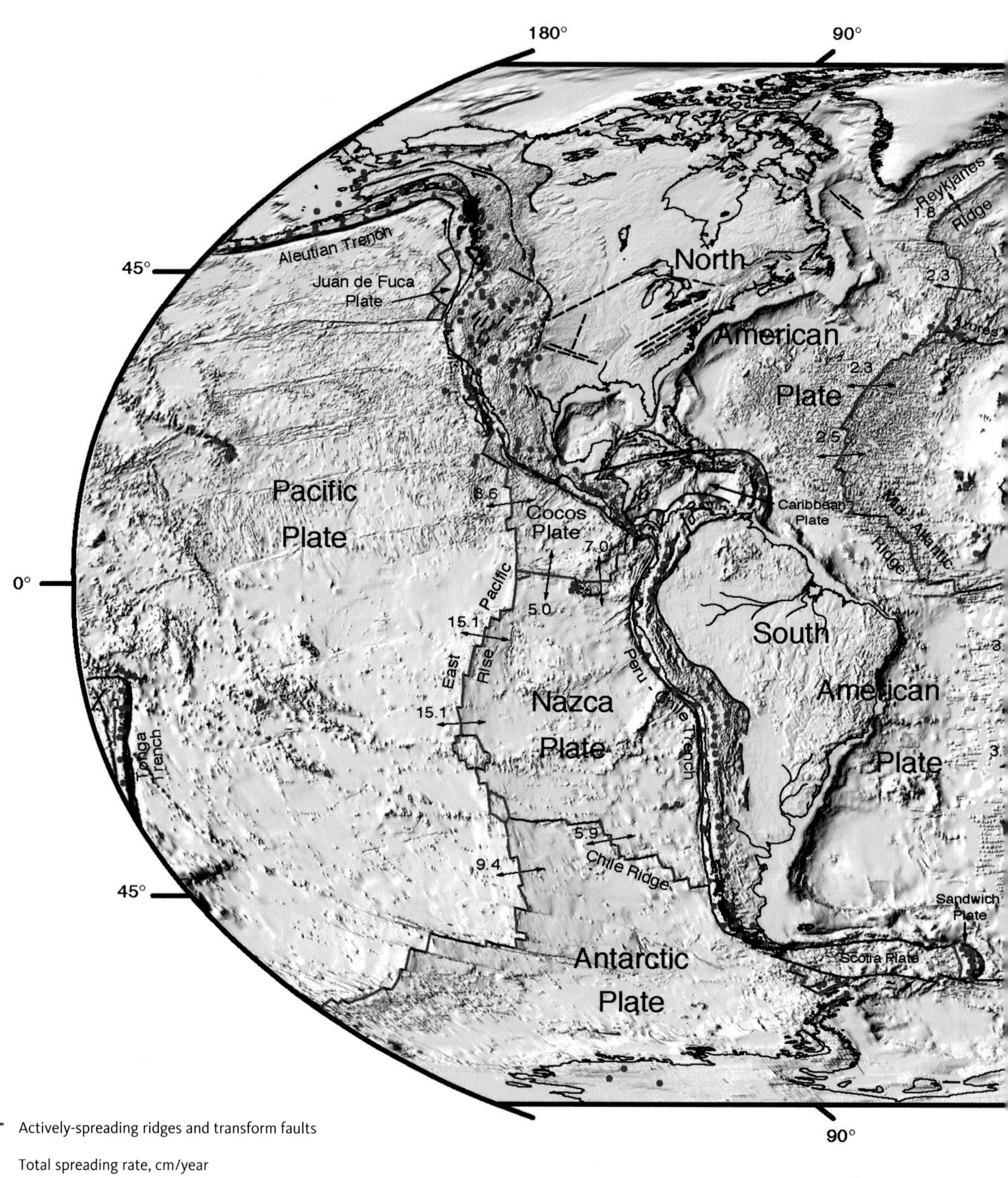

Actively-spreading ridges and transform faults

Total spreading rate, cm/year

Major active fault or fault zone; dashed where nature, location, or activity uncertain

Normal fault or rift; hachures on downthrown side

Reverse fault (overthrust, subduction zones); generalized; barbs on upthrown side

Volcanic centers active within the last one million years; generalized. Minor basaltic centers and seamounts omitted.

This NASA map summarizes the tectonic and volcanic activity of Earth during th past 1 million years. The reconstructions at far right indicate positions of Earth's major land masses through time.

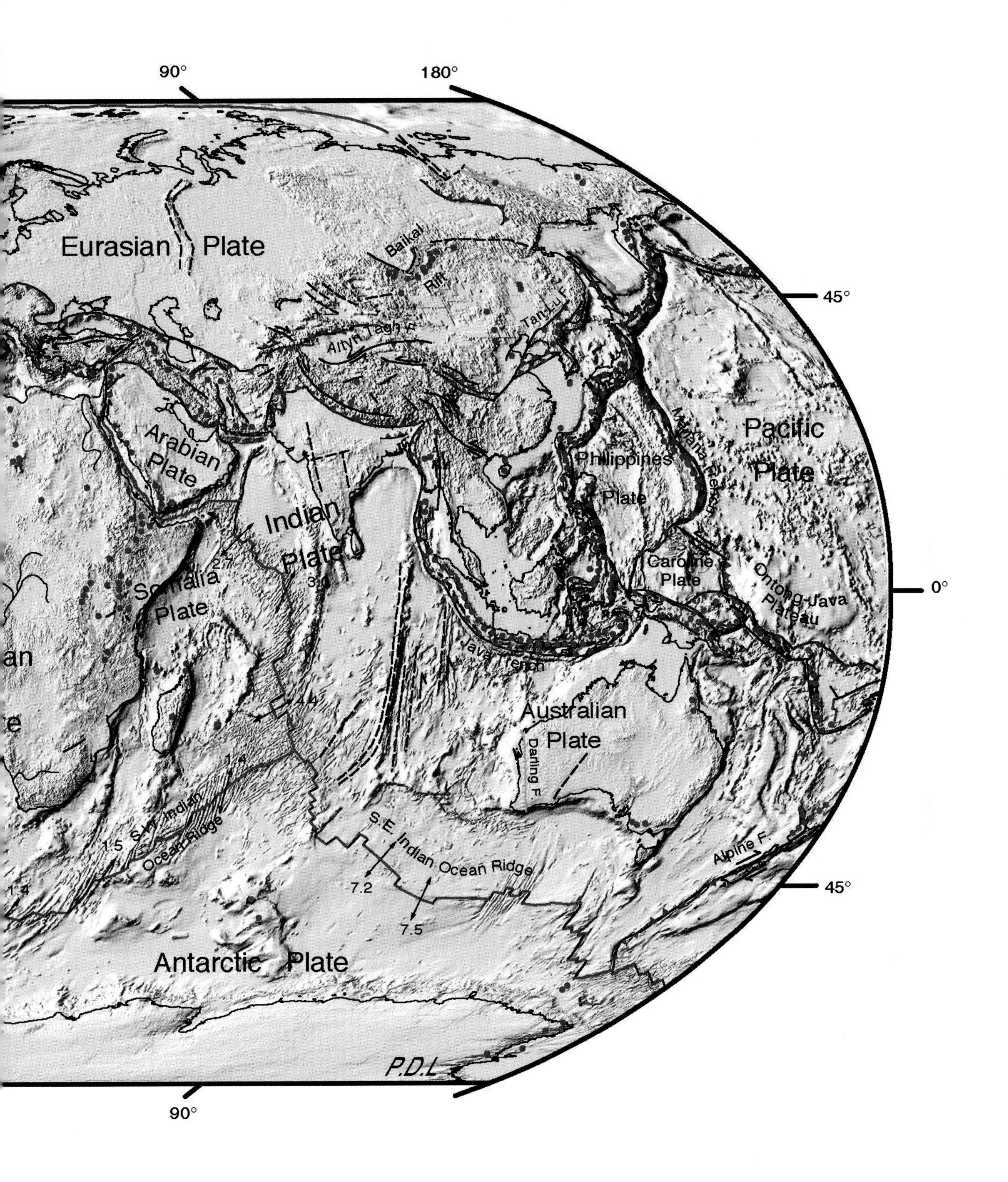

90°
180°
Eurasian Plate
Baikal
Rift
Tan-Lu F.
Altyn Tagh F.
45°
Arabian Plate
Indian Plate
Philippines Plate
Pacific Plate
Mariana Trench
Caroline Plate
Ontong-Java Plateau
0°
Somalia Plate
2.7
3.0
Java Trench
4.4
Australian Plate
Darling F.
1.4
1.5
S.W. Indian Ocean Ridge
S.E. Indian Ocean Ridge
7.2
7.5
Alpine F.
45°
Antarctic Plate
P.D.L.
90°

Appendix VI. Units of Measure

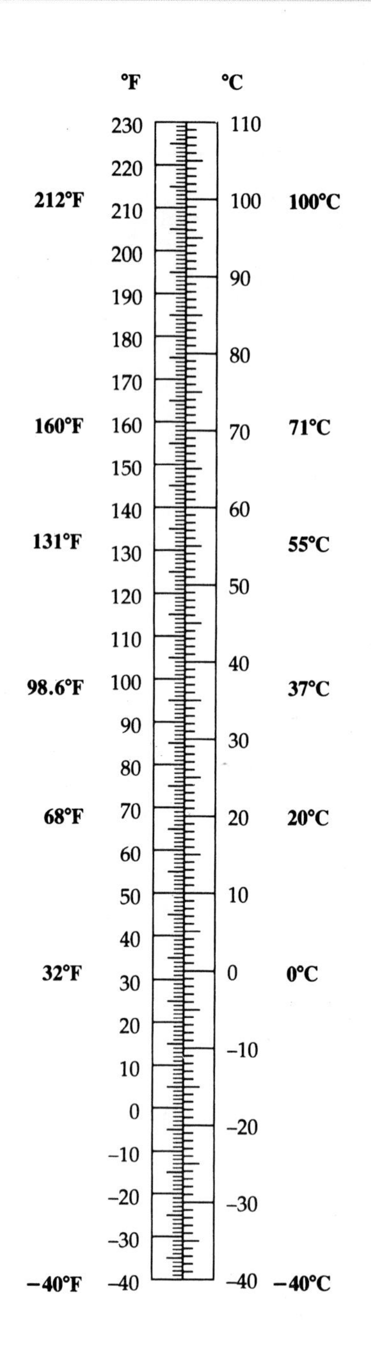

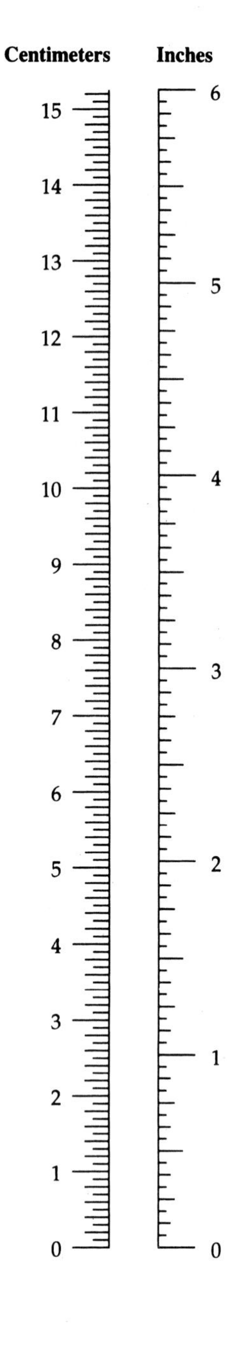

LENGTH

1 kilometer (km) = 0.62 miles (mi)
1 meter (m) = 39.37 inches (in)
1 centimeter (cm) = 0.39 inches

To convert	*multiply by*	*to obtain*
inches	2.25	centimeters
feet	30.48	centimeters
centimeters	0.39	inches
millimeters	0.039	inches

AREA

1 square kilometer = 0.386 square miles
1 square meter = 1.196 square yards
1 square centimeter = 0.155 square inches

VOLUME

1 cubic meter = 35.31 cubic feet
1 liter = 1.06 quarts
1 milliliter = 0.034 fluid ounces = 1/5 teaspoon

To convert	*multiply by*	*to obtain*
quarts	0.95	liters
fluid ounces	28.41	milliliters
liters	1.06	quarts
milliliters	0.03	fluid ounces

WEIGHT

1 metric ton (mt) = 2,205 pounds (lb) = 1.1 tons (t)
1 kilogram (kg) = 2.205 pounds (lb)
1 gram (g) = 0.035 ounces (oz)

To convert	*multiply by*	*to obtain*
pounds	0.454	kilograms
pounds	454	grams
ounces	28.35	grams
kilograms	2.205	pounds
grams	0.035	ounces

TEMPERATURE

Celcius (°C) to Fahrenheit (°F): $°F = 1.8\,(°C) + 32$

Fahrenheit (°F) to Celsius: $°C = \frac{(°F - 32)}{1.8}$

	°C	°F
Water boils	100	212
Human body temperature	37	98.6
Water freezes	0	32

Appendix VII. Answers to Self-Quizzes and Genetics Problems

CHAPTER 1

Answer	Section
1. a	1.1
2. c	1.1
3. energy, nutrients	1.2
4. homeostasis	1.2
5. d	1.2
6. reproduction	1.2
7. d	1.2
8. a, d, e	1.1–1.4
9. Animals	1.3
10. a, b	1.1, 1.3
11. domains	1.4
12. b	1.5
13. b	1.7
14. b	1.7
15. c	1.1
e	1.7
b	1.4
d	1.5
a	1.5
f	1.2

CHAPTER 2

Answer	Section
1. a	2.1
2. b	2.1
3. d	2.2
4. d	2.1
5. b	2.2
6. a	2.3
7. a	2.3
8. c	2.3
9. c	2.4
10. c	2.4
11. d	2.2, 2.5
12. a	2.5
13. c	2.5
14. b	2.4
15. c	2.4
b	2.1
d	2.4
a	2.1, 2.2
f	2.4
e	2.1, 2.2

CHAPTER 3

Answer	Section
1. c	3.1
2. four	3.1
3. b	3.1, 3.3, 3.4
4. e	3.2, 3.6
5. c	3.3
6. False	3.3, 3.7
7. b	3.3
8. starch, cellulose, glycogen	3.2
9. e	3.3
10. d	3.4, 3.6
11. d	3.5
12. d	3.6
13. a amino acid	3.4
b carbohydrate	3.2
c polypeptide	3.4
d fatty acid	3.3
14. c	3.3
a	3.2
b	3.3
15. g	3.4
a	3.3
b	3.4
c	3.3
d	3.6
j	3.3
f	3.6
i	3.4
h	3.2
e	3.2

CHAPTER 4

Answer	Section
1. c	4.1
2. c	4.1
3. c	4.1, 4.4
4. b	4.1
5. b	4.3
6. c	4.3
7. a	4.3
8. b	4.5
9. a	4.6
10. c	4.7
11. a	4.6
12. c, b, d, a	4.6
13. d	4.10
14. a	4.10
15. c	4.7
g	4.8
e	4.4, 4.6
d	4.5
a	4.10
b	4.4, 4.9
f	4.1, 4.5

CHAPTER 5

Answer	Section
1. c	5.1
2. b	5.1
3. d	5.1
4. a	5.2
5. c	5.2
6. c	5.2
7. temperature, pH, salt, pressure	5.3
8. d	5.4
9. a	5.5
10. more/less	5.6
11. c	5.6, 5.7
12. b	5.7
13. a	5.6
14. d	5.8
15. c	5.2
e	5.8
f	5.1
b	5.2
a	5.5
g	5.6
h	5.7
d	5.7

CHAPTER 6

Answer	Section
1. autotroph: weed; heterotrophs: cat, bird, caterpillar	6.1
2. c	6.1
3. a	6.3
4. d	6.4
5. b	6.3, 6.4
6. b	6.4
7. c	6.4
8. c	6.3, 6.5
9. b	6.5
10. b	6.5
11. a	6.1
12. b	6.5, 6.6
13. f	6.5
h	6.5
g	6.4
d	6.4
e	6.5
b	6.1, 6.3
a	6.1
c	6.1

CHAPTER 7

Answer	Section
1. False	7.1
2. d	7.1, 7.3
3. a	7.1, 7.6
4. c	7.3
5. b	7.1, 7.4, 7.5
6. d	7.1, 7.6
7. e	7.4
8. b	7.4
9. c	7.5
10. c	7.5
11. c	7.6
12. d	7.7
13. f	7.6
14. c	7.5
15. b	7.4
d	7.3
a	7.3, 7.6
c	7.4, 7.7
e	7.3
f	7.2

CHAPTER 8

Answer	Section
1. c	8.2
2. c	8.2
3. b	8.2
4. b	8.2
5. a	8.3
6. b	8.3
7. b	8.3
8. a	8.4
9. d	8.4
10. c	8.2, 8.4
11. b	8.4
12. d	8.3, 8.4
13. d	8.5
14. d	8.6
15. d	8.1
b	8.6
a	8.2
g	8.3
e	8.4
h	8.4
c	8.3
f	8.5

CHAPTER 9

Answer	Section
1. c	9.1
2. b	9.2
3. a	9.1
4. c	9.1
5. a	9.1
6. b	9.1, 9.3
7. b	9.2
8. c	9.3
9. a	9.3
10. a	9.3
11. a	9.2
12. a	9.2, 9.4
13. b	9.2, 9.4
14. c	9.4
15. c	9.3
b	9.2
e	9.4
a	9.2
f	9.3
d	9.2

CHAPTER 10

Answer	Section
1. d	10.1
2. d	10.1, 10.2
3. b	10.1
4. b	10.1
5. h	10.1
6. b	10.1
7. c	10.2
8. c	10.2
9. d	10.2
10. b	10.2
11. b	10.3
12. b	10.3
13. c	10.3
14. b	10.4
15. f	10.2
a	10.3
b	10.1, 10.4
e	10.3
c	10.1
d	10.5

CHAPTER 11

Answer	Section
1. e	11.1
2. b	11.1
3. d	11.1
4. e	11.1
5. c	11.2
6. c	11.1
7. a	11.1
8. c	11.1
9. a	11.1
10. d	11.3
11. interphase, prophase, metaphase, anaphase, telophase	11.2
12. d	11.5
13. a	11.5
14. c	11.3
f	11.2
a	11.5
g	11.3
b	11.3
e	11.5
d	11.2
h	11.4
15. d	11.2
b	11.2
c	11.2
e	11.1
a	11.2
f	11.3

CHAPTER 12

Answer	Section
1. b	12.1
2 b	12.2
3. c	12.1, 12.4, 12.6
4. d	12.2
5. b	12.2
6. a	12.2
7. Sister chromatids are still attached	12.3
8. c	12.3
9. b	12.4
10. a	12.4
11. e	12.1, 12.4
12. b	12.4, 12.5
13. c	12.3
d	12.3
a	12.1
f	12.2
e	12.2
b	12.1, 12.6
g	12.4

CHAPTER 13

Answer	Section
1. b	13.1
2. a	13.1
3. b	13.2
4. b	13.2
5. c	13.2
6. a	13.2, 13.3
7. b	13.2
8. d	13.3
9. c	13.3
10. c	13.4
11. b	13.4
12. Continuous variation	13.6
13. b	13.3
d	13.2
a	13.1
c	13.1

CHAPTER 14

Answer	Section
1. b	14.1
2. b	14.1
3. a	14.1
4. b	14.2
5. False	14.3
6. d (could be due to both parents carrying an autosomal recessive allele, or the mom carrying an x-linked recessive allele)	14.2, 14.3
7. d	14.3
8. X from mom, Y from dad	14.3
9. d	14.2
10. Y-linked inheritance (this is a critical thinking question)	14.3
11. d	14.5
12. b	14.5
13. True	14.5
14. c	14.5
15. c	14.5
e	14.4
f	14.5
b	14.4
a	14.1
d	14.4

CHAPTER 15

Answer	Section
1. c	15.1
2. a	15.1
3. b	15.1
4. b	15.2
5. c	15.2
6. b	15.3
7. b	15.3
8. d	15.2, 15.4
9. d	15.4
10. True	15.6
11. b	15.6
d	15.6
e	15.5
f	15.1
12. d	15.5
13. b	15.6
14. a	15.2
d	15.1

Appendix VII. Answers to Self-Quizzes and Genetics Problems *(continued)*

c 15.1
e 15.3
b 15.1
15. c 15.4
f 15.5
d 15.6
b 15.4
a 15.5
e 15.5

CHAPTER 16

1. b 16.1
2. c 16.1, 16.7
3. d 16.2
4. b 16.2
5. b 16.3
6. d 16.4
7. Gondwana 16.6
8. a 16.1–16.3
c 16.6
d 16.5
e 16.3, 16.5
f 16.9
9. morphological divergence 16.7
10. d 16.7
11. b 16.8
12. 66 16.9
13. e 16.7, 16.8
14. b 16.2
d 16.1, 16.3
g 16.2
e 16.4
c 16.7
f 16.7
a 16.3

CHAPTER 17

1. a 17.1
2. c 17.1, 17.7
3. a, c, b 17.3–17.5
4. d 17.6
5. d 17.6
6. b 17.7
7. f 17.1, 17.2, 17.7
8. allopatric speciation 17.9
9. d 17.6, 17.8
10. d 17.12
11. c 17.12
(this is a critical thinking question)
12. c 17.12
13. b 17.12
14. c 17.7
e 17.6
g 17.11
b 17.7
d 17.12
f 17.11
a 17.11
h 17.12

CHAPTER 18

1. c 18.1
2. a 18.1
3. c 18.2
4. a 18.2
5. a 18.2
6. c 18.4
7. b 18.5
8. b 18.4
9. c 18.2
10. a 18.4
11. d 18.5, 18.6
12. c 18.5
13. d 18.6
14. a 18.1
c 18.5
b 18.3
e 18.5
f 18.6
d 18.6

CHAPTER 19

1. c 19.1
2. b 19.3
3. c 19.1
4. b 19.1
5. d 19.2
6. c 19.3
7. one 19.4
8. c 19.5
9. d 19.5
10. d 19.5
11. c 19.6
12. d 19.2, 19.7
13. a 19.4
14. b 19.6
a 19.1
a 19.2
b 19.6
b 19.6
15. d 19.5
e 19.7
a 19.3
f 19.1
b 19.8
c 19.8

CHAPTER 20

1. c 20.2
2. d 20.3
3. b 20.2
4. c 20.4
5. b 20.5
6. a 20.7
7. d 20.7
8. c 20.4
9. c 20.6
10. d 20.6
11. c 20.7
12. a 20.8
13. a 20.2
14. a 20.7
15. d 20.9
g 20.5
a 20.3
b 20.6
f 20.6
c 20.7
e 20.6

CHAPTER 21

1. c 21.1, 21.2
2. a 21.1, 21.3
(ferns produce spores, not seeds)
3. b 21.2
4. c 21.3
5. a 21.4
6. e 21.2, 21.3, 21.5
7. b 21.5, 21.6
8. True 21.3
9. c 21.6
10. a 21.2
11. a 21.1
12. c 21.7
13. b 21.5
14. c 21.2
d 21.3
a 21.5
b 21.6
15. c 21.4
i 21.1
a 21.1
b 21.1
e 21.1
f 21.6
d 21.3
g 21.3
h 21.4

CHAPTER 22

1. c 22.1
2. a 22.1
3. b 22.1
4. a 22.3
5. d 22.2
6. c 22.2
7. c 22.2
8. d 22.2
9. b 22.4
10. d 22.3
11. a 22.3
12. c 22.4
13. b 22.5
14. a 22.3
15. d 22.1
b 22.1
a 22.1
f 22.2
g 22.2
c 22.3
e 22.3

CHAPTER 23

1. False 23.1
2. d 23.1
3. a 23.4
4. a 23.5
5. b 23.11
6. a 23.7
7. b 23.2
8. a 23.1, 23.9
9. c 23.9, 23.11
10. b 23.12
11. b 23.6, 23.7
12. d 23.11
13. b 23.12
f 23.7
d 23.3
h 23.4
c 23.5
a 23.8
g 23.6
e 23.9

CHAPTER 24

1. c 24.1
2. c 24.1
3. a 24.2
4. b 24.2
5. d 24.3
6. c 24.4
7. f 24.4, 24.5
8. a 24.5
9. b 24.9
10. c 24.6
11. c 24.11
12. b 24.1
i 24.2
g 24.3
f 24.8
c 24.6
d 24.7
a 24.7
h 24.7
e 24.10

CHAPTER 25

1. b 25.2
2. c 25.2
3. a, b 25.2
4. b 25.3
5. False 25.5
6. c 25.2, 25.6
7. c 25.4
8. a 25.3, 25.5
9. b, c 25.3
10. d 25.5
11. b 25.5
12. a 25.6
13. b 25.6
14. b 25.7
15. c 25.6
d 25.2, 25.4
e 25.6
a 25.3
b 25.3
f 25.1, 25.4
g 25.2, 25.4

CHAPTER 26

1. f 26.1
2. b 26.1
3. e 26.2
4. b 26.2
5. b 26.3
6. c 26.4
7. d 26.4
8. d 26.4
9. c 26.6
10. c 26.5
11. c 26.5
12. a 26.5
13. b 26.6
14. a 26.6
15. c 26.5
g 26.1
e 26.6
b 26.2
d 26.4
h 26.4
f 26.6
a 26.4

CHAPTER 27

1. pollination 27.1
2. a, b, c 27.1, 27.13
3. b 27.1
4. b 27.1
5. c 27.3
6. c 27.3
7. d 27.4
8. d 27.3, 27.4
9. c 27.5
10. a 27.5
11. e 27.6–27.9
12. c 27.11
13. b 27.10
d 27.10
a 27.10
f 27.11
c 27.11
e 27.10
14. c 27.9
15. c 27.9
e 27.7
b 27.10
a 27.8
d 27.9

CHAPTER 28

1. a 28.3
2. a 28.3
3. a 28.3
4. b 28.4
5. b 28.4
6. c 28.4
7. c 28.5
8. a 28.5
9. d 28.6
10. a 28.8
11. a 28.7
12. b 28.9
13. a 28.8
14. c 28.3
15. b 28.3
j 28.3
a 28.4
c 28.8
d 28.5
h 28.3
f 28.8
i 28.3
e 28.4
g 28.3

CHAPTER 29

1. a 29.2
2. c 29.3
3. b 29.4
4. a 29.5
5. a 29.4
6. c 29.7
7. a 29.7
8. b 29.11
9. c 29.12
10. c 29.1
11. b 29.9
12. a 29.8
13. j 29.9
d 29.5
g 29.11
b 29.9
h 29.10
a 29.9
e 29.1
i 29.7
c 29.8
f 29.8

CHAPTER 30

1. b 30.2
2. c 30.1
3. c 30.2
4. e 30.3
5. d 30.2
6. b 30.8

7. b	30.7
8. a	30.2
9. b	30.7
10. c	30.5
11. d	30.4
12. b	30.4
13. a	30.8
14. c	30.3
15. d	30.5
g	30.7
f	30.5
a	304.
h	30.5
e	30.3
b	30.8
i	30.7
c	30.3

CHAPTER 31

1. a	31.1
2. b	31.3
3. a	31.3
4. d	31.1
5. c	31.8
6. a	31.5
7. b	31.8
8. d	31.5
9. c	31.5
10. c	31.5
11. c	31.7
12. d	31.6
f	31.5
c	31.5
e	31.8
a	31.4
b	31.3
13. b	31.9
14. d	31.4
c	31.5
b	31.6
g	31.7
f	31.7
e	31.8
a	31.8

CHAPTER 32

1. a	32.2
2. c	32.7
3. b	32.4
4. a	32.3
5. a	32.3
6. biceps	32.4
7. a	32.4
8. b	32.5
9. d	32.5
10. d	32.6
11. c	32.6
12. d	32.6
13. a	32.8
14. c	32.1
15. h	32.4
f	32.5
g	32.8
e	32.3
i	32.5
c	32.3
b	32.2
d	32.3
a	32.7

CHAPTER 33

1. a	33.1
2. d	33.1
3. pulmonary	33.1
4. c	33.4
5. b	33.4
6. a	33.4
7. b	33.3
8. d	33.3
9. a	33.6
10. b	33.8
11. d	33.3
12. a	33.2
13. b	33.10
14. b	33.9
15. f	33.1
a	33.10
e	33.3
g	33.3
b	33.3
c	33.8
d	33.2

CHAPTER 34

1. d	34.1
2. e	34.2, 34.3
3. g	34.3
4. h	34.3–34.8
5. d	34.3
6. Immediate, fixed, general, and no antigen memory are all correct.	34.1, 34.3
7. Self/nonself discrimination, diversity, specificity, and memory are all correct.	34.5
8. d	34.4, 34.6
9. a	34.5
10. b	34.5, 34.6
11. e	34.5, 34.8
12. b	34.5, 34.7
13. c	34.9
14. a	34.9
15. b	34.6, 34.8
e	34.4
c	34.6
d	34.8
a	34.8
16. f	34.9
e	34.5
d	34.9
c	34.3
b	34.9
a	34.4
g	34.5

CHAPTER 35

1. d	35.1
2. b	35.2
3. a	35.3
4. c	35.4
5. d	35.5
6. a	35.5
7. iron	35.6
8. b	35.8
9. d	35.6
10. c	35.5
11. d	35.5
12. True	35.5
13. a	35.4
14. bacteria	35.7
15. d	35.4
g	35.4
e	35.4
f	35.4
c	35.4
b	35.4
a	35.4

CHAPTER 36

1. d	36.1
2. b	36.3
3. c	36.5
4. b	36.5
5. a	36.5
6. c	36.6
7. a	36.3, 36.5, 36.6
8. b	36.5
9. d	36.8
10. d	36.7
11. c	36.4
12. c	36.9
13. b	36.10
14. c	36.8
15. g	36.5
b	36.6
a	36.5
d	36.4
h	36.3
c	36.5
f.	36.2
e.	36.6

CHAPTER 37

1. c	37.2
2. d	37.2
3. c	37.2
4. a	37.3
5. b	37.4
6. a	37.4
7. b	37.4
8. a	37.4
9. water	37.4
10. dialysis	37.5
11. c	37.3
a	37.3
b	37.3
e	37.4
d	37.4
12. c	37.6
13. b	37.6
a	37.6
d	37.6
c	37.6
e	37.6

CHAPTER 38

1. a	38.1
2. b	38.1
3. a	38.6
4. c	38.8
5. b	38.7
6. c	38.11
7. c	38.6
d	38.10
g	38.7
e	38.6
b	38.6
a	38.7
f	38.8
8. a	38.3
9. False	38.2
10. endoderm, mesoderm, ectoderm	38.2
11. c	38.12
12. b	38.15
a	38.6
d	38.8
c	38.8
e	38.8
g	38.12
f	38.15
13. a, d, c, f	38.12
b, e	38.13

CHAPTER 39

1. d	39.1
2. b	39.1
3. a	39.1
4. c	39.3
5. a	39.3
6. birds	39.4
7. cooperation in predator detection, defense, rearing young, learning by imitation, defending territory	39.5
8. d	39.6
9. d	39.6
10. a	39.4
11. e	39.6
h	39.5
c	39.2
b	39.4
d	39.4
a	39.2
f	39.5
g	39.2

CHAPTER 40

1. clumped	40.1
2. f	40.2
3. 400	40.1
4. 1,600	40.2
5. a	40.2
6. d	40.3
7. d	40.4
8. b	40.6
9. d	40.7
10. a	40.6
11. a	40.4
12. c	40.4
13. c	40.3
d	40.2
a	40.2
e	40.3
b	40.3

CHAPTER 41

1. b	41.1
2. d	41.1, 41.3
3. e	41.3
4. d	41.3
5. b	41.2
d	41.5
c	41.1
a	41.4
e	41.3
6. b	41.5
7. b	41.6
8. Pioneer	41.6
9. c	41.6
10. a	41.1
11. b	41.3
12. c	41.5
13. a	41.4
14. c	41.8
b	41.6
d	41.6
a	41.7
f	41.7
e	41.3

CHAPTER 42

1. b	42.1
2. d	42.1
3. d	42.1
4. d	42.1
5. a	42.1
6. c	42.5
7. b	42.5
8. d	42.6
9. d	42.6, 42.7
10. a	42.8
11. c	42.8
12. b	42.6
13. a	42.7
14. d	42.6
15. d	42.6
c	42.6
b	42.7
a	42.7

CHAPTER 43

1. b	43.1
2. d	43.1
3. d	43.1
4. a	43.2
5. b	43.2
6. b	43.2
7. d	43.3
8. c	43.6
9. a	43.8
10. b	43.9
11. d	43.12
12. c	43.11
13. c	43.3
14. b	43.3
15. d	43.8
e	43.6
f	43.7
c	43.6
b	43.10
h	43.5
i	43.6
a	43.4
g	43.12

CHAPTER 44

1. b	44.1
2. d	44.1
3. a	44.2
4. c	44.4
5. c	44.5
6. a	44.5
7. b	44.3
8. b	44.6
9. b	44.5
10. c	44.7
11. d	44.7
12. c	44.7
13. d	44.8
14. g	44.7
a	44.5
i	44.7
e	44.4
d	44.1
f	44.4
h	44.6
b	44.2
c	44.2

CHAPTER 13: GENETICS PROBLEMS

1. Yellow is recessive. Because F_1 plants have a green phenotype and must be heterozygous, green must be dominant over the recessive yellow.

2. a. *AB*
   b. *AB*, *aB*
   c. *Ab*, *ab*
   d. *AB*, *Ab*, *aB*, *ab*

3. a. All offspring will be *AaBB*.

4. A mating of two M^L cats yields 1/4 *MM*, 1/2 M^LM, and 1/4 M^LM^L. Because M^LM^L is lethal, the probability that any one kitten among the survivors will be heterozygous is 2/3.

5. Because both parents are heterozygous (Hb^AHb^S), each child has a 1 in 4 possibility of inheriting two Hb^S alleles and having sickle cell anemia.

6. The data reveal that these genes do not assort independently because the observed ratio is very far from the 9:3:3:1 ratio expected with independent assortment. Instead, the results can be explained if the genes are located close to each other on the same chromosome.

CHAPTER 14: GENETICS PROBLEMS

1. Autosomal recessive. If the allele was inherited in a dominant pattern, individuals in the last generation would all have the phenotype. If it was X-linked, offspring of the first generation would all have the phenotype.

2. a. Human males (XY) inherit their X chromosome from their mother.
   b. A male with an X-linked allele produces two kinds of gametes: one with an X chromosome (and the X-linked allele), and the other with a Y chromosome.
   c. A female homozygous for an X-linked allele produces one type of gamete, which carries the X-linked allele.
   d. A female heterozygous for an X-linked allele produces two types of gametes: one that carries the X-linked allele, and another that carries the partnered allele on the homologous chromosome.

3. As a result of translocation, chromosome 21 may get attached to the end of chromosome 14. The new individual's chromosome number would still be 46, but its somatic cells would have the translocated chromosome 21 in addition to two normal chromosomes 21.

4. 50 percent

Glossary

abscisic acid (ABA) Plant hormone involved in stomata function and stress responses; inhibits germination. **466**

abscission Process by which plant parts are shed. **466**

acid Substance that releases hydrogen ions when it dissolves in water. **32**

acid rain Low-pH rain that forms when sulfur dioxide and nitrogen oxides mix with water vapor in the atmosphere. **789**

actin Globular protein that plays a role in cell movements; main component of thin filaments in myofibrils. **561**

action potential Abrupt reversal of the charge difference across a plasma membrane. **501**

activation energy Minimum amount of energy required to start a reaction. **80**

activator Transcription factor that increases the rate of transcription. **164**

active site Of an enzyme, pocket in which substrates bind and a reaction occurs. **82**

active transport Energy-requiring mechanism in which a transport protein pumps a solute across a cell membrane against its concentration gradient. **90**

adaptation (adaptive trait) A heritable trait that enhances an individual's fitness in a particular environment. **257**

adaptive immunity Of vertebrate animals, a set of immune defenses that can be tailored to specific pathogens encountered by an organism during its lifetime. Characterized by self/nonself recognition, specificity, diversity, and memory. **590**

adaptive radiation Macroevolutionary pattern in which a burst of genetic divergences from a lineage gives rise to many new species. **292**

adaptive trait *See* adaptation.

adhering junction Cell junction composed of adhesion proteins that connect to cytoskeletal elements. Fastens animal cells to each other and basement membrane. **69**

adhesion protein Plasma membrane protein that helps cells stick to one another and (in animals) to extracellular matrix. **57**

adipose tissue Connective tissue that specializes in fat storage. **485**

adrenal cortex Outer portion of an adrenal gland; secretes aldosterone and cortisol. **544**

adrenal gland Endocrine gland located atop the kidney. **544**

adrenal medulla Inner portion of an adrenal gland; secretes epinephrine and norepinephrine. **544**

aerobic Involving or occurring in the presence of oxygen. **115**

aerobic respiration Oxygen-requiring metabolic pathway that breaks down sugars to produce ATP. Includes glycolysis, acetyl–CoA formation, the Krebs cycle, and electron transfer phosphorylation. **114**

age structure Of a population, the distribution of its members among various age categories. **723**

agglutination The clumping together of foreign cells bound by antibodies; the clumps attract phagocytic cells. Basis of blood typing tests. **601**

AIDS Acquired immunodeficiency syndrome. A secondary immune deficiency that develops as the result of infection by the HIV virus. **606**

alcoholic fermentation Anaerobic sugar breakdown pathway that produces ATP, CO_2, and ethanol. **122**

aldosterone Adrenal hormone that makes kidney tubules more permeable to sodium; encourages sodium reabsorption, which in turn increases water reabsorption and concentrates the urine. **655**

allantois Extraembryonic membrane that, in mammals, becomes part of the umbilical cord. **685**

allele frequency Abundance of a particular allele in a population's gene pool. **275**

alleles Forms of a gene with slightly different DNA sequences; may encode slightly different versions of the gene's product. **192**

allergen A normally harmless substance that provokes an immune response in some people. **604**

allergy Sensitivity to an allergen. **604**

allopatric speciation Speciation pattern in which a physical barrier ends gene flow between populations. **288**

allosteric regulation Control of enzyme activity by a regulatory molecule or ion that binds to a region outside the enzyme's active site. **84**

alternation of generations Of land plants and some algae, a life cycle that includes haploid and diploid multicelled bodies. **338**

alternative splicing Post-translational RNA modification in which some exons are removed or joined in various combinations. **151**

altruistic behavior Behavior that benefits others at the expense of the individual. **706**

alveoli Air sacs in the lung; gas exchange occurs across their lining. **619**

amino acid Small organic compound that is a subunit of proteins. Consists of a carboxyl group, an amine group, and a characteristic side group (R), all typically bonded to the same carbon atom. **44**

amino acid–derived hormone An amine (modified amino acid), peptide, or protein that functions as a hormone. **536**

ammonia Nitrogen-containing compound (NH_3) formed by breakdown of amino acids. **650**

amnion Extraembryonic membrane that encloses an amniote embryo and the amniotic fluid. **684**

amniote Vertebrate whose egg has waterproof membranes that allow it to develop away from water; a reptile, bird, or mammal. **401**

amoeba Single-celled protist that extends pseudopods to move and to capture prey. **340**

amoebozoan Shape-shifting heterotrophic protist with no pellicle or cell wall; an amoeba or slime mold. **340**

amphibian Tetrapod with scaleless skin; develops in water, then lives on land as a carnivore with lungs. For example, a frog or salamander. **404**

anaerobic Occurring in the absence of oxygen. **115**

analogous structures Similar body structures that evolved separately in different lineages (by morphological convergence). **267**

anaphase Stage of mitosis during which sister chromatids separate and move toward opposite spindle poles. **181**

aneuploid Having too many or too few copies of a particular chromosome. **230**

angiosperms Highly diverse seed plant lineage; only plants that make flowers and fruits. **356**

animal A eukaryotic heterotroph that is made up of unwalled cells and develops through a series of stages. Most ingest food, reproduce sexually, and move. **9**, **376**

animal hormone Intercellular signaling molecule that is secreted by an endocrine gland or cell and travels in the blood. **536**

annelid Segmented worm with a coelom, complete digestive system, and closed circulatory system. **384**

antenna Of some arthropods, sensory structure on the head that detects touch and odors. **389**

anther Of a flower, the part of the stamen that produces pollen. **454**

anthropoid primate Humanlike primate; monkey, ape, or human. **411**

antibody Y-shaped antigen receptor protein, made only by B cells; antibody bound to antigenic particles activates complement and triggers effector cell function such as phagocytosis. **596**

antibody-mediated immune response Immune response in which antibodies are produced in response to an antigen. **598**

anticodon In a tRNA, set of three nucleotides that base-pairs with an mRNA codon. **153**

antidiuretic hormone Pituitary hormone that encourages water reabsorption in the kidney, thus concentrating the urine. **654**

antigen A molecule or particle that the immune system recognizes as nonself. Its presence in the body triggers an immune response. **590**

antioxidant Substance that prevents oxidation of other molecules. **86**

anus Body opening that serves solely as the exit for wastes from a tubular digestive tract. **630**

aorta Large artery that receives oxygenated blood pumped out of the heart's left ventricle. **572**

ape Common name for a tailless nonhuman primate; a gibbon, orangutan, gorilla, chimpanzee, or bonobo. **411**

apical dominance In plants, effect in which a lengthening shoot tip inhibits the growth of lateral buds. **464**

apical meristem Meristem in the tip of a shoot or root; gives rise to primary growth (lengthening) in a plant. **432**

apicomplexan Parasitic protist that reproduces inside cells of its host; for example, the protist that causes malaria. **336**

apoptosis Mechanism of cell suicide. **671**

appendicular skeleton Of vertebrates, bones of the limbs or fins and bones that connect these structures to the axial skeleton. **557**

appendix Wormlike projection from the first part of the large intestine; serves as a reservoir for bacteria. **638**

aquifer Porous rock layer that holds some groundwater. **752**

arachnids Land-dwelling arthropods with no antennae and four pairs of walking legs; spiders, scorpions, mites, and ticks. **390**

archaea Singular **archaean**. Group of single-celled organisms that lack a nucleus but are more closely related to eukaryotes than to bacteria. **8**

arctic tundra Highest-latitude Northern Hemisphere biome, where low, cold-tolerant plants survive with only a brief growing season. **773**

area effect Larger islands have more species than small ones. **742**

arteriole Blood vessel that conveys blood from an artery to capillaries. **578**

artery Large-diameter vessel that carries blood away from the heart. **570**

arthropod Invertebrate with jointed legs and a hard exoskeleton that is periodically molted. **389**

asexual reproduction Reproductive mode of eukaryotes by which offspring arise from a single parent only. **178, 666**

atmospheric cycle Biogeochemical cycle in which a gaseous form of an element plays a significant role. For example, the carbon cycle. **754**

atom Fundamental building block of all matter. Consists of varying numbers of protons, neutrons, and electrons. **4**

atomic number Number of protons in the atomic nucleus; determines the element. **24**

ATP Adenosine triphosphate. Nucleotide that consists of an adenine base, a ribose sugar, and three phosphate groups. Functions as a subunit of RNA and as a coenzyme in many reactions. Important energy carrier in cells. **46**

ATP/ADP cycle Process by which cells regenerate ATP. ADP forms when a phosphate group is removed from ATP, then ATP forms again as ADP gains a phosphate group. **87**

atrioventricular (AV) node Clump of cells that conveys excitatory signals between the atria and ventricles. **575**

atrium Heart chamber that receives blood from veins. **574**

australopith Extinct African hominins in the genus *Australopithecus*; some are considered likely human ancestors. **413**

autoimmune response Immune response that inappropriately targets one's own tissues. **605**

autonomic nervous system Division of the peripheral nervous system that relays signals to and from internal organs and glands. **506**

autosome A chromosome that is the same in males and females. **137**

autotroph Producer. An organism that makes its own food using energy from the environment and carbon from inorganic molecules such as CO_2. **100**

auxin Plant hormone that causes lengthening; also has a central role in growth by coordinating the effects of other hormones. **464**

AV node *See* atrioventricular node.

axial skeleton Bones of the main body axis; skull, backbone, and rib cage. **557**

axon Of a neuron, a cytoplasmic extension that transmits electrical signals along its length and secretes chemical signals at its endings. **500**

bacteria Singular bacterium. The most diverse and well-known group of prokaryotes. **8**

bacteriophage Virus that infects bacteria. **133, 316**

balanced polymorphism Maintenance of two or more alleles for a trait at high frequency in a population. **283**

bark In woody plants, all living and dead tissue that lies outside of the vascular cambium. **433**

Barr body Inactivated X chromosome in a cell of a female mammal. The other X chromosome is active. **168**

basal body Organelle that develops from a centriole; occurs at base of cilium or flagellum. **67**

basal metabolic rate Rate at which a body uses energy when at rest. **644**

base Substance that accepts hydrogen ions when it dissolves in water. **32**

basement membrane Secreted layer that attaches an epithelium to an underlying tissue. **482**

base-pair substitution Type of mutation in which a single base pair changes. **156**

basophil Circulating white blood cell that releases the contents of its granules in response to antigen or injury. **591**

B cell B lymphocyte. White blood cell that can make antibodies. Central to antibody-mediated immune responses. **591**

B cell receptor Antigen receptor on the surface of a B cell; an antibody that has not been released from the B cell's plasma membrane. **597**

bell curve Bell-shaped curve; typically results from graphing frequency versus distribution for a trait that varies continuously. **216**

benthic province The ocean's sediments and rocks. **778**

bilateral symmetry Having paired structures so the right and left halves are mirror images. **376**

bile Mix of salts, pigments, and cholesterol produced in the liver, then stored and concentrated in the gallbladder; emulsifies fats when secreted into the small intestine. **636**

binary fission Method of asexual reproduction that divides one bacterial or archaeal cell into two identical descendant cells. **320**

bioaccumulation The concentration of a chemical pollutant in the tissues of an organism rises over the course of the organism's lifetime. **759**

biodiversity Scope of variation among living organisms; the genetic variation within species, variety of species, and variety of ecosystems. **8, 792**

biofilm Community of microorganisms living within a shared mass of secreted slime. **59**

biogeochemical cycle A nutrient moves among environmental reservoirs and into and out of food webs. **752**

biogeography Study of patterns in the geographic distribution of species and communities. **254**

biological magnification A chemical pollutant becomes increasingly concentrated as it moves up through food chains. **759**

biological pest control Use of a pest's natural enemies to control its population size. **737**

biology The scientific study of life. **4**

bioluminescence Light emitted by a living organism. **335**

biomarker Substance found only or mainly in cells of one type. **307**

biome A region (often discontinuous) characterized by its climate and dominant vegetation. **768**

biosphere All regions of Earth where organisms live. **5**

biotic potential Maximum possible population growth rate under optimal conditions. **715**

bipedalism Habitual upright walking. **412**

bird Feathered reptile of a lineage in which the body became adapted for flight. **408**

bivalve Mollusk with a hinged two-part shell. For example, a clam. **386**

blastocyst Mammalian blastula. **684**

blastula Hollow ball of cells that forms as a result of cleavage. **668**

blood Circulatory fluid of a closed circulatory system. In vertebrates, a fluid connective tissue consisting of plasma and cellular components (red cells, white cells, platelets) that form in bones. **485, 570**

blood–brain barrier Protective mechanism that prevents unwanted substances from entering cerebrospinal fluid. **508**

blood pressure Pressure exerted by blood against a vessel wall. **579**

bone tissue Connective tissue made up of cells surrounded by a mineral-hardened matrix of their own secretions. **485**

boreal forest Extensive high-latitude forest of the Northern Hemisphere; conifers are the predominant vegetation. **770**

bottleneck Reduction in population size so severe that it reduces genetic diversity. **284**

Bowman's capsule Region of a nephron tubule that forms a cup around the glomerulus; coveys filtrate to the proximal tubule. **652**

brain Central control organ of a nervous system; receives and integrates sensory information, regulates internal conditions, and sends out signals that result in movements. **498**

bronchiole A small airway that leads from a bronchus to alveoli. **619**

bronchus Airway connecting the trachea to a lung. **619**

brood parasitism One egg-laying species benefits by having another raise its offspring. **737**

brown alga Multicelled marine protist with a brown accessory pigment in its chloroplasts. **337**

brush border cell In the lining of the small intestine, an epithelial cell with microvilli at its surface. **635**

bryophyte Nonvascular plant; a moss, liverwort, or hornwort. **348**

buffer Set of chemicals that can keep the pH of a solution stable by alternately donating and accepting ions that contribute to pH. **32**

C3 plant Type of plant that uses only the Calvin–Benson cycle to fix carbon. **106**

C4 plant Type of plant that minimizes photorespiration by fixing carbon twice, in two cell types. **107**

calcitonin Thyroid hormone that encourages bone to take up and incorporate calcium. **543**

Calvin–Benson cycle Cyclic carbon-fixing pathway that builds sugars from CO_2; the light-independent reactions of photosynthesis. **106**

camera eye Eye with an adjustable opening and a single lens that focuses light on a retina. **524**

camouflage Coloration or body form that helps an organism blend in with its surroundings and escape detection. **735**

CAM plant Type of plant that conserves water by fixing carbon twice, at different times of day. **107**

cancer Disease that occurs when a malignant neoplasm physically and metabolically disrupts body tissues. **185**

capillary Small-diameter blood vessel; exchanges with interstitial fluid occur across its wall. **570**

carbohydrate Molecule that consists primarily of carbon, hydrogen, and oxygen atoms in a 1:2:1 ratio. Complex kinds (e.g., cellulose, starch, glycogen) are polymers of simple kinds (sugars). **40**

carbon cycle Movement of carbon, mainly between the oceans, atmosphere, and living organisms. **754**

carbon fixation Process by which carbon from an inorganic source such as carbon dioxide becomes incorporated (fixed) into an organic molecule. **106**

cardiac cycle Sequence of contraction and relaxation of heart chambers that occurs with each heartbeat. **574**

cardiac muscle tissue Muscle of the heart wall. **486**

carpel Floral reproductive organ that produces female gametophytes; consists of an ovary, stigma, and often a style. **356, 454**

carrying capacity (*K*) Of a species, the maximum number of individuals that a particular environment can sustain; can change over time. **716**

cartilage Connective tissue that consists of cells surrounded by a rubbery matrix of their own secretions. **484**

cartilaginous fish Jawed fish with a skeleton of cartilage; a shark, ray, or skate. **402**

Casparian strip Waxy band between the plasma membranes of abutting root endodermal cells; forms a seal that prevents soil water from seeping through cell walls into the vascular cylinder. **442**

catalysis The acceleration of a reaction rate by a molecule that is unchanged by participating in the reaction. **82**

cDNA Complementary strand of DNA synthesized from an RNA template by the enzyme reverse transcriptase. **239**

cell Smallest unit of life; at minimum, consists of plasma membrane, cytoplasm, and DNA. **4**

cell cortex Reinforcing mesh of microfilaments under a plasma membrane. **66**

cell cycle A series of events from the time a cell forms until its cytoplasm divides. **178**

cell junction Structure that connects a cell to another cell or to extracellular matrix; e.g., tight junction, adhering junction, or gap junction (of animals); plasmodesmata (of plants). **69**

cell-mediated immune response Immune response involving cytotoxic T cells and NK cells that destroy infected or cancerous body cells. **598**

cell plate A disk-shaped structure that forms during cytokinesis in a plant cell; matures as a cross-wall between the two new nuclei. **182**

cell theory Theory that all organisms consist of one or more cells, which are the basic unit of life; all cells come from division of preexisting cells; and all cells pass hereditary material to offspring. **52**

cellular slime mold Amoeba-like protist that feeds as a single predatory cell; under unfavorable conditions, it joins with others to form a multicellular spore-bearing structure. **340**

cell wall Rigid but permeable structure that surrounds the plasma membrane of some cells. **59**

cellulose Tough, insoluble carbohydrate that is the major structural material in plants. **40**

central nervous system Of vertebrates, the brain and spinal cord. **499**

central vacuole Large, fluid-filled vesicle in many plant cells. **62**

centriole Barrel-shaped organelle from which microtubules grow. **67**

centromere Of a duplicated eukaryotic chromosome, constricted region where sister chromatids attach to each other. **136**

cephalization Evolutionary trend whereby nerve cells and sensory structures become concentrated in the head of a bilateral animal. **376**, **498**

cephalopod Predatory mollusk that has a closed circulatory system and moves by jet propulsion. For example an octopus or squid. **387**

cerebellum Hindbrain region that coordinates voluntary movements. **510**

cerebral cortex Outer gray matter layer of the cerebrum; region responsible for most complex behavior. **512**

cerebrospinal fluid Fluid that surrounds the brain and spinal cord and fills spaces (ventricles) within the brain. **508**

cerebrum Forebrain region that controls higher functions. **510**

cervix Narrow part of uterus that connects to the vagina. **676**

chaparral Biome of dry shrubland in regions with hot, dry summers and cool, rainy winters. **771**

character Quantifiable, heritable characteristic or trait. **294**

character displacement Evolutionary process in which two competing species become less similar in their resource requirements over time. **733**

charge Electrical property; opposite charges attract, and like charges repel. **24**

chelicerates Arthropod group with specialized feeding structures (chelicerae) and no antennae; arachnids and horseshoe crabs. **390**

chemical bond An attractive force that arises between two atoms when their electrons interact; joins atoms as molecules. *See* covalent bond, ionic bond. **28**

chemoautotroph Organism that uses carbon dioxide as its carbon source and obtains energy by oxidizing inorganic molecules. **321**

chemoheterotroph Organism that obtains energy and carbon by breaking down organic compounds. **321**

chemoreceptor Sensory receptor that responds to a chemical. **520**

chlorophyll *a* Main photosynthetic pigment in plants. **100**

chloroplast Organelle of photosynthesis in the cells of plants and photosynthetic protists. Has two outer membranes enclosing semifluid stroma. Light-dependent reactions occur at its inner thylakoid membrane; light-independent reactions, in the stroma. Stores excess sugars as starch. **65**

choanoflagellate Heterotrophic freshwater protist with a flagellum and a food-capturing "collar." May be solitary or colonial. **341**

chordate Animal with an embryo that has a notochord, dorsal nerve cord, pharyngeal gill slits, and a tail that extends beyond the anus. A lancelet, tunicate, or vertebrate. **400**

chorion Outermost extraembryonic membrane of amniotes; major component of the placenta in placental mammals. **685**

chromosome A structure that consists of DNA and associated proteins; carries part or all of a cell's genetic information. **136**

chromosome number The total number of chromosomes in a cell of a given species. **136**

chyme Mix of food and gastric fluid. **634**

chytrid Fungus that makes flagellated spores. **365**

ciliate Single-celled, heterotrophic protist with many cilia. **335**

cilium Plural, **cilia** Short, movable structure that projects from the plasma membrane of some eukaryotic cells. **66**

circadian rhythm A biological activity that is repeated about every 24 hours. **470**

circulatory system Organ system consisting of a heart or hearts and vessels that distribute circulatory fluid through a body. May be closed or open. **570**

clade A group whose members share one or more defining derived traits. **294**

cladistics Making hypotheses about evolutionary relationships among clades. **295**

cladogram Evolutionary tree diagram that shows evolutionary connections among a group of clades. **295**

cleavage Mitotic division of an animal cell. **668**

cleavage furrow In a dividing animal cell, the indentation where cytoplasmic division will occur. **182**

climate Average weather conditions in a region. **764**

cloaca Body opening that serves as the exit for digestive waste and urine; also functions in reproduction. **402**, **630**

cloning vector A DNA molecule that can accept foreign DNA and be replicated inside a host cell. **239**

closed circulatory system Circulatory system in which blood flows through a continuous network of vessels; all materials are exchanged across the walls of those vessels. **384**, **570**

club fungus Fungus that produces spores in club-shaped structures during sexual reproduction. **365**

cnidarian Radially symmetrical invertebrate with two tissue layers; uses tentacles with stinging cells to capture food. For example, a jelly or a sea anemone. **380**

cnidocyte Stinging cell unique to cnidarians. **380**

coal Fossil fuel formed over millions of years by compaction and heating of plant remains. **352**

cochlea Coiled, fluid-filled structure in the inner ear; holds the mechanoreceptors (hair cells) involved in hearing. **528**

codominance Effect in which the full and separate phenotypic effects of two alleles are apparent in heterozygous individuals. **212**

codon In an mRNA, a nucleotide base triplet that codes for an amino acid or stop signal during translation. **152**

coelom A fluid-filled body cavity between the gut and body wall; it is lined with tissue derived from mesoderm. **377**

coenzyme An organic molecule that functions as a cofactor; e.g., NAD. **86**

coevolution The joint evolution of two closely interacting species; macroevolutionary pattern in which each species is a selective agent for traits of the other. **293**

cofactor A metal ion or organic molecule that associates with an enzyme and is necessary for its function. **86**

cohesion Property of a substance that arises from the tendency of its molecules to resist separating from one another. **31**

cohesion–tension theory Explanation of how transpiration creates a tension that pulls a cohesive column of water upward through xylem, from roots to shoots. **445**

cohort Group of individuals born during the same interval. **718**

coleoptile Rigid sheath that protects a growing embryonic shoot of monocots. **461**

collecting tubule Kidney tubule that receives fluid from several nephrons and delivers it to the renal pelvis. **653**

collenchyma Simple plant tissue composed of living cells with unevenly thickened walls; provides flexible support. **424**

colonial organism Organism composed of many similar cells, each capable of living and reproducing on its own. **332**

colonial theory of animal origins Hypothesis that the first animals evolved from a colonial protist. **378**

commensalism Species interaction that benefits one species and neither helps nor harms the other. **730**

community All populations of all species in a given area. **5, 730**

compact bone Dense bone that makes up the shaft of long bones. **558**

companion cell In phloem, specialized parenchyma cell that provides a partnered sieve element with metabolic support. **446**

comparative morphology The scientific study of similarities and differences in body plans. **255**

competitive exclusion Process whereby two species compete for a limiting resource, and one drives the other to local extinction. **732**

complement A set of proteins that circulate in inactive form in blood; activated complement proteins attract phagocytic white blood cells, coat antigenic particles, and puncture lipid bilayers. **590**

complete digestive tract Tubelike digestive system; food enters through one opening and wastes leave through another. **630**

compound Molecule that has atoms of more than one element. **28**

compound eye Of some arthropods, a motion-sensitive eye made up of many image-forming units, each with its own lens. **389, 524**

concentration Amount of solute per unit volume of a solution. **32**

condensation Chemical reaction in which an enzyme builds a large molecule from smaller subunits; water also forms. **39**

cone cell Photoreceptor that provides sharp vision and allows detection of color. **526**

conjugation Mechanism of horizontal gene transfer in which one prokaryote passes a plasmid to another. **320**

connective tissue Animal tissue with an extensive extracellular matrix; provides structural and functional support. **484**

conservation biology Field of applied biology that surveys biodiversity and seeks ways to maintain and use it nondestructively. **792**

consumer Organism that obtains energy and carbon by feeding on tissues, wastes, or remains of other organisms; a heterotroph. **6, 748**

continuous variation Range of small differences in a shared trait. **216**

contractile vacuole In freshwater protists, an organelle that collects and expels excess water. **333**

control group In an experiment, group of individuals identical to an experimental group except for the independent variable under investigation. **13**

coral bleaching A coral expels its photosynthetic dinoflagellate symbionts in response to stress and becomes colorless. **777**

coral reef Highly diverse marine ecosystem centered around reefs built by living corals that secrete calcium carbonate. **777**

cork Plant tissue that waterproofs, insulates, and protects the surfaces of woody stems and roots. **433**

cork cambium Lateral meristem that gives rise to periderm in plants. **433**

cornea Clear, protective covering at the front of a vertebrate eye; helps focus light on the retina. **525**

corpus luteum Hormone-secreting structure that forms from follicle cells left behind after ovulation. **677**

cortisol Adrenal cortex hormone that influences metabolism and immunity; secretions rise with stress. **544**

cotyledon Seed leaf of a flowering plant embryo. **357**

countercurrent exchange Exchange of substances between two fluids moving in opposite directions. **616**

covalent bond Chemical bond in which two atoms share a pair of electrons. **28**

critical thinking The act of judging information before accepting it. **12**

crossing over Process by which homologous chromosomes exchange corresponding segments of DNA during prophase I of meiosis. **198**

crustaceans Mostly marine arthropods with a calcium-hardened cuticle and two pairs of antennae; for example lobsters, crabs, krill, and barnacles. **390**

cuticle Secreted covering at a body surface. **69, 346**

cyanobacteria Photosynthetic, oxygen-producing bacteria. **322**

cytokines Signaling molecules secreted by white blood cells to coordinate their activities during immune responses. **591**

cytokinesis Cytoplasmic division; process in which a eukaryotic cell divides in two after mitosis or meiosis. **182**

cytokinin Plant hormone that promotes cell division in shoot apical meristem and cell differentiation in root apical meristem. Often interacts antagonistically with auxin. **464**

cytoplasm Semifluid substance enclosed by a cell's plasma membrane. **52**

cytoskeleton Network of interconnected protein filaments that support, organize, and move eukaryotic cells and their parts. *See* microtubules, microfilaments, intermediate filaments. **66**

data Experimental results. **13**

decomposer Organism that feeds on wastes and remains; breaks organic material down into its inorganic subunits. **323, 748**

deductive reasoning Using a general idea to make a conclusion about a specific case. **12**

deletion Mutation in which one or more nucleotides are lost from DNA. **156**

demographics Statistics that describe a population. **712**

demographic transition model Model describing the changes in human birth and death rates that occur as a region becomes industrialized. **724**

denature To unravel the shape of a protein or other large biological molecule. **46**

dendrite Of a motor neuron or interneuron, a cytoplasmic extension that receives chemical signals sent by other neurons and converts them to electrical signals. **500**

dendritic cell Phagocytic white blood cell that patrols solid tissues; important antigen-presenting cell in adaptive immune responses. **591**

denitrification Conversion of nitrates or nitrites to nitrogen gas. **757**

dense, irregular connective tissue Connective tissue that consists of randomly arranged fibers and scattered fibroblasts. **484**

dense, regular connective tissue Connective tissue that consists of fibroblasts arrayed between parallel arrangements of fibers. **484**

density-dependent limiting factor Factor that limits population growth and has a greater effect in dense populations; for example, competition for a limited resource. **716**

density-independent limiting factor Factor that limits population growth and acts regardless of population size; for example a flood. **717**

dental plaque On teeth, a thick biofilm composed of bacteria, their extracellular products, and saliva proteins. **592**

dependent variable In an experiment, a variable that is presumably affected by an independent variable being tested. **13**

derived trait A novel trait present in a clade but not in the clade's ancestors. **294**

dermal tissues Tissues that cover and protect the plant body. *See* epidermis, periderm. **422**

dermis Deep layer of skin that consists of connective tissue with nerves and blood vessels running through it. **491**

desert Biome with little rain and low humidity; plants that have water-storing and water-conserving adaptations predominate. **772**

desertification Conversion of dry grassland to desert. **786**

detrital food chain Food chain in which energy is transferred directly from producers to detritivores. **750**

detritivore Consumer that feeds on small bits of organic material. **748**

deuterostomes Lineage of bilateral animals in which the second opening on the embryo surface develops into a mouth; includes echinoderms and chordates. **377**

development Multistep process by which the first cell of a new multicelled organism gives rise to an adult. **7**

diaphragm Muscle between the thoracic and abdominal cavities; contracts during inhalation. **619**

diastole Relaxation phase of the cardiac cycle. **574**

diastolic pressure Blood pressure when ventricles are relaxed. **579**

diatom Single-celled photosynthetic protist with a brown accessory pigment in its chloroplasts and a two-part silica shell. **337**

differentiation Process by which cells become specialized during development; occurs as different cells in an embryo begin to use different subsets of their DNA. **142**

diffusion Spontaneous spreading of molecules or ions. **88**

dihybrid cross Cross between two individuals identically heterozygous for two genes; for example $AaBb \times AaBb$. **210**

dikaryotic Having two genetically distinct nuclei in a cell ($n + n$). **365**

dinoflagellate Single-celled, aquatic protist that moves with a whirling motion; may be heterotrophic or photosynthetic. **335**

dinosaur Group of reptiles that includes the ancestors of birds; became extinct at the end of the Cretaceous. **406**

diploid Having two of each type of chromosome characteristic of the species ($2n$). **137**

directional selection Mode of natural selection that shifts an allele's frequency in a consistent direction, so phenotypes at one end of a range of variation are favored. **278**

disruptive selection Mode of natural selection in which traits at the extremes of a range of variation are adaptive, and intermediate forms are not. **281**

distal tubule Region of a kidney tubule that delivers filtrate to a collecting tubule. **653**

distance effect Islands close to a mainland have more species than those farther away. **742**

DNA Deoxyribonucleic acid. Carries hereditary information that guides development and other activities; consists of two chains of nucleotides (adenine, guanine, thymine, and cytosine) twisted into a double helix. **7, 46**

DNA cloning Set of methods that uses living cells to make many identical copies of a DNA fragment. **238**

DNA library Collection of cells that host different fragments of foreign DNA, often representing an organism's entire genome. **240**

DNA ligase Enzyme that seals gaps in double-stranded DNA. **138**

DNA polymerase DNA replication enzyme. Uses one strand of DNA as a template to assemble a complementary strand of DNA from nucleotides. **138**

DNA profiling Identifying an individual by analyzing the unique parts of his or her DNA. **244**

DNA replication Process by which a cell duplicates its DNA before it divides. **138**

DNA sequence Order of nucleotides in a strand of DNA. **135**

DNA sequencing *See* sequencing.

dominance hierarchy Social system in which resources and mating opportunities are unequally distributed within a group. **705**

dominant Refers to an allele that masks the effect of a recessive allele paired with it in heterozygous individuals. **207**

dormancy Period of temporarily suspended metabolism. **456**

dosage compensation Mechanism in which X chromosome inactivation equalizes gene expression between males and females. **168**

double fertilization Mode of fertilization in flowering plants in which one sperm cell fuses with the egg, and a second sperm cell fuses with the endosperm mother cell. **456**

duplication Repeated section of a chromosome. **228**

echinoderms Invertebrates with a water–vascular system and hardened plates and spines embedded in the skin or body. Radials as adults, but bilateral as larvae. For example, a sea star. **394**

ECM *See* extracellular matrix.

ecological footprint Area of Earth's surface required to sustainably support a particular level of development and consumption. **725**

ecological niche All of a species' requirements and roles in an ecosystem. **732**

ecological restoration Actively altering an area in an effort to restore an ecosystem that has been damaged or destroyed. **793**

ecology Study of interactions among organisms, and among organisms and their environment. **712**

ecosystem A community interacting with its environment through a one-way flow of energy and cycling of materials. **5, 748**

ectoderm Outermost tissue layer of an animal embryo. **376, 668**

ectotherm Animal whose body temperature varies with that of its environment; controls its internal temperature by altering its behavior; for example, a fish or a lizard. **406, 658**

effector cell Antigen-sensitized B cell or T cell that forms in an immune response and acts immediately. **598**

egg Female gamete. **666**

electron Negatively charged subatomic particle. **24**

electronegativity Measure of the ability of an atom to pull electrons away from other atoms. **28**

electron transfer chain Array of enzymes and other molecules in a cell membrane that accept and give up electrons in sequence, thus releasing the energy of the electrons in small, usable steps. **85**

electron transfer phosphorylation Process in which electron flow through electron transfer chains sets up a hydrogen ion gradient that drives ATP formation. **104**

electrophoresis Laboratory technique that separates DNA fragments by size. **242**

element A pure substance that consists only of atoms with the same number of protons. **24**

embryo In animals, a developing individual from first cleavage until hatching or birth; in humans, usually refers to an individual in weeks 2 to 8 of development. **673**

embryonic induction Embryonic cells produce signals that alter the behavior of neighboring cells. **670**

emergent property A characteristic of a system that does not appear in any of the system's component parts. **4**

emerging disease A disease that was previously unknown or has recently begun spreading to a new region. **318**

emigration Movement of individuals out of a population. **714**

emulsification Dispersion of fat droplets in a fluid. **636**

endangered species A species that faces extinction in all or a part of its range. **784**

endemic species Species that remains restricted to the area where it evolved. **784**

endergonic Describes a reaction that requires a net input of free energy to proceed. **80**

endocrine gland Ductless gland; aggregation of epithelial cells that secrete a hormone or hormones into the blood. **483, 538**

endocytosis Process by which a cell takes in a small amount of extracellular fluid (and its contents) by the ballooning inward of the plasma membrane. **92**

endoderm Innermost tissue layer of an animal embryo. **376, 668**

endodermis Outer layer of the vascular cylinder in a plant root; sheet of cells just outside the pericycle. **430**

endomembrane system Series of interacting organelles (endoplasmic reticulum, Golgi bodies, vesicles) between nucleus and plasma membrane; produces lipids, proteins. **62**

endoplasmic reticulum (ER) Organelle that is a continuous system of sacs and tubes extending from the nuclear envelope. Smooth ER makes lipids and breaks down carbohydrates and fatty acids; rough ER modifies polypeptides made by ribosomes on its surface. **63**

endorphin One type of natural painkiller molecule. **521**

endoskeleton Internal skeleton made up of hardened components such as bones. **401, 556**

endosperm Nutritive tissue in the seeds of flowering plants. **357, 456**

endospore Resistant resting stage of some soil bacteria. **323**

endosymbiont hypothesis Theory that mitochondria and chloroplasts evolved from bacteria that entered and lived in a host cell. **308**

endotherm Animal that maintains its temperature by adjusting its production of metabolic heat; for example, a bird or mammal. **406, 658**

energy The capacity to do work. **78**

enhancer In eukaryotic cells, a binding site in DNA for an activator. **164**

entropy Measure of how much the energy of a system is dispersed. **78**

enzyme Protein or RNA that speeds up a chemical reaction without being changed by it. **39**

eosinophil Circulating white blood cell with granules; specialized to combat multicelled parasites that are too large for phagocytosis. **591**

epidermis Outermost tissue layer. In young plants, dermal tissue. In animals, the epithelial layer of skin. **424, 490**

epididymis Duct where sperm mature; empties into a vas deferens. **674**

epigenetic Refers to heritable changes in gene expression that are not the result of changes in DNA sequence. **172**

epiglottis Tissue flap that folds down to prevent food from entering the trachea during swallowing. **619**

epistasis Polygenic inheritance, in which a trait is influenced by multiple genes. **213**

epithelial tissue Sheetlike animal tissue that covers outer body surfaces and lines internal tubes and cavities. **482**

equilibrium model of island biogeography Model that predicts the number of species on an island based on the island's area and distance from the mainland. **742**

ER *See* endoplasmic reticulum.

erosion *See* soil erosion.

esophagus Muscular tube that connects the pharynx (throat) to the stomach. **633**

essential amino acid Amino acid that the body cannot make and must obtain from food. **639**

essential fatty acid Fatty acid that the body cannot make and must obtain from food. **639**

estrogens Sex hormones that function in reproduction and cause development of female secondary sexual characteristics; secreted by the ovaries. **545, 676**

estrous cycle Reproductive cycle in which the uterine lining thickens, and, if pregnancy does not occur, is reabsorbed. **679**

estuary A highly productive ecosystem where nutrient-rich water from a river mixes with seawater. **776**

ethylene Gaseous plant hormone that participates in germination, abscission, ripening, and stress responses. **467**

eudicot Flowering plant in which the embryo has two seed leaves (cotyledons). For example, a tomato, cherry, or cactus. **357**

eugenics Idea of deliberately improving the genetic qualities of the human race. **248**

euglenoid Flagellated protozoan with multiple mitochondria; may be heterotrophic or have chloroplasts descended from green algae. **333**

eukaryote Organism whose cells characteristically have a nucleus; a protist, fungus, plant, or animal. **8**

eusocial animal Animal that lives in a multigenerational group in which many sterile workers cooperate in all tasks essential to the group's welfare, while a few members of the group produce offspring. **706**

eutrophication Nutrient enrichment of an aquatic habitat. **774**

evaporation Transition of a liquid to a vapor. **31**

evolution Change in a line of descent. **256**

evolutionary tree Diagram showing evolutionary connections. **295**

exaptation Evolutionary adaptation of an existing structure for a completely new purpose. **292**

exergonic Describes a reaction that ends with a net release of free energy. **80**

exocrine gland Gland that secretes milk, sweat, saliva, or some other substance through a duct. **483**

exocytosis Process by which a cell expels a vesicle's contents to extracellular fluid. **92**

exon Nucleotide sequence that remains in an RNA after post-transcriptional modification. **151**

exoskeleton Of some invertebrates, hard external parts that muscles attach to and move. **389, 556**

exotic species A species that evolved in one community and later became established in a different one. **740**

experiment A test designed to support or falsify a prediction. **12**

experimental group In an experiment, a group of individuals who have a certain characteristic or receive a certain treatment as compared with a control group. **13**

exponential growth A population grows by a fixed percentage in successive time intervals; the size of each increase is determined by the current population size. **714**

external fertilization Sperm and eggs unite in the external environment. **667**

extinct Refers to a species that no longer has living members. **292**

extracellular fluid Of a multicelled organism, body fluid outside of cells; serves as the body's internal environment. **480**

extracellular matrix (ECM) Complex mixture of cell secretions; its composition and function vary by cell type. E.g., basement membrane of epithelial tissue. **68**

extreme halophile Organism adapted to life in a highly salty environment. **326**

extreme thermophile Organism adapted to life in a very high-temperature environment. **326**

facilitated diffusion Passive transport mechanism in which a solute follows its concentration gradient across a membrane by moving through a transport protein. **90**

fat Lipid that consists of a glycerol molecule with one, two, or three fatty acid tails. *See* saturated fat, unsaturated fat. **42**

fatty acid Organic compound that consists of a chain of carbon atoms with an acidic carboxyl group at one end. Carbon chain of saturated types has single bonds only; that of unsaturated types has one or more double bonds. **42**

feces Unabsorbed food material and cellular waste that is expelled from the digestive tract. **638**

feedback inhibition Regulatory mechanism in which a change that results from some activity decreases or stops the activity. **84**

fermentation A metabolic pathway that breaks down sugars to produce ATP and does not require oxygen. E.g., lactate fermentation. **114**

fertilization Fusion of two gametes to form a zygote; part of sexual reproduction. **195**

fetus Developing human from about 9 weeks until birth. **673**

fever A temporary, internally induced rise in core body temperature above the normal set point. **595**

fibrin Threadlike protein formed during blood clotting from the soluble plasma protein fibrinogen. **577**

fibrous root system Root system composed of an extensive mass of similar-sized adventitious roots; typical of monocots. **430**

first law of thermodynamics Energy cannot be created or destroyed. **78**

fitness Degree of adaptation to an environment, as measured by an individual's relative genetic contribution to future generations. **257**

fixed Refers to an allele for which all members of a population are homozygous. **284**

flagellated protozoan Protist belonging to an entirely or mostly heterotrophic lineage with no cell wall and one or more flagella. **333**

flagellum Long, slender cellular structure used for locomotion through fluid surroundings. **59**

flatworm Bilaterally symmetrical invertebrate with organs but no body cavity; for example, a planarian or tapeworm. **382**

flower Specialized reproductive structure of a flowering plant. **356, 454**

fluid mosaic Model of a cell membrane as a two-dimensional fluid of mixed composition. **56**

follicle-stimulating hormone (FSH) Anterior pituitary hormone with roles in ovarian follicle maturation and sperm production. **678**

food chain Description of who eats whom in one path of energy flow through an ecosystem. **749**

food web Set of cross-connecting food chains. **750**

foraminifera Heterotrophic single-celled protists with a porous calcium carbonate shell and long cytoplasmic extensions. **334**

fossil Physical evidence of an organism that lived in the ancient past. **255**

founder effect After a small group of individuals found a new population, allele frequencies in the new population differ from those in the original population. **284**

fovea Retinal region where cone cells are most concentrated. **526**

free radical Atom with an unpaired electron; most are highly reactive and can damage biological molecules. **27**

frequency-dependent selection Mode of natural selection in which a trait's adaptive value depends on its frequency in a population. **283**

fruit Mature ovary of a flowering plant; often with accessory parts; encloses a seed or seeds. **357, 458**

FSH *See* follicle-stimulating hormone.

functional group An atom (other than hydrogen) or a small molecular group bonded to a carbon of an organic compound; imparts a specific chemical property. **39**

fungus Single-celled or multicellular eukaryotic consumer that digests food outside its body, then absorbs the resulting breakdown products. Has chitin-containing cell walls. **9, 364**

gallbladder Organ that stores and concentrates bile produced by the liver. **636**

gamete Mature, haploid reproductive cell; e.g., an egg or a sperm. **194**

gametophyte Multicelled, haploid, gamete-producing body that forms in the life cycle of land plants and some multicelled algae. **338, 346**

ganglion Cluster of nerve cell bodies. **498**

gap junction Cell junction that forms a closable channel across the plasma membranes of adjoining animal cells. **69**

gastric fluid Fluid secreted by the stomach lining; contains digestive enzymes, acid, and mucus. **634**

gastropod Mollusk in which the lower body consists of a broad "foot"; for example, a snail or slug. **386**

gastrovascular cavity Saclike gut that also functions in gas exchange. **380, 630**

gastrula Three-layered structure formed by gastrulation during animal development. **668**

gastrulation Animal developmental process by which cell movements produce a three-layered gastrula. **668**

gene A part of a chromosome that encodes an RNA or protein product in its DNA sequence. Unit of hereditary information. **148**

gene expression Process by which the information in a gene guides assembly of an RNA or protein product. **149**

gene flow The movement of alleles into and out of a population. **285**

gene pool All the alleles of all the genes in a population; a pool of genetic resources. **275**

gene therapy Treating a genetic defect or disorder by transferring a normal or modified gene into the affected individual. **248**

genetically modified organism (GMO) Organism whose genome has been modified by genetic engineering. **246**

genetic code Complete set of sixty-four mRNA codons. **152**

genetic drift Change in allele frequency due to chance alone. **284**

genetic engineering Process by which deliberate changes are introduced into an individual's genome. **246**

genetic equilibrium Theoretical state in which an allele's frequency never changes in a population's gene pool. **276**

genome An organism's complete set of genetic material. **240**

genomics The study of genomes. **244**

genotype The particular set of alleles that is carried by an individual's chromosomes. **207**

genus plural **genera** A group of species that share a unique set of traits; first part of a species name. **10**

geologic time scale Chronology of Earth's history; correlates geologic and evolutionary events. **264**

germ cell Immature reproductive cell that gives rise to haploid gametes when it divides. **194**

germinate To resume metabolic activity after dormancy. **456**

germ layer One of three primary layers in an early embryo. **668**

GH *See* growth hormone.

gibberellin Plant hormone that induces stem elongation; also helps seeds break dormancy. **465**

gill Of an aquatic animal, a folded or filamentous respiratory organ in which blood or hemolymph exchanges gases with water. **615**

global climate change A currently ongoing rise in average temperature that is altering climate patterns around the world. **755**

glomeromycete Fungus that partners with plant roots; fungal hyphae grow inside the cell walls of root cells. **365**

glomerular filtration Protein-free plasma forced out of glomerular capillaries by blood pressure enters Bowman's capsule. **654**

glomerulus In a kidney, a ball of leaky capillaries enclosed by Bowman's capsule. **653**

glottis Opening formed when the vocal cords relax. **618**

glucagon Pancreatic hormone that raises blood glucose level. **546**

glycolysis Set of reactions in which a six-carbon sugar (such as glucose) is broken down to two pyruvate for a net yield of two ATP. First part of carbohydrate-breakdown pathways. **116**

GMO *See* genetically modified organism.

Golgi body Membrane-enclosed organelle that modifies proteins and lipids, then packages the finished products into vesicles. **63**

gonad Gamete-forming organ of an animal; testes or ovaries. **666**

Gondwana Supercontinent that existed before Pangea, more than 500 million years ago. **263**

Gram-positive bacteria Bacteria with thick cell walls that are colored purple when prepared for microscopy by Gram staining. **323**

grassland Biome in the interior of continents; perennial grasses and other nonwoody plants adapted to grazing and fire predominate. **771**

gravitropism Plant growth in a direction influenced by gravity. **468**

gray matter Central nervous system tissue that consists of neuron axon terminals, cell bodies, and dendrites, along with some neuroglial cells. **508**

grazing food chain Food chain in which energy is transferred from producers to grazers (herbivores). **750**

green alga Single-celled, colonial, or multicelled photosynthetic protist that has chloroplasts containing chlorophylls *a* and *b*. **338**

greenhouse effect Warming of Earth's lower atmosphere and surface as a result of heat trapped by greenhouse gases. **755**

ground tissues Tissues that make up the bulk of the plant body; all plant tissues other than vascular and dermal tissues. **422**

groundwater Soil water and water in aquifers. **752**

growth In multicelled species, an increase in the number, size, and volume of cells. **7**

growth factor Molecule that stimulates mitosis and differentiation. **184**

growth hormone (GH) Anterior pituitary hormone that regulates growth and metabolism. **541**

guard cell One of a pair of cells that define a stoma across the epidermis of a plant leaf or stem. **446**

gymnosperm Seed plant whose seeds are not enclosed within a fruit; a conifer, cycad, ginkgo, or gnetophyte. **354**

habitat Type of environment in which a species typically lives. **730**

habituation Learning not to respond to a repeated neutral stimulus. **699**

half-life Characteristic time it takes for half of a quantity of a radioisotope to decay. **260**

haploid Having one of each type of chromosome characteristic of the species. **194**

heart Muscular organ that pumps blood through a body. **570**

hemolymph Fluid that circulates in an open circulatory system. **570**

hemostasis Process by which blood clots in response to injury. **577**

herbivory An animal feeds on plants or plant parts. **735**

hermaphrodite Animal that has both male and female gonads, either simultaneously or at different times in its life. **379, 666**

heterotherm Animal that sometimes maintains its temperature by producing metabolic heat, and at other times allows its temperature to fluctuate with the environment. **658**

heterotroph Consumer. An organism that obtains carbon from organic compounds assembled by other organisms. **100**

heterozygous Having two different alleles of a gene; describes genotype of a diploid organism. **207**

hippocampus Brain region essential to formation of declarative memories. **513**

histone Type of protein that structurally organizes eukaryotic chromosomes. **136**

HIV (human immunodeficiency virus) Virus that causes AIDS. **317**

homeostasis Process in which an organism keeps its internal conditions within tolerable ranges by sensing and responding to change. **7**

homeotic gene Type of master gene; its expression controls formation of specific body parts during development. **166**

hominin Human or an extinct primate species more closely related to humans than to any other primates. **412**

Homo erectus Extinct hominin that arose about 1.8 million years ago in East Africa; migrated out of Africa. **414**

Homo habilis Extinct hominin; earliest named *Homo* species; known only from Africa, where it arose 2.3 million years ago. **414**

homologous chromosomes Chromosomes with the same length, shape, and genes. In sexual reproducers, one member of a homologous pair is paternal and the other is maternal. **179**

homologous structures Body structures that are similar in different lineages because they evolved in a common ancestor. **266**

Homo neanderthalensis Extinct hominin; closest known relative of *H. sapiens*; lived in Africa, Europe, Asia. **414**

homozygous Having identical alleles of a gene; describes genotype of a diploid organism. **207**

horizontal gene transfer Transfer of genetic material between existing individuals. **320**

hormone *See* animal hormone or plant hormone.

hot spot Threatened region that is habitat for species not found elsewhere and is considered a high priority for conservation efforts. **792**

human immunodeficiency virus *See* HIV.

humus Decaying organic matter in soil. **440**

hybrid The heterozygous offspring of a cross or mating between two individuals that breed true for different forms of a trait. **207**

hydrocarbon Compound or region of one that consists only of carbon and hydrogen atoms. **38**

hydrogen bond Attraction between a covalently bonded hydrogen atom and another atom taking part in a separate covalent bond. Collectively, they impart special properties to liquid water and stabilize the structure of biological molecules. **30**

hydrolysis Water-requiring chemical reaction in which an enzyme breaks a molecule into smaller subunits. **39**

hydrophilic Describes a substance that dissolves easily in water. **30**

hydrophobic Describes a substance that resists dissolving in water. **30**

hydrostatic skeleton Of soft-bodied invertebrates, a fluid-filled chamber that muscles exert force against, redistributing the fluid. **380, 556**

hydrothermal vent Underwater opening where hot, mineral-rich water streams out from an underwater opening in Earth's crust. **302, 778**

hypertonic Describes a fluid that has a high solute concentration relative to another fluid separated by a semipermeable membrane. **88**

hypha Component of a fungal mycelium; a filament made up of cells arranged end to end. **364**

hypothalamus Forebrain region that controls processes related to homeostasis and has endocrine functions. **511, 540**

hypothesis Testable explanation of a natural phenomenon. **12**

hypotonic Describes a fluid that has a low solute concentration relative to another fluid separated by a semipermeable membrane. **88**

immigration Movement of individuals into a population. **714**

immunity The body's ability to resist and fight infections. **590**

immunization Any procedure designed to induce immunity to a specific disease. **607**

imprinting Learning that can occur only during a specific interval in an animal's life. **698**

inbreeding Mating among close relatives. **285**

incomplete dominance Effect in which one allele is not fully dominant over another, so the heterozygous phenotype is an intermediate blend between the two homozygous phenotypes. **212**

independent variable In an experiment, variable that is controlled by an experimenter in order to explore its relationship to a dependent variable. **13**

indicator species Species whose presence and abundance in a community provides information about conditions in the community. **739**

induced-fit model Substrate binding to an active site improves the fit between the two. **82**

inductive reasoning Drawing a conclusion based on observation. **12**

inferior vena cava Vein that delivers blood from the lower body to the heart. **573**

inflammation A local response to tissue damage or infection; characterized by redness, warmth, swelling, and pain. **594**

inheritance Transmission of DNA to offspring. **7**

innate immunity In all multicelled organisms, set of immediate, general defenses against infection. **590**

inner ear Fluid-filled cochlea and vestibular apparatus. **528**

insect Most diverse arthropod group; members have six legs, two antennae, and, in some groups, wings. **392**

insertion Mutation in which one or more nucleotides become inserted into DNA. **156**

instinctive behavior An innate response to a simple stimulus. **698**

insulin Pancreatic hormone that lowers blood glucose level. **546**

intermediate disturbance hypothesis Species richness is greatest in communities with moderate levels of disturbance. **739**

intermediate filament Stable cytoskeletal element that structurally supports cell membranes and tissues. **66**

internal fertilization Sperm fertilize eggs inside a female's body. **667**

interneuron Neuron that both receives signals from and sends signals to other neurons. Located mainly in the brain and spinal cord. **500**

interphase In a eukaryotic cell cycle, the interval between mitotic divisions when a cell grows, roughly doubles the number of its cytoplasmic components, and replicates its DNA. **178**

interspecific competition Competition between members of different species. **732**

interstitial fluid Fluid in spaces between body cells. **480**

intervertebral disk Cartilage disk between two vertebrae. **556**

intraspecific competition Competition for resources among members of the same species. **716**

intron Nucleotide sequence that intervenes between exons and is removed during post-transcriptional modification. **151**

inversion Structural rearrangement of a chromosome in which part of the DNA becomes oriented in the reverse direction. **228**

invertebrate Animal that does not have a backbone. **376**

ion Charged atom. **27**

ionic bond Type of chemical bond in which a strong mutual attraction links ions of opposite charge. **28**

iris Circular muscle that adjusts the shape of the pupil to regulate how much light enters the eye. **525**

isotonic Describes two fluids with identical solute concentrations and separated by a semipermeable membrane. **88**

isotopes Forms of an element that differ in the number of neutrons their atoms carry. **24**

jawless fish Fish with a skeleton of cartilage, no fins or jaws; a lamprey or hagfish. **402**

joint Region where bones meet. **559**

karyotype Image of an individual's set of chromosomes arranged by size, length, shape, and centromere location. **137**

key innovation An evolutionary adaptation that gives its bearer the opportunity to exploit a particular environment much more efficiently or in a new way. **292**

keystone species A species that has a disproportionately large effect on community structure relative to its abundance. **740**

kidney Organ of the vertebrate urinary system that filters blood, adjusts its composition, and forms urine. **651**

kinetic energy The energy of motion. **78**

knockout An experiment in which a gene is deliberately inactivated in a living organism; also, an organism that carries a knocked-out gene. **166**

Krebs cycle Cyclic pathway that, along with acetyl–CoA formation, breaks down pyruvate to carbon dioxide in aerobic respiration's second stage. **118**

***K*-selection** Selection favoring traits that allow their bearers to outcompete others for limited resources; occurs when a population is near its environment's carrying capacity. **719**

labor Process of giving birth; expulsion of a placental mammal from its mother's uterus by muscle contractions. **690**

lactate fermentation Anaerobic sugar breakdown pathway that produces ATP and lactate. **122**

lactation Milk production by a female mammal. **690**

lancelet Invertebrate chordate that has a fishlike shape and retains the defining chordate traits into adulthood. **400**

large intestine Organ that receives digestive waste from the small intestine and concentrates it as feces. **633**

larva Sexually immature stage in some animal life cycles. **379**

larynx Short airway containing the vocal cords (voice box). **618**

lateral meristem Vascular cambium or cork cambium; cylindrical sheet of meristem that runs lengthwise through shoots and roots; gives rise to secondary growth (thickening) in a plant. **432**

law of independent assortment During meiosis, members of a pair of genes on homologous chromosomes tend to be distributed into gametes independently of other gene pairs. **210**

law of nature Generalization that describes a consistent natural phenomenon for which there is incomplete scientific explanation. **18**

law of segregation The two members of each pair of genes on homologous chromosomes end up in different gametes during meiosis. **209**

leaching Process by which water moving through soil removes nutrients from it. **440**

learned behavior Behavior that is modified by experience. **698**

lek Of some birds, a communal mating display area for males. **703**

lens Disk-shaped structure that bends light rays so they fall on an eye's photoreceptors. **524**

lethal mutation Mutation that alters phenotype so drastically that it causes death. **274**

LH *See* luteinizing hormone.

lichen Composite organism consisting of a fungus and green algae or cyanobacteria. **368**

life history A set of traits related to growth, survival, and reproduction such as life span, age-specific mortality, age at first reproduction, and number of breeding events. **718**

ligament Strap of dense connective tissue that holds bones together at a joint. **559**

light-dependent reactions First stage of photosynthesis; metabolic pathway that converts light energy to chemical energy. A noncyclic pathway produces oxygen; a cyclic pathway does not. **103**

light-independent reactions Second stage of photosynthesis; metabolic pathway that uses ATP and NADPH to assemble sugars from water and CO_2. E.g., Calvin–Benson cycle in C3 plants. **103**

lignin Material that strengthens the cell walls of vascular plants. **68, 346**

limbic system Group of structures deep in the brain that function in expression of emotion. **513**

lineage Line of descent. **256**

linkage group All genes on a chromosome. **211**

lipid A fat, steroid, or wax. **42**

lipid bilayer Double layer of lipids (mainly phospholipids) arranged tail-to-tail; structural foundation of all cell membranes. **43**

loam Soil with roughly equal amounts of sand, silt, and clay. **440**

lobe-finned fish Jawed fish with fleshy fins that contain bones; a coelacanth or lungfish. **403**

locomotion Self-propelled movement from place to place. **554**

locus Location of a gene on a chromosome. **206**

logistic growth A population grows exponentially at first, then growth slows as population size approaches the environment's carrying capacity for that species. **716**

loop of Henle U-shaped portion of a kidney tubule; connects the proximal and distal regions of the tubule. **653**

loose connective tissue Connective tissue that consists of fibroblasts and fibers scattered in a gel-like matrix. **484**

lung Internal respiratory organ that exchanges gases with the air. **615**

luteinizing hormone (LH) Anterior pituitary hormone; with roles in ovulation, corpus luteum formation, and sperm production. **678**

lymph Fluid in the lymph vascular system. **584**

lymph node Small mass of lymphatic tissue through which lymph filters; contains many lymphocytes (B and T cells). **584**

lymph vascular system System of vessels that takes up interstitial fluid and carries it (as lymph) to the blood. **584**

lysogenic pathway Bacteriophage replication path in which viral DNA becomes integrated into the host's chromosome and is passed to the host's descendants. **317**

lysosome Enzyme-filled vesicle that breaks down cellular wastes and debris. **62**

lysozyme Antibacterial enzyme in body secretions such as saliva and mucus. **593**

lytic pathway Bacteriophage replication pathway in which a virus immediately replicates in its host and kills it. **316**

macroevolution Large-scale evolutionary patterns and trends; e.g., adaptive radiation, exaptation. **292**

macrophage Phagocytic white blood cell that patrols tissues and tissue fluids. **591**

Malpighian tubules Of insects and spiders, tubular organs that take up waste solutes and deliver them to the gut for excretion. **650**

mammal Animal with hair or fur; females secrete milk from mammary glands. **409**

mark–recapture sampling Method of estimating population size of mobile animals by marking individuals, releasing them, then checking the proportion of marks among individuals recaptured at a later time. **713**

marsupial Mammal in which young are born at an early stage and complete development in a pouch on the mother's surface. **409**

mass number Of an isotope, the total number of protons and neutrons in the atomic nucleus. **24**

mast cell Stationary white blood cell that releases the contents of its granules in response to antigen as well as signaling molecules from the endocrine and nervous system. Factor in inflammation. **591**

master gene Gene encoding a product that affects the expression of many other genes. **166**

mechanoreceptor Sensory receptor that responds to pressure, position, or acceleration. **520**

medulla oblongata Hindbrain region that influences breathing and controls reflexes such as coughing and vomiting. **510**

megaspore Of seed plants, haploid spore that forms in an ovule and gives rise to an egg-producing gametophyte. **352, 456**

meiosis Nuclear division process that halves the chromosome number. Basis of sexual reproduction. **194**

melatonin Pineal gland hormone that regulates sleep–wake cycles and seasonal changes. **542**

membrane potential Voltage difference across a cell membrane; arises from differences in charge on opposite sides of the membrane. **501**

memory cell Long-lived, antigen-sensitized B cell or T cell that forms in a primary response and is held in reserve to act in a secondary response. **598**

meninges Membranes that enclose the brain and spinal cord. **508**

menopause Permanent cessation of menstrual cycles. **679**

menstrual cycle Reproductive cycle in which the uterus lining thickens and then, if pregnancy does not occur, is shed. **678**

menstruation Flow of shed uterine tissue out of the vagina. **678**

meristem In a plant, a zone of undifferentiated cells; all plant growth arises from divisions of meristem cells. **432**

mesoderm Middle tissue layer of a three-layered animal embryo. **376, 668**

mesophyll Photosynthetic parenchyma. **425**

messenger RNA (mRNA) RNA that has a protein-building message. **148**

metabolic pathway Series of enzyme-mediated reactions by which cells build, remodel, or break down an organic molecule. **84**

metabolism All of the enzyme-mediated chemical reactions by which cells acquire and use energy as they build and break down organic molecules. **39**

metamorphosis Dramatic remodeling of body form during the transition from larva to adult. **389**

metaphase Stage of mitosis at which all chromosomes are aligned midway between spindle poles. **181**

metastasis The process in which malignant cells of a neoplasm spread from one part of the body to another. **185**

methanogen Organism that produces methane gas (CH_4) as a metabolic by-product. **326**

MHC markers Self-proteins on the surface of human body cells. **596**

microevolution Change in allele frequency. **275**

microfilament Cytoskeletal element composed of actin subunits. Reinforces cell membranes; functions in movement and muscle contraction. **66**

microspore Of seed plants, a haploid spore formed in pollen sacs; gives rise to a pollen grain. **352, 456**

microtubule Hollow cytoskeletal element composed of tubulin subunits. Involved in movement of a cell or its parts. **66**

microvilli Thin projections that increase the surface area of some epithelial cells. **482, 635**

middle ear Eardrum and the tiny bones that transfer sound to the inner ear. **528**

mimicry An evolutionary pattern in which one species becomes more similar in appearance to another. **735**

mineral In the diet, an inorganic substance that is required in small amounts for normal metabolism. **641**

mitochondrion Double-membraned organelle that produces ATP by aerobic respiration in eukaryotes. **64**

mitosis Nuclear division mechanism that maintains the chromosome number. Basis of body growth and tissue repair in multicelled eukaryotes; also asexual reproduction in some multicelled eukaryotes and many single-celled ones. **178**

model Analogous system used to test an object or event that cannot be tested directly. **12**

mold Fungus that grows as a mass of asexually reproducing hyphae. **365**

molecule Two or more atoms joined by chemical bonds. **4**

mollusk Invertebrate with a reduced coelom and a mantle. For example, a bivalve, gastropod, or cephalopod. **386**

molting Periodic shedding of an outer body layer or part. **388**

monocot Flowering plant with one seed leaf (cotyledon). For example, a grass, orchids, or palm. **357**

monohybrid cross Cross between two individuals identically heterozygous for one gene; e.g., $Aa \times Aa$. **208**

monomers Molecules that are subunits of polymers. **39**

monophyletic group An ancestor in which a derived trait evolved, together with all of its descendants. **294**

monotreme Egg-laying mammal. **409**

monsoon Wind that reverses direction seasonally. **766**

morphogen Substance that regulates development by affecting cells in a concentration-dependent manner. **670**

morphological convergence Evolutionary pattern in which similar body parts (analogous structures) evolve separately in different lineages. **266**

morphological divergence Evolutionary pattern in which a body part of an ancestor changes in its descendants. **266**

motor neuron Neuron that controls a muscle or gland. **500**

motor protein Type of energy-using protein that interacts with cytoskeletal elements to move the cell's parts or the whole cell. **66**

motor unit One motor neuron and the muscle fibers it controls. **563**

mRNA *See* messenger RNA.

multicellular organism Organism that consists of interdependent cells of multiple types. **310, 332**

multiple allele system Gene for which three or more alleles persist in a population at relatively high frequency. **212**

muscle tension Force exerted by a contracting muscle. **563**

muscle tissue Tissue that consists mainly of contractile cells. **486**

mutation Permanent change in the nucleotide sequence of DNA. **140**

mutualism Species interaction that benefits both species. **731**

mycelium Mass of threadlike filaments (hyphae) that make up the body of a multicelled fungus. **364**

mycorrhiza Mutually beneficial partnership between a fungus and a plant root. **368**

myelin Fatty material produced by neuroglial cells; insulates axons and thus speeds conduction of action potentials. **503**

myofibril Within a muscle fiber, a threadlike contractile component made up of sarcomeres arranged end to end. **561**

myoglobin Muscle protein that reversibly binds oxygen. **564**

myosin ATP-dependent motor protein; makes up the thick filaments in a sarcomere. **561**

myriapod Long-bodied terrestrial arthropod with one pair of antennae and many similar segments; a centipede or millipede. **390**

natural killer cell *See* NK cell.

natural selection Differential survival and reproduction of individuals of a population based on differences in shared, heritable traits. Driven by environmental pressures. **257**

nectar Sweet fluid exuded by some flowers; attracts animal pollinators. **455**

negative feedback mechanism A change causes a response that reverses the change; important mechanism of homeostasis. **492**

neoplasm An accumulation of abnormally dividing cells. **184**

nephridium Of some invertebrates, an organ that takes up body fluid and expels excess water and solutes through a pore at the body surface. **650**

nephron A kidney tubule and associated capillaries; filters blood and forms urine. **652**

nerve Neuron fibers bundled inside a sheath of connective tissue. **498**

nerve cord Bundle of nerve fibers running the length of a body. **498**

nerve net Of cnidarians, a mesh of interacting neurons with no central control organ. **380, 498**

nervous tissue Animal tissue composed of neurons and supporting cells; detects stimuli and controls responses to them. **487**

neuroglial cell Cell that supports neurons. **498**

neuromuscular junction Synapse between a neuron and a muscle. **504**

neuron One of the cells that make up communication lines of nervous systems; transmits electrical signals along its plasma membrane and sends chemical messages to other cells. **487, 498**

neurosecretory cell Specialized neuron that secretes a hormone into the blood in response to an action potential. **540**

neurotransmitter Chemical signal released by axon terminals of a neuron. **504**

neutral mutation A mutation that has no effect on survival or reproduction. **274**

neutron Uncharged subatomic particle. **24**

neutrophil Most abundant circulating phagocytic white blood cell. **591**

niche *See* ecological niche.

nitrification Conversion of ammonium to nitrate. **756**

nitrogen cycle Movement of nitrogen among the atmosphere, soil, and water, and into and out of food webs. **756**

nitrogen fixation Incorporation of nitrogen gas (N_2) into ammonia (NH_3). **322, 756**

NK cell Natural killer cell. Lymphocyte that can kill cancer cells undetectable by cytotoxic T cells. **591**

node A region of stem where new shoots form. **426**

nondisjunction Failure of sister chromatids or homologous chromosomes to separate during nuclear division. **230**

nonvascular plant Plant that does not have xylem and phloem; a bryophyte such as a moss. **346**

normal flora Microorganisms that typically live on human surfaces, including the interior tubes and cavities of the digestive and respiratory tracts. **324, 592**

notochord Stiff rod of connective tissue that runs the length of the body in chordate larvae or embryos. **400**

nuclear envelope A double membrane that constitutes the outer boundary of the nucleus. Pores in the membrane control which substances can cross. **61**

nucleic acid Polymer of nucleotides; DNA or RNA. **46**

nucleic acid hybridization Convergence of complementary nucleic acid strands. Arises because of base-pairing interactions. **138**

nucleoid Of a bacterium or archaeon, region of cytoplasm where the DNA is concentrated. **59**

nucleolus In a cell nucleus, a dense, irregularly shaped region where ribosomal subunits are assembled. **61**

nucleoplasm Viscous fluid enclosed by the nuclear envelope. **61**

nucleosome A length of chromosomal DNA wound twice around a spool of histone proteins. **136**

nucleotide Small organic compound that is a subunit of nucleic acids. Consists of a five-carbon sugar, nitrogen-containing base, and one or more phosphate groups. E.g., adenine, guanine, cytosine, thymine, uracil. **46**

nucleus Of a eukaryotic cell, organelle with a double membrane that holds, protects, and controls access to the cell's DNA. **8, 52** Of an atom, core region occupied by protons and neutrons. **24**

nutrient Substance that an organism needs for growth and survival but cannot make for itself. **6**

olfactory receptor Chemoreceptor involved in the sense of smell. **522**

oncogene Gene that helps transform a normal cell into a tumor cell. **184**

oocyte Immature egg. **676**

open circulatory system Circulatory system in which hemolymph leaves vessels and flows among tissues before returning to the heart. **386, 570**

operator In prokaryotes, a binding site in DNA for a repressor. **164**

operon Group of genes together with a promoter–operator DNA sequence that controls their transcription. **170**

organ In multicelled organisms, a structure that consists of tissues engaged in a collective task. **5**

organelle Structure that carries out a specialized metabolic function inside a cell; e.g., a mitochondrion. **52**

organic Describes a molecule that consists mainly of carbon and hydrogen atoms. **38**

organism Individual that consists of one or more cells. **4**

organs of equilibrium Sensory organs that respond to body position and motion; function in sense of balance. **530**

organ system In multicelled organisms, set of organs that interact closely in a collective task. **5**

osmosis Diffusion of water across a selectively permeable membrane; occurs when the fluids on either side of the membrane are not isotonic. **89**

osmotic pressure Amount of turgor that prevents osmosis into cytoplasm or other hypertonic fluid. **89**

outer ear External ear (pinna) and the air-filled auditory canal. **528**

ovarian follicle In animals, immature egg and surrounding cells. **677**

ovary In flowering plants, the enlarged base of a carpel, inside which one or more ovules form and eggs are fertilized. Matures as a fruit. **356, 454** In animals, an egg-producing gonad. **668**

oviduct Duct between an ovary and the uterus. **676**

ovulation Release of a secondary oocyte from an ovary. **677**

ovule Of seed plants, reproductive structure in which egg-bearing gametophyte develops; after fertilization, it matures into a seed. **352, 454**

ovum Mature animal egg. **681**

ozone layer Upper atmospheric region with a high concentration of ozone (O_3) that screens out incoming UV radiation. **307, 790**

pain Perception of tissue injury. **521**

pain receptor Sensory receptor that responds to tissue damage. **520**

pancreas Organ that secretes digestive enzymes into the small intestine, and the hormones insulin and glucagon into the blood. **546**

Pangea Supercontinent that formed about 270 million years ago. **262**

parapatric speciation Speciation pattern in which populations speciate while in contact along a common border. **291**

parasitism Relationship in which one species withdraws nutrients from another species, without immediately killing it. **736**

parasitoid An insect that lays eggs in another insect, and whose young devour their host from the inside. **736**

parasympathetic neurons Neurons of the autonomic system that encourage digestion and other "housekeeping" tasks. **506**

parathyroid glands Four small endocrine glands on the rear of the thyroid that secrete parathyroid hormone. **543**

parathyroid hormone Hormone that regulates the concentration of calcium ions in the blood. **543**

parenchyma Simple tissue composed of living cells with different functions depending on location; main component of ground tissue. **424**

passive transport Membrane-crossing mechanism that requires no energy input. **90**

pathogen Disease-causing agent. **318**

PCR *See* polymerase chain reaction.

pedigree Chart showing the pattern of inheritance of a trait through generations in a family. **222**

pelagic province The ocean's open waters. **778**

pellicle Layer of proteins that gives shape to many unwalled, single-celled protists. **333**

penis Male organ of intercourse. **674**

peptide Short chain of amino acids linked by peptide bonds. **44**

peptide bond A bond between the amine group of one amino acid and the carboxyl group of another. Joins amino acids in proteins. **44**

per capita growth rate (*r*) Of a population, the change in individuals added over some time interval, divided by the number of individuals in the population. **714**

perception The meaning a brain derives from a sensation. **520**

pericycle In a plant, layer of cells just inside the endodermis of a root vascular cylinder. **430**

periderm Plant dermal tissue that replaces epidermis during secondary growth of woody stems and roots. **433**

periodic table Tabular arrangement of all known elements by their atomic number. **24**

peripheral nervous system Of vertebrates, nerves that carry signals between the central nervous system and the rest of the body. **499**

peristalsis Wavelike smooth muscle contractions that propel food through the digestive tract. **633**

peritubular capillaries Capillaries that surround a kidney tubule and exchange substances with it during urine formation. **653**

permafrost Continually frozen soil layer that lies beneath arctic tundra and prevents water from draining. **773**

peroxisome Enzyme-filled vesicle that breaks down amino acids, fatty acids, and toxic substances. **62**

pH Measure of the number of hydrogen ions in a fluid. Decreases with increasing acidity. **32**

phagocytosis "Cell eating"; an endocytic pathway by which a cell engulfs particles such as microbes or cellular debris. **92**

pharynx Throat; opens to airways and digestive tract. **618**

phenotype An individual's observable traits. **207**

pheromone Chemical that serves as a communication signal among members of an animal species. **522, 700**

phloem Complex vascular tissue of plants; its living sieve elements compose sieve tubes that distribute sugars. Each sieve element has an associated companion cell that provides it with metabolic support. **346, 425**

phospholipid A lipid with a phosphate group in its hydrophilic head, and two nonpolar tails typically derived from fatty acids. Major component of cell membranes. **43**

phosphorus cycle Movement of phosphorus among Earth's rocks and waters, and into and out of food webs. **758**

phosphorylation A phosphate-group transfer. **87**

photoautotroph Organism that obtains carbon from carbon dioxide and energy from light. **321**

photoheterotroph Organism that obtains carbon from organic compounds and energy from light. **321**

photolysis Process by which light energy breaks down a molecule. **104**

photoperiodism Biological response to seasonal changes in the relative lengths of day and night. **470**

photoreceptor Sensory receptor that responds to light. **520**

photorespiration Reaction in which rubisco attaches oxygen instead of carbon dioxide to ribulose bisphosphate. **106**

photosynthesis Metabolic pathway by which most autotrophs use light energy to make sugars from carbon dioxide and water. Converts light energy into chemical bond energy. **6**

photosystem Cluster of pigments and proteins that converts light energy to chemical energy in photosynthesis. **104**

phototropism Plant growth in a direction influenced by light. **468**

phylogeny Evolutionary history of a species or group of species. **294**

phytochrome A light-sensitive pigment that helps set plant circadian rhythms based on length of night. **470**

pigment An organic molecule that selectively absorbs light of certain wavelengths. Reflected light imparts a characteristic color. E.g., chlorophyll. **100**

pilus A protein filament that projects from the surface of some bacterial cells. **59**

pineal gland Endocrine gland in the brain; secretes melatonin under low-light or dark conditions. **542**

pioneer species Species that can colonize a new habitat. **738**

pituitary gland Endocrine gland in the forebrain; interacts closely with the adjacent hypothalamus. **540**

placenta Of placental mammals, organ that forms during pregnancy and allows diffusion of substances between the maternal and embryonic bloodstreams. **667**

placental mammal Mammal in which maternal and embryonic bloodstreams exchange materials by means of a placenta. **409**

placozoans Group of tiny marine animals having a simple asymmetrical body and a small genome; considered an ancient lineage. **378**

plankton Community of tiny drifting or swimming organisms. **334**

plant Multicelled, typically photosynthetic eukaryote; develops from an embryo that forms on the parent and is nourished by it. **9, 346**

plant hormone Extracellular signaling molecule of plants that exerts its effect at very low concentration. E.g., auxin, gibberellin. **463**

plaque *See* dental plaque.

plasma Fluid portion of blood. **576**

plasma membrane A cell's outermost membrane; controls movement of substances into and out of the cell. **52**

plasmid Of many prokaryotes, a small ring of nonchromosomal DNA. **59**

plasmodesmata Cell junctions that form an open channel between the cytoplasm of adjacent plant cells. **69**

plasmodial slime mold Protist that feeds as a multinucleated mass and forms a spore-bearing structure when environmental conditions become unfavorable. **340**

plastid One of several types of double-membraned organelles in plants and algal cells; for example, a chloroplast or amyloplast. **65**

platelet Cell fragment that helps blood clot. **577**

plate tectonics theory Theory that Earth's outer layer of rock is cracked into plates, the slow movement of which rafts continents to new locations over geologic time. **262**

pleiotropy Effect in which a single gene affects multiple traits. **213**

plot sampling Using demographics observed in sample plots to estimate demographics of a population as a whole. **712**

polar body Tiny cell produced by unequal cytoplasmic division during egg production. **677**

polarity Separation of charge into positive and negative regions. **28**

pollen grain Walled, immature male gametophyte of a seed plant. Forms in an anther. **347, 454**

pollen sac Of seed plants, reproductive structure in which pollen grains develop. **352**

pollination Arrival of pollen on a receptive stigma of a seed plant. **352, 454**

pollination vector Environmental agent that moves pollen grains from one plant to another. **454**

pollinator An animal that facilitates pollination by moving pollen from one plant to another. **358, 454**

pollutant A substance that is released into the environment by human activities and interferes with the function of organisms that evolved in the absence of the substance or with lower levels. **789**

polymer Molecule that consists of multiple monomers. **39**

polymerase chain reaction (PCR) Method that rapidly generates many copies of a specific section of DNA. **240**

polypeptide Long chain of amino acids linked by peptide bonds. **44**

polyploid Having three or more of each type of chromosome characteristic of the species. **230**

pons Hindbrain region that influences breathing and serves as a bridge to the adjacent midbrain. **510**

population A group of organisms of the same species who live in a specific location and breed with one another more often than they breed with members of other populations. **5, 274**

population density Number of individuals per unit area. **712**

population distribution Location of population members relative to one another; clumped, uniformly dispersed, or randomly dispersed. **712**

population size Total number of individuals in a population. **712**

positive feedback mechanism A response intensifies the conditions that caused its occurrence. **502**

potential energy Stored energy. **79**

Precambrian Period from 4.6 billion to 542 million years ago. **310**

predation One species captures, kills, and eats another. **734**

prediction Statement, based on a hypothesis, about a condition that should exist if the hypothesis is correct. **12**

pressure flow theory Explanation of how a difference in turgor between sieve elements in source and sink regions pushes sugar-rich fluid through a sieve tube in a plant. **447**

primary endosymbiosis Over generations, evolution of an organelle from bacteria that entered a host cell and lived inside it. **332**

primary growth Of a plant, lengthening of young shoots and roots; originates at apical meristems. **432**

primary motor cortex Region of frontal lobe that controls voluntary movement. **512**

primary production The rate at which an ecosystem's producers capture and store energy. **748**

primary succession A new community becomes established in an area where there was previously no soil. **738**

primary wall The first cell wall of young plant cells. **68**

primate Mammal having grasping hands with nails and a body adapted to climbing; for example, a lemur, monkey, ape, or human. **410**

primer Short, single strand of DNA that base-pairs with a targeted DNA sequence. **138**

prion Infectious protein. **46**

probability The chance that a particular outcome of an event will occur; depends on the total number of outcomes possible. **16**

probe Short fragment of DNA labeled with a tracer; designed to hybridize with a nucleotide sequence of interest. **240**

producer Organism that makes its own food using energy and nonbiological raw materials from the environment; an autotroph. **6, 748**

product A molecule that is produced by a reaction. **80**

progesterone Sex hormone secreted by ovaries; prepares a female body for pregnancy and helps maintain a pregnancy. **545, 676**

prokaryote Informal name for a single-celled organism without a nucleus; a bacterium or archaean. **8**

promoter In DNA, a sequence to which RNA polymerase binds. **150**

prophase Stage of mitosis during which chromosomes condense and become attached to a newly forming spindle. **181**

prostate gland Exocrine gland that contributes to semen. **675**

protein Organic molecule that consists of one or more polypeptides. **44**

proteobacteria Most diverse bacterial lineage. **322**

protist Eukaryote that is not a plant, fungus, or animal. **8, 332**

protocell Membranous sac that contains interacting organic molecules; hypothesized to have formed prior to the earliest life forms. **304**

proton Positively charged subatomic particle. **24**

proto-oncogene Gene that, by mutation, can become an oncogene. **184**

protostomes Lineage of bilateral animals in which the first opening on the embryo surface develops into a mouth. **377**

proximal tubule Region of kidney tubule nearest Bowman's capsule. **652**

pseudocoelom Unlined body cavity around the gut. **377**

pseudopod A temporary protrusion that helps some eukaryotic cells move and engulf prey. **67**

puberty Period when reproductive organs begin to function. **673**

pulmonary artery Vessel that carries oxygen-poor blood from the heart to a lung. **572**

pulmonary circuit Circuit through which blood flows from the heart to the lungs and back. **571**

pulmonary vein Vessel carrying oxygen-rich blood from a lung to the heart. **572**

pulse Brief expansion of artery walls that occurs when ventricles contract. **578**

Punnett square Diagram used to predict the genetic and phenotypic outcome of a breeding or mating. **208**

pupil Adjustable opening through which light enters the eye. **525**

pyruvate Three-carbon end product of glycolysis. **116**

radial symmetry Having parts arranged around a central axis, like the spokes of a wheel. **376**

radioactive decay Process by which atoms of a radioisotope emit energy and/or subatomic particles when their nucleus spontaneously breaks up. **25**

radioisotope Isotope with an unstable nucleus. **24**

radiolaria Heterotrophic single-celled protists with a porous shell of silica and long cytoplasmic extensions. **334**

radiometric dating Method of estimating the age of a rock or fossil by measuring the content and proportions of a radioisotope and its daughter elements. **260**

rain shadow Dry region downwind of a coastal mountain range. **767**

ray-finned fish Jawed fish with fins supported by thin rays derived from skin; member of most diverse lineage of fishes. **403**

reabsorption *See* tubular reabsorption.

reactant A molecule that enters a reaction and is changed by participating in it. **80**

reaction Process of molecular change, in which reactants become products. **39**

receptor protein Plasma membrane protein that triggers a change in cell activity after binding to a particular substance. **57**

recessive Refers to an allele with an effect that is masked by a dominant allele on the homologous chromosome. **207**

recognition protein Plasma membrane protein that identifies a cell as belonging to self (one's own body or species). **57**

recombinant DNA A DNA molecule that contains genetic material from more than one organism. **238**

rectum Region of the large intestine in which feces are stored prior to excretion. **633**

red alga Photosynthetic protist; typically multicelled, with chloroplasts containing red accessory pigments (phycobilins). **338**

red blood cell Hemoglobin-filled blood cell that carries oxygen. **576**

red marrow Bone marrow that makes blood cells. **558**

redox reaction Oxidation–reduction reaction, in which one molecule accepts electrons (it becomes reduced) from another molecule (which becomes oxidized). Also called electron transfer. **85**

reflex Automatic response that occurs without conscious thought or learning. **508**

replacement fertility rate Number of children a woman must bear to replace herself with one daughter of reproductive age. **723**

repressor Transcription factor that reduces the rate of transcription. **164**

reproduction Processes by which parents produce offspring. *See* sexual reproduction, asexual reproduction. **7**

reproductive base Of a population, members of the reproductive and pre-reproductive age categories. **723**

reproductive cloning Technology that produces genetically identical individuals. **142**

reproductive isolation The end of gene flow between populations. **286**

reptile Amniote subgroup that includes lizards, snakes, turtles, crocodilians, and birds. **406**

resource partitioning Evolutionary process whereby species become adapted in different ways to access different portions of a limited resource; allows species with similar needs to coexist. **733**

respiration Physiological process by which an animal body supplies cells with oxygen and disposes of their waste carbon dioxide. **614**

respiratory cycle One inhalation and one exhalation. **620**

respiratory membrane Membrane consisting of alveolar epithelium, capillary endothelium, and their fused basement membranes. **622**

respiratory protein A protein that reversibly binds oxygen when the oxygen concentration is high and releases it when oxygen concentration is low. For example, hemoglobin. **614**

respiratory surface Moist surface across which gases are exchanged between animal cells and the external environment. **614**

resting potential Membrane potential of a neuron at rest. **501**

restriction enzyme Type of enzyme that cuts DNA at a specific nucleotide sequence. **238**

retina Photoreceptor-containing layer of tissue in an eye. **524**

retrovirus RNA virus that uses the enzyme reverse transcriptase to produce viral DNA in a host cell. **317**

reverse transcriptase An enzyme that uses mRNA as a template to make a strand of cDNA. **239**

rhizoid Threadlike structure that holds a nonvascular plant in place. **348**

rhizome Stem that grows horizontally along or under the ground. **350**

ribosomal RNA (rRNA) RNA that is part of ribosomes. **148**

ribosome Organelle of protein synthesis. An intact ribosome has two subunits, each composed of rRNA and proteins. Ribosomes are not enclosed by membranes. **59**

ribozyme RNA that functions as an enzyme. **305**

RNA Ribonucleic acid. Nucleic acid with roles in gene expression; consists of a single-stranded chain of nucleotides (adenine, guanine, cytosine, and uracil). *See* messenger RNA, transfer RNA, ribosomal RNA. **46**

RNA polymerase Enzyme that carries out transcription. **150**

RNA world Hypothetical early interval when RNA served as the genetic information. **305**

rod cell Photoreceptor that is active in dim light; provides coarse perception of image and detects motion. **526**

root hairs Hairlike, absorptive extensions of an epidermal cell on the surface of a plant root. **430**

root nodules Of some plant roots, swellings that contain nitrogen-fixing bacteria. **443**

roundworm Cylindrical worm with a pseudocoelom. **388**

rRNA *See* ribosomal RNA.

***r*-selection** Selection that favors traits that allow their bearers to produce many offspring quickly; occurs when population density is low and resources are abundant. **719**

rubisco Ribulose bisphosphate carboxylase. Carbon-fixing enzyme of the Calvin–Benson cycle. **106**

ruminant Hoofed mammal with a multiple-chamber stomach that adapts it to a cellulose-rich diet. **631**

runoff Water that flows over soil into streams. **752**

sac fungi Fungi that form spores in a sac-shaped structure during sexual reproduction. **365**

salivary gland Exocrine gland that secretes saliva into the mouth. **633**

salt Compound that releases ions other than H+ and OH– when it dissolves in water. **30**

sampling error Difference between results derived from testing an entire group of events or individuals, and results derived from testing a subset of the group. **16**

SA node *See* sinoatrial node.

sarcomere Contractile unit of skeletal and cardiac muscle. **561**

sarcoplasmic reticulum Specialized endoplasmic reticulum in muscle cells; stores and releases calcium ions. **563**

saturated fat Triglyceride that has three saturated fatty acid tails. **42**

savanna Biome dominated by perennial grasses with a few scattered shrubs and trees. **771**

science Systematic study of the observable world. **12**

scientific method Systematically making, testing, and evaluating hypotheses about the natural world. **13**

scientific theory Hypothesis that has not been disproven after many years of rigorous testing. **18**

sclerenchyma Simple plant tissue composed of cells that die when they are mature; their lignin-reinforced cell walls remain and structurally support plant parts. Includes fibers, sclereids. **424**

SCNT *See* somatic cell nuclear transfer.

seamount An undersea mountain. **778**

secondary endosymbiosis Evolution of an organelle from a protist that itself contains organelles that arose by primary endosymbiosis. **332**

secondary growth Of a plant, thickening of older stems and roots; originates at lateral meristems. **432**

secondary sexual characteristics Traits that differ between the sexes but do not play a direct role in reproduction. **545**

secondary succession A new community develops in a site where a community previously existed. **738**

secondary wall Lignin-reinforced wall that forms inside the primary wall of a plant cell. **68**

second law of thermodynamics Energy disperses spontaneously. **78**

second messenger Molecule that forms inside a cell when a hormone binds to a receptor in the plasma membrane; sets in motion reactions that alter activity inside the cell. **537**

sedimentary cycle Biochemical cycle in which the atmosphere plays little role and rocks are the major reservoir. **758**

seed Embryo sporophyte of a seed plant packaged with nutritive tissue inside a protective coat. **347, 458**

seedless vascular plant Plant that disperses by releasing spores and has xylem and phloem. For example, a club moss or fern. **350**

segmentation Having a body composed of similar units that repeat along its length. **377**

selfish herd Temporary group that forms when individuals cluster to minimize their individual risk of predation. **704**

semen Sperm mixed with secretions from exocrine glands. **674**

semicircular canals Organs of equilibrium that respond to rotation and angular movement of the head; part of the vestibular apparatus. **530**

semiconservative replication Describes the process of DNA replication, which produces two copies of a DNA molecule: one strand of each copy is new, and the other is parental. **139**

seminal vesicles Exocrine glands that add sugary fluid to semen. **675**

seminiferous tubules In testes, tiny tubes in which sperm form. **674**

sensation Detection of a stimulus. **520**

sensory adaptation Slowing or cessation of a sensory receptor's response to an ongoing stimulus. **520**

sensory neuron Neuron that is activated when its receptor endings detect a specific stimulus, such as light or pressure. **500**

sensory receptor Cell or cell component that responds to a specific stimulus, such as temperature or light. **492**

sequencing Method of determining the order of nucleotides in DNA. **242**

sex chromosome Member of a pair of chromosomes that differs between males and females. **137**

sex hormone Steroid hormone produced by gonads; functions in reproduction and may affect secondary sexual characteristics. **545**

sexual dimorphism Difference in appearance between males and females of a species. **282, 702**

sexual reproduction Reproductive mode by which offspring arise from two parents and inherit genes from both. **192, 666**

sexual selection Mode of natural selection in which some individuals outreproduce others because they are better at securing mates. **282**

shell model Model of electron distribution in an atom. **26**

short tandem repeat In chromosomal DNA, sequences of a few nucleotides repeated multiple times in a row. Used in DNA profiling. **216**

sieve elements Living plant cells that compose sugar-conducting sieve tubes of phloem. Each sieve tube consists of a stack of sieve elements that meet end to end at sieve plates. **446**

sieve tube Sugar-conducting tube of phloem; consists of stacked sieve elements. **446**

single-nucleotide polymorphism (SNP) One-nucleotide DNA sequence variation carried by a measurable percentage of a population. **244**

sink Region of plant tissue where sugars are being used or stored. **447**

sinoatrial (SA) node Cardiac pacemaker; group of cells that spontaneously emits rhythmic action potentials that result in contraction of cardiac muscle. **575**

sister chromatids The two attached DNA molecules of a duplicated eukaryotic chromosome; attachs at the centromere. **136**

sister groups The two lineages that emerge from a node on a cladogram. **295**

skeletal muscle fiber Multinucleated contractile cell that runs the length of a skeletal muscle. **561**

skeletal muscle tissue Muscle that interacts with skeletal elements to move body parts; under voluntary control. **486**

sliding-filament model Explanation of how interactions among actin and myosin filaments shorten a sarcomere and bring about muscle contraction. **562**

small intestine Longest portion of the digestive tract, and the site of most digestion and absorption. **633**

smooth muscle tissue Muscle that lines blood vessels and forms the wall of hollow organs. **486**

SNP *See* single-nucleotide polymorphism.

social animal Animal that lives in a multigenerational group in which members, who are usually relatives, cooperate in some tasks. **704**

soil erosion Loss of soil under the force of wind and water. **440**

soil water Water between soil particles. **752**

solute A substance dissolved in a solvent. **30**

solution Uniform mixure of solute completely dissolved in solvent. **30**

solvent Liquid that can dissolve other substances. **30**

somatic cell nuclear transfer (SCNT) Reproductive cloning method in which the DNA of an adult donor's body cell is transferred into an unfertilized egg. **142**

somatic nervous system Division of the peripheral nervous system that controls skeletal muscles and relays sensory signals about movements and external conditions. **506**

somatic sensations Sensations such as touch and pain that arise when sensory neurons in skin, muscle, or joints are activated. **521**

sorus Cluster of spore-producing capsules on a fern leaf. **350**

source Region of a plant tissue where sugars are being produced or released from storage. **446**

speciation Evolutionary process in which new species arise. **286**

species Unique type of organism designated by genus name and specific epithet. Of sexual reproducers, often defined as one or more groups of individuals that can potentially interbreed, produce fertile offspring, and do not interbreed with other groups. **10**

specific epithet Second part of a species name. **10**

sperm Male gamete. **666**

sphincter Ring of muscle that controls passage through a tubular organ or body opening. **560**

spinal cord Portion of the central nervous system that connects peripheral nerves with the brain. **508**

spindle Temporary structure that moves chromosomes during nuclear division; consists of microtubules. **181**

spirochetes Bacteria that resemble a stretched-out spring. **323**

spleen Large lymphoid organ that functions in immunity and filters pathogens, old red blood cells, and platelets from the blood. **584**

sponge Aquatic invertebrate that has no tissues or organs and filters food from the water. **379**

spongy bone Lightweight bone with many internal spaces; contains red marrow. **558**

sporophyte Spore-forming diploid body that forms in the life cycle of land plants and some multicelled algae. **338, 346**

stabilizing selection Mode of natural selection in which an intermediate form of a trait is adaptive, and extreme forms are not. **280**

stamen Floral reproductive organ that produces male gametophytes; typically consists of an anther on the tip of a filament. **356, 454**

stasis Evolutionary pattern in which a lineage persists with little or no change over evolutionary time. **292**

statistically significant Refers to a result that is statistically unlikely to have occurred by chance. **16**

stem cell Cell that can divide to produce more stem cells or differentiate into specialized cell types. **493**

steroid Type of lipid with four carbon rings and no fatty acid tails. **43**

steroid hormone Lipid-soluble hormone derived from cholesterol. **537**

stigma Of a flower, the upper part of the carpel. Adapted to receive pollen. **454**

stoma Plural **stomata**. Opening across a plant's cuticle and epidermis; can be opened for gas exchange or closed to prevent water loss. **346**

stomach Digestive organ that mixes food with enzymes and acid. **633**

strobilus Of some nonflowering plants, a spore-forming, cone-shaped structure composed of modified leaves. **351**

stroma The cytoplasm-like fluid between the thylakoid membrane and the two outer membranes of a chloroplast. Site of light-independent reactions of photosynthesis. **103**

stromatolite Rocky structures composed of layers of bacterial cells and sediments. **307**

substrate Of an enzyme, a reactant that is specifically acted upon by the enzyme. **82**

substrate-level phosphorylation The formation of ATP by the direct transfer of a phosphate group from a substrate to ADP. **116**

superior vena cava Vein that delivers blood from the upper body to the heart. **573**

surface-to-volume ratio A relationship in which the volume of an object increases with the cube of the diameter, and the surface area increases with the square. Limits cell size. **53**

survivorship curve Graph showing how many members of a cohort remain alive over time. **718**

suspension feeder Animal that filters food from water around it. **379**

symbiosis One species lives in or on another in a commensal, mutualistic, or parasitic relationship. **730**

sympathetic neurons Neurons of the autonomic system that are activated during stress and danger. **506**

sympatric speciation Speciation pattern in which speciation occurs within a population, in the absence of a physical barrier to gene flow. **290**

synapse Region where a neuron's axon terminals transmit signaling molecules (neurotransmitter) to another cell. **504**

synaptic integration The summation of excitatory and inhibitory signals by a postsynaptic cell. **505**

systemic acquired resistance In plants, inducible whole-body resistance to a wide range of pathogens and abiotic stressors. **473**

systemic circuit Circulatory circuit through which blood flows from the heart to the body tissues and back. **571**

systole Contractile phase of the cardiac cycle. **574**

systolic pressure Blood pressure when ventricles are contracting. **579**

T cell T lymphocyte. White blood cell central to adaptive immunity. E.g., helper T cell, cytotoxic T cell. **591**

T cell receptor (TCR) Antigen receptor on the surface of a T cell. **596**

taiga *See* boreal forest.

taproot system An enlarged primary root together with all of the lateral roots that branch from it. Typical of eudicots. **430**

taste receptor Chemoreceptor involved in the sense of taste. **523**

taxonomy The science of naming and classifying species. **10**

taxon, plural taxa Group of organisms that share a unique set of traits. **10**

telomere Noncoding, repetitive DNA sequence at the end of eukaryotic chromosomes; protects the coding sequences from degradation. **183**

telophase Stage of mitosis during which chromosomes arrive at opposite spindle poles and decondense, and two new nuclei form. **181**

temperate deciduous forest Northern Hemisphere biome in which the main plants are broadleaf trees that lose their leaves in fall and become dormant during cold winters. **770**

temperature Measure of molecular motion. **31**

tendon Strap of dense connective tissue that connects a skeletal muscle to bone. **560**

territory Area an animal or group of animals occupies and defends. **702**

testcross Method of determining genotype by tracking a trait in the offspring of a cross between an individual of unknown genotype and an individual known to be homozygous recessive. **208**

testis Sperm-producing animal gonad. **668**

testosterone Hormone secreted by testes; functions in sperm formation and development of male secondary sex characteristics. **545, 674**

tetrapod Vertebrate with four legs, or a descendant thereof. **401**

thalamus Forebrain region that relays signals to the cerebrum; affects sleep–wake cycles. **511**

theory of inclusive fitness Alleles associated with altruism can be advantageous if the expense of this behavior to the altruist is outweighed by increases in the reproductive success of relatives. **706**

therapeutic cloning The use of SCNT to produce human embryos for research purposes. **142**

thermoreceptor Temperature-sensitive sensory receptor. **520**

thermoregulation Maintaining body temperature within a limited range. **658**

thigmotropism Plant growth in a direction influenced by contact. **469**

threatened species Species likely to become endangered in the near future. **784**

threshold potential Neuron membrane potential at which gated sodium channels open, causing an action potential to occur. **502**

thylakoid membrane A chloroplast's highly folded inner membrane system; forms a continuous compartment in the stroma. Site of light reactions of photosynthesis. **103**

thymus Hormone-producing gland in which T lymphocytes (T cells) mature. **584**

thyroid gland Endocrine gland in the base of the neck that secretes thyroid hormone and calcitonin. **542**

thyroid hormone Iodine-containing hormones that collectively increase metabolic rate and play a role in development. **542**

tight junctions In animals, arrays of adhesion proteins that join epithelial cells and collectively prevent fluids from leaking between them. **69**

tissue In multicelled organisms, specialized cells organized in a pattern that allows them to perform a collective function. **4**

tissue culture propagation Laboratory method in which individual plant cells (typically from meristem) are induced to form embryos. **462**

topsoil Uppermost soil layer; contains the most organic matter and nutrients for plant growth. **440**

total fertility rate Expected number of children a women will bear over the course of a lifetime. **722**

tracer A molecule with a detectable component; researchers can track it after delivery into a body or other system. **25**

trachea Airway to the lungs; windpipe. **619**

tracheal system Of insects, tubes that convey gases between the body surface and internal tissues. **615**

tracheids Tapered cells of xylem that die when mature; their interconnected, pitted walls remain and form water-conducting tubes in plants. **444**

tract Bundle of axons in the central nervous system. **508**

trait An observable characteristic of an organism or species. **10**

transcription Process by which enzymes assemble an RNA using the nucleotide sequence of a gene as a template. **148**

transcription factor Protein that influences transcription by binding directly to DNA; for example, an activator or repressor. **164**

transfer RNA (tRNA) RNA that delivers amino acids to a ribosome during translation. **148**

transgenic Refers to a genetically modified organism that carries a gene from a different species. **246**

translation Process by which a polypeptide chain is assembled from amino acids in the order specified by an mRNA. **149**

translocation Of a chromosome, major structural change in which a broken piece gets reattached in the wrong location. In a plant, movement of organic compounds through phloem. **228, 446**

transpiration Evaporation of water from aboveground plant parts. **445**

transport protein Protein that allows specific ions or molecules to cross a membrane. Those that function in active transport require an energy input, as from ATP; those that function in facilitated difussion require no energy input. **57**

transposable element Segment of DNA that can move spontaneously within or between chromosomes. **228**

triglyceride A fat with three fatty acid tails. **42**

tRNA *See* transfer RNA.

trophic level Position of an organism in a food chain. **748**

tropical rain forest Highly productive and species-rich biome in which year-round rains and warmth support continuous growth of evergreen broadleaf trees. **769**

tropism In plants, directional growth response to an environmental stimulus. **468**

trypanosome Parasitic flagellated protist with a single mitochondrion and a flagellum that runs along the back of the cell. **333**

tubular reabsorption Water and solutes move from the filtrate inside a kidney tubule into the peritubular capillaries. **654**

tubular secretion Ions and breakdown products of organic molecules move out of peritubular capillaries and into filtrate. **654**

tumor A neoplasm that forms a lump. **184**

tunicate Invertebrate chordate that loses most of its defining chordate traits during the transition to adulthood. **400**

turgor Pressure that a fluid exerts against a wall, membrane, or other structure that contains it. **89**

unsaturated fat Triglyceride that has one or more unsaturated fatty acid tails. **42**

urea Main nitrogen-containing compound in urine of mammals. **651**

ureter Tube that carries urine from a kidney to the bladder. **652**

urethra Tube through which urine from the bladder exits the body. **652**

uric acid Main nitrogen-containing compound in the urine of insects, as well as birds and other reptiles. **650**

urinary bladder Hollow, muscular organ that stores urine. **652**

urine Fluid that consists of water and soluble wastes; formed and excreted by the vertebrate kidneys. **651**

uterus Muscular chamber where offspring develop; womb. **676**

vaccine A preparation introduced into the body in order to elicit immunity to a specific antigen. **607**

vacuole A membrane-enclosed, fluid-filled organelle that isolates or disposes of waste, debris, or toxic materials. **62**

vagina Female organ of intercourse and birth canal. **676**

variable In an experiment, a characteristic or event that differs among individuals or over time. **12**

vascular bundle In a stem or leaf, multistranded bundle formed by xylem, phloem, and sclerenchyma fibers. **426**

vascular cambium Lateral meristem that produces secondary xylem and phloem in plants. **433**

vascular cylinder Central column of vascular tissue in a plant root. **430**

vascular plant Plant with xylem and phloem. **346**

vascular tissues Tissues that distribute water and nutrients through a plant body. *See* xylem, phloem. **422**

vas deferens One of a pair of long ducts that carry mature sperm to the ejaculatory duct. **674**

vasoconstriction Narrowing of a blood vessel when smooth muscle that rings it contracts. **578**

vasodilation Widening of a blood vessel when smooth muscle that rings it relaxes. **578**

vector Of a disease, an animal that carries the pathogen from one host to the next. **318**

vegetative reproduction Growth of new roots and shoots from extensions or fragments of a parent plant; form of asexual reproduction in plants. **462**

vein In plants, a vascular bundle in a leaf or other structure. In animals with a circulatory system, a large-diameter vessel that returns blood to the heart. **428, 570**

ventricle Heart chamber that pumps blood into arteries. **574**

venule Blood vessel that conveys blood from capillaries to a vein. **581**

vernalization In plants, stimulation of flowering in spring by prolonged exposure to low temperature in winter. **471**

vertebrae Bones of the backbone, or vertebral column. **556**

vertebral column Backbone. **556**

vertebrate Animal with a backbone. **400**

vesicle Small, membrane-enclosed organelle; different kinds store, transport, or break down their contents; e.g., a peroxisome or lysosome. **62**

vessel elements Of xylem, cells that form in stacks and die when mature; their pitted walls remain to form water-conducting tubes in plants. Each tube consists of a stack of vessel elements that meet end to end at perforation plates. **444**

vestibular apparatus System of fluid-filled sacs and canals in the inner ear; contains organs of equilibrium. **530**

villi Singular **villus.** Multicelled projections from the lining of the small intestine. **635**

viral recombination Multiple strains of virus infect a host simultaneously and swap genes. **319**

viroid Small noncoding RNA that can infect plants. **319**

virus Noncellular, infectious particle of protein and nucleic acid; replicates only in a host cell. **316**

visceral sensations Sensations that arise when sensory neurons associated with organs inside body cavities are activated. **521**

visual accommodation Process of making adjustments to lens shape so light from an object falls on the retina. **525**

vital capacity Maximum amount of air moved in and out of lungs with forced inhalation and exhalation. **620**

vitamin Organic substance required in the diet in small amounts for normal metabolism. **640**

vomeronasal organ Pheromone-detecting organ of vertebrates. **523**

water cycle Movement of water among Earth's oceans, atmosphere, and the freshwater reservoirs on land. **752**

water mold Heterotrophic protist that grows as a mesh of nutrient-absorbing filaments. **337**

water–vascular system Of echinoderms, a system of fluid-filled tubes and tube feet that function in locomotion. **394**

wavelength Distance between the crests of two successive waves. **100**

wax Water-repellent mixture of lipids with long fatty acid tails bonded to long-chain alcohols or carbon rings. **43**

white blood cell Blood cell with a role in housekeeping and defense. **576**

white matter Central nervous system tissue consisting mainly of myelinated axons. **508**

wood Accumulated secondary xylem inside the cylinder of vascular cambium in an older plant stem or root. **433**

X chromosome inactivation Developmental shutdown of one of the two X chromosomes in the cells of female mammals. **168**

xylem Complex vascular tissue of plants; its dead tracheids and vessel elements distribute water and dissolved minerals through the plant body. **346, 425**

yeast Fungus that lives as single cell. **364**

yellow marrow Bone marrow that is mostly fat; fills cavity in most long bones. **558**

yolk Nutritious material in many animal eggs. **667**

zero population growth Interval in which births equal deaths. **714**

zygote Cell formed by fusion of two gametes at fertilization; the first cell of a new individual. **195**

zygote fungi Fungi that live in damp places and form a thick-walled zygospore during sexual reproduction. **365**

Index

Figures and tables are indicated by f and t. Glossary terms are indicated with bold page numbers. Applications related to human health are indicated with green bullets and environmental applications with red bullets.

A

C

E

J

K

L

R

S